MW01284481

OPTIMAL CONTROL

OPTIMAL CONTROL

Third Edition

FRANK L. LEWIS
Department of Electrical Engineering
Automation & Robotics Research Institute
University of Texas at Arlington
Arlington, Texas

DRAGUNA L. VRABIE
United Technologies Research Center
East Hartford, Connecticut

VASSILIS L. SYRMOS
Department of Electrical Engineering
University of Hawaii at Manoa
Honolulu, Hawaii

WILEY

JOHN WILEY & SONS, INC.

For general information about our other products and services, please contact our Customer Care Department within the United States at (800) 762-2974, outside the United States at (317) 572-3993 or fax (317) 572-4002.

Wiley publishes in a variety of print and electronic formats and by print-on-demand. Some material included with standard print versions of this book may not be included in e-books or in print-on-demand. If this book refers to media such as a CD or DVD that is not included in the version you purchased, you may download this material at http://booksupport.wiley.com. For more information about Wiley products, visit www.wiley.com.

Library of Congress Cataloging-in-Publication Data:

Lewis, Frank L.
 Optimal control / Frank L. Lewis, Draguna L. Vrabie, Vassilis L. Syrmos.—3rd ed.
  p. cm.
 Includes bibliographical references and index.
 ISBN 978-0-470-63349-6 (cloth); ISBN 978-1-118-12263-1 (ebk); ISBN 978-1-118-12264-8 (ebk); ISBN 978-1-118-12266-2 (ebk); ISBN 978-1-118-12270-9 (ebk); ISBN 978-1-118-12271-6 (ebk); ISBN 978-1-118-12272-3 (ebk)
 1. Control theory. 2. Mathematical optimization. I. Vrabie, Draguna L. II. Syrmos, Vassilis L. III. Title.
 QA402.3.L487 2012
 629.8'312–dc23

2011028234

Printed in the United States of America
10 9 8 7 6 5 4 3 2 1

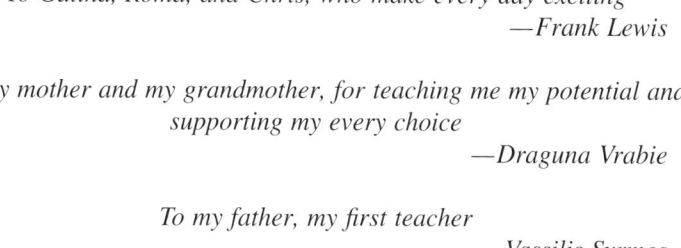

To Galina, Roma, and Chris, who make every day exciting
—*Frank Lewis*

To my mother and my grandmother, for teaching me my potential and supporting my every choice
—*Draguna Vrabie*

To my father, my first teacher
—*Vassilis Syrmos*

CONTENTS

PREFACE

This book is intended for use in a second graduate course in modern control theory. A background in the state-variable representation of systems is assumed. Matrix manipulations are the basic mathematical vehicle and, for those whose memory needs refreshing, Appendix A provides a short review.

The book is also intended as a reference. Numerous tables make it easy to find the equations needed to implement optimal controllers for practical applications.

Our interactions with nature can be divided into two categories: observation and action. While observing, we process data from an essentially uncooperative universe to obtain knowledge. Based on this knowledge, we act to achieve our goals. This book emphasizes the control of systems assuming perfect and complete knowledge. The dual problem of estimating the state of our surroundings is briefly studied in Chapter 9. A rigorous course in optimal estimation is required to conscientiously complete the picture begun in this text.

Our intention is to present optimal control theory in a clear and direct fashion. This goal naturally obscures the more subtle points and unanswered questions scattered throughout the field of modern system theory. What appears here as a completed picture is in actuality a growing body of knowledge that can be interpreted from several points of view and that takes on different personalities as new research is completed.

We have tried to show with many examples that computer simulations of optimal controllers are easy to implement and are an essential part of gaining an intuitive feel for the equations. Students should be able to write simple programs as they progress through the book, to convince themselves that they have confidence in the theory and understand its practical implications.

Relationships to classical control theory have been pointed out, and a root-locus approach to steady-state controller design is included. Chapter 9 presents some multivariable classical design techniques. A chapter on optimal control of polynomial systems is included to provide a background for further study in the field of adaptive control. A chapter on robust control is also included to expose the reader to this important area. A chapter on differential games shows how to extend the optimality concepts in the book to multiplayer optimization in interacting teams.

Optimal control relies on solving the matrix design equations developed in the book. These equations can be complicated, and exact solution of the Hamilton-Jacobi equations for nonlinear systems may not be possible. The last chapter, on optimal adaptive control, gives practical methods for solving these matrix design equations. Algorithms are given for finding approximate solutions online in real-time using adaptive learning techniques based on data measured along the system trajectories.

The first author wants to thank his teachers: J. B. Pearson, who gave him the initial excitement and passion for the field; E. W. Kamen, who tried to teach him persistence and attention to detail; B. L. Stevens, who forced him to consider applications to real situations; R. W. Newcomb, who gave him self-confidence; and A. H. Haddad, who showed him the big picture and the humor behind it all. We owe our main thanks to our students, who force us daily to take the work seriously and become a part of it.

Acknowledgments

This work was supported by NSF grant ECCS-0801330, ARO grant W91NF-05-1-0314, and AFOSR grant FA9550-09-1-0278.

1

STATIC OPTIMIZATION

In this chapter we discuss optimization when time is not a parameter. The discussion is preparatory to dealing with time-varying systems in subsequent chapters. A reference that provides an excellent treatment of this material is Bryson and Ho (1975), and we shall sometimes follow their point of view.

Appendix A should be reviewed, particularly the section that discusses matrix calculus.

1.1 OPTIMIZATION WITHOUT CONSTRAINTS

A scalar *performance index* $L(u)$ is given that is a function of a *control* or *decision vector* $u \in R^m$. It is desired to determine the value of u that results in a minimum value of $L(u)$.

We proceed to solving this optimization problem by writing the Taylor series expansion for an increment in L as

$$dL = L_u^T du + \frac{1}{2} du^T L_{uu} du + O(3), \qquad (1.1\text{-}1)$$

where $O(3)$ represents terms of order three. The gradient of L with respect to u is the column vector

$$L_u \triangleq \frac{\partial L}{\partial u}, \qquad (1.1\text{-}2)$$

and the Hessian matrix is

$$L_{uu} = \frac{\partial^2 L}{\partial u^2}. \qquad (1.1\text{-}3)$$

1

L_{uu} is called the *curvature matrix*. For more discussion on these quantities, see Appendix A.

Note. The gradient is defined throughout the book as a *column* vector, which is at variance with some authors, who define it as a row vector.

A *critical* or *stationary point* is characterized by a zero increment dL to first order for all increments du in the control. Hence,

$$L_u = 0 \tag{1.1-4}$$

for a critical point.

Suppose that we are at a critical point, so $L_u = 0$ in (1.1-1). For the critical point to be a local minimum, it is required that

$$dL = \frac{1}{2} du^{\mathrm{T}} L_{uu}\, du + O(3) \tag{1.1-5}$$

is positive for all increments du. This is guaranteed if the curvature matrix L_{uu} is positive definite,

$$L_{uu} > 0. \tag{1.1-6}$$

If L_{uu} is negative definite, the critical point is a local maximum; and if L_{uu} is indefinite, the critical point is a saddle point. If L_{uu} is semidefinite, then higher terms of the expansion (1.1-1) must be examined to determine the type of critical point.

The following example provides a tangible meaning to our initial mathematical developments.

Example 1.1-1.   Quadratic Surfaces

Let $u \in R^2$ and

$$L(u) = \frac{1}{2} u^{\mathrm{T}} \begin{bmatrix} q_{11} & q_{12} \\ q_{12} & q_{22} \end{bmatrix} u + [s_1 \quad s_2] u \tag{1}$$

$$\triangleq \frac{1}{2} u^{\mathrm{T}} Q u + S^{\mathrm{T}} u. \tag{2}$$

The critical point is given by

$$L_u = Qu + S = 0 \tag{3}$$

and the optimizing control is

$$u^* = -Q^{-1} S. \tag{4}$$

By examining the Hessian

$$L_{uu} = Q \tag{5}$$

one determines the type of the critical point.

The point u^* is a minimum if $L_{uu} > 0$ and it is a maximum if $L_{uu} < 0$. If $|Q| < 0$, then u^* is a saddle point. If $|Q| = 0$, then u^* is a *singular point* and in this case L_{uu} does not provide sufficient information for characterizing the nature of the critical point.

By substituting (4) into (2) we find the extremal value of the performance index to be

$$L^* \triangleq L(u^*) = \frac{1}{2} S^T Q^{-1} Q Q^{-1} S - S^T Q^{-1} S$$

$$= -\frac{1}{2} S^T Q^{-1} S. \tag{6}$$

Let

$$L = \frac{1}{2} u^T \begin{bmatrix} 1 & 1 \\ 1 & 2 \end{bmatrix} u + [0 \quad 1] u. \tag{7}$$

Then

$$u^* = -\begin{bmatrix} 2 & -1 \\ 1 & 1 \end{bmatrix} \begin{bmatrix} 0 \\ 1 \end{bmatrix} = \begin{bmatrix} 1 \\ -1 \end{bmatrix} \tag{8}$$

is a minimum, since $L_{uu} > 0$. Using (6), we see that the minimum value of L is $L^* = -\frac{1}{2}$.

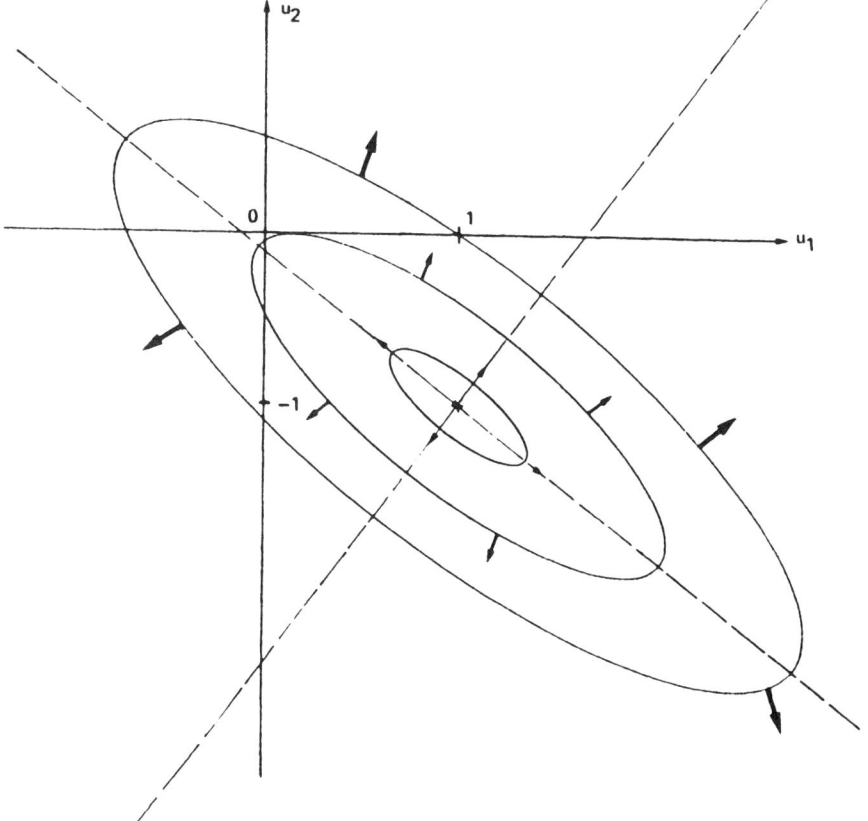

FIGURE 1.1-1   Contours and the gradient vector.

The contours of the $L(u)$ in (7) are drawn in Fig. 1.1-1, where $u = [u_1 \quad u_2]^T$. The arrows represent the gradient

$$L_u = Qu + S = \begin{bmatrix} u_1 + u_2 \\ u_1 + 2u_2 + 1 \end{bmatrix}. \tag{9}$$

Note that the gradient is always perpendicular to the contours and pointing in the direction of increasing $L(u)$.

We shall use an asterisk to denote optimal values of u and L when we want to be explicit. Usually, however, the asterisk will be omitted.   ∎

Example 1.1-2.   Optimization by Scalar Manipulations

We have discussed optimization in terms of vectors and the gradient. As an alternative approach, we could deal entirely in terms of scalar quantities. To demonstrate, let

$$L(u_1, u_2) = \frac{1}{2}u_1^2 + u_1 u_2 + u_2^2 + u_2, \tag{1}$$

where u_1 and u_2 are scalars. A critical point is present where the derivatives of L with respect to *all arguments* are equal to zero:

$$\frac{\partial L}{\partial u_1} = u_1 + u_2 = 0,$$

$$\frac{\partial L}{\partial u_2} = u_1 + 2u_2 + 1 = 0. \tag{2}$$

Solving this system of equations yields

$$u_1 = 1, \quad u_2 = -1; \tag{3}$$

thus, the critical point is $(1, -1)$. Note that (1) is an expanded version of (7) in Example 1.1-1, so we have just derived the same answer by another means.

Vector notation is a tool that simplifies the bookkeeping involved in dealing with multidimensional quantities, and for that reason it is very attractive for our purposes.   ∎

1.2   OPTIMIZATION WITH EQUALITY CONSTRAINTS

Now let the scalar performance index be $L(x, u)$, a function of the control vector $u \in R^m$ and an *auxiliary* (state) *vector* $x \in R^n$. The optimization problem is to determine the control vector u that minimizes $L(x, u)$ and at the same time satisfies the *constraint equation*

$$f(x, u) = 0. \tag{1.2-1}$$

The auxiliary vector x is determined for a given u by the relation (1.2-1). For a given u, (1.2-1) defines a set of n scalar equations.

To find necessary and sufficient conditions for a local minimum that also satisfies $f(x, u) = 0$, we proceed exactly as we did in the previous section, first expanding dL in a Taylor series and then examining the first- and second-order terms. Let us first gain some insight into the problem, however, by considering it from three points of view (Bryson and Ho 1975, Athans and Falb 1966).

Lagrange Multipliers and the Hamiltonian

Necessary Conditions   At a stationary point, dL is equal to zero in the first-order approximation with respect to increments du when df is zero. Thus, at a critical point the following equations are satisfied:

$$dL = L_u^T \, du + L_x^T \, dx = 0 \tag{1.2-2}$$

and

$$df = f_u \, du + f_x \, dx = 0. \tag{1.2-3}$$

Since (1.2-1) determines x for a given u, the increment dx is determined by (1.2-3) for a given control increment du. Thus, the Jacobian matrix f_x is nonsingular and one can write

$$dx = -f_x^{-1} f_u \, du. \tag{1.2-4}$$

Substituting this into (1.2-2) yields

$$dL = \left(L_u^T - L_x^T f_x^{-1} f_u \right) du. \tag{1.2-5}$$

The derivative of L with respect to u holding f constant is therefore given by

$$\left. \frac{\partial L}{\partial u} \right|_{df=0} = \left(L_u^T - L_x^T f_x^{-1} f_u \right)^T = L_u - f_u^T f_x^{-T} L_x, \tag{1.2-6}$$

where f_x^{-T} means $(f_x^{-1})^T$. Note that

$$\left. \frac{\partial L}{\partial u} \right|_{dx=0} = L_u. \tag{1.2-7}$$

Thus, for dL to be zero in the first-order approximation with respect to arbitrary increments du when $df = 0$, we must have

$$L_u - f_u^T f_x^{-T} L_x = 0. \tag{1.2-8}$$

This is a necessary condition for a minimum. Before we derive a sufficient condition, let us develop some more insight by examining two more ways to obtain (1.2-8). Write (1.2-2) and (1.2-3) as

$$\begin{bmatrix} dL \\ df \end{bmatrix} = \begin{bmatrix} L_x^T & L_u^T \\ f_x & f_u \end{bmatrix} \begin{bmatrix} dx \\ du \end{bmatrix} = 0. \tag{1.2-9}$$

This set of linear equations defines a stationary point, and it must have a solution $[dx^T \; du^T]^T$. The critical point is obtained only if the $(n + 1) \times (n + m)$ coefficient matrix has rank less than $n + 1$. That is, its rows must be linearly dependent so there exists an n vector λ such that

$$[1 \; \lambda^T] \begin{bmatrix} L_x^T & L_u^T \\ f_x & f_u \end{bmatrix} = 0. \tag{1.2-10}$$

Then

$$L_x^T + \lambda^T f_x = 0, \tag{1.2-11}$$

$$L_u^T + \lambda^T f_u = 0. \tag{1.2-12}$$

Solving (1.2-11) for λ gives

$$\lambda^T = -L_x^T f_x^{-1}, \tag{1.2-13}$$

and substituting in (1.2-12) again yields the condition (1.2-8) for a critical point.

Note. The left-hand side of (1.2-8) is the transpose of the Schur complement of L_u^T in the coefficient matrix of (1.2-9) (see Appendix A for more details).

The vector $\lambda \in R^n$ is called a *Lagrange multiplier*, and it will turn out to be an extremely useful tool for us. To give it some additional meaning now, let $du = 0$ in (1.2-2), (1.2-3) and eliminate dx to get

$$dL = L_x^T f_x^{-1} \, df. \tag{1.2-14}$$

Therefore,

$$\left. \frac{\partial L}{\partial f} \right|_{du=0} = \left(L_x^T f_x^{-1} \right)^T = -\lambda, \tag{1.2-15}$$

so that $-\lambda$ is the partial of L with respect to the constraint holding the control u constant. It shows the effect on the performance index of holding the control constant when the constraints are changed.

As a third method of obtaining (1.2-8), let us develop the approach we shall use for our analysis in subsequent chapters. Include the constraints in the performance index to define the *Hamiltonian* function

$$H(x, u, \lambda) = L(x, u) + \lambda^T f(x, u), \tag{1.2-16}$$

where $\lambda \in R^n$ is an as yet undetermined Lagrange multiplier. To determine x, u, and λ, which result in a critical point, we proceed as follows.

Increments in H depend on increments in x, u, and λ according to

$$dH = H_x^T\, dx + H_u^T\, du + H_\lambda^T d\lambda. \qquad (1.2\text{-}17)$$

Note that

$$H_\lambda = \frac{\partial H}{\partial \lambda} = f(x, u), \qquad (1.2\text{-}18)$$

so suppose we choose some value of u and demand that

$$H_\lambda = 0. \qquad (1.2\text{-}19)$$

Then x is determined for the given u by $f(x, u) = 0$, which is the constraint relation. In this situation the Hamiltonian equals the performance index:

$$H|_{f=0} = L. \qquad (1.2\text{-}20)$$

Recall that if $f = 0$, then dx is given in terms of du by (1.2-4). We should rather not take into account this coupling between du and dx, so it is convenient to choose λ so that

$$H_x = 0. \qquad (1.2\text{-}21)$$

Then, by (1.2-17), increments dx do not contribute to dH. Note that this yields a value for λ given by

$$\frac{\partial H}{\partial x} = L_x + f_x^T \lambda = 0 \qquad (1.2\text{-}22)$$

or (1.2-13).

If (1.2-19) and (1.2-21) hold, then

$$dL = dH = H_u^T\, du, \qquad (1.2\text{-}23)$$

since $H = L$ in this situation. To achieve a stationary point, we must therefore finally impose the *stationarity condition*

$$H_u = 0. \qquad (1.2\text{-}24)$$

In summary, necessary conditions for a minimum point of $L(x, u)$ that also satisfies the constraint $f(x, u) = 0$ are

$$\frac{\partial H}{\partial \lambda} = f = 0, \qquad (1.2\text{-}25a)$$

$$\frac{\partial H}{\partial x} = L_x + f_x^{\mathrm{T}}\lambda = 0, \qquad (1.2\text{-}25b)$$

$$\frac{\partial H}{\partial u} = L_u + f_u^{\mathrm{T}}\lambda = 0, \qquad (1.2\text{-}25c)$$

with $H(x, u, \lambda)$ defined by (1.2-16). The way we shall often use them, these three equations serve to determine x, λ, and u in that respective order. The last two of these equations are (1.2-11) and (1.2-12). In most applications determining the value of λ is not of interest, but this value is required, since it is an intermediate variable that allows us to determine the quantities of interest, u, x, and the minimum value of L.

The usefulness of the Lagrange-multiplier approach can be summarized as follows. In reality dx and du are not independent increments, because of (1.2-4). By introducing an undetermined multiplier λ, however, we obtain an extra degree of freedom, and λ can be selected to make dx and du behave *as if* they were independent increments. Therefore, setting independently to zero the gradients of H with respect to *all arguments* as in (1.2-25) yields a critical point. By introducing Lagrange multipliers, the problem of minimizing $L(x, u)$ subject to the constraint $f(x, u) = 0$ is replaced with the problem of *minimizing the Hamiltonian $H(x, u, \lambda)$ without constraints*.

Sufficient Conditions   Conditions (1.2-25) determine a stationary (critical) point. We are now ready to derive a test that guarantees that this point is a minimum. We proceed as we did in Section 1.1.

Write Taylor series expansions for increments in L and f as

$$dL = \begin{bmatrix} L_x^{\mathrm{T}} & L_u^{\mathrm{T}} \end{bmatrix} \begin{bmatrix} dx \\ du \end{bmatrix} + \frac{1}{2}\begin{bmatrix} dx^{\mathrm{T}} & du^{\mathrm{T}} \end{bmatrix}\begin{bmatrix} L_{xx} & L_{xu} \\ L_{ux} & L_{uu} \end{bmatrix}\begin{bmatrix} dx \\ du \end{bmatrix} + O(3), \quad (1.2\text{-}26)$$

$$df = \begin{bmatrix} f_x & f_u \end{bmatrix} \begin{bmatrix} dx \\ du \end{bmatrix} + \frac{1}{2}\begin{bmatrix} dx^{\mathrm{T}} & du^{\mathrm{T}} \end{bmatrix}\begin{bmatrix} f_{xx} & f_{xu} \\ f_{ux} & f_{uu} \end{bmatrix}\begin{bmatrix} dx \\ du \end{bmatrix} + O(3), \quad (1.2\text{-}27)$$

where

$$f_{xu} \triangleq \frac{\partial^2 f}{\partial u\, \partial x}$$

and so on. (What are the dimensions of f_{xu}?) To introduce the Hamiltonian, use these equations to see that

$$\begin{bmatrix} 1 & \lambda^{\mathrm{T}} \end{bmatrix}\begin{bmatrix} dL \\ df \end{bmatrix} = \begin{bmatrix} H_x^{\mathrm{T}} & H_u^{\mathrm{T}} \end{bmatrix}\begin{bmatrix} dx \\ du \end{bmatrix} + \frac{1}{2}\begin{bmatrix} dx^{\mathrm{T}} & du^{\mathrm{T}} \end{bmatrix}\begin{bmatrix} H_{xx} & H_{xu} \\ H_{ux} & H_{uu} \end{bmatrix}\begin{bmatrix} dx \\ du \end{bmatrix} + O(3).$$
$$(1.2\text{-}28)$$

A critical point requires that $f = 0$, and also that dL is zero in the first-order approximation for all increments dx, du. Since f is held equal to zero, df is also zero. Thus, these conditions require $H_x = 0$ and $H_u = 0$ exactly as in (1.2-25).

To find sufficient conditions for a minimum, let us examine the second-order term. First, it is necessary to include in (1.2-28) the dependence of dx on du. Hence, let us suppose we are at a critical point so that $H_x = 0$, $H_u = 0$, and $df = 0$. Then by (1.2-27)

$$dx = -f_x^{-1} f_u \, du + O(2). \tag{1.2-29}$$

Substituting this relation into (1.2-28) yields

$$dL = \frac{1}{2} du^{\mathrm{T}} \begin{bmatrix} -f_u^{\mathrm{T}} f_x^{-\mathrm{T}} & I \end{bmatrix} \begin{bmatrix} H_{xx} & H_{xu} \\ H_{ux} & H_{uu} \end{bmatrix} \begin{bmatrix} -f_x^{-1} f_u \\ I \end{bmatrix} du + O(3). \tag{1.2-30}$$

To ensure a minimum, dL in (1.2-30) should be positive for all increments du. This is guaranteed if the *curvature matrix with constant f equal to zero*

$$L_{uu}^f \triangleq L_{uu}|_f = \begin{bmatrix} -f_u^{\mathrm{T}} f_x^{-\mathrm{T}} & I \end{bmatrix} \begin{bmatrix} H_{xx} & H_{xu} \\ H_{ux} & H_{uu} \end{bmatrix} \begin{bmatrix} -f_x^{-1} f_u \\ I \end{bmatrix}$$

$$= H_{uu} - f_u^{\mathrm{T}} f_x^{-\mathrm{T}} H_{xu} - H_{ux} f_x^{-1} f_u + f_u^{\mathrm{T}} f_x^{-\mathrm{T}} H_{xx} f_x^{-1} f_u \tag{1.2-31}$$

is positive definite. Note that if the constraint $f(x, u)$ is identically zero for all x and u, then (1.2-31) reduces to L_{uu} in (1.1-6). If (1.2-31) is negative definite (indefinite), then the stationary point is a constrained maximum (saddle point).

Examples

To gain a feel for the theory we have just developed, let us consider some examples. The first example is a geometric problem that allows easy visualization, while the second involves a quadratic performance index and linear constraint. The second example is representative of the case that is used extensively in controller design for linear systems.

Example 1.2-1.   *Quadratic Surface with Linear Constraint*

Suppose the performance index is as given in Example 1.1-1:

$$L(x, u) = \frac{1}{2} [x \quad u] \begin{bmatrix} 1 & 1 \\ 1 & 2 \end{bmatrix} \begin{bmatrix} x \\ u \end{bmatrix} + [0 \quad 1] \begin{bmatrix} x \\ u \end{bmatrix}, \tag{1}$$

where we have simply renamed the old scalar components u_1, u_2 as x, u, respectively. Let the constraint be

$$f(x, u) = x - 3 = 0. \tag{2}$$

The Hamiltonian is

$$H = L + \lambda^T f = \frac{1}{2}x^2 + xu + u^2 + u + \lambda(x - 3), \tag{3}$$

where λ is a scalar. The conditions for a stationary point are (1.2-25), or

$$H_\lambda = x - 3 = 0, \tag{4}$$
$$H_x = x + u + \lambda = 0, \tag{5}$$
$$H_u = x + 2u + 1 = 0. \tag{6}$$

Solving in the order (4), (6), (5) yields $x = 3$, $u = -2$, and $\lambda = -1$. The stationary point is therefore

$$(x, u)^* = (3, -2). \tag{7}$$

To verify that (7) is a minimum, find the constrained curvature matrix (1.2-31):

$$L_{uu}^f = 2. \tag{8}$$

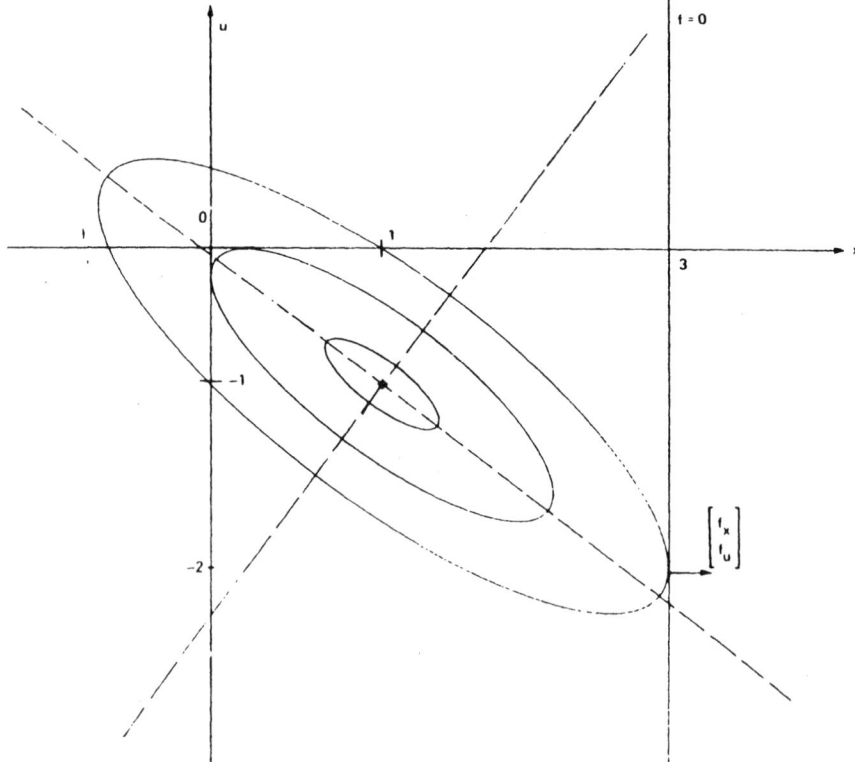

FIGURE 1.2-1   Contours of $L(x, u)$, and the constraint $f(x, u)$.

This is positive, so (7) is a minimum. The contours of $L(x, u)$ and the constraint (2) are shown in Fig. 1.2-1.

It is worthwhile to make an important point. The gradient of $f(x, u)$ in the (x, u) plane is

$$\begin{bmatrix} f_x \\ f_u \end{bmatrix} = \begin{bmatrix} 1 \\ 0 \end{bmatrix}, \tag{9}$$

as shown in Fig. 1.2-1. The gradient of $L(x, u)$ in the plane is

$$\begin{bmatrix} L_x \\ L_u \end{bmatrix} = \begin{bmatrix} x + u \\ x + 2u + 1 \end{bmatrix} \tag{10}$$

(cf. (9) in Example 1.1-1). At the constrained minimum $(3, -2)$, this has a value of

$$\begin{bmatrix} L_x \\ L_u \end{bmatrix} = \begin{bmatrix} 1 \\ 0 \end{bmatrix}. \tag{11}$$

Note that the gradients of f and L are *parallel* at the stationary point. This means that the constrained minimum occurs where the constraint (2) is tangent to an elliptical contour of L. Moving in either direction along the line $f = 0$ will then increase the value of L. The value of L at the constrained minimum is found by substituting $x = 3$, $u = -2$ into (1) to be $L^* = 0.5$. Since $\lambda = -1$, holding u constant at -2 and changing the constraint by df (i.e., moving the line in Fig. 1.2-1 to the right by df) will result in an *increase* in the value of $L(x, u)$ of $dL = -\lambda \, df = df$ (see (1.2-15)).   ∎

Example 1.2-2.   Quadratic Performance Index with Linear Constraint

Consider the quadratic performance index

$$L(x, u) = \frac{1}{2}x^{\mathrm{T}}Qx + \frac{1}{2}u^{\mathrm{T}}Ru \tag{1}$$

with linear constraint

$$f(x, u) = x + Bu + c = 0, \tag{2}$$

where $x \in R^n$, $u \in R^m$, $f \in R^n$, $\lambda \in R^n$, Q, R, and B are matrices, and c is an n vector. We assume $Q > 0$ and $R > 0$ (with both symmetric). This *static linear quadratic* (LQ) *problem* will be further generalized in Chapters 2 and 3 to apply to time-varying systems.

The contours of $L(x, u)$ are hyperellipsoids, and $f(x, u) = 0$ defines a hyperplane intersecting them. The stationary point occurs where the gradients of f and L are parallel.

The Hamiltonian is

$$H = \frac{1}{2}x^{\mathrm{T}}Qx + \frac{1}{2}u^{\mathrm{T}}Ru + \lambda^{\mathrm{T}}(x + Bu + c) \tag{3}$$

and the conditions for a stationary point are

$$H_\lambda = x + Bu + c = 0, \tag{4}$$

$$H_x = Qx + \lambda = 0, \tag{5}$$

$$H_u = Ru + B^{\mathrm{T}}\lambda = 0. \tag{6}$$

To solve these, first use the stationarity condition (6) to find an expression for u in terms of λ,

$$u = -R^{-1}B^{\mathrm{T}}\lambda. \tag{7}$$

According to (5)

$$\lambda = -Qx, \tag{8}$$

and taking into account (4) results in

$$\lambda = QBu + Qc. \tag{9}$$

Using this in (7) yields

$$u = -R^{-1}B^{\mathrm{T}}(QBu + Qc) \tag{10}$$

or

$$(I + R^{-1}B^{\mathrm{T}}QB)u = -R^{-1}B^{\mathrm{T}}Qc,$$

$$(R + B^{\mathrm{T}}QB)u = -B^{\mathrm{T}}Qc. \tag{11}$$

Since $R > 0$ and $B^{\mathrm{T}}QB > 0$, we can invert $R + B^{\mathrm{T}}QB$ and so the optimal control is

$$u = -(R + B^{\mathrm{T}}QB)^{-1}B^{\mathrm{T}}Qc. \tag{12}$$

Using (12) in (4) and (9) gives the optimal-state and multiplier values of

$$x = -(I - B(R + B^{\mathrm{T}}QB)^{-1}B^{\mathrm{T}}Q)c, \tag{13}$$

$$\lambda = (Q - QB(R + B^{\mathrm{T}}QB)^{-1}B^{\mathrm{T}}Q)c. \tag{14}$$

By the matrix inversion lemma (see Appendix A)

$$\lambda = (Q^{-1} + BR^{-1}B^{\mathrm{T}})^{-1}c \tag{15}$$

if $|Q| \neq 0$.

To verify that control (12) results in a minimum, use (1.2-31) to determine that the constrained curvature matrix is

$$L_{uu}^{f} = R + B^{\mathrm{T}}QB, \tag{16}$$

which is positive definite by our restrictions on R and Q. Using (12) and (13) in (1) yields the optimal value

$$L^{*} = \frac{1}{2}c^{\mathrm{T}}\left[Q - QB(R + B^{\mathrm{T}}QB)^{-1}B^{\mathrm{T}}Q\right]c, \tag{17}$$

$$L^{*} = \frac{1}{2}c^{\mathrm{T}}\lambda, \tag{18}$$

so that

$$\frac{\partial L^{*}}{\partial c} = \lambda. \tag{19}$$

∎

Effect of Changes in Constraints

Equation (1.2-28) expresses the increment dL in terms of df, dx, and du. In the discussion following that equation we let $df = 0$, found dx in terms of du, and expressed dL in terms of du. That gave us conditions for a stationary point ($H_x = 0$ and $H_u = 0$) and led to the second-order coefficient matrix L_{uu}^f in (1.2-31), which provided a test for the stationary point to be a constrained minimum.

In this subsection we are interested in dL as a function of an increment df in the constraint. We want to see how the performance index L changes in response to changes in the constraint f *if we remain at a stationary point*. We are therefore trying to find stationary points *near* a given stationary point. See Fig. 1.2-2, which shows how the stationary point moves with changes in f.

At the stationary point $(u, x)^*$ defined by $f(x, u) = 0$, the conditions $H_\lambda = 0$, $H_x = 0$, and $H_u = 0$ are satisfied. If the constraint changes by an increment so that $f(x, u) = df$, then the stationary point moves to $(u + du, x + dx)$. The partials in (1.2-25) change by

$$dH_\lambda = df = f_x\, dx + f_u\, du, \tag{1.2-32a}$$

$$dH_x = H_{xx}\, dx + H_{xu}\, du + f_x^T d\lambda, \tag{1.2-32b}$$

$$dH_u = H_{ux}\, dx + H_{uu}\, du + f_u^T d\lambda. \tag{1.2-32c}$$

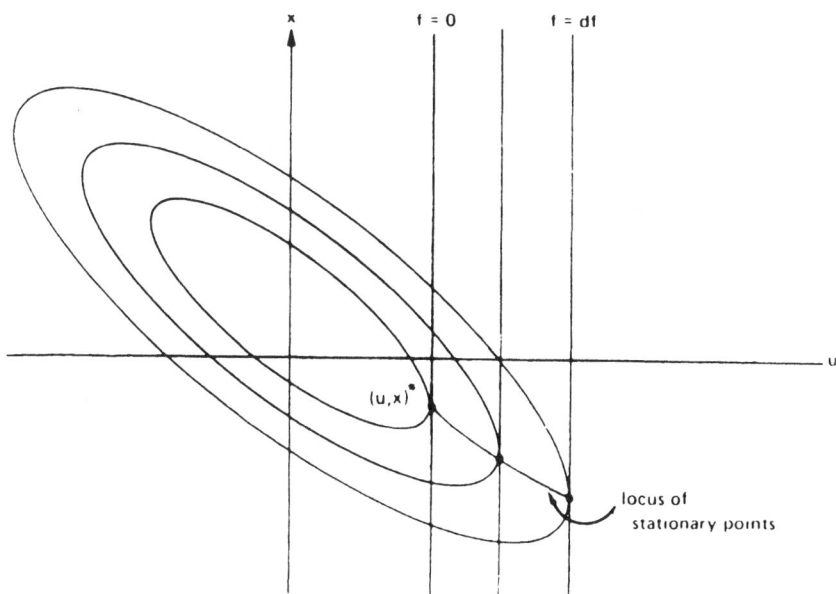

FIGURE 1.2-2 Locus of stationary points as the constraint varies.

In order that we remain at a stationary point, the increments dH_x and dH_u should be zero. This requirement imposes certain relations between the changes dx, du, and df, which we shall use in (1.2-28) to determine dL as a function of df.

To find dx and du as functions of df with the requirement that we remain at an optimal solution, use (1.2-32a) to find

$$dx = f_x^{-1} df - f_x^{-1} f_u \, du, \tag{1.2-33}$$

and set (1.2-32b) to zero to find

$$d\lambda = -f_x^{-T}(H_{xx} \, dx + H_{xu} \, du). \tag{1.2-34}$$

Now use these relations in (1.2-32c) to obtain

$$dH_u = \left(H_{uu} - H_{ux} f_x^{-1} f_u - f_u^T f_x^{-T} H_{xu} + f_u^T f_x^{-T} H_{xx} f_x^{-1} f_u \right) du$$
$$+ \left(H_{ux} - f_u^T f_x^{-T} H_{xx} \right) f_x^{-1} df = 0$$

so that

$$du = - \left(L_{uu}^f \right)^{-1} \left(H_{ux} - f_u^T f_x^{-T} H_{xx} \right) f_x^{-1} df \triangleq -C \, df. \tag{1.2-35}$$

Using (1.2-35) in (1.2-33) yields

$$dx = \left[I + f_x^{-1} f_u \left(L_{uu}^f \right)^{-1} \left(H_{ux} - f_u^T f_x^{-1} H_{xx} \right) \right] f_x^{-1} df$$
$$= f_x^{-1} \left(I + f_u C \right) df. \tag{1.2-36}$$

Equations (1.2-35) and (1.2-36) are the required expressions for the increments in the stationary values of control and state as functions of df. If $|L_{uu}^f| \neq 0$, then dx and du can be determined in terms of df, and the existence of neighboring optimal solutions as f varies is guaranteed.

To determine the increment dL in the optimal performance index as a function of df, substitute (1.2-35) and (1.2-36) into (1.2-28), using $H_x = 0$, $dH_u = 0$, since we began at a stationary point $(u, x)^*$. The result is found after some work to be

$$dL = -\lambda^T df + \frac{1}{2} df^T \left(f_x^{-T} H_{xx} f_x^{-1} - C^T L_{uu}^f C \right) df + O(3). \tag{1.2-37}$$

From this we see that the first and second partial derivatives of $L^*(x, u)$ with respect to $f(x, u)$ under the restrictions $dH_x = 0$, $dH_u = 0$ are

$$\left. \frac{\partial L^*}{\partial f} \right|_{H_x, H_u} = -\lambda, \tag{1.2-38}$$

$$\left. \frac{\partial^2 L^*}{\partial f^2} \right|_{H_x, H_u} = f_x^{-T} H_{xx} f_x^{-1} - C^T L_{uu}^f C. \tag{1.2-39}$$

Equation (1.2-38) allows a further interpretation of the Lagrange multiplier; it indicates the rate of change of the *optimal* value of the performance index with respect to the constraint.

1.3   NUMERICAL SOLUTION METHODS

Analytic solutions for the stationary point $(u, x)^*$ and minimal value L^* of the performance index cannot be found except for simple functions $L(x, u)$ and $f(x, u)$. In most practical cases, numerical optimization methods must be used. Many methods exist, but *steepest descent* or *gradient* (Luenberger 1969, Bryson and Ho 1975) methods are probably the simplest.

The steps in constrained minimization by the method of steepest descent are (Bryson and Ho 1975)

1. Select an initial value for u.
2. Determine x from $f(x, u) = 0$.
3. Determine λ from $\lambda = -f_x^{-T} L_x$.
4. Determine the gradient vector $H_u = L_u + f_u^T \lambda$.
5. Update the control vector by $\Delta u = -\alpha H_u$, where K is a positive scalar constant (to find a maximum use $\Delta u = \alpha H_u$).
6. Determine the predicted change in the value of L, $\Delta L = H_u^T \Delta u = -\alpha H_u^T H_u$. If ΔL is sufficiently small, stop. Otherwise, go to step 2.

There are many variations to this procedure. If the step-size constant K is too large, then the algorithm may overshoot the stationary point $(u, x)^*$ and convergence may not occur. The step size should usually be reduced as $(u, x)^*$ is approached, and several of the existing variations differ in the approach to adapting K.

Many software routines are available for unconstrained optimization. The numerical solution of the constrained optimization problem of minimizing $L(x, u)$ subject to $f(x, u) = 0$ can be obtained using the MATLAB function *constr.m* available under the Optimization Toolbox. This function takes in the user-defined subroutine *funct.m*, which computes the value of the function, the constraints, and the initial conditions.

PROBLEMS

Section 1.1

1.1-1. Find the critical points u^* (classify them) and the value of $L(u^*)$ in Example 1.1-1 if

a. $Q = \begin{bmatrix} -1 & 1 \\ 1 & -2 \end{bmatrix}$, $S^T = \begin{bmatrix} 0 & 1 \end{bmatrix}$.

b. $Q = \begin{bmatrix} -1 & 1 \\ 1 & 2 \end{bmatrix}$, $S^T = [0 \quad 1]$.

Sketch the contours of L and find the gradient L_u.

1.1-2. Find the minimum value of

$$L(x_1, x_2) = x_1^2 - x_1 x_2 + x_2^2 + 3x_1. \tag{1}$$

Find the curvature matrix at the minimum. Sketch the contours, showing the gradient at several points.

1.1-3. Failure of test for minimality. The function $f(x, y) = x^2 + y^4$ has a minimum at the origin.
a. Verify that the origin is a critical point.
b. Show that the curvature matrix is singular at the origin.
c. Prove that the critical point is indeed a minimum.

Section 1.2

1.2-1. Ship closest point of approach. A ship is moving at 10 miles per hour on a course of $30°$ (measured clockwise from north, which is $0°$). Find its closest point of approach to an island that at time $t = 0$ is 20 miles east and 30 miles north of it. Find the distance to the island at this point. Find the time of closest approach.

1.2-2. Shortest distance between two points. Let $P_1 = (x_1, y_1)$ and $P_2 = (x_2, y_2)$ be two given points. Find the third point $P_3 = (x_3, y_3)$ such that $d_1 = d_2$ is minimized, where d_1 is the distance from P_3 to P_1 and d_2 is the distance from P_3 to P_2.

1.2-3. Meteor closest point of approach. A meteor is in a hyperbolic orbit described with respect to the earth at the origin by

$$\frac{x^2}{a^2} - \frac{y^2}{b^2} = 1. \tag{1}$$

Find its closest point of approach to a satellite that is in such an orbit that it has a constant position of (x_1, y_1). Verify that the solution indeed yields a minimum.

1.2-4. Shortest distance between a parabola and a point. A meteor is moving along the path

$$y = x^2 + 3x - 6. \tag{1}$$

A space station is at the point $(x, y) = (2, 2)$.
a. Use Lagrange multipliers to find a cubic equation for x at the closest point of approach.

b. Find the closest point of approach (x, y), and the distance from this point to $(2, 2)$.

1.2-5. Rectangles with maximum area, minimum perimeter

a. Find the rectangle of maximum area with perimeter p. That is, maximize

$$L(x, y) = xy \tag{1}$$

subject to

$$f(x, y) = 2x + 2y - p = 0. \tag{2}$$

b. Find the rectangle of minimum perimeter with area a^2. That is, minimize

$$L(x, y) = 2x + 2y \tag{3}$$

subject to

$$f(x, y) = xy - a^2 = 0. \tag{4}$$

c. In each case, sketch the contours of $L(x, y)$ and the constraint. Optimization problems related like these two are said to be *dual*.

1.2-6. Linear quadratic case. Minimize

$$L = \frac{1}{2}x^{\mathrm{T}}\begin{bmatrix} 1 & 0 \\ 0 & 2 \end{bmatrix}x + \frac{1}{2}u^{\mathrm{T}}\begin{bmatrix} 2 & 1 \\ 1 & 1 \end{bmatrix}u$$

if

$$x = \begin{bmatrix} 1 \\ 3 \end{bmatrix} = \begin{bmatrix} 2 & 2 \\ 1 & 0 \end{bmatrix}u.$$

Find x^*, u^*, λ^*, L^*.

1.2-7. Linear quadratic case. In the LQ problem define the *Kalman gain*

$$K \triangleq (B^{\mathrm{T}}QB + R)^{-1}B^{\mathrm{T}}Q \tag{1}$$

a. Express u^*, λ^*, x^*, and L^* in terms of K.

b. Let

$$S_0 \triangleq Q - QB(B^{\mathrm{T}}QB + R)^{-1}B^{\mathrm{T}}Q \tag{2}$$

so that $L^* = c^{\mathrm{T}}S_0c/2$. Show that

$$S_0 = Q(I - BK) = (I - BK)^{\mathrm{T}}Q(I - BK) + K^{\mathrm{T}}RK. \tag{3}$$

Hence, factor L^* as a perfect square. (Let $\sqrt{Q}$ and $\sqrt{R}$ be the square roots of Q and R.)

c. Show that

$$S_0 = (Q^{-1} + BR^{-1}B^T)^{-1}. \tag{4}$$

1.2-8. Geometric mean less than or equal to arithmetic mean

a. Show that the minimum value of $x^2y^2z^2$ on the sphere $x^2 + y^2 + z^2 = r^2$ is $(r^2/3)^3$.

b. Show that the maximum value of $x^2 + y^2 + z^2$ on the sphere $x^2y^2z^2 = (r^2/3)^3$ is r^2.

c. Generalize part a or b and so deduce that, for $a_i > 0$,

$$(a_1a_2 \cdots a_n)^{1/n} \le (a_1 + a_2 + \cdots + a_n)/n.$$

Note: The problems in parts a and b are dual (Fulks 1967).

1.2-9. Find the point nearest the origin on the line $3x + 2y + z = 1$, $x + 2y - 3z = 4$.

1.2-10. Rectangle inside Ellipse

a. Find the rectangle of maximum perimeter that can be inscribed inside an ellipse. That is, maximize $4(x + y)$ subject to constraint $x^2/a^2 + y^2/b^2 = 1$.

b. Find the rectangle of maximum area $4xy$ that can be inscribed inside an ellipse.

2

OPTIMAL CONTROL
OF DISCRETE-TIME SYSTEMS

We are now ready to extend the methods of Chapter 1 to the optimization of a performance index associated with a system developing dynamically through time. It is important to realize that we shall be making a fairly subtle change of emphasis. In Chapter 1, the focus of our attention was initially on the performance index, and we introduced the notion of constraints as the discussion proceeded. In this and subsequent chapters we are forced to begin with the constraint equations, since these represent the dynamics of the system. These constraint relations are *fixed by the physics of the problem*. The performance index is *selected by the engineer* as it represents the desired behavior of the dynamical system.

In Section 2.1 we derive the general solution of the optimization problem for discrete-time systems. In Section 2.2 we discuss the very important special case of linear systems with a quadratic performance index. We first discuss the case of *fixed final state*, which yields an open-loop control, followed by the situation of *free final state*, which yields a closed-loop control. In Section 2.3 we show how to apply these results to the digital control of continuous-time systems.

Some connections with classical root-locus design are given in Section 2.5.

2.1 SOLUTION OF THE GENERAL DISCRETE-TIME OPTIMIZATION PROBLEM

Problem Formulation

Let the plant be described by the very general nonlinear discrete-time dynamical equation

$$x_{k+1} = f^k(x_k, u_k) \qquad (2.1\text{-}1)$$

19

with initial condition x_0. The superscript on function f indicates that, in general, the system, and thus its model, can have time-varying dynamics. Let the state x_k be a vector of size n and the control input u_k be a vector of size m. Equation (2.1-1) represents the constraint, since it determines the state at time $k + 1$ given the control and state at time k. Clearly, f is a vector of n functions.

Let an associated scalar performance index, specified by the engineer, be given in the general form

$$J_i = \phi(N, x_N) + \sum_{k=i}^{N-1} L^k(x_k, u_k),$$ (2.1-2)

where $[i, N]$ is the time interval, on a discrete time scale with a fixed sample step, over which we are interested in the behavior of the system. $\phi(N, x_N)$ is a function of the final time N and the state at the final time, and $L^k(x_k, u_k)$ is a generally time-varying function of the state and control input at each intermediate time k in $[i, N]$.

The *optimal control problem* is to find the control u_k^* on the interval $[i, N]$ (i.e., $u_k^*, \forall k \in [i, N]$) that drives the system (2.1-1) along a trajectory x_k^* such that the value of the performance index (2.1-2) is optimized.

Here we note that relative to the meaning that it is attached to the performance index, the optimization problem can be either a minimization or a maximization problem. For the case that the performance index represents the costs accrued during the operation of the system over the time interval $[i, N]$, the optimal control input is determined to minimize the performance index, while in the situation related to accumulation of value over the time interval $[i, N]$, the optimal control input is determined to minimize the performance index (2.1-2). As in most industrial applications the optimal control problem deals with minimization of control errors as well as of control effort, without reducing the generality of the formulation, herein we will treat the optimal control problem as a minimization problem.

Example 2.1-1.   Some Useful Performance Indices

To clarify the problem formulation, it is worthwhile to discuss some common performance indices that we can select for the given system (2.1-1).

a. Minimum-time Problems

Suppose we want to find the control u_k to drive the system from the given initial state x_0 to a desired final state $x \in R^n$ in minimum time. Then we could select the performance index

$$J = N = \sum_{k=0}^{N-1} 1$$ (1)

and specify the *boundary condition*

$$x_N = x.$$ (2)

In this case one can consider either $\phi = N$ and $L = 0$, or equivalently $\phi = 0$ and $L = 1$.

b. Minimum-fuel Problems

To find the scalar control u_k to drive the system from x_0 to a desired final state x at a fixed time N using minimum fuel, we could use

$$J = \sum_{k=0}^{N-1} |u_k|, \tag{3}$$

since the fuel burned is proportional to the magnitude of the control vector. Then $\phi = 0$ and $L_k = |u_k|$. The boundary condition $x_N = x$ would again apply.

c. Minimum-energy Problems

Suppose we want to find u_k to minimize the energy of the final state and all intermediate states, and also that of the control used to achieve this. Let the final time N again be fixed. Then we could use

$$J = \frac{1}{2} s x_N^T x_N + \frac{1}{2} \sum_{k=0}^{N-1} \left(q x_k^T x_k + r u_k^T u_k \right), \tag{4}$$

where q, r, and s are scalar weighting factors. Then $\phi = \frac{1}{2} s x_N^T x_N$ and $L = \frac{1}{2} (q x_k^T x_k + r u_k^T u_k)$ are quadratic functions.

Minimizing the energy corresponds in some sense to keeping the state and the control close to zero. If it is more important to us that the intermediate state be small, then we should choose q *large* to weight it heavily in J, which we are trying to minimize. If it is more important that the control energy be small, then we should select a large value of r. If we are more interested in a small final state, then s should be large.

For more generality, we could select weighting *matrices* Q, R, S instead of scalars. The performance index can in this case be represented as

$$J = \frac{1}{2} x_N^T S x_N + \frac{1}{2} \sum_{k=0}^{N-1} \left(x_k^T Q x_k + u_k^T R u_k \right). \tag{5}$$

∎

At this point, several things should be clearly understood. First, the system dynamics (2.1-1) are *given* by the physics of the problem, while the performance index (2.1-2) is what we *choose* to achieve the desired system response. Second, to achieve different control objectives, different types of performance indices J are selected. Finally, the optimal control problem is characterized by *compromises and trade-offs*, with different weighting factors in J resulting in different balances between conformability with performance objectives and magnitude of the required optimal controls.

In practice, it is usually necessary to do a control design with a trial performance index J, compute the optimal control u_k^*, and then run a computer simulation to see how the system responds to this u_k^*. If the response is not acceptable, the entire process is repeated using another J with different state and control weightings. After several repetitions have been done to find an acceptable u_k^*, this final version of u_k^* is applied to the actual system.

Problem Solution

Let us now solve the optimal control problem for the general nonlinear system (2.1-1) with associated performance index (2.1-2). To determine the optimal control sequence $u_i^*, u_{i+1}^*, \ldots, u_{N-1}^*$ minimizing J, we proceed basically as we did in Chapter 1, using the powerful Lagrange-multiplier approach. Since there is a constraint function $f^k(x_k, u_k)$ specified at each time k in the interval of interest $[i, N]$, we also require a Lagrange multiplier at each time. *Each constraint has an associated Lagrange multiplier.*

Thus, let $\lambda_k \in R^n$, and append the constraint (2.1-1) to the performance index (2.1-2) to define an augmented performance index J' by

$$ J' = \phi(N, x_N) + \sum_{k=i}^{N-1} \left[L^k(x_k, u_k) + \lambda_{k+1}^{\mathrm{T}} \left(f^k(x_k, u_k) - x_{k+1} \right) \right]. \qquad (2.1\text{-}3) $$

Note that we have associated with f^k the multiplier λ_{k+1}, not λ_k. This is done with the benefit of hindsight, as it makes the solution neater.

Defining the Hamiltonian function as

$$ H^k(x_k, u_k) = L^k(x_k, u_k) + \lambda_{k+1}^{\mathrm{T}} f^k(x_k, u_k), \qquad (2.1\text{-}4) $$

we can write

$$ J' = \phi(N, x_N) - \lambda_N^{\mathrm{T}} x_N + H^i(x_i, u_i) + \sum_{k=i+1}^{N-1} \left[H^k(x_k, u_k) - \lambda_k^{\mathrm{T}} x_k \right], \qquad (2.1\text{-}5) $$

where some minor manipulations with indices have been performed. Note that the Hamiltonian is defined slightly differently than in Chapter 1, since we did not include x_{k+1} in H^k. Furthermore, a Hamiltonian is defined at each time k.

We now want to examine the increment in J' due to increments in all the variables x_k, λ_k, and u_k. We assume the final time N is fixed. According to the Lagrange-multiplier theory, at a constrained minimum this increment dJ' should be zero. Therefore, write

$$ dJ' = (\phi_{x_N} - \lambda_N)^{\mathrm{T}} dx_N + \left(H_{x_i}^i \right)^{\mathrm{T}} dx_i + \left(H_{u_i}^i \right)^{\mathrm{T}} du_i $$

$$ + \sum_{k=i+1}^{N-1} \left[\left(H_{x_k}^k - \lambda_k \right)^{\mathrm{T}} dx_k + \left(H_{u_k}^k \right)^{\mathrm{T}} du_k \right] + \sum_{k=i+1}^{N} \left(H_{\lambda_k}^{k-1} - x_k \right)^{\mathrm{T}} d\lambda_k, \qquad (2.1\text{-}6) $$

where

$$ H_{x_k}^k \triangleq \frac{\partial H^k}{\partial x_k} $$

and so on. Necessary conditions for a constrained minimum are thus given by

$$ x_{k+1} = \frac{\partial H^k}{\partial \lambda_{k+1}}, \quad k = i, \ldots, N-1, \qquad (2.1\text{-}7a) $$

$$\lambda_k = \frac{\partial H^k}{\partial x_k}, \quad k = i, \ldots, N-1, \tag{2.1-7b}$$

$$0 = \frac{\partial H^k}{\partial u_k}, \quad k = i, \ldots, N-1, \tag{2.1-7c}$$

which arise from the terms inside the summations and the coefficient of du_i, and

$$\left(\frac{\partial \phi}{\partial x_N} - \lambda_N\right)^{\mathrm{T}} dx_N = 0, \tag{2.1-8a}$$

$$\left(\frac{\partial H^i}{\partial x_i}\right)^{\mathrm{T}} dx_i = 0. \tag{2.1-8b}$$

Examining (2.1-3) and (2.1-4) one can see that λ_i does not appear in J'. We have defined it in such a manner that the lower index in (2.1-7b) can be taken as i, instead of $i + 1$, solely as a matter of neatness.

These conditions are certainly not intuitively obvious, so we should discuss them a little to see what they mean. First, compare (2.1-7) with the conditions for a static minimum (1.2-25). They are very similar, except that our new conditions must hold at each time k in the interval of interest $[i, N-1]$, since x_k, u_k, and λ_k are now sequences. Equation (2.1-7c) is called the *stationarity condition*.

Writing (2.1-7) explicitly in terms of L^k and f^k using (2.1-4) yields the formulation in Table 2.1-1. Equality (2.1-9a) is just the constraint, or system, equation. It is a recursion for the state x_k that develops *forward* in time. Evidently, (2.1-9b) is a recursion for λ_k that develops *backward* in time! The (fictitious) Lagrange multiplier is thus a variable that is determined by its own dynamical equation. It is called the *costate* of the system, and (2.1-9b) is called the *adjoint system*. The system (2.1-9a) and the adjoint system (2.1-9b) are coupled difference equations. They define a *two-point boundary-value problem*, since the boundary conditions required for solution are the *initial* state x_i and the *final* costate λ_N. These problems are, in general, extremely difficult to solve. We consider some examples later.

The stationarity condition (2.1-9c) allows the optimal control u_k to be expressed in terms of the costate. We therefore have a rather curious situation: we do not really care what λ_k is, but this method of solution requires us to find λ_k as an intermediate step in finding the optimal control.

We have not yet discussed (2.1-8). The first of these equations holds only at final time $k = N$, whereas the second holds only at initial time $k = i$. They are not dynamical recursions like (2.1-7a) and (2.1-7b). These two equations specify the *split boundary conditions* needed to solve the recursions (2.1-9). Two possibilities exist for each of these equations.

If the *initial state* x_i *is fixed*, then $dx_i = 0$, so that (2.1-8b) holds regardless of the value of $H^i_{x_i}$. In the case of *free initial state*, dx_i is not zero, so (2.1-8b) demands that

$$\frac{\partial H^i}{\partial x_i} = 0. \tag{2.1-10}$$

TABLE 2.1-1   Discrete Nonlinear Optimal Controller

System model:

$$x_{k+1} = f^k(x_k, u_k), \quad k > i$$

Performance index:

$$J_i = \phi(N, x_N) + \sum_{k=i}^{N-1} L^k(x_k, u_k)$$

Hamiltonian:

$$H^k = L^k + \lambda_{k+1}^{\mathrm{T}} f^k$$

Optimal controller
  State equation:

$$x_{k+1} = \frac{\partial H^k}{\partial \lambda_{k+1}} = f^k(x_k, u_k) \tag{2.1-9a}$$

  Costate equation:

$$\lambda_k = \frac{\partial H_k}{\partial x_k} = \left(\frac{\partial f^k}{\partial x_k}\right)^{\mathrm{T}} \lambda_{k+1} + \frac{\partial L^k}{\partial x_k} \tag{2.1-9b}$$

  Stationarity condition:

$$0 = \frac{\partial H^k}{\partial u_k} = \left(\frac{\partial f^k}{\partial u_k}\right)^{\mathrm{T}} \lambda_{k+1} + \frac{\partial L^k}{\partial u_k} \tag{2.1-9c}$$

  Boundary conditions:

$$\left(\frac{\partial L^i}{\partial x_i} + \left(\frac{\partial f^i}{\partial x_i}\right)^{\mathrm{T}} \lambda_{i+1}\right)^{\mathrm{T}} dx_i = 0$$

$$\left(\frac{\partial \phi}{\partial x_N} - \lambda_N\right)^{\mathrm{T}} dx_N = 0$$

In our applications the system starts at a known initial state x_i. Thus, the first case holds, $dx_i = 0$, and there is no constraint on the value of $H_{x_i}^i$. We therefore ignore (2.1-8b) and use as the initial condition the given value of x_i.

We need to deal with two possibilities for the final state x_N. In the case of a *fixed final state* we use the desired value of x_N as the terminal condition. Then x_N is not free to be varied in determining the optimal solution and $dx_N = 0$, so that (2.1-8a) holds. On the other hand, if we are not interested in a particular value for the final state, then x_N can be varied in determining the optimal solution. In this case dx_N is not zero. For this *free-final-state* situation, (2.1-8a) demands that

$$\lambda_N = \frac{\partial \phi}{\partial x_N}. \tag{2.1-11}$$

Then, the terminal condition is the value (2.1-11) of the final costate λ_N.

In summary, the initial condition for the two-point boundary-value problem (2.1-9) is the known value of x_i. The final condition is either a desired value of x_N or the value (2.1-11) of λ_N. These comments will become clearer as we proceed.

An Example

To develop some feel for the theory we have just derived, let us consider an example. We shall see that the solution of the optimal control problem is not straightforward even in the simplest cases, because of the two-point nature of the state and costate equations, but that once the solution is obtained it imparts a great deal of intuition about the control of the system.

We also show how to run software simulations to test our optimal control designs.

Example 2.1-2.   Optimal Control for a Scalar Linear System

Consider the simple linear dynamical system

$$x_{k+1} = ax_k + bu_k, \tag{1}$$

where lowercase a and b are used to emphasize that we are dealing with the scalar case. Let the given initial condition be x_0. Suppose the interval of interest is $[0, N]$ and that we are concerned with minimizing control energy so that

$$J_0 = \frac{r}{2} \sum_{k=0}^{N-1} u_k^2 \tag{2}$$

for some scalar weighting factor r.

Let us discuss two cases, corresponding to two ways in which we might want the system to behave.

a. Fixed Final State

First, suppose we want to make the system end up at time $k = N$ in exactly the particular (reference) state r_N:

$$x_N = r_N. \tag{3}$$

To find the optimal control sequence $u_0, u_1, \ldots, u_{N-1}$ (note that x_N does not depend on u_N) that drives (1) from the given x_0 to the desired $x_N = r_N$ while minimizing (2), we can use Table 2.1-1. First, let us compute (2.1-9). The Hamiltonian is

$$H^k = L^k + \lambda_{k+1}^T f^k = \frac{r}{2} u_k^2 + \lambda_{k+1}(ax_k + bu_k), \tag{4}$$

so the conditions (2.1-9) are

$$x_{k+1} = \frac{\partial H^k}{\partial \lambda_{k+1}} = ax_k + bu_k, \tag{5}$$

$$\lambda_k = \frac{\partial H^k}{\partial x_k} = a\lambda_{k+1}, \tag{6}$$

$$0 = \frac{\partial H^k}{\partial u_k} = ru_k + b\lambda_{k+1}. \tag{7}$$

Solving the stationarity condition (7) for u_k in terms of the costate yields

$$u_k = -\frac{b}{r}\lambda_{k+1}. \tag{8}$$

If we can find the optimal λ_k, we can therefore use (8) to find the optimal control. To find λ_k, eliminate u_k in (5) using (8). Then

$$x_{k+1} = ax_k - \frac{b^2}{r}\lambda_{k+1} = ax_k - \gamma\lambda_{k+1}, \tag{9}$$

where

$$\gamma \triangleq \frac{b^2}{r} \tag{10}$$

is the ratio of "control effect" to control weighting.

Equation (6) is a simple homogeneous difference equation, with solution given by

$$\lambda_k = a^{N-k}\lambda_N. \tag{11}$$

This is all well and good, but we do not know λ_N. To find it, proceed as follows. Use (11) in (9) to get

$$x_{k+1} = ax_k - \gamma a^{N-k-1}\lambda_N. \tag{12}$$

This can be viewed as a difference equation with a forcing function of $-\gamma\lambda_N a^{N-k-1}$, so the solution in terms of x_0 is

$$x_k = a^k x_0 - \sum_{i=0}^{k-1} a^{k-i-1}\left(\gamma\lambda_N a^{N-i-1}\right)$$

$$= a^k x_0 - \gamma\lambda_N a^{N+k-2}\sum_{i=0}^{k-1} a^{-2i}. \tag{13}$$

Using the formula for the sum of a geometric series we have

$$x_k = a^k x_0 - \gamma\lambda_N a^{N+k-2}\frac{1-a^{-2k}}{1-a^{-2}}$$

$$= a^k x_0 - \gamma\lambda_N a^{N-k}\frac{1-a^{2k}}{1-a^2}. \tag{14}$$

The state at time k is thus a linear combination of the known initial state and the as yet unknown final costate. To find λ_N, we need to make use of the boundary conditions (2.1-8).

Since x_0 is fixed, $dx_0 = 0$ and (2.1-8b) is satisfied. Since we are also demanding the fixed final state (3), we have $dx_N = 0$ so that (2.1-8a) is satisfied. In words, the algorithm cannot vary either x_0 or x_N in determining the constrained minimum for this problem.

According to (14), the final state is expressed in terms of the unknown λ_N as

$$x_N = a^N x_0 - \frac{\gamma(1-a^{2N})}{1-a^2}\lambda_N = a^N x_0 - \Lambda\lambda_N, \tag{15}$$

where

$$\Lambda \triangleq \frac{\gamma(1 - a^{2N})}{1 - a^2} = \frac{b^2(1 - a^{2N})}{r(1 - a^2)}.$$ (16)

(Defining auxiliary variables is a good trick for making our results look neater than they actually are!) Solving for λ_N in terms of the given x_0 and the known desired $x_N = r_N$ yields

$$\lambda_N = -\frac{1}{\Lambda}(r_N - a^N x_0).$$ (17)

Note that $a^N x_0$ is the final state of the plant (1) in the case of zero input. The final costate λ_N is thus proportional to the desired final state r_N minus the final state $a^N x_0$, which the system would reach by itself with no control; it makes sense that the control required to drive x_0 to $x_N = r_N$ should depend on this difference!

At this point, we can determine the costate to be, using (11),

$$\lambda_k = -\frac{1}{\Lambda}\left(r_N - a^N x_0\right)a^{N-k}$$ (18)

and the optimal control to be, using (8),

$$u_k^* = \frac{b}{r\Lambda}\left(r_N - a^N x_0\right)a^{N-k-1}.$$ (19)

This is the solution to our problem, and u_k^*, for $k = 0, 1, \ldots, N - 1$ will drive x_0 to $x_N = r_N$ while minimizing the control energy (2).

It is worthwhile to examine u_k^* a little bit. Note that (19) can be written

$$u_k^* = \frac{1 - a^2}{b(1 - a^{2N})}\left(r_N - a^N x_0\right)a^{N-k-1},$$ (20)

so that in the case of fixed final state the optimal control is independent of the control weighting r. It should also be stressed that u_k^* given by (19) is an *open-loop control*. It depends on the initial and final states, but not on intermediate values x_k of the state. This is discussed further in Section 2.2.

For completeness, let us determine the optimal state trajectory x_k^* and performance index J_0^* under the influence of u_k^*. Substituting (20) in (1) yields

$$x_{k+1}^* = ax_k^* + \frac{1 - a^2}{1 - a^{2N}}\left(r_N - a^N x_0\right)a^{N-k-1}.$$ (21)

The first observation worthy of note is that the optimal state trajectory x_k^* is independent of both r and b!

Equation (21) is a dynamical system with forcing function given by the second term, so its solution is

$$x_k^* = a^k x_0 + \frac{1 - a^2}{1 - a^{2N}}\left(r_N - a^N x_0\right)\sum_{i=0}^{k-1} a^{k-i-1}a^{N-i-1}.$$ (22)

Using the formula for the sum of a geometric series and simplifying yields

$$x_k^* = a^k x_0 + \left(r_N - a^N x_0\right)\frac{1 - a^{2k}}{1 - a^{2N}}a^{N-k}$$ (23)

or

$$x_k^* = \frac{(1 - a^{2(N-k)})a^k x_0 + (1 - a^{2k})a^{N-k} r_N}{1 - a^{2N}}. \tag{24}$$

Note that $x_0^* = x_0$ and $x_N^* = r_N$ as required. In fact, x_k^* is a time-varying linear combination of x_0 and r_N containing proportionately less of x_0 and more of r_N as k increases from 0 to N.

The optimal performance index is found by using (20) in (2):

$$J_0^* = \frac{r}{2} \frac{(1 - a^2)^2}{b^2 (1 - a^{2N})^2} (r_N - a^N x_0)^2 \sum_{k=0}^{N-1} a^{2(N-k-1)}. \tag{25}$$

Using the formula for the sum of a geometric series and simplifying results in

$$J_0^* = \frac{1}{2\Lambda} (r_N - a^N x_0)^2. \tag{26}$$

Thus, the farther the plant zero-input response $a^N x_0$ is from the desired final state r_N, the larger the cost J_0^*.

b. Free Final State

Suppose that we still desire the system to end up in state r_N at time x_N, but we decide to choose quite a different method for ensuring this. We do not need x_N to be exactly equal to r_N, we only need x_N to be *close to* r_N. Let us, therefore, make the difference $x_N - r_N$ small by including it in the performance index, so that (2) becomes

$$J_0 = \frac{1}{2}(x_N - r_N)^2 + \frac{r}{2} \sum_{k=0}^{N-1} u_k^2. \tag{27}$$

Now the optimal control will attempt to make $|x_N - r_N|$ small while also using low control energy. (As we shall see, this will not guarantee that x_N will exactly equal r_N.) In this case

$$\phi = \tfrac{1}{2}(x_N - r_N)^2, \tag{28}$$

but f_k and L_k are not changed. The Hamiltonian is still given by (4), and conditions (2.1-9) are still (5)–(7). This means that all of our work in part a up through (15) is unchanged by adding to J the final-state weighting term. The only change is in the boundary conditions (2.1-8).

Since x_N is not constrained to take on an exact value, it can be varied in determining the optimal control. Hence, $dx_N \neq 0$, and so according to (2.1-8a) we must have

$$\lambda_N = \frac{\partial \phi}{\partial x_N} = x_N - r_N. \tag{29}$$

The final costate is now related to the final state x_N and the desired final state r_N by (29); this is our new terminal condition for the solution of (6) and (9). The initial condition is still the given value of x_i.

Returning to part a and picking up at equation (15), we must now use (29) (instead of (3)) to see that at the final time

$$x_N = a^N x_0 - \Lambda(x_N - r_N). \tag{30}$$

Solving for x_N gives

$$x_N = \frac{\Lambda r_N + a^N x_0}{1 + \Lambda},$$

(31)

which should be compared to (17). Thus, fixing x_N by (3) allowed us to solve for the final *costate* in part a, while here the terminal relation (29) has allowed us to solve for the final *state* in terms of x_0 and the desired r_N.

According to (29) and (31),

$$\lambda_N = \frac{-(r_N - a^N x_0)}{1 + \Lambda},$$

(32)

so the costate is given by (11) as

$$\lambda_k = \frac{-(r_N - a^N x_0)}{1 + \Lambda} a^{N-k}$$

(33)

(cf. (18)), and the optimal control is given by (8) as

$$u_k^* = \frac{b}{r(1 + \Lambda)}(r_N - a^N x_0) a^{N-k-1}.$$

(34)

This is the optimal control that solves our free-final-state problem. Note that, unlike (19), the control (34) does depend on r.

In the limit as $r \to 0$, we are concerned less and less about the control energy we use since u_k^2 is weighted less and less heavily in J_0. In this case the free-final-state control (34) tends to the fixed-final-state control (19) (which is independent of r). Therefore, the less we are concerned about control energy (i.e., the smaller we make r in (27)), the closer the final state x_N comes to the desired r_N. This illustrates quite nicely the characteristic trade-off of optimal control that we discussed earlier.

As we let r go to infinity, meaning that we are expressing more concern about the control energy by weighting it more heavily in J_0, the optimal control goes to zero. For completeness, let us determine the optimal state trajectory and performance index under the influence of (34). Substitute (34) into (1) to get

$$x_{k+1}^* = ax_k^* + \frac{b^2}{r(1 + \Lambda)}(r_N - a^N x_0) a^{N-k-1}.$$

(35)

Solving and manipulating yields

$$x_k^* = \frac{[(1 - a^2)/\gamma + (1 - a^{2(N-k)})]a^k x_0 + (1 - a^{2k})a^{N-k} r_N}{(1 - a^2)/\gamma + (1 - a^{2N})}.$$

(36)

Note that $x_0^* = x_0$, but

$$x_N^* = \frac{[(1 - a^2)/\gamma]a^N x_0 + (1 - a^{2N}) r_N}{(1 - a^2)/\gamma + (1 - a^{2N})}.$$

(37)

As $r \to 0$, we have $\gamma \to \infty$, so that x_N^* approaches the desired r_N. In fact, if $\gamma \to \infty$, then the entire optimal state trajectory (36) approaches (24).

To find the optimal value of the performance index, use (34) in (2) and simplify to obtain

$$J_0^* = \frac{\Lambda}{2(1+\Lambda)^2}(r_N - a^N x_0)^2. \tag{38}$$

As $r \to 0$, we have $\Lambda \to \infty$, so that this approaches the fixed-final-state cost (26).

c. Computer Simulation

In practical situations, the optimal control should be simulated to ensure that it results in acceptable system behavior before it is applied to the physical system. This is easy to do. Figure 2.1-1 shows a computer program in MATLAB to simulate the plant (1) with fixed-final-state optimal control (19). We are using an initial state of $x_0 = 0$ and a desired final state of $r_N = 10$. Figure 2.1-2 shows a computer program in MATLAB to simulate the plant (1) with free-final-state optimal control (19). Figure 2.1-3 shows the optimal state trajectory x_k^* for the fixed-final-state (i.e., $r = 0$) and for the free final state control with several values of weighting r, system parameters $a = .99$, $b = .1$, and $N = 100$. Figure 2.1-4 shows the corresponding optimal control sequences u_k^*. As expected, x_N approaches r_N and the control energy increases as r becomes small.

```
function [x, u] =scoptco_fixed (a, b, r, N, x0, rN)
% Simulation of Optimal Control for Scalar Systems
% Fixed Final State Case
x(1) =x0;
alam= (1-a^ (2*N)) / (1-a^2); alam=alam*b^2;
u(1) =b*(rN-x(1)*a^N)*a^N/ (alam);
u(1) =u(1) / a;
for k=1:N
% Update the Plant State
x(k+1)=a*x(k) +b*u(k);
% Update the Optimal Control Input
u(k+1) =u(k) /a;
end
```

FIGURE 2.1-1   MATLAB simulation of fixed-final-state optimal control.

```
function [x, u]=scoptco_free (a, b, r, N, x0, rN)
% Simulation of Optimal Control for Scalar Systems
% Free Final State Case
x(1) =x0;
alam=(1-a^ (2*N)) / (1-a^ 2); alam=alam*b^2/r;
u(1) =b*(rN-x(1)*a^N)*a^N/ (r*(alam+1));
u(1) =u(1) /a;
for k=1:N
% Update the Plant State
x(k+1) =a*x(k) +b*u(k);
% Update the Optimal Control Input
u(k+1) =u(k) /a;
end
```

FIGURE 2.1-2   MATLAB simulation of free-final-state optimal control.

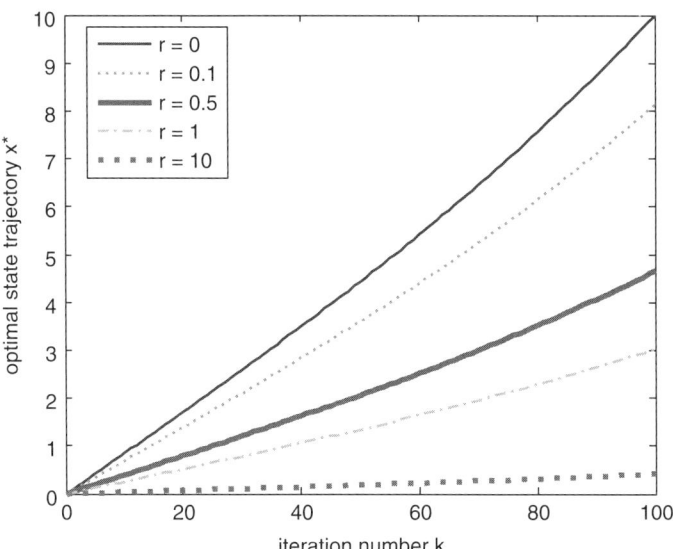

FIGURE 2.1-3 State trajectories for the fixed-final-state problem ($r = 0$) and for the free-final-state problem for several values of r.

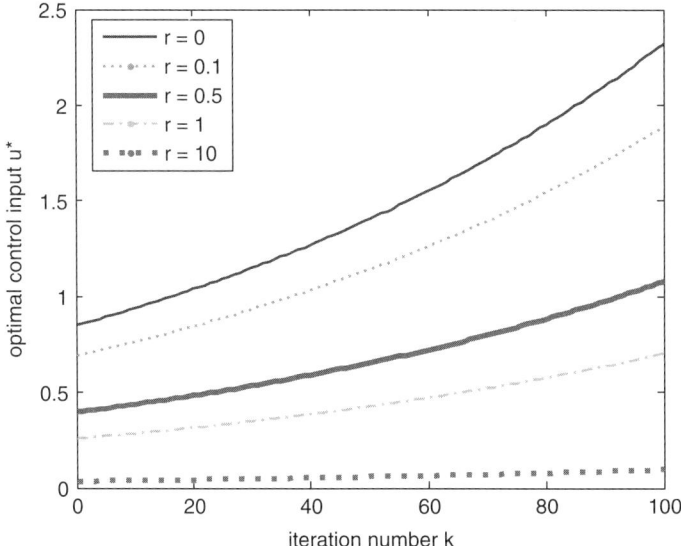

FIGURE 2.1-4 Optimal control functions corresponding to Fig. 2.1-3.

Using these simulation results, we can select r and the corresponding control u_k^* to yield an acceptable state trajectory and control energy for our particular application. This control is then used on the actual system. ∎

2.2 DISCRETE-TIME LINEAR QUADRATIC REGULATOR

Table 2.1-1 provides the solution to the optimal control problem for nonlinear systems with very general performance indices, but explicit expressions for the optimal control are difficult to deduce. In this section we consider the extremely important special case of linear systems with quadratic performance indices. These performance indices can be considered as $(n + m)$-dimensional quadratic surfaces, where n and m are the dimensions of the state and control vectors. The linear system is a hyperplane that intersects the quadratic surface and moves through the space as a function of time. This section is therefore the culmination of the natural progression begun by Examples 1.2-1 and 1.2-2.

We shall discover that very refined solutions can be given in two instances: the fixed-final-state situation, which leads to an open-loop control strategy, and the free-final-state situation, which leads to a closed-loop strategy.

The State and Costate Equations

Let the plant to be controlled be described by the linear equation

$$x_{k+1} = A_k x_k + B_k u_k, \tag{2.2-1}$$

with $x_k \in R^n$ and $u_k \in R^m$. The associated performance index is the quadratic function

$$J_i = \frac{1}{2} x_N^T S_N x_N + \frac{1}{2} \sum_{k=i}^{N-1} (x_k^T Q_k x_k + u_k^T R_k u_k), \tag{2.2-2}$$

defined over the time interval of interest $[i, N]$. Note that both the plant and the cost-weighting matrices can, in general, be time-varying. The initial plant state is given as x_i. We assume that Q_k, R_k, and S_N are symmetric positive semidefinite matrices and that $|R_k| \neq 0$ for all k.

The objective is to find the control sequence $u_i, u_{i+1}, \ldots, u_{N-1}$ that minimizes J_i. To solve this *linear quadratic* (LQ) *regulator problem*, we begin with the Hamiltonian function

$$H^k = \frac{1}{2}\left(x_k^T Q_k x_k + u_k^T R_k u_k\right) + \lambda_{k+1}^T (A_k x_k + B_k u_k). \tag{2.2-3}$$

Then Table 2.2-1 presents the state and costate equations

$$x_{k+1} = \frac{\partial H_k}{\partial \lambda_{k+1}} = A_k x_k + B_k u_k \tag{2.2-4}$$

$$\lambda_k = \frac{\partial H_k}{\partial x_k} = Q_k x_k + A_k^T \lambda_{k+1} \tag{2.2-5}$$

TABLE 2.2-1   Discrete Linear Quadratic Regulator (Final State Free)

System model:
$$x_{k+1} = A_k x_k + B_k u_k, \quad k > i$$

Performance index:
$$J_i = \frac{1}{2} x_N^T S_N x_N + \frac{1}{2} \sum_{k=i}^{N-1} (x_k^T Q_k x_k + u_k^T R_k u_k)$$

Assumptions:
$$S_N \geq 0, \quad Q_k \geq 0, \quad R_k > 0, \quad \text{and all three are symmetric}$$

Optimal feedback control:
$$S_k = A_k^T [S_{k+1} - S_{k+1} B_k (B_k^T S_{k+1} B_k + R_k)^{-1} B_k^T S_{k+1}] A_k + Q_k,$$
$$k < N, \; S_N \text{ given}$$
$$K_k = (B_k^T S_{k+1} B_k + R_k)^{-1} B_k^T S_{k+1} A_k, \quad k < N$$
$$u_k = -K_k x_k, \quad k < N$$
$$J_i^* = \tfrac{1}{2} x_i^T S_i x_i$$

and the stationarity condition

$$0 = \frac{\partial H_k}{\partial u_k} = R_k u_k + B_k^T \lambda_{k+1}. \tag{2.2-6}$$

According to (2.2-6),
$$u_k = -R_k^{-1} B_k^T \lambda_{k+1}, \tag{2.2-7}$$

so the optimal control sequence is determined if we can find the costate sequence. A block diagram of the optimal controller appears in Fig. 2.2-1. We see, however, that it cannot be implemented in this form, since it is not causal.

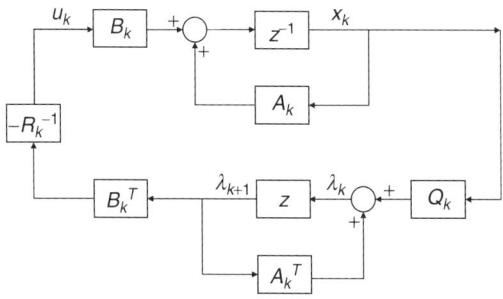

FIGURE 2.2-1   State–costate formulation of the discrete linear quadratic optimal controller.

Using (2.2-7) to eliminate u_k in (2.2-4) gives

$$x_{k+1} = A_k x_k - B_k R_k^{-1} B_k^T \lambda_{k+1}. \tag{2.2-8}$$

At this point we drop the subscripts on the plant and weighting matrices to simplify the notation.

The coupled state and costate equations can be written as the single unforced system

$$\begin{bmatrix} x_{k+1} \\ \lambda_k \end{bmatrix} = \begin{bmatrix} A & -BR^{-1}B^T \\ Q & A^T \end{bmatrix} \begin{bmatrix} x_k \\ \lambda_{k+1} \end{bmatrix}. \tag{2.2-9}$$

Relative to the coefficient matrix describing (2.2-9) we must make an important observation. For this we introduce the following definition. A $2n \times 2n$ matrix H is said to be of Hamiltonian type if it satisfies

$$JHJ = H^T$$

where $J = \begin{bmatrix} 0 & -I_n \\ I_n & 0 \end{bmatrix}$.

It can be easily verified that the coefficient matrix in (2.2-9) is of Hamiltonian type and we shall refer to it as the *discrete Hamiltonian matrix*. The unforced system (2.2-9) will be known as the *discrete Hamiltonian system*. This system is difficult to solve since part of it develops forward and part backward in time.

If $|A| \neq 0$ (which is the case whenever it is obtained by sampling a continuous system), then we can write (2.2-8) as the backward recursion

$$x_k = A^{-1} x_{k+1} + A^{-1} BR^{-1} B^T \lambda_{k+1}. \tag{2.2-10}$$

Using this in (2.2-5) allows us to write (2.2-9) in the modified form

$$\begin{bmatrix} x_k \\ \lambda_k \end{bmatrix} = \begin{bmatrix} A^{-1} & A^{-1}BR^{-1}B^T \\ QA^{-1} & A^T + QA^{-1}BR^{-1}B^T \end{bmatrix} \begin{bmatrix} x_{k+1} \\ \lambda_{k+1} \end{bmatrix}. \tag{2.2-11}$$

This equation develops purely backward in time, so if we can determine x_N and λ_N, then we can find x_k and λ_k and hence the optimal control. Unfortunately, we are given x_0, not λ_N.

Although we do not use equation (2.2-11) here, we discuss it again in Section 2.4, where we see that, in the time-invariant case, optimal controls can be computed from the eigenvectors of its coefficient matrix!

The costate equation (2.2-5) is an *adjoint system* for the plant (2.2-4). Let us discuss this notion to develop some more intuition on the relation between the state and the costate.

First suppose that $u_k = 0$ and $Q_k = 0$, and for simplicity let $i = 0$. Then the state and costate equations are

$$x_{k+1} = A x_k$$
$$\lambda_k = A^T \lambda_{k+1},$$

with solutions

$$x_k = A^k x_0, \tag{2.2-12}$$

$$\lambda_k = \left(A^{\mathrm{T}}\right)^{N-k} \lambda_N. \tag{2.2-13}$$

Therefore,

$$\lambda_k^{\mathrm{T}} x_k = \lambda_N^{\mathrm{T}} (A)^{N-k} A^k x_0 = \lambda_N^{\mathrm{T}} A^N x_0.$$

According to (2.2-12) and (2.2-13) this means that for all $k \in [0, N]$

$$\lambda_k^{\mathrm{T}} x_k = \lambda_0^{\mathrm{T}} x_0 = \lambda_N^{\mathrm{T}} x_N, \tag{2.2-14}$$

so that in the case of zero intermediate-state weighting ($Q_k = 0$) and zero input u_k, the inner product of the state and the costate is invariant with time. Considering x_k and λ_k as vectors in n space, we have

$$\lambda_k^{\mathrm{T}} x_k = |\lambda_k| \cdot |x_k| \cdot \cos \theta_k, \tag{2.2-15}$$

where $| \cdot |$ represents magnitude and θ_k is the angle between the vectors. As the state x_k develops through time, the magnitude and angle of the costate vary so that (2.2-15) is constant for all k.

To solve the optimal control problem, we need to solve for the costate λ_k and then use (2.2-7). To do this, we must use the boundary conditions (2.1-8). We know the initial state. We shall consider two special cases for the final condition (2.1-8a).

Before we do so, it is quite instructive to determine the value of the cost J_i when the control input u_k is zero.

Zero-input Cost and the Lyapunov Equation

Let the input u_k to the plant (2.2-1) be zero. We want to determine the value of the cost J_i in this uncontrolled situation.

Up to this point we have taken the initial time i as fixed, and k has been the variable time index, but it is now necessary to make a subtle shift of emphasis. In this subsection we want to find J_i as a function of i when $u_k = 0$ over $[i, N]$. To do this, we first let i equal the final time N and determine J_N. Then we increment i *backward* and find J_{N-1}, J_{N-2}, and so on. This amounts to taking i as our variable time index, considering successively longer time intervals with the final time N as the fixed quantity. The following discussion will make it clear why we wrote J_i and S_N in (2.2-2) as explicit functions of i and N, respectively.

To begin, note that

$$J_N = \tfrac{1}{2} x_N^{\mathrm{T}} S_N x_N. \tag{2.2-16}$$

Now, let $i = N - 1$, $u_k = 0$ and write

$$J_{N-1} = \tfrac{1}{2} x_N^{\mathrm{T}} S_N x_N + \tfrac{1}{2} x_{N-1}^{\mathrm{T}} Q_{N-1} x_{N-1}. \tag{2.2-17}$$

Use the plant dynamics (2.2-1) to see that

$$J_{N-1} = \tfrac{1}{2} x_{N-1}^T (A_{N-1}^T S_N A_{N-1} + Q_{N-1}) x_{N-1}. \qquad (2.2\text{-}18)$$

To make this look like (2.2-16), define a new intermediate variable (an $n \times n$ matrix) by

$$S_{N-1} = A_{N-1}^T S_N A_{N-1} + Q_{N-1}. \qquad (2.2\text{-}19)$$

Then

$$J_{N-1} = \tfrac{1}{2} x_{N-1}^T S_{N-1} x_{N-1}. \qquad (2.2\text{-}20)$$

It is clear that we can repeat this procedure for $i = N - 2, N - 3, \ldots$ because (2.2-16) and (2.2-20) have the same form. The result is that we define an entirely new intermediate sequence of $n \times n$ matrices by the backward recursion

$$S_k = A_k^T S_{k+1} A_k + Q_k, \qquad k < N, \qquad (2.2\text{-}21)$$

with boundary condition S_N given as the final-state weighting matrix in (2.2-2). (Note that we have shifted from dummy index i to k.) This is a *discrete Lyapunov equation* for S_k, also known as the *observability Lyapunov equation*.

In terms of the new quantities S_k, the zero-input performance index over the interval $[k, N]$ is equal to

$$J_k = \tfrac{1}{2} x_k^T S_k x_k. \qquad (2.2\text{-}22)$$

We call S_k the performance index *kernel* sequence.

The result (2.2-22) is quite interesting. The kernel S_k can be computed offline, *before* we know the system state trajectory, since it depends only on the plant and cost-weighting matrices. According to (2.2-22), then, we can calculate the zero-input cost over the interval $[k, N]$ by knowing the precomputed S_k and only the *initial* state x_k! If we know where we start out, then we know what the ride will cost. Note that we can interpret $2J_k$ as the squared seminorm of x_k with respect to S_k, since $S_k \geq 0$. If $Q_k > 0$, then S_k is positive definite for all k, and $2J_k$ becomes a squared norm of x_k.

In the time-invariant case, we know that the solution of the Lyapunov equation (2.2-21) is

$$S_k = (A^T)^{N-k} S_N A^{N-k} + \sum_{i=k}^{N-1} (A^T)^{N-i-1} Q A^{N-i-1}. \qquad (2.2\text{-}23)$$

By the Lyapunov stability theory, as $(N - k) \to \infty$ this converges to the *steady-state value*

$$S_\infty = \sum_{i=0}^{\infty} (A^T)^i Q A^i \qquad (2.2\text{-}24)$$

if the plant is asymptotically stable. Figure 2.2-2 shows the limiting behavior of S_k for a scalar plant for the stable and unstable cases.

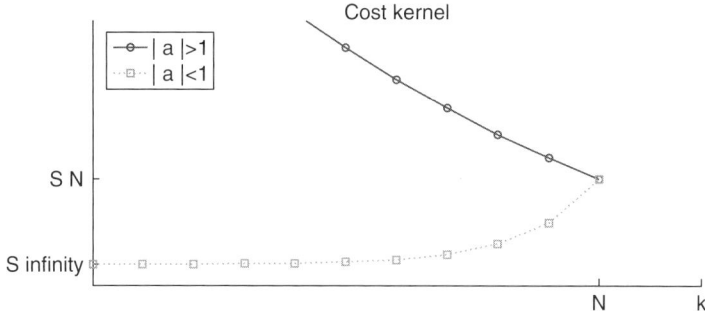

FIGURE 2.2-2   Zero-input cost kernel for a stable and an unstable plant.

If the plant is stable, the cost over the interval $[-\infty, N]$, or equivalently $[0, \infty]$, is given by the *steady-state cost*

$$J_\infty = \tfrac{1}{2} x_0^T S_\infty x_0. \tag{2.2-25}$$

It should be clearly understood that if A is stable, the uncontrolled cost over an infinite interval is *finite*. This is true because the state goes to zero with time.

If A is not stable, the zero-input steady-state cost can be infinite (depending on Q) since the norm of x_k is unbounded. In the steady-state case, $S \triangleq S_k = S_{k+1}$ for large k, so that (2.2-21) becomes the *algebraic Lyapunov equation*

$$S = A^T S A + Q. \tag{2.2-26}$$

By the Lyapunov theory, this equation has a positive semidefinite solution S_∞ if A is stable, and this solution is given by (2.2-24).

If $\sqrt{Q}$ is defined by

$$Q = \sqrt{Q}^T \sqrt{Q}, \tag{2.2-27}$$

then S_∞ is the *unique* positive *definite* solution if A is stable and $(A, \sqrt{Q})$ is observable. This latter condition means that Q was selected so that the plant state is "observable by the performance index," so that variations in any direction of the state have an effect on J. If $(A, \sqrt{Q})$ is not observable, then the unobservable state components can tend to infinity with k, but the performance index may still be finite since these components have no effect on J. On the other hand, if $(A, \sqrt{Q})$ is observable and A is unstable, then J_k will grow without bound as $k \to -\infty$.

We shall see later that these results generalize to the case of nonzero input and provide a means for computing the optimal control sequence.

Fixed-final-state and Open-loop Control

Here we return to the problem of determining the optimal control for the plant (2.2-1) with the cost (2.2-2). The state and costate equations are given, once

u_k has been expressed in terms of λ_{k+1}, by (2.2-8) and (2.2-5). (For ease of notation, assume the time-invariant case; the results generalize to the time-varying situation.) To solve these, we need to determine the terminal conditions.

For simplicity, let the initial time be $i = 0$. The initial state x_0 is given. In this subsection our terminal objective will be to make x_N match *exactly* the desired final reference state r_N. The final condition is thus

$$x_N = r_N. \tag{2.2-28}$$

Since in this fixed-final-state case $dx_N = 0$, condition (2.1-8a) is automatically satisfied.

Since we are demanding that x_N be equal to a known desired r_N, the final-state contribution to J_0 in (2.2-2) always has a fixed value of $\frac{1}{2}r_N^T S_N r_N$. It is therefore redundant to include a final-state weighting term in J_0. Accordingly, we may as well set $S_N = 0$.

Let the cost function be

$$J_0 = \frac{1}{2} \sum_{k=0}^{N-1} u_k^T R u_k, \tag{2.2-29}$$

and so we are asking for a control that drives x_0 exactly to $x_N = r_N$ using minimum control energy. The state and costate equations are now

$$x_{k+1} = Ax_k - BR^{-1}B^T\lambda_{k+1}, \tag{2.2-30}$$

$$\lambda_k = A^T\lambda_{k+1}. \tag{2.2-31}$$

Since $Q = 0$, the costate equation is decoupled from the state equation, and the problem has an easy solution. To find it, write the solution of (2.2-31) in terms of the, as yet unknown, final costate as

$$\lambda_k = (A^T)^{N-k}\lambda_N. \tag{2.2-32}$$

Use this to eliminate λ_{k+1} in (2.2-30) to get

$$x_{k+1} = Ax_k - BR^{-1}B^T(A^T)^{N-k-1}\lambda_N. \tag{2.2-33}$$

Considering this as a first-order difference equation with the second term as the input, we get ✓ *i = k-1 + A° which makes sense*

$$x_k = A^k x_0 - \sum_{i=0}^{k-1} A^{k-i-1}BR^{-1}B^T(A^T)^{N-i-1}\lambda_N. \tag{2.2-34}$$

To find λ_N, evaluate (2.2-34) at $k = N$,

$$x_N = A^N x_0 - \sum_{i=0}^{N-1} A^{N-i-1}BR^{-1}B^T(A^T)^{N-i-1}\lambda_N.$$

Therefore, the final costate, with $x_N = r_N$, is

$$\lambda_N = -G_{0,N}^{-1}(r_N - A^N x_0), \tag{2.2-35}$$

where

$$G_{0,N} = \sum_{i=0}^{N-1} A^{N-i-1} BR^{-1} B^{\mathrm{T}} (A^{\mathrm{T}})^{N-i-1}. \tag{2.2-36}$$

Using (2.2-32) the costate is

$$\lambda_k = -(A^{\mathrm{T}})^{N-k} G_{0,N}^{-1}(r_N - A^N x_0), \tag{2.2-37}$$

and so by (2.2-7) the optimal control sequence is

$$u_k^* = R^{-1} B^{\mathrm{T}} (A^{\mathrm{T}})^{N-k-1} G_{0,N}^{-1}(r_N - A^N x_0). \tag{2.2-38}$$

This is the minimum-control-energy solution to the fixed-final-state LQ regulator problem; the problem is solved.

We can easily demonstrate that u_k^* is a control that drives x_0 to $x_N = r_N$. The solution to state equation (2.2-1) is

$$x_k = A^k x_0 + \sum_{i=0}^{k-1} A^{k-i-1} Bu_i. \tag{2.2-39}$$

Evaluating the state at $k = N$ and using $u_i = u_i^*$ as given by (2.2-38) yields

$$x_N = A^N x_0 + \sum_{i=0}^{N-1} A^{N-i-1} BR^{-1} B^{\mathrm{T}} (A^{\mathrm{T}})^{N-i-1} G_{0,N}^{-1}(r_N - A^N x_0),$$

but $G_{0,N}^{-1}(r_N - A^N x_0)$ does not depend on i, and the remaining portion of the sum is just $G_{0,N}$. This means that

$$x_N = A^N x_0 + G_{0,N} G_{0,N}^{-1}(r_N - A^N x_0) = r_N \tag{2.2-40}$$

as desired!

To gain an understanding of our optimal control result, let us discuss it a little. First, note that in the absence of an input, the solution to the state equation (2.2-1) is

$$x_k = A^k x_0, \tag{2.2-41}$$

so that $x_N = A^N x_0$ is the final state with zero input. Thus, $r_N - A^N x_0$ is the difference between the desired and undriven final states; it makes sense that u_k^* should depend on this quantity.

Now we examine $G_{0,N}$. This is the *weighted reachability gramian* of the system. In terms of the system reachability matrix $U_k = [B \ \ AB \ \ \cdots \ \ A^{k-1}B]$, it can be written as

$$G_{0,N} = U_N \begin{bmatrix} R^{-1} & & 0 \\ & \ddots & \\ 0 & & R^{-1} \end{bmatrix} U_N^{\mathrm{T}}. \tag{2.2-42}$$

If $R = I$, then $G_{0,N} = U_N U_N^{\mathrm{T}}$. The optimal control exists if and only if $|G_{0,N}| \neq 0$. Since we assumed $|R| \neq 0$, this is equivalent to U_N having full rank n, where n is the state dimension. Therefore, we can drive any given x_0 to any desired $x_N = r_N$ for some N if and only if the system is *reachable*! Since reachability implies that U_{n+j} has full rank for all $j \geq 0$, if the system is reachable, we can guarantee the existence of a control to drive x_0 to $x_N = r_N$ for any r_N by selecting $N \geq n$.

The following point should be clearly understood. The optimal control (2.2-38) is an *open-loop* control. It can be precomputed knowing only the given x_0 and the desired r_N, and it is independent of intermediate values of x_k within the interval $[0, N]$. This means that if we apply u_k^* as calculated by (2.2-38) to the actual system, all is well as long as (2.2-1) is an exact description of the dynamics and nothing occurs to cause x_k to deviate from the optimal state trajectory. In practice, however, nature is seldom cooperative. Modeling uncertainties and noise cause errors in x_k, and the control (2.2-38) does not take these errors into account. Open-loop control schemes are not robust in most actual applications.

To compute the reachability gramian, there is an attractive alternative to (2.2-36) or (2.2-42). The solution to the Lyapunov equation

$$P_{k+1} = AP_k A^{\mathrm{T}} + BR^{-1}B^{\mathrm{T}}, \quad k > 0 \tag{2.2-43}$$

is

$$P_k = A^k P_0 (A^k)^{\mathrm{T}} + \sum_{i=0}^{k-1} A^{k-i-1} BR^{-1} B^{\mathrm{T}} (A^{\mathrm{T}})^{k-i-1}, \tag{2.2-44}$$

so if we solve this equation with $P_0 = 0$, then $G_{0,k} = P_k$ for each k. First this recursion is solved to obtain $G_{0,N}$ for the final time of interest N, and then (2.2-38) is used to compute the optimal control u_k^* for each k in $[0, N]$. This *reachability Lyapunov equation* should be compared with the *observability Lyapunov equation* (2.2-21).

The next example illustrates the use of these results.

Example 2.2-1. Open-loop Control of a Scalar System

Consider the scalar plant

$$x_{k+1} = ax_k + bu_k \tag{1}$$

with cost

$$J_0 = \frac{r}{2} \sum_{k=0}^{N-1} u_k^2. \tag{2}$$

The system is reachable since

$$U_1 = b \neq 0. \tag{3}$$

The reachability gramian (2.2-36) is

$$G_{0,N} = \sum_{i=0}^{N-1} \frac{b^2}{r} a^{2(N-i-1)} = \frac{b^2}{r} \frac{1-a^{2N}}{1-a^2}, \tag{4}$$

which is exactly Λ from Example 2.1-2.

To drive any given x_0 to a desired r_N for any $N \geq n = 1$, we should use the optimal control sequence u_k^* for $k \in [0, N-1]$ given by (2.2-38), so that

$$
\begin{aligned}
u_k^* &= \frac{a^{N-k-1}b}{r} \frac{r(1-a^2)}{b^2(1-a^{2N})}(r_N - a^N x_0) \\
&= \frac{1-a^2}{b(1-a^{2N})}(r_N - a^N x_0)a^{N-k-1}.
\end{aligned}
\tag{5}
$$

This is exactly the result obtained by solving the state and costate equations in Example 2.1-2a.

Note that the optimal control at times k and $k+1$ are related by

$$u_{k+1}^* = u_k^*/a, \tag{6}$$

which makes the control sequence easy to calculate. See Fig. 2.1-1.   ∎

Free-final-state and Closed-loop Control

We have just found the minimum-energy optimal control for system (2.2-1) in the case where x_N is required to have a fixed given value. It is now desired to find the optimal control sequence that drives the system (2.2-1) along the trajectory, beginning at a given x_i, resulting in a minimum value of (2.2-2). We shall make here *no restriction* on the value of the final state x_N. This *free-final-state* problem will result in a radically different sort of control.

The state and costate equations with the input u_k eliminated are (2.2-8) and (2.2-5), which are reproduced here:

$$x_{k+1} = A_k x_k - B_k R_k^{-1} B_k^T \lambda_{k+1}, \tag{2.2-45}$$

$$\lambda_k = Q_k x_k + A_k^T \lambda_{k+1}. \tag{2.2-46}$$

The control is given as (2.2-7) or

$$u_k = -R_k^{-1} B_k^T \lambda_{k+1}. \tag{2.2-47}$$

The initial condition is given as x_i, and the final state x_N is free. This means that x_N can be varied in determining the constrained minimum. Hence, $dx_N \neq 0$.

According to (2.1-8), then, it is required that

$$\lambda_N = \frac{\partial \phi}{\partial x_N}. \tag{2.2-48}$$

The final state weighting function is $\phi = \frac{1}{2} x_N^T S_N x_N$, so that

$$\lambda_N = S_N x_N. \tag{2.2-49}$$

This relation between the final costate and state is the new terminal condition; in the fixed-final-state problem the terminal condition was (2.2-28).

To solve this two-point boundary-value problem, we shall use the *sweep method* (Bryson and Ho 1975). Thus, assume that a linear relation like (2.2-49) holds for all times $k \leq N$:

$$\lambda_k = S_k x_k \tag{2.2-50}$$

for some intermediate sequence of $n \times n$ matrices S_k. If we can find a consistent formula for these postulated S_k, then evidently (2.2-50) is a valid assumption. To do this, use (2.2-50) in (2.2-45) to get

$$x_{k+1} = A_k x_k - B_k R_k^{-1} B_k^T S_{k+1} x_{k+1}.$$

Solving for x_{k+1} yields

$$x_{k+1} = (I + B_k R_k^{-1} B_k^T S_{k+1})^{-1} A_k x_k, \tag{2.2-51}$$

which is a forward recursion for the state.

Now substitute (2.2-50) into costate equation (2.2-46) to see that

$$S_k x_k = Q_k x_k + A_k^T S_{k+1} x_{k+1},$$

or by (2.2-51)

$$S_k x_k = Q_k x_k + A_k^T S_{k+1} (I + B_k R_k^{-1} B_k^T S_{k+1})^{-1} A_k x_k.$$

Since x_k is generally nonzero, and this equation holds for all state sequences given any x_i, evidently

$$S_k = A_k^T S_{k+1} (I + B_k R_k^{-1} B_k^T S_{k+1})^{-1} A_k + Q_k, \tag{2.2-52}$$

or, using the matrix inversion lemma (Appendix A)

$$S_k = A_k^T [S_{k+1} - S_{k+1} B_k (B_k^T S_{k+1} B_k + R_k)^{-1} B_k^T S_{k+1}] A_k + Q_k. \tag{2.2-53}$$

This is a backward recursion for the postulated S_k, which completely specifies it in terms of S_{k+1} and the known system and weighting matrices. The boundary condition is known: it is just the final-state weighting matrix S_N. Therefore, we have been able to discover an intermediate sequence such that the relation (2.2-50) holds.

The matrix quadratic equation (2.2-53) is called a *Riccati equation* in honor of the Italian mathematician Count Jacopo Riccati, who investigated a scalar version of its differential counterpart in 1724. If $|S_k| \neq 0$ for all k, then we can use the matrix inversion lemma to rewrite the Riccati equation as

$$S_k = A_k^{\mathrm{T}}(S_{k+1}^{-1} + B_k R_k^{-1} B_k^{\mathrm{T}})^{-1} A_k + Q_k. \tag{2.2-54}$$

The intermediate sequence S_k can be computed offline knowing only the plant and parameters of the performance index. Then (2.2-51), with the given initial state x_i, yields the optimal state trajectory. We do not yet know, however, the control required to move the plant along that trajectory! To determine the optimal control write

$$u_k = -R_k^{-1} B_k^{\mathrm{T}} \lambda_{k+1} = -R_k^{-1} B_k^{\mathrm{T}} S_{k+1} x_{k+1}. \tag{2.2-55}$$

At first glance it appears that the problem is now solved, since the Riccati equation gives sequence the S_k, (2.2-51) gives sequence x_k, and we can use this to find sequence u_k. This is, however, quite inconvenient. Let us manipulate (2.2-55) to find a more satisfactory expression for the optimal control. To this end, use the plant equation $x_{k+1} = A_k x_k + B_k u_k$ in (2.2-55) to get

$$u_k = -R_k^{-1} B_k^{\mathrm{T}} S_{k+1}(A_k x_k + B_k u_k)$$

or

$$(I + R_k^{-1} B_k^{\mathrm{T}} S_{k+1} B_k)u_k = -R_k^{-1} B_k^{\mathrm{T}} S_{k+1} A_k x_k.$$

Premultiply by R_k and then solve for the control to obtain

$$u_k = -(B_k^{\mathrm{T}} S_{k+1} B_k + R_k)^{-1} B_k^{\mathrm{T}} S_{k+1} A_k x_k. \tag{2.2-56}$$

Defining the Kalman gain sequence

$$K_k = (B_k^{\mathrm{T}} S_{k+1} B_k + R_k)^{-1} B_k^{\mathrm{T}} S_{k+1} A_k, \tag{2.2-57}$$

the control takes the form

$$u_k = -K_k x_k. \tag{2.2-58}$$

Note that in these equations we could use the notation u_k^*, since they give the optimal control.

The form of (2.2-58) makes it particularly clear what we now have on our hands. The Kalman gain is given in terms of the Riccati equation solution S_k and the system and weighting matrices. It can therefore be computed and stored in computer memory *before* the control is ever applied to the plant. It does not depend on the state trajectory. Therefore, (2.2-58) is a *time-varying state-variable*

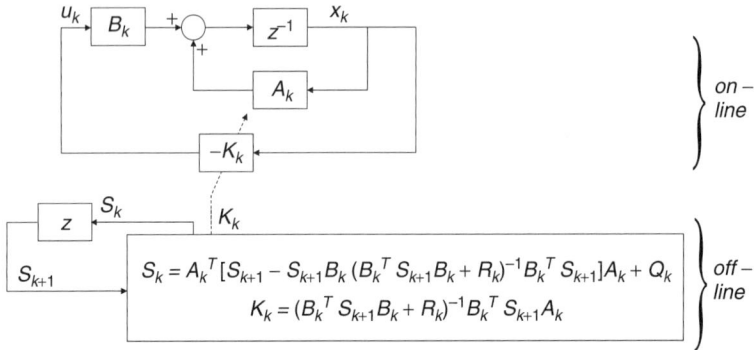

FIGURE 2.2-3 Free-final-state LQ regulator optimal control scheme.

feedback, which expresses the current required control in terms of the current state. In the free-final-state linear quadratic (LQ) regulator, the optimal control is thus given by a *closed-loop control law*. Closed-loop control is inherently more robust than open-loop control, because any deviations from the optimal state trajectory x_k^*, which is given by (2.2-51), are automatically accounted for.

The problem is solved. To determine the optimal control, we need only solve the Riccati equation for S_k and (2.2-57) for K_k, both of which can be done offline, and then use the feedback (2.2-58). These equations are summarized in Table 2.2-1, and a block diagram of the control scheme appears in Fig. 2.2-3.

The closed-loop system with the optimal feedback (2.2-58) is

$$x_{k+1} = (A_k - B_k K_k)x_k, \tag{2.2-59}$$

which provides an alternative to (2.2-51) for computing the optimal state trajectory. It is worth noting that an alternative more efficient way of computing S_k and K_k is to use the recursion

$$K_k = (B_k^T S_{k+1} B_k + R_k)^{-1} B_k^T S_{k+1} A_k, \tag{2.2-60}$$

$$S_k = A_k^T S_{k+1}(A_k - B_k K_k) + Q_k. \tag{2.2-61}$$

It is not difficult to show (see the problems) that equation (2.2-61) is equivalent to the *Joseph stabilized version* of the Riccati equation

$$S_k = (A_k - B_k K_k)^T S_{k+1}(A_k - B_k K_k) + K_k^T R_k K_k + Q_k. \tag{2.2-62}$$

This equation has better numerical properties when it comes to computation and will soon be useful to us.

If the system and weighting matrices A, B, Q, R are all time invariant, then the feedback gain K_k is still a function of time, since, in general, the Riccati-equation solution S_k is time varying. Thus, even for time-invariant systems, the optimal control is a *time-varying* state feedback. This means that the optimal closed-loop system $(A - BK_k)$ is time varying, and, of course, accounts for the fact that it cannot be found by classical frequency-domain methods.

For completeness, let us determine the optimal value of the performance index under the influence of the optimal control (2.2-58). First, observe that

$$\frac{1}{2} \sum_{k=i}^{N-1} \left(x_{k+1}^T S_{k+1} x_{k+1} - x_k^T S_k x_k \right) = \frac{1}{2} x_N^T S_N x_N - \frac{1}{2} x_i^T S_i x_i. \tag{2.2-63}$$

We can therefore add zero, in the form of the left-hand side of (2.2-63) minus its right-hand side, to the performance index (2.2-2). The result is

$$J_i = \frac{1}{2} x_i^T S_i x_i + \frac{1}{2} \sum_{k=i}^{N-1} \left[x_{k+1}^T S_{k+1} x_{k+1} + x_k^T (Q_k - S_k) x_k + u_k^T R_k u_k \right]. \tag{2.2-64}$$

If we take into account the state equation (2.2-1), this is equivalent to

$$J_i = \frac{1}{2} x_i^T S_i x_i + \frac{1}{2} \sum_{k=i}^{N-1} \left[x_k^T \left(A_k^T S_{k+1} A_k + Q_k - S_k \right) x_k \right. \tag{2.2-65}$$
$$\left. + x_k^T A_k^T S_{k+1} B_k u_k + u_k^T B_k^T S_{k+1} A_k x_k + u_k^T \left(B_k^T S_{k+1} B_k + R_k \right) u_k \right].$$

According to the Riccati equation (2.2-53), this is

$$J_i = \frac{1}{2} x_i^T S_i x_i + \frac{1}{2} \sum_{k=i}^{N-1} \left[x_k^T A_k^T S_{k+1} B_k \left(B_k^T S_{k+1} B_k + R_k \right)^{-1} B_k^T S_{k+1} A_k x_k \right.$$
$$\left. + x_k^T A_k^T S_{k+1} B_k u_k + u_k^T B_k^T S_{k+1} A_k x_k + u_k^T (B_k S_{k+1} B_k + R_k) u_k \right]. \tag{2.2-66}$$

The summand can be written as the perfect square of a norm with respect to $(B_k^T S_{k+1} B_k + R_k)$ (McReynolds 1966):

$$J_i = \frac{1}{2} x_i^T S_i x_i$$
$$+ \frac{1}{2} \sum_{k=i}^{N-1} \left\| \left(B_k^T S_{k+1} B_k + R_k \right)^{-1} B_k^T S_{k+1} A_k x_k + u_k \right\|_{\left(B_k^T S_{k+1} B_k + R_k \right)}^2. \tag{2.2-67}$$

If we now select the optimal control (2.2-56), then the optimal value of the performance index is seen to be

$$J_i^* = \frac{1}{2} x_i^T S_i x_i. \tag{2.2-68}$$

This result deserves some discussion. The sequence S_k can be computed offline before the optimal control is applied, so that S_i is known a priori. Given any initial state x_i, then, we can use (2.2-68) to compute the optimal cost of applying the control before we ever apply it! In general, we can treat any time k in $[i, N]$

as the initial time of a subinterval, so that

$$J_k^* = \tfrac{1}{2} x_k^T S_k x_k. \tag{2.2-69}$$

What this means is that given the current state x_k we can compute the cost to go (i.e., remaining cost) of applying the optimal control from times k through N.

Because of (2.2-69), we call S_k the performance index *kernel* matrix. The optimal value of the performance index J_k is simply one-half of the semi-norm squared of the current state x_k with respect to S_k. (If $|S_k| \neq 0$, the semi-norm becomes a norm.)

It is worth noting that, according to (2.2-67), $B_k^T S_{k+1} B_k + R_k$ is equal to $\partial^2 J_i / \partial u_k^2$, the second derivative of J_i with respect to the kth input under the constraint (2.2-1). Compare this with the constrained curvature matrix L_{uu}^f in Example 1.2-3. Evidently, $B_k^T S_{k+1} B_k + R_k$ is a time-varying curvature matrix.

To contribute to the development of an intuitive grasp of the Riccati-equation-based control law, let us point out a few links with our previous work. First, use the matrix inversion lemma to write the optimal state trajectory recursion (2.2-51) as

$$x_{k+1} = \left[I - B_k (B_k^T S_{k+1} B_k + R_k)^{-1} B_k^T S_{k+1} \right] A_k x_k \tag{2.2-70}$$

(which is, in fact, the same as (2.2-59)). It is now clear that this equation and (2.2-69) (with S_k given by (2.2-53)) are simply generalizations to the case where time is a parameter of our static results in Example 1.2-3.

We can also link our present results to the zero-input cost discussion, for if control matrix B_k is zero, then the (quadratic) Riccati equation reduces to the (linear) Lyapunov equation (2.2-21)! The control-dependent term in the Riccati equation makes the value of S_k *smaller* than $A_k^T S_{k+1} A_k + Q_k$ at each step, and expresses the decrease in the value of the performance index that results if we are allowed to control the plant.

There is a second way to reduce a Riccati equation to a Lyapunov equation, and it provides the connection between the fixed-final-state and free-final-state control laws. Suppose $Q_k = 0$, so that there is no intermediate-state weighting. For simplicity, assume time-invariant plant and cost-weighting matrices. Suppose also that $|A| \neq 0$. Under these circumstances, (2.2-54) can be written

$$S_k^{-1} = A^{-i} S_{k+1}^{-1} A^{-T} + A^{-1} B R^{-1} B^T A^{-T}, \tag{2.2-71}$$

which is a backward-developing Lyapunov recursion for S_k^{-1}.

Now, if we want to ensure that x_N approaches exactly a desired final value of $r_N = 0$, in the performance index we can let the final-state weighting matrix S_N go to infinity. This tells the optimal control to make $x_N = 0$ in order to keep J_i finite. In this limit, $S_N^{-1} = 0$, which provides the terminal condition for (2.2-71). It is easy to show (by writing out explicitly the first few steps in the recursion and then using induction) that the solution to (2.2-71) with $S_N^{-1} = 0$ is just

$$S_{N-k}^{-1} = A^{-k} G_{0,k} (A^T)^{-k}, \tag{2.2-72}$$

where $G_{0,k}$ is the weighted reachability gramian (2.2-36).

It takes a few more steps to prove it, but it is fairly evident at this point that if $Q = 0$ and $S_N \to \infty$, then the free-final-state closed-loop control (2.2-56) reduces to the fixed-final-state open-loop control (2.2-38) in the case $r_N = 0$! See the related discussion in Example 2.1-2. (Note that $S_N \to \infty$ is equivalent to $R \to 0$ if $Q = 0$.)

As a final connection, examine part b of Example 2.1-2. It can be demonstrated that the control law given in equation (34) of that example is just an alternative formulation of (2.2-56). In practical applications the latter closed-loop formulation would be used.

As a final comment on the closed-loop LQ regulator, we make the following very important point. In the fixed-final-state problem, reachability of the system was required to be able to solve for the optimal control to drive any given initial state to any desired final state. In our discussion of the free-final-state problem, however, reachability never came up. In fact, it is *not* necessary for the system to be reachable for the control in Table 2.2-1 to exist. If the system is not reachable, the control will still do its best to minimize J_i. Clearly, if the system is, in fact, reachable, the control will do a better job of minimizing J_i, since then all vectors in R^n are candidates for the optimal x_N.

We shall see later that reachability does become important in the *steady-state* control problem, where N tends to infinity. Let us now work through an example.

Example 2.2-2.   Optimal Feedback Control of a Scalar System

The plant to be controlled is the time-invariant scalar system

$$x_{k+1} = ax_k + bu_k \tag{1}$$

with performance index

$$J_i = \frac{1}{2}s_N x_N^2 + \frac{1}{2}\sum_{k=i}^{N-1}(qx_k^2 + ru_k^2). \tag{2}$$

In Example 2.1-2 we considered a special case ($q = 0$) of this problem, and we found the optimal controls for two different terminal conditions by direct solution of the state and costate equations. Here we shall find the optimal control in the state feedback formulation using the results we have just derived.

In this scalar case the Riccati equation is

$$s_k = a^2 s_{k+1} - \frac{a^2 b^2 s_{k+1}^2}{b^2 s_{k+1} + r} + q \tag{3}$$

or

$$s_k = \frac{a^2 r s_{k+1}}{b^2 s_{k+1} + r} + q. \tag{4}$$

The Kalman gain is

$$K_k = \frac{abs_{k+1}}{b^2 s_{k+1} + r} = \frac{a/b}{1 + r/(b^2 s_{k+1})} \tag{5}$$

and the optimal control is

$$u_k = -K_k x_k. \tag{6}$$

The optimal value of the performance index is

$$J_k = \tfrac{1}{2} s_k x_k^2. \tag{7}$$

The optimal closed-loop system (2.2-59) is

$$x_{k+1} = (a - bK_k)x_k = \frac{a}{1 + (b^2/r)s_{k+1}} x_k. \tag{8}$$

Even in this simple case a closed-form solution to (3) is hard to find. Let us therefore consider three special cases. Then we shall demonstrate that even if we cannot solve (3) analytically, for particular values of a, b, q, r, s_N, it is very easy to compute the optimal control sequence u_k^* and to simulate applying it to the plant on a digital computer.

a. No Control Weighting

Let $r = 0$, meaning that we do not care how much control is used (i.e., u_k is not weighted in J_i so that the optimal solution will make no attempt to keep it small). Then (4) is

$$s_k = q, \tag{9}$$

the feedback gain (5) is $K_k = a/b$, and the optimal control becomes

$$u_k = -\frac{a}{b} x_k. \tag{10}$$

Under the influence of this control, the performance index is

$$J_k = \tfrac{1}{2} q x_k^2, \tag{11}$$

and the closed-loop system (8) is $x_{k+1} = 0$!

We can understand this as follows. If we have a given value x_k for the state at time k, then a naive approach to minimizing the magnitude of the state vector (which is all we require since $r = 0$) is to solve the state equation (1) for the u_k required to make x_{k+1} equal to zero, so that $0 = x_{k+1} = ax_k + bu_k$. This yields the control (10).

b. Very Large Control Weighting

If we are very concerned not to use too much control energy, we can let $r \to \infty$. Then (4) becomes

$$s_k = a^2 s_{k+1} + q. \tag{12}$$

The solution to this (Lyapunov) difference equation is

$$s_k = s_N a^{2(N-k)} + \sum_{i=k}^{N-1} q a^{2(N-i-1)} = s_N a^{2(N-k)} + \left(\frac{1 - a^{2(N-k)}}{1 - a^2} \right) q. \tag{13}$$

The Kalman gain is $K_k = 0$, and so the optimal control is $u_k = 0$. The closed-loop system is $x_{k+1} = ax_k$.

If we are very concerned about using too much control, the best policy is to use none at all!

c. No Intermediate-state Weighting

Let us set $q = 0$. Then we are concerned only about making x_N^2 small without using too much control energy. It is easier in this case to deal with the inverse of the cost kernel, s_k^{-1}, so use (2.2-71) to rewrite (4) as

$$s_k^{-1} = \frac{s_{k+1}^{-1}}{a^2} + \frac{b^2}{a^2 r}. \tag{14}$$

The solution to this (Lyapunov) difference equation is

$$s_k^{-1} = a^{-2(N-k)} s_N^{-1} + \sum_{i=k}^{N-1} \frac{b^2}{a^2 r} a^{-2(N-i-1)}, \tag{15}$$

or, changing variables,

$$s_k^{-1} = a^{-2(N-k)} s_N^{-1} + \frac{b^2}{a^2 r} \sum_{i=0}^{N-k-1} a^{-2i} = s_N^{-1} a^{-2(N-k)} + \frac{b^2}{r a^{2(N-k)}} \frac{1 - a^{2(N-k)}}{1 - a^2}. \tag{16}$$

After a few lines of work we get

$$s_k = \frac{s_N a^{2(N-k)}}{1 + s_N (b^2/r)[(1 - a^{2(N-k)})/(1 - a^2)]}. \tag{17}$$

This is the Riccati-equation solution for $q = 0$.

d. Implementation of Optimal Control

To actually compute and implement the control described by (4)–(6), none of the analysis subsequent to equation (6) is needed. The implementation of optimal controls has two phases: computing the control sequence and applying it to the plant. We show here how to compute u_k^* and then simulate its application using a digital computer.

```
function [x, u, K, S]=scaopt (a, b, q, r, s, x0, N)
% Program to Compute and Simulate Optimal Feedback Control
% Compute and Store Optimal Feedback Sequence
% (Backward Iteration)
S(N+1) =s;
for k=N:-1:1
K(k)=(a*b*s)/(r+s*b^2);
s=q+(r*s*a^2)/(r+s*b^2);
S(k)=s;
end
% Apply Optimal Control to Plant (Forward Iteration)
x(1)=x0;
for k=1:N
% Compute Optimal Control
u(k)=-K(k)*x(k);
% Update the Plant State
x(k+1)=a*x(k)+b*u(k);
end
```

FIGURE 2.2-4   MATLAB code to compute and simulate optimal feedback control.

A routine implementing (4) and (5), for $a = 1.05$, $b = 0.01$, $q = r = 1$, $x_0 = 10$, $s_N = 5$, and $N = 100$, is shown in Fig. 2.2-4. This must solve the Riccati equation *backward* on the desired interval $[0, N]$, using as a starting value s_N. The optimal feedback gains K_k are also computed for all $k \in [0, N]$, and they are stored in memory. (The s_k need not be stored. They are stored in this example *only* so they can be plotted for illustrative purposes. In an application the storage of s_k can be avoided.) This completes the computation of the control law.

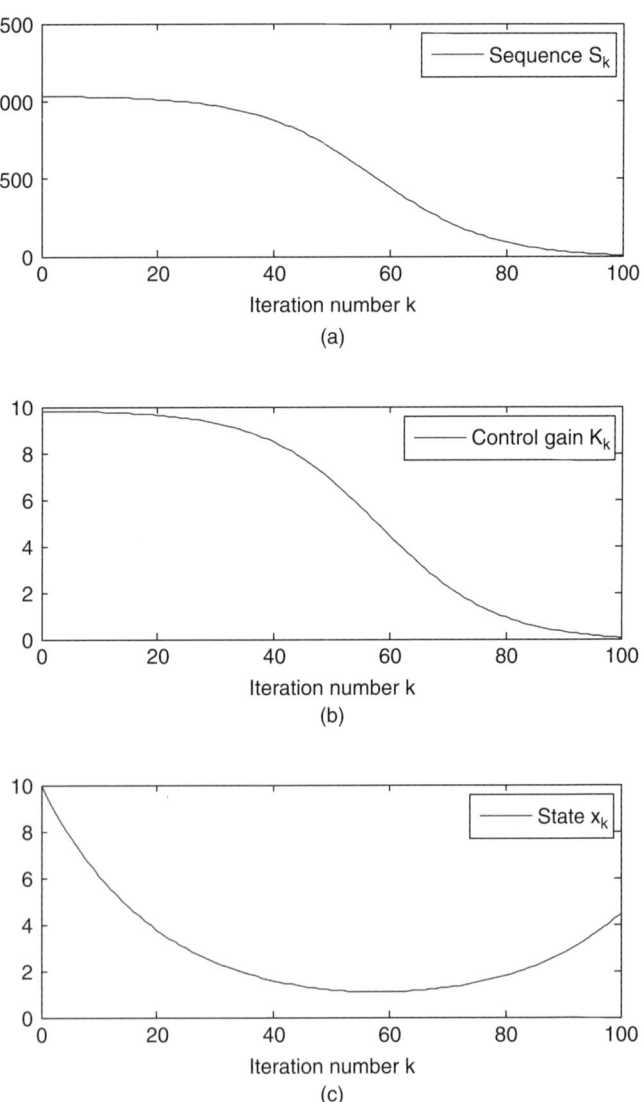

FIGURE 2.2-5 Optimal control simulations for $s_N = 5$. (a) Cost kernel s_k. (b) Optimal feedback gains K_k. (c) Optimal trajectory x_k^*.

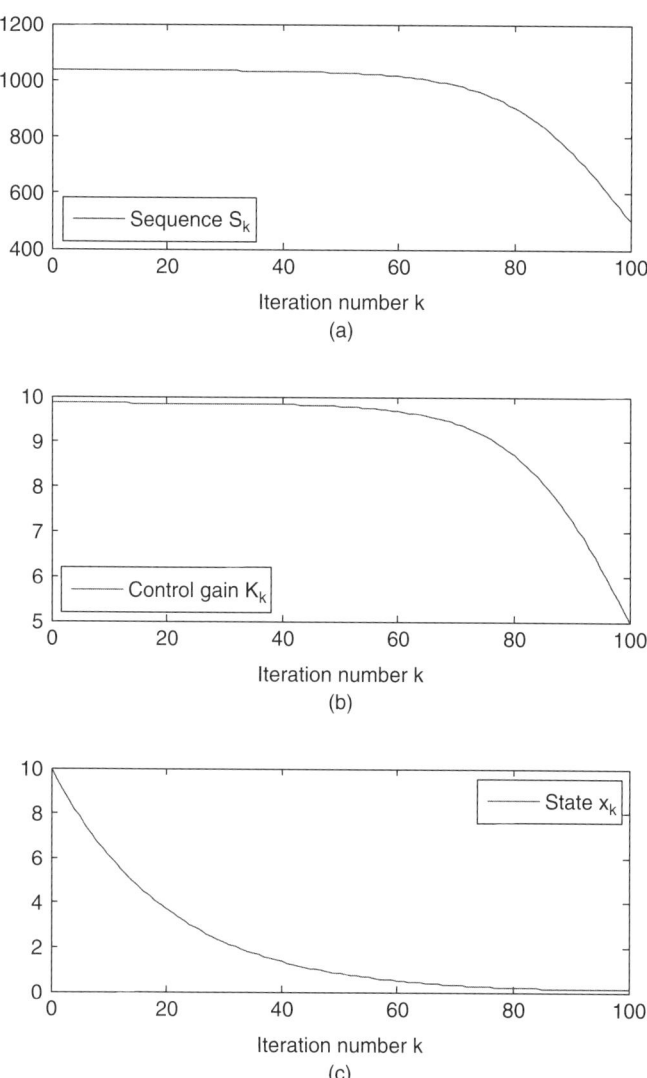

FIGURE 2.2-6 Optimal control simulations for $s_N = 500$. (a) Cost kernel s_k. (b) Optimal feedback gains K_k. (c) Optimal trajectory x_k^*.

A MATLAB routine simulating the application of the feedback control to the plant (1) is also shown in Fig. 2.2-4. Beginning at the given x_0, it steps *forward* in time using (1) and (6), with precomputed stored values of K_k, to calculate the resulting trajectory x_k^*. The backward-computed kernel and gain sequences s_k and K_k and the forward-computed state sequence x_k^* are plotted in Fig. 2.2-5. Note that if s_N is finite, x_N will not be exactly equal to zero, but that as $s_N \to \infty$, the final state x_N approaches zero more closely. The value of s_N for Fig. 2.2-5 was 5. Figure 2.2-6 is the simulation using $s_N = 500$.   ■

Software for the efficient solution of the Riccati equation is included in Bierman (1977). However, this example has made the point that a little *preliminary analysis* can result in simplified algorithms (i.e., we used (4) in the example, not (3), for implementation). This is especially true in the case where the number of states n is greater than one and the matrices are sparse.

To see why, note that the Riccati equation is symmetric (the transpose of the right-hand side is equal to itself if S_{k+1} is symmetric). Therefore, S_k is symmetric for all k if $S_N^T = S_N$, which we have assumed. This means that S_k has n^2 elements, only $n(n+1)/2$ of which are distinct. Some preliminary analysis can yield $n(n+1)/2$ *scalar* recursions for the components of S_k, which are easier to use than one $n \times n$ *matrix* recursion. We shall see examples of this in Section 2.3.

Exercise 2.2-3. *LQ Regulator with Weighting of State–input Inner Product*

Let the plant be given by

$$x_{k+1} = A_k x_k + B_k u_k, \tag{1}$$

but consider the modified performance index

$$J_i = \frac{1}{2} x_N^T S_N x_N + \frac{1}{2} \sum_{k=i}^{N-1} [x_k^T \quad u_k^T] \begin{bmatrix} Q_k & T_k \\ T_k^T & R_k \end{bmatrix} \begin{bmatrix} x_k \\ u_k \end{bmatrix}, \tag{2}$$

where the block coefficient matrix in the sum is positive definite. This index allows us to weight products of state and input components to keep them small.

a. Show that the Hamiltonian system is

$$\begin{bmatrix} x_{k+1} \\ \lambda_k \end{bmatrix} = \begin{bmatrix} \overline{A}_k & -B_k R_k^{-1} B_k^T \\ \overline{Q}_k & A_k^T \end{bmatrix} \begin{bmatrix} x_k \\ \lambda_{k+1} \end{bmatrix}, \tag{3}$$

where

$$\overline{A}_k = A_k - B_k R_k^{-1} T_k^T, \tag{4}$$

$$\overline{Q}_k = Q_k - T_k R_k^{-1} T_k^T. \tag{5}$$

b. Show that the optimal control is given by

$$u_k = -\left(B_k^T S_{k+1} B_k + R_k\right)^{-1} (B_k^T S_{k+1} A_k + T_k^T) x_k, \tag{6}$$

where sequence S_k is give by the Riccati equation

$$S_k = \overline{A}_k^T \left[S_{k+1} - S_{k+1} B_k (B_k^T S_{k+1} B_k + R_k)^{-1} B_k^T S_{k+1} \right] \overline{A}_k + \overline{Q}_k, \qquad k < N \tag{7}$$

with boundary condition of S_N.

c. Show that (7) can alternatively be written

$$S_k = A_k^T S_{k+1} A_k - K_k^T (B_k^T S_{k+1} B_k + R_k) K_k + Q_k \tag{8}$$

where the Kalman gain is

$$K_k = (B_k^T S_{k+1} B_k + R_k)^{-1} (B_k^T S_{k+1} A_k + T_k^T). \qquad (9)$$

d. Show that the optimum cost to go on the subinterval [k, N] is given in terms of the state x_k as

$$J_k = \tfrac{1}{2} x_k^T S_k x_k. \qquad (10)$$

e. Show that the Riccati equation (2.2-53) can be written as (8) with K_k given by (2.2-57). Hence, the only changes introduced by the off-diagonal weighting term T_k are that the Kalman gain must be modified as in (9) and that the Riccati equation formulation (8) should be used.

■

2.3   DIGITAL CONTROL OF CONTINUOUS-TIME SYSTEMS

With the increasing sophistication of microprocessors, more and more control schemes are being implemented digitally. In these schemes, the control input is switched to new values at discrete time steps, with a zero-order hold usually used between switchings so that the control is constant during these intervals. Such controls must be designed using a discretized version of the continuous plant.

Design of Digital Controls

The design of digital control laws is very straightforward. If the continuous time-invariant plant is given by

$$\dot{x}(t) = Ax(t) + Bu(t), \qquad (2.3\text{-}1)$$

then the discretized version of the plant, using a sampling period of T, is

$$x_{k+1} = A^s x_k + B^s u_k, \qquad (2.3\text{-}2)$$

where the sampled plant and control matrices are

$$A^s = e^{AT}, \qquad (2.3\text{-}3a)$$

$$B^s = \int_0^T e^{A\tau} B \, d\tau. \qquad (2.3\text{-}3b)$$

This discretization process assumes that the control input $u(t)$ to the continuous plant is switched only at times kT, and that it is held constant (with a zero-order hold) between switchings so that $u(t) = u_k$, for $kT \leq t < (k + 1)T$. This is within our power to guarantee, since we select the control input. If we do manufacture $u(t)$ from u_k in this fashion, then the continuous state is related to the discrete state according to $x(kT) = x_k$.

To guarantee that the *samples* of the state, $x_k = x(kT)$, and the control, $u_k = u(kT)$, display a desired behavior, it is necessary only to apply our optimal control law computation methods to the sampled system (2.3-2) to compute the optimal sequence u_k^*. The actual continuous input that is sent to the plant is then manufactured from u_k^*.

The values of $x(t)$ between sampling instants kT cannot be specified using a control law design based on (2.3-2), but they can be determined in a very easy manner, for between the sampling instants kT and $(k+1)T$ the state propagates according to the plant dynamics with a constant input:

$$\dot{x}(t) = Ax(t) + Bu_k, \quad kT \leq t < (k+1)T. \tag{2.3-4}$$

The solution to this equation is

$$x(t) = e^{A(t-kT)}x_k + \int_{kT}^{T} e^{A(t-\tau)} B \, d\tau \, u_k, \quad kT \leq t < (k+1)T. \tag{2.3-5}$$

If the continuous plant is time varying, then the discretization process yields a time-varying discrete plant

$$A_k^s = \phi((k+1)T, kT), \tag{2.3-6a}$$

$$B_k^s = \int_{kT}^{(k+1)T} \phi((k+1)T, \tau)B(\tau) \, d\tau, \tag{2.3-6b}$$

where $\phi(t, t_0)$, is the state transition matrix of the time invariant plant described by (2.3-1). The optimal control approach just described still applies.

Simulation of Digital Controls

Once the optimal control sequence u_k^* has been designed based on (2.3-2), we should like to simulate the application of the resulting digital control to the plant (2.3-1) to verify that it has a satisfactory behavior. For completeness, we should like to observe the behavior of the state *between* sampling instants as well as at the sampling instants.

To do this, we can use the simulation scheme shown in Fig. 2.3-1. The continuous state equation (2.3-1) is represented in the figure as

$$F(t, X, \dot{X}) \tag{2.3-7}$$

and is integrated using the MATLAB routine *lsim.m*. The control input $u(t)$ is updated at each time kT and then held until time $(k+1)T$. It is important to realize that *two* sampling time intervals are involved, the plant sampling interval T and the simulation sampling interval T_s. The states are evaluated by the routine at the sampling interval and hence we choose T_s to be a divisor of T.

Note that we are using discrete system (2.3-2) to design the control law, but that, to find $x(t)$ at all values of t, we are using the continuous plant (2.3-1) to *simulate* the control law.

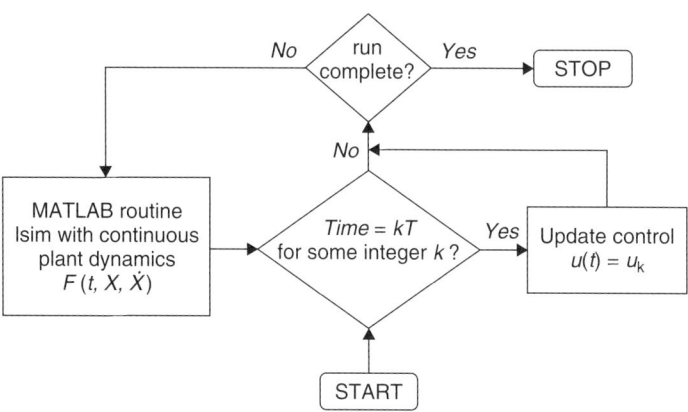

FIGURE 2.3-1   Digital control simulation scheme.

Some Examples

Let us demonstrate how easy this procedure is by considering some examples. The examples also illustrate how to simplify the coding for the computation of optimal control u_k by doing some preliminary analysis; by tailoring the Riccati equation to each problem, no matrix manipulations are used.

Example 2.3-1.   Digital Control of an RC Circuit

The electric circuit in Fig. 2.3-2 provides a scalar example that nicely illustrates our approach. The continuous state equation is

$$\dot{x} = \frac{-1}{\tau}x + \frac{1}{\tau}u, \tag{1}$$

with time constant

$$\tau = \frac{1}{RC}. \tag{2}$$

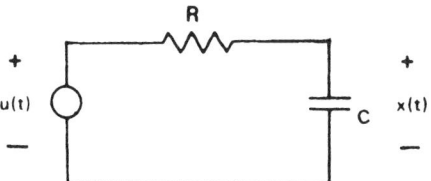

FIGURE 2.3-2   RC circuit.

Let $\tau = 5$, so that

$$\dot{x} = -0.2x + 0.2u. \tag{3}$$

It is desired to control the capacitor voltage $x(t)$ by a scheme in which input $u(t)$ is switched only at discrete instants kT by a microprocessor. The microprocessor also samples $x(t)$ at each sampling period T. Let $T = 0.5$ sec. (For good control, we should select $T < \tau/10$.)

a. Design of Digital Control Law for Free Final State

The discretized system is

$$x_{k+1} = e^{-T/\tau} x_k + \int_0^T e^{-\lambda/\tau} \frac{1}{\tau} d\lambda \cdot u_k$$

$$= e^{-T/\tau} x_k + (1 - e^{-T/\tau}) u_k, \tag{4}$$

or

$$x_{k+1} = a x_k + b u_k, \tag{5}$$

with $a = 0.9048$ and $b = 0.0952$. Suppose that we want the control and state samples u_k and x_k to be small over a 5-sec interval for any initial voltage $x(0)$. Then $N = 5/T = 10$. To express these control objectives mathematically, select the performance index

$$J = \frac{1}{2} s_N x_N^2 + \frac{1}{2} \sum_{k=0}^{N-1} (a x_k^2 + r u_k^2). \tag{6}$$

This is the same system and cost we examined in Example 2.2-2, so the optimal control is just given by the Riccati equation as

$$K_k = \frac{a b s_{k+1}}{b^2 s_{k+1} + r}, \tag{7}$$

$$s_k = \frac{a^2 r s_{k+1}}{b^2 s_{k+1} + r} + q, \tag{8}$$

$$u_k = K_k x_k. \tag{9}$$

By making s_N large, we can force the final state $x_N = x(5)$ to be small.

b. Simulation of Digital Control Law for Free Final State

It is quite simple to simulate the digital control law being applied to the continuous plant (3). First notice that equations (7), (8), and (9) can be computed by using the *scaopt.m* function. The overall driver program for the implementation of the digital control law is given in Fig. 2.3-3, where $a_c = -0.2$, $b_c = 0.2$, $a_d = 0.9048$, $b_d = 0.0952$, $q = r = 1$, $s = 100$, $N = 10$, and $x_0 = 10$.

Results of the simulation run are shown in Fig. 2.3-4, where the state $x(t)$ with no control is shown along with the state resulting on application of our digital control. Also shown in Fig. 2.3-5 is $u(t)$, the digital control manufactured from u_k in (9).

c. Design of Control Law for Fixed Final State

To achieve a fixed final value of $x(5) = 0$, we can use a very large value of s_N in parts a and b. To achieve a nonzero value r_N for $x(5)$, however, we must use the open-loop control of (2.2-38). (When we discuss the *function of final-state-fixed problem* in Section 4.5, we shall see how to achieve a nonzero fixed final state using a Riccati-equation-based design.)

```
function ex2_3_1b(a_c, b_c, a_d, b_d, q, r, s, x0, N)
% Compute Optimal Control Input
[x, u, K, S] =scaopt(a_d, b_d, q, r, s, x0, N);
% Define the time interval T
T=0:0.05:5;
% Expand the input u to the specified interval T
U=kron(u, ones(1, 10));
U=[U u(length(u))];
% Simulate the plant dynamics
system=ss(a_c,b_c,1,0);
figure(1)
[Y,T,X]=lsim(system, U, T, x0); plot(T,Y); hold;
%Simulate the plant dynamics with zero input
[Y,T,X]=lsim(system, [0 kron(u, zeros(1, 10))], T, x0); plot(T,Y);
legend('x(t) with zero control input','x(t) with control');
      xlabel('Time [s]');
% Plot the input u(t)
figure(2)
T=0:0.05:5;
plot(T, U); legend('Control input'); xlabel('Time [s]');
end
```

FIGURE 2.3-3 Driver program to compute u_k^* and simulate the resulting digital control scheme.

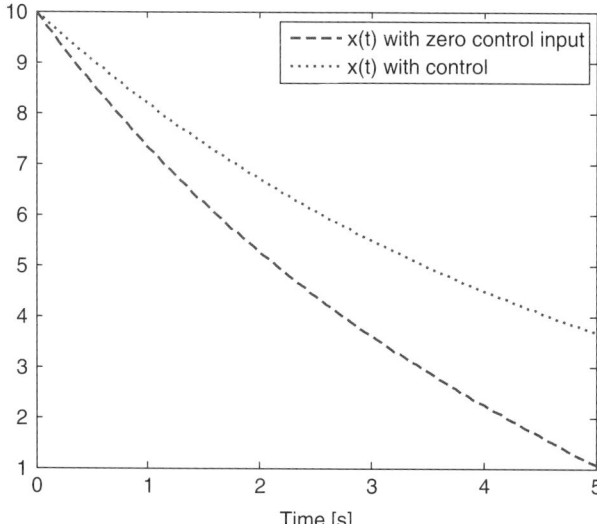

FIGURE 2.3-4 Simulation of continuous plant dynamics comparing zero-input response to controlled response.

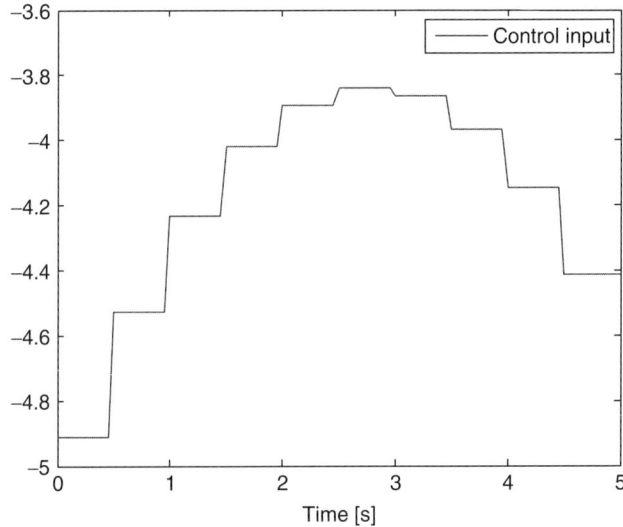

FIGURE 2.3-5   The control input obtained from u_k.

Thus, suppose we want to drive the capacitor voltage from $x(0) = x_0 = 10$ V exactly to $x(5) = r_N = 20$ V while minimizing the energy

$$J_0 = \frac{r}{2} \sum_{k=0}^{N-1} u_k^2 \qquad (10)$$

of the control samples. The design of the control to make the sampled system (5) achieve this objective is exactly the problem we solved in Example 2.2-1, so the optimal discrete control sequence is

$$u_k = \frac{(1-a)^2}{b(1-a^{2N})}(r_N - a^N x_0)a^{N-k-1}. \qquad (11)$$

d. Simulation of Digital Control for Fixed Final State

In this open loop scheme, the simulation is as easy as before. We just use the driver program as before with the exception that in this case we cannot use *scaopt.m* to compute u_k; instead, we wrote a simple script to implement (11). The driver program is shown in Fig. 2.3-6.

The simulation results are shown in Fig. 2.3-7. Note that $-u(t)$ (not $u(t)$) is plotted there.

While the open-loop control is easier to implement, it is not as robust as the closed-loop control. If at a time of 2 sec, for example, a noise source drives the state off the optimal trajectory shown in Fig. 2.3-7, the final state $x(5)$ will no longer be $20V$.

```
function ex2_3_1d (a_c, b_c, a_d, b_d, q, r, s, rN, x0, N)
% Compute Optimal Control Input
for k=1:N
u(k)=((1-a_d^2)/(b_d*(1-a_d^(2*N))))*(rN-x0*a_d^N)*a_d^ N-k);
end
% Define the time interval T
T=0.05:0.05:5;
% Expand the input u to the specified interval T
U=kron(u,ones(1,10));
% Simulate the plant dynamics
lsim(a_c, b_c, 1, 0, U, T, x0); hold
%Simulate the plant dynamics with zero input
lsim (a_c, b_c, 1, 0, kron (u, zeros (1, 10)), T, x0);
% Plot the input -u(t)
plot (T, U);
```

FIGURE 2.3-6 Driver program for on-line computation of open-loop control.

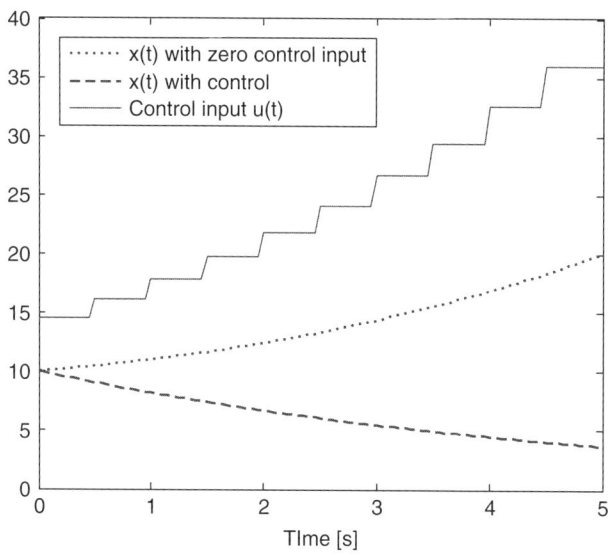

FIGURE 2.3-7 Simulation of continuous plant dynamics comparing zero-input response to controlled response. ■

Example 2.3-2. Digital Control of Systems Obeying Newton's Laws

Newton's laws $m\ddot{d} = F$ can be expressed in state-variable form as

$$\dot{x} = \begin{bmatrix} 0 & 1 \\ 0 & 0 \end{bmatrix} x + \begin{bmatrix} 0 \\ 1 \end{bmatrix} u, \tag{1}$$

where $x \triangleq [d \quad v]^T$, with $d(t)$ and $v(t)$ representing position and velocity. Input $u(t)$ is an acceleration; to find the input in units of force, we can multiply $u(t)$ by m, the mass of the body.

The optimal control law we shall derive applies to any system obeying (1). To lend more interest to our example, however, let us formulate a particular problem to solve.

a. The Rendezvous Problem

See Fig. 2.3-8. A target aircraft A_t is moving in the y_1 direction with a constant velocity of V_t. Its initial y_1 coordinate is Y_t. Our aircraft A is moving in the y_1 direction with a constant velocity of $V > V_t$. Our initial y_1 coordinate is 0. Thus, our velocity relative to A_t is $(V - V_t)$. Clearly, at time

$$t_f = \frac{Y_t}{V - V_t} \tag{2}$$

the two aircraft A and A_t will be abreast of each other (i.e., have the same y_1 coordinate). The y_1 velocities V and V_t are fixed throughout the problem, and hence the final time t_f is known.

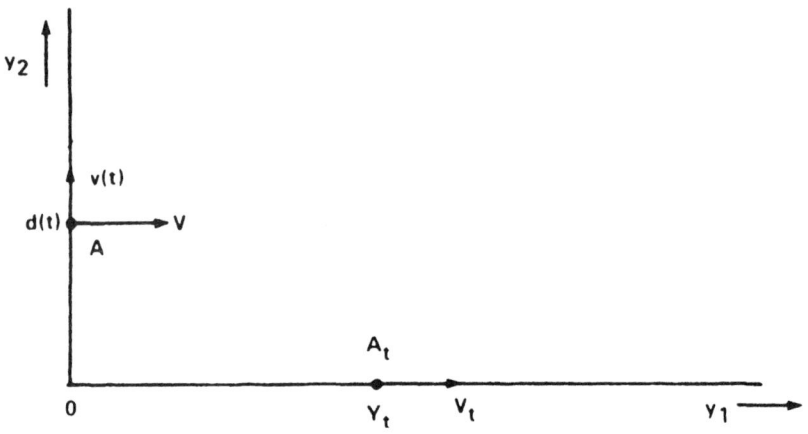

FIGURE 2.3-8   Rendezvous problem geometry.

The optimal control problem is as follows. The y_2 position and velocity of aircraft A relative to A_t are $d(t)$ and $v(t)$, and the y_2 dynamics of A are described by (1). It is required to determine the control acceleration $u(t)$ needed in the y_2 direction so that aircraft A will rendezvous with the target A_t at time t_f. This means that we must determine $u(t)$ so that $d(t_f)$ and $v(t_f)$ are both zero.

b. Design of Digital Control Law

To find such a control $u(t)$, discretize the y_2 dynamics (1) to get (note that $e^{AT} = I + AT$ since $A^2 = 0$.):

$$x_{k+1} = \begin{bmatrix} 1 & T \\ 0 & 1 \end{bmatrix} x_k + \begin{bmatrix} T^2/2 \\ T \end{bmatrix} u_k \tag{3}$$

for some sampling period T. Suppose that $t_f = 5$ sec. Then $T = 0.5$ sec is reasonable.

Associate with (3) the performance index

$$J_0 = \frac{1}{2}x_N^T \begin{bmatrix} s_d & 0 \\ 0 & s_v \end{bmatrix} x_N + \frac{1}{2}\sum_{k=0}^{N-1}\left(x_k^T\begin{bmatrix} q_d & 0 \\ 0 & q_v \end{bmatrix}x_k + ru_k^2\right). \tag{4}$$

Select control weighting $r = 1$, position weighting $q_d = 1$, and velocity weighting $q_v = 1$. To ensure that the final y_2 position d_N and velocity V_N are very small, select the final-state component weights as $s_d = 100$, $s_v = 100$. The number of iterations is

$$N = t_f/T = 10. \tag{5}$$

Now the optimal control is given by (2.2-58), (2.2-60), and (2.2-61). Let us do some preliminary analysis on these equations, defining a few intermediate variables to simplify things.

Since we know that S_k is symmetric for all k, let

$$S_k \overset{\Delta}{=} \begin{bmatrix} s_1 & s_2 \\ s_2 & s_3 \end{bmatrix}. \tag{6}$$

(We shall not require time subscripts on the cost kernel components s_i since they will be updated at each k by MATLAB replacement.) Then the feedback gain is updated by

$$\delta = B^T S_{k+1} B + r$$

$$= r + \frac{s_1 T^4}{4} + s_2 T^3 + s_3 T^2, \tag{7}$$

$$K_k = B^T S_{k+1} A / \delta$$

$$= \frac{1}{\delta}\left[\frac{s_1 T^2}{2} + s_2 T \frac{s_1 T^3}{2} + \frac{3s_2 T^2}{2} + s_3 T\right]. \tag{8}$$

Letting

$$K_k \overset{\Delta}{=} [k_1 \quad k_2] \tag{9}$$

we can write

$$k_1 = \left(\frac{s_1 T^2}{2} + s_2 T\right)/\delta, \tag{10a}$$

$$k_2 = \left(\frac{s_1 T^3}{2} + \frac{3s_2 T^2}{2} + s_3 T\right)/\delta. \tag{10b}$$

The closed-loop plant matrix is

$$A_k^{cl} \overset{\Delta}{=} A - BK_k = \begin{bmatrix} 1 - k_1 T^2/2 & T - k_2 T^2/2 \\ -k_1 T & 1 - k_2 T \end{bmatrix}. \tag{11}$$

Defining the components of A_k^{cl} as

$$A_k^{cl} \overset{\Delta}{=} \begin{bmatrix} a_{11}^{cl} & a_{12}^{cl} \\ a_{21}^{cl} & a_{22}^{cl} \end{bmatrix}, \tag{12}$$

we have the four scalar updates:

$$a_{11}^{cl} = 1 - k_1 T^2/2, \tag{13a}$$

$$a_{12}^{cl} = T - k_2 T^2/2, \tag{13b}$$

$$a_{21}^{cl} = -k_1 T, \tag{13c}$$

$$a_{22}^{cl} = 1 - k_2 T. \tag{13d}$$

The updated cost kernel is

$$S_k = A^T S_{k+1} A_k^{cl} + Q$$

$$= \begin{bmatrix} s_1 a_{11}^{cl} + s_2 a_{21}^{cl} + q_d & s_1 a_{12}^{cl} + s_2 a_{22}^{cl} \\ (s_1 T + s_2) a_{11}^{cl} + (s_2 T + s_3) a_{21}^{cl} & (s_1 T + s_2) a_{12}^{cl} + (s_2 T + s_3) a_{22}^{cl} + q_v \end{bmatrix}, \tag{14}$$

which yields the scalar updates:

$$s_1 = s_1 a_{11}^{cl} + s_2 a_{21}^{cl} + q_d, \tag{15a}$$

$$s_2 = s_1 a_{12}^{cl} + s_2 a_{22}^{cl}, \tag{15b}$$

$$s_3 = (s_1 T + s_2) a_{12}^{cl} + (s_2 T + s_3) a_{22}^{cl} + q_v. \tag{15c}$$

Note that (15b) and

$$s_2 = (s_1 T + s_2) a_{11}^{cl} + (s_2 T + s_3) a_{21}^{cl} \tag{16}$$

are evidently equivalent since S_k is symmetric. (Prove this.) For numerical stability in the face of computer roundoff error, we could use the average of (15b) and (16) for s_2.

The optimal feedback gains are therefore found by iterating (7), (10), (13), and (15) for time index $k = N - 1, N - 2, \ldots, 0$. Software for this is contained in subroutine *ex2_3_2c* in Fig. 2.3-10.

c. Simulation of Digital Control

Suppose that the initial y_2 position and velocity of aircraft A are $d(0) = 10$, $v(0) = 10$. By using subroutine *lsim* to simulate the dynamics (1) and driving with a zero input as follows

```
T=0:0.05:5;
U=zeros(1,101);
A=[0 1;0 0]; b=[0;1]; c=eye(2); d=zeros(2, 1);
system=ss(A,b,c,d);
ic=[10 10];
[Y,T,X]=lsim(system, U, T, ic);
plot(T,Y)
axis([0 5 0 60]);
legend('d(t)','v(t)'); xlabel('Time [s]');
```

the uncontrolled state plot in Fig. 2.3-9 was obtained.

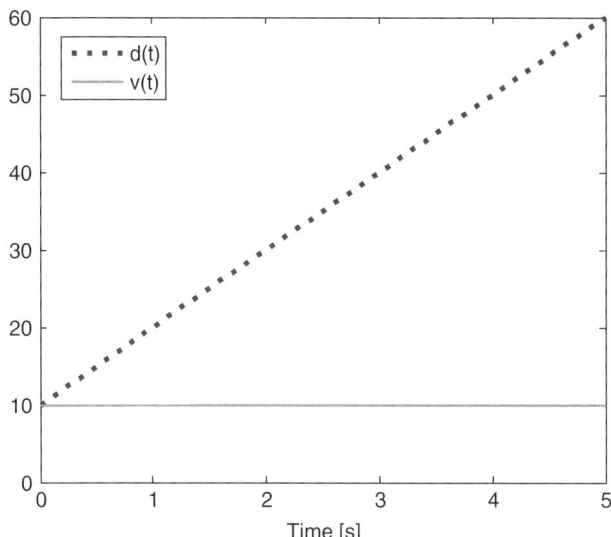

FIGURE 2.3-9 Uncontrolled response of plant.

This plot represents the zero input solution

$$d(t) = d(0) + v(0)t,$$

$$v(t) = v(0).$$

Clearly, this behavior is not exactly what we had in mind!

Let us examine the behavior of (1) manufactured under the influence of a digital control law manufactured from the optimal discrete control found in part b according to

$$u(t) = u_k, \quad kT \le t < (k+1)T.$$

To do this, we can use the driver program shown in Fig. 2.3-10.

Since we computed the digital control law, we can use *lsim* for simulating the dynamics of the continuous plant. The resulting state trajectory is shown in Fig. 2.3-11. The first thing the control input does is go negative to -20 in order to decrease the y_2 velocity $v(t)$ to zero by a time of 0.5 sec. Then velocity $v(t)$ becomes negative so that our aircraft A begins to approach the target (i.e., $d(t)$ begins to decrease). Velocity $v(t)$ is then gradually returned to zero to achieve a rendezvous at $t_f = 5$ sec.

Apparently, our choice of final weighting $s_d = 100$ and $s_v = 100$ was satisfactory. The simulation shows that at $t = 5$ sec, $d(t)$ and $v(t)$ are indeed very close to zero as required.

Preliminary simplification of the Riccati equation can lead to some nice, simple implementations, as this example shows. For greater than $n = 2$ states, however, it can be more trouble than it is worth. It should be clearly realized that preliminary analysis is only for convenience; it is never actually required, since the software in Bierman (1977) can be used to solve the Riccati equation in the general case.

```
function u = ex2_3_2c(A_d, b_d, q, r, s, N, T, x0)
% Backward iteration for Cost Kernel and FB Gains
k=N;
K=zeros(N,2);
while k>0,
T2=T^2;
div=r+(s(1)*T2^2)/4 + s(2)*T*T2 + s(3)*T2;
% Feedback Gains
K(k, 1) = (s(1)*T2/2+s(2)*T)/div;
K(k, 2) = (s(1)*T2*T/2 +3*s(2)*T2/2 + s(3)*T)/div;
% Closed-loop Plant matrix
Acl=[ 1-K(k,1)*T2/2  T-K(k,2)*T2/2; -K(k,1)*T  1-K(k,2)*T]
% Cost Kernel Update
s(3)=(s(1)*T+s(2))*Acl(1, 2) + (s(2)*T+s(3))*Acl(2, 2)+q(2);
temp=s(2);
s(2)=s(1)*Acl(1,2)+temp*Acl(2,2);
s(1)=s(1)*Acl(1,1)+temp*Acl(2,1)+q(1);
k=k-1;
end
% Apply Optimal Control (Forward Iteration)
x(:,1)=x0;
for k=1:N
% Compute Optimal Control Law
u(k)=-K(k,:)*x(:,k);
% Update the Plant State
x(:,k+1)=A_d*x(:,k)+b_d*u(k);
end
end
```

FIGURE 2.3-10   Driver program to compute u_k^* and simulate the resulting control scheme.

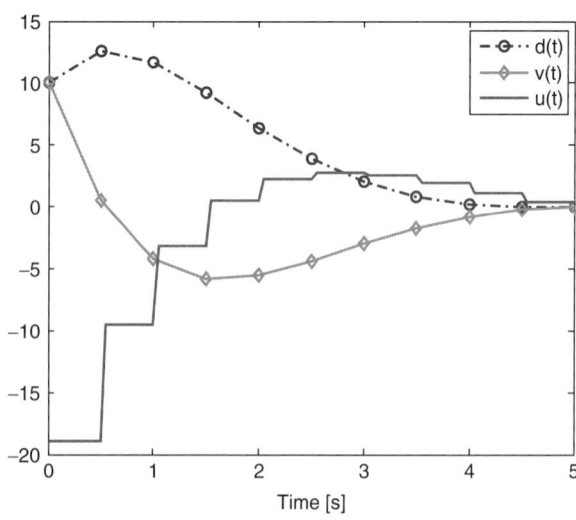

FIGURE 2.3-11   Optimal state and control trajectories for rendezvous problem.

2.4   STEADY-STATE CLOSED-LOOP CONTROL AND SUBOPTIMAL FEEDBACK

We have seen that the solution to the LQ optimal control problem is a state feedback of the form

$$u_k = -K_k x_k, \tag{2.4-1}$$

with gain sequence K_k given in terms of the solution sequence S_k to the Riccati equation as

$$S_k = A^{\mathrm{T}}[S_{k+1} - S_{k+1}B(B^{\mathrm{T}}S_{k+1}B + R)^{-1}B^{\mathrm{T}}S_{k+1}]A + Q, \tag{2.4-2}$$

$$K_k = (B^{\mathrm{T}}S_{k+1}B + R)^{-1}B^{\mathrm{T}}S_{k+1}A. \tag{2.4-3}$$

In most of this section we assume the time-invariant case. Even in this situation, the closed-loop system

$$x_{k+1} = (A - BK_k)x_k \tag{2.4-4}$$

is time varying since the optimal feedback gains K_k are time varying.

This time-varying feedback is not always convenient to implement; it requires the storage of an entire sequence of $m \times n$ matrices. We might be interested in using instead a suboptimal feedback gain that does not actually minimize the performance index but is a *constant* so that

$$u_k = -K x_k. \tag{2.4-5}$$

Such a feedback is certainly easier to implement than (2.4-1).

As one candidate for a constant feedback gain, we might consider the limit of the optimal K_k as the final time N goes to infinity (or equivalently as $k \to -\infty$). We shall see that when this limit exists, it provides a constant feedback that is often satisfactory.

Let us first consider the effect of using an *arbitrary* feedback to control the plant.

Suboptimal Feedback Gains

The plant, which we shall assume time invariant in this subsection only for notational simplicity, is

$$x_{k+1} = Ax_k + Bu_k. \tag{2.4-6}$$

Let us use as a control the state feedback (2.4-1) for some arbitrary given matrix sequence K_k. We are not yet concerned about how to select K_k; all we want to know is the resulting value of the performance index

$$J_i = \frac{1}{2}x_N^{\mathrm{T}}S_N x_N + \frac{1}{2}\sum_{k=i}^{N-1}(x_k^{\mathrm{T}}Qx_k + u_k^{\mathrm{T}}Ru_k). \tag{2.4-7}$$

We can find this value by using a derivation like the one leading to (2.2-66). Thus, add the left-hand side minus the right-hand side of (2.2-63) (i.e., add zero) to (2.4-7) and use (2.4-1) to get

$$J_i = \frac{1}{2}x_i^T S_i x_i + \frac{1}{2}\sum_{k=i}^{N-1}\left[x_{k+1}^T S_{k+1} x_{k+1} + x_k^T (Q - S_k + K_k^T RK_k)x_k\right]. \quad (2.4\text{-}8)$$

The sequence S_k is at this point undefined. Taking into account the state equation (2.4-6) with the control (2.4-1) yields

$$J_i = \frac{1}{2}x_i^T S_i x_i + \frac{1}{2}\sum_{k=i}^{N-1} x_k^T\left[(A - BK_k)^T S_{k+1}(A - BK_k) + Q + K_k^T RK_k - S_k\right]x_k.$$

$$(2.4\text{-}9)$$

Now suppose the sequence S_k satisfies the matrix equation (2.2-62). The sum is then zero, so that finally $J_i = \frac{1}{2}x_i^T S_i x_i$.

We can summarize this result as follows. Let the feedback (2.4-1) for any given K_k be applied to the plant. Then the resulting cost on $[k, N]$ is given for each time k by

$$J_k = \frac{1}{2}x_k^T S_k x_k, \quad (2.4\text{-}10)$$

where the kernel is the solution to

$$S_k = (A - BK_k)^T S_{k+1}(A - BK_k) + K_k^T RK_k + Q \quad (2.4\text{-}11)$$

with boundary condition S_N.

We should be sure we know exactly what is going on here. Equation (2.4-11) is *not* the Joseph-stabilized Riccati equation! It becomes the Joseph-Riccati equation *only if* the optimal gain sequence (2.4-3) is used. If K_k is an arbitrary given gain sequence, then (2.4-11) is simply a (linear) *Lyapunov* equation in terms of the known closed-loop plant matrix $(A - BK_k)$.

Note in particular that K_k can be a *constant* feedback matrix K, as in (2.4-5). If K_k is not the optimal gain, then J_k given by (2.4-10) and (2.4-11) is, in general, greater than the optimal cost J_k^*.

The next example illustrates the use of these new results.

Example 2.4-1.   Suboptimal Feedback Control of a Scalar System

Let us reconsider the system of Example 2.2-2. The plant is

$$x_{k+1} = ax_k + bu_k \quad (1)$$

with performance index

$$J_0 = \frac{1}{2}s_N x_N^2 + \frac{1}{2}\sum_{k=0}^{N-1}(qx_k^2 + ru_k^2). \quad (2)$$

The optimal control is a time-varying state feedback

$$u_k = -K_k x_k, \tag{3}$$

with gain determined by the Riccati equation as

$$s_k = \frac{a^2 r s_{k+1}}{b^2 s_{k+1} + r} + q, \tag{4}$$

$$K_k = \frac{abs_{k+1}}{b^2 s_{k+1} + r}. \tag{5}$$

For parameters of $a = 1.05$, $b = 0.01$, $q = r = s_N = 5$, with final time $N = 100$, a simulation was run to obtain the Kalman gain sequence shown in Fig. 2.2-5b. For $N - k = 100$, a steady-state value of $K_\infty \triangleq K_0 = 9.808$ has been reached. The corresponding Riccati-equation solution s_k^* is shown in Fig. 2.4-1a. If we apply the feedback (3) to the plant with $x_0 = 10$, the optimal state trajectory x_k^* in Fig. 2.4-1b results. The optimal cost

$$J_k^* = \tfrac{1}{2} s_k^* x_k^{*2} \tag{6}$$

along this trajectory is shown in Fig. 2.4-1c.

Now, let us suppose we want a simpler feedback control than (3) with K_k as in Fig. 2.2-5b. Suppose we try to use the *constant* state feedback

$$u_k = -K_\infty x_k = -9.808 x_k. \tag{7}$$

Then the cost is given by

$$J_k = \tfrac{1}{2} s_k x_k^2, \tag{8}$$

where s_k is the solution to the Lyapunov equation (2.4-11), which becomes

$$s_k = s_{k+1}(a - bK_\infty)^2 + rK_\infty^2 + q, \tag{9}$$

with boundary condition s_N. This suboptimal kernel sequence is shown in Fig. 2.4-1a. Note that it is an upper bound for the optimal sequence s_k^*. Thus, for any state x_k at any time k, the cost to go J_k using feedback (7) satisfies

$$J_k^* = \tfrac{1}{2} s_k^* x_k^2 \le \tfrac{1}{2} s_k x_k^2 = J_k. \tag{10}$$

Simulating the plant (1) with the input (7) yields the suboptimal state trajectory x_k shown in Fig. 2.4-1b. The associated suboptimal cost J_k is shown in Fig. 2.4-1c. It is greater than J_k^*. The suboptimal trajectory x_k has less energy than x_k^*, but J_k is larger than J_k^* because of the larger control energy required to achieve x_k.

Note that initially, at times well removed from the final time N, x_k and x_k^* are about the same. As k approaches N, K_k^* deviates more from K_∞, and the trajectories differ markedly.

Using the constant gain K_∞, the closed-loop system is

$$x_{k+1} = (a - bK_\infty)x_k = 0.952 x_k, \tag{11}$$

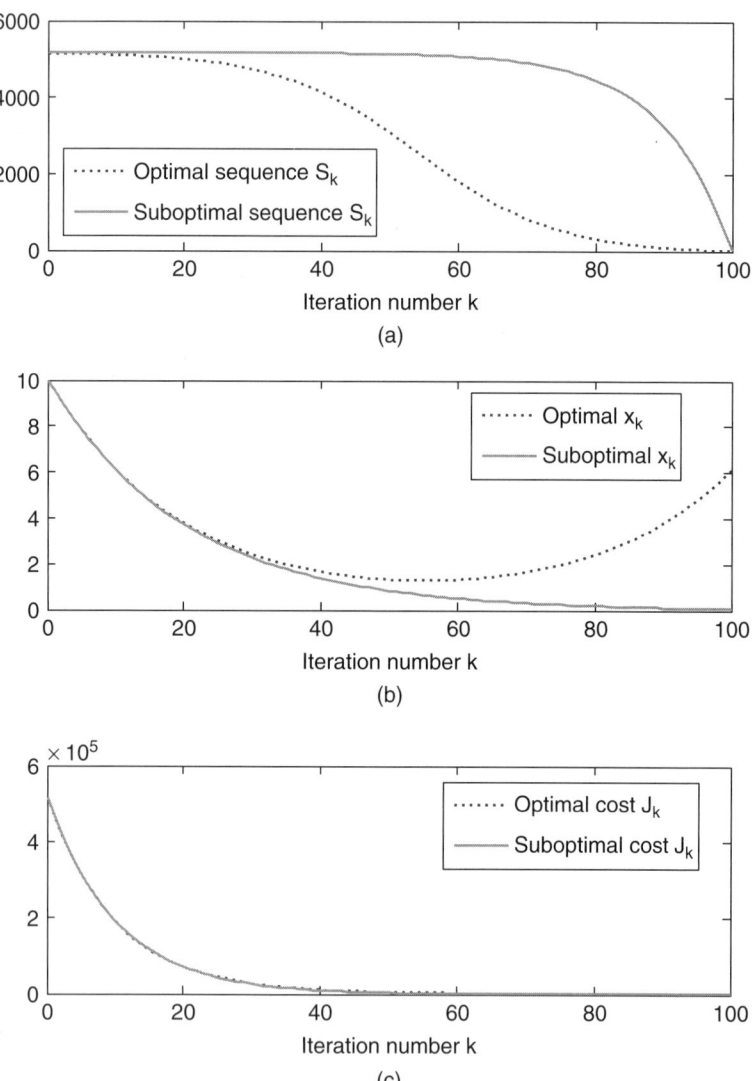

FIGURE 2.4-1   Optimal and suboptimal closed-loop behavior. (a) Cost kernels. (b) State trajectories. (c) Costs.

so the suboptimal trajectory is simply given by

$$x_k = 0.952^k x_0 = 10(0.952)^k. \tag{12}$$

By examining these graphs, we can determine whether the behavior under the influence of the simplified gain (7) is satisfactory and whether we are willing to use the extra control energy required. If so, we can go ahead and use this simplified feedback on the actual physical plant.   ∎

The Algebraic Riccati Equation

The comments in this subsection apply only for time-invariant plant and cost matrices. Equation (2.4-2) is solved backward in time beginning at time N. As $k \to -\infty$, the sequence S_k can have several types of behavior, as symbolized in Fig. 2.4-2. It can converge to a steady-state matrix S_∞, which may be zero, positive semi-definite, or positive definite. It can also fail to converge to a finite matrix.

If S_k does converge, then for large negative k, evidently $S \triangleq S_k = S_{k+1}$. Thus, in the limit, (2.4-2) becomes the *algebraic Riccati equation* (ARE)

$$S = A^{\mathrm{T}}[S - SB(B^{\mathrm{T}}SB + R)^{-1}B^{\mathrm{T}}S]A + Q, \qquad (2.4\text{-}12)$$

which has no time dependence. The limiting solution S_∞ to (2.4-2) is clearly a solution of (2.4-12).

Note that if S_N is symmetric and positive semidefinite, then the solution S_k to (2.4-2) is also symmetric and positive semidefinite for all k (transpose both sides of the equation). The algebraic equation (2.4-12), on the other hand, can have nonpositive semidefinite, nonsymmetric, and even complex solutions. Thus, all solutions to the algebraic Riccati equation are not also limiting solutions to the (time-varying) Riccati equation for some S_N.

If the limiting solution to (2.4-2) exists and is denoted by S_∞, then the corresponding *steady-state Kalman gain* is

$$K_\infty = (B^{\mathrm{T}}S_\infty B + R)^{-1}B^{\mathrm{T}}S_\infty A. \qquad (2.4\text{-}13)$$

This is a *constant* feedback gain. Under some circumstances it may be acceptable to use the time-invariant feedback law

$$u_k = -K_\infty x_k \qquad (2.4\text{-}14)$$

instead of the optimal control (2.4-1)–(2.4-3). The cost associated with such a control strategy is given by (2.4-10), where S_k satisfies (2.4-11) with $K_k = K_\infty$.

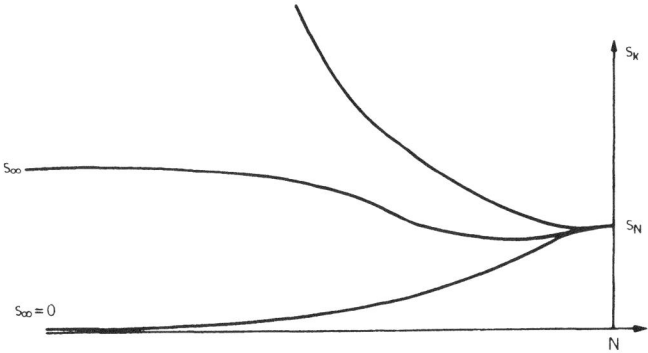

FIGURE 2.4-2   Limiting behavior of Riccati-equation solution.

To examine the consequences of using this steady-state feedback, let us discuss the limiting behavior of the closed-loop system (2.4-4) using the optimal time-varying feedback (2.4-1)–(2.4-3).

Limiting Behavior of the Riccati-equation Solution

This subsection applies only for time-invariant plant and cost matrices. We are interested in answering three questions here:

1. When does there exist a bounded limiting solution S_∞ to the Riccati equation for all choices of S_N?
2. In general, the limiting solution S_∞ depends on the boundary condition S_N. When is S_∞ the same for all choices of S_N?
3. When is the closed-loop plant (2.4-4) asymptotically stable?

Question 3 is particularly important. We have designed the feedback K_k to minimize a performance index, but as $k \to -\infty$ (or equivalently, as final time $N \to \infty$), we should certainly like for the closed-loop system to be stable! In some circumstances this can be guaranteed.

We can answer these questions in terms of the dynamical properties of the original system (2.4-6) with associated performance index (2.4-7). Recall that (A, B) is *reachable* if the eigenvalues of $(A - BK)$ can be arbitrarily assigned by appropriate choice of the feedback matrix K. (A, B) is *stabilizable* if there exists a matrix K such that $(A - BK)$ is asymptotically stable. This is equivalent to the reachability of all the unstable modes of A. Recall also that (A, C) is *observable* if the eigenvalues of $(A - LC)$ can be arbitrarily assigned by appropriate choice of the output injection matrix L. (A, C) is *detectable* if $(A - LC)$ can be made asymptotically stable by some matrix L. This is equivalent to the observability of the unstable modes of A.

Up to this point, we have been interested only in the state equation and not in an output, because our performance index was expressed in terms of the state and the input. Let us now define a *fictitious output* for our system. Let C and D denote the square roots of Q and R, such that $Q = C^T C$ and $R = D^T D$. In addition, let $S_N = C_N^T C_N$. Then define a fictitious output by the output equations

$$y_k = \begin{bmatrix} C \\ 0 \end{bmatrix} x_k + \begin{bmatrix} 0 \\ D \end{bmatrix} u_k, \qquad k = 0, \ldots, N - 1,$$

$$y_N = C_N x_N.$$

(2.4-15)

In terms of this "output," the performance index can be written

$$J_i = \frac{1}{2} y_N^T y_N + \frac{1}{2} \sum_{k=i}^{N-1} y_k^T y_k.$$

(2.4-16)

We shall see that the answers to our three questions can be given in terms of the stabilizability of (A, B) and the observability of (A, C). The following theorem tells us when there is a finite limiting cost kernel S_∞.

Theorem 2.4-1.   Let (A, B) be stabilizable. Then for every choice of S_N, there is a bounded limiting solution S_∞ to (2.4-2). Furthermore, S_∞ is a positive semidefinite solution to the algebraic Riccati equation (2.4-12).

Proof: Since (A, B) is stabilizable, there exists a constant feedback L so that

$$u_k = -Lx_k$$

and

$$x_{k+1} = (A - BL)x_k$$

is asymptotically stable. Thus, x_k is bounded and goes to zero as $k \to \infty$. Therefore, the associated cost

$$J_i = \frac{1}{2}x_N^T S_N x_N + \frac{1}{2}\sum_{k=i}^{N-1}(x_k^T Q x_k + u_k^T R u_k)$$

is finite as $i \to \infty$. It is given by (2.4-10) with $k = i$, where S_i satisfies the Lyapunov equation (2.4-11) (with index k replaced by i) with K_i there equal to L, and using S_N as boundary condition.

The *optimal* cost, however, is given for all i by

$$J_i^* = \tfrac{1}{2}x_i^T S_i^* x_i,$$

where S_i^* is the solution to (2.4-2) with S_N as boundary condition. Since $J_i^* \leq J_i$ for all initial states x_i, S_i provides an upper bound for S_i^* for all i ($S_i^* \leq S_i$ means $(S_i - S_i^*) \geq 0$). Hence, the solution sequence to (2.4-2) is bounded above by a finite sequence.

It can be shown that the solution S_i^* to (2.4-2) is smooth, so that if it is bounded above by a finite S_i, then it converges to a constant limit S_∞. For details, see Casti (1977) or Kwakernaak and Sivan (1972).

Since equation (2.4-2) is symmetric, then so is S_i for all i if S_N is symmetric, which we have assumed. The structure of the equation and the assumptions on Q and R also imply positive semidefiniteness.

Clearly, S_∞ is a solution to the limiting equation (2.4-12).   ∎

An important point should be noted here. In presenting the solution in Table 2.2-1 to the free-final-state LQ regulator and in the subsequent examples we discussed, we did not make any controllability assumptions on the plant. Regardless of the controllability properties of the plant, the optimal control will do the best it can to minimize the performance index. We have just shown that if the plant is, in fact, stabilizable, then there is a finite limiting solution S_∞ to the Riccati equation. This means that as the time interval $[i, N]$ goes to infinity, the optimal cost J_i^* stays bounded. Since $R > 0$, this in turn guarantees that the optimal control u_k^* itself does not go to infinity.

We can often show the stronger condition of reachability, which implies stabilizability, and for which there is a simple test based on the full rank of the reachability matrix

$$U_n = [B \quad AB \quad \cdots \quad A^{n-1}B], \tag{2.4-17}$$

where $x \in R^n$.

Theorem 2.4-1 is based on system stabilizability, and it makes intuitive sense because we are dealing with the optimal control problem. The next result initially seems very strange; it provides answers to our second and third questions in terms of the observability of the plant through the fictitious output!

Theorem 2.4-2. Let C be a square root of the intermediate-state weighting matrix, so that $Q = C^T C \geq 0$, and suppose $R > 0$. Suppose that (A, C) is observable; then (A, B) is stabilizable if and only if:

a. There is a unique positive definite limiting solution S_∞ to the Riccati equation (2.4-2). Furthermore, S_∞ is the unique positive definite solution to the algebraic Riccati equation (2.4-12).

b. The closed-loop plant

$$x_{k+1} = (A - BK_\infty)x_k$$

is asymptotically stable, where K_∞ is given by (2.4-13).

Proof:

Necessity

Define D by $R = D^T D$. Then $|D| \neq 0$, so that $B = MD$ for some M. One, therefore, has

$$\text{rank} \begin{bmatrix} zI - A \\ C \\ DK \end{bmatrix} = \text{rank} \begin{bmatrix} I & 0 & M \\ 0 & I & 0 \\ 0 & 0 & I \end{bmatrix} \begin{bmatrix} zI - A \\ C \\ DK \end{bmatrix} = \text{rank} \begin{bmatrix} zI - (A - BK) \\ C \\ DK \end{bmatrix}. \tag{1}$$

If (A, C) is observable, then by the PBH (Popov–Belevitch–Hautus) rank test (Kailath 1980),

$$\text{rank} \begin{bmatrix} zI - A \\ C \end{bmatrix} = n \quad \text{for every } z, \tag{2}$$

so that by (1)

$$\left((A - BK), \begin{bmatrix} C \\ DK \end{bmatrix} \right)$$

is observable for any K.

Now, stabilizability implies the existence of a feedback control $u_k = -Lx_k$, so that

$$x_{k+1} = (A - BL)x_k \tag{3}$$

is asymptotically stable. The cost of such a control on $[i, \infty]$ is

$$J_i = \tfrac{1}{2}x_i^{\mathrm{T}} S x_i \tag{4}$$

with S the limiting solution to (2.4-11) with $K_i = L$. The optimal closed-loop system (2.4-2)–(2.4-4) has an associated cost on $[i, \infty]$ of

$$J_i^* = \frac{1}{2}\sum_{k=i}^{\infty}\left(x_k^{\mathrm{T}} Q x_k + u_k^{\mathrm{T}} R u_k\right) = \frac{1}{2}x_i^{\mathrm{T}} S * x_i \le J_i. \tag{5}$$

with S^* the limiting solution to (2.4-2). Therefore, $Cx_k \to 0$ and, since $|R| \neq 0$, $u_k \to 0$. Select an N so that Cx_k and u_k are negligible for $k > N$. Then for $k > N$,

$$0 = \begin{bmatrix} Cx_k \\ Cx_{k+1} \\ \vdots \\ Cx_{k+n-1} \end{bmatrix} = \begin{bmatrix} C \\ CA \\ \vdots \\ CA^{n-1} \end{bmatrix} x_k, \tag{6}$$

and so observability of (A, C) requires $x_k \to 0$. Hence, the optimal closed-loop system (2.4-4) is asymptotically stable.

Write (2.4-12) as

$$S = (A - BK_\infty)^{\mathrm{T}} S(A - BK_\infty) + \begin{bmatrix} C \\ DK_\infty \end{bmatrix}^{\mathrm{T}} \begin{bmatrix} C \\ DK_\infty \end{bmatrix} \tag{7}$$

with K_∞ the optimal feedback. (Prove that this can be done.) Then (7) is a Lyapunov equation with

$$\left((A - BK_\infty), \begin{bmatrix} C \\ DK_\infty \end{bmatrix} \right)$$

observable and $(A - BK_\infty)$ stable. Therefore, there is a unique positive definite solution S^* to (2.4-12).

Sufficiency

If $x_{k+1} = (A - BK_\infty)x_k$ is asymptotically stable, then (A, B) is certainly stabilizable. ∎

The structure of this result should be examined. All it is, is a restatement of Theorem 2.4-1 under the observability hypothesis. The observability condition has made our previous theorem quite a bit stronger!

Let us discuss two aspects of this theorem: what it does for us, and how we can guarantee that it holds in a particular problem. Part a of the theorem says that if interval $[i, N]$ is a large enough, then the optimal cost of control is

$$J_i^* = \tfrac{1}{2} x_i^\mathrm{T} S_\infty x_i,$$ (2.4-18)

which is both finite and independent of the value we selected for the final-state weighting S_N. Thus, whether we weight the final state heavily or lightly in J_i, the cost of optimal control over $[i, N]$ is the same. Since $R > 0$, the finiteness of J_i means that the optimal control is also finite; all our objectives can be achieved with finite control energy.

The theorem also guarantees the existence of a steady-state gain K_∞ that stabilizes the plant. This means two things. First, at the beginning of a control run, far from the final time N, the closed-loop plant $(A - BK_k)$ is nearly $(A - BK_\infty)$, so it starts out stable. Second, if we should decide to use the constant suboptimal gain $u_k = -K_\infty x_k$ for all k instead of the harder to implement optimal time-varying feedback, we are guaranteed at least closed-loop stability.

These properties are all quite desirable. To guarantee that they hold, we need only to ensure that the plant is stabilizable and to be judicious in our choice of the state weighting matrix Q. We should select $Q = C^\mathrm{T} C$ for some C such that (A, C) is observable. Note that this is always the case if Q is selected to be positive *definite*, for (A, C) is observable for any C of rank n.

Intuitively, all of this means the following. If the plant is observable through the fictitious output, then motions in all of the directions of the state space R^n have an influence on the performance index. If any state component begins to increase, then so does the cost J_k. Hence, if J_k is small, necessarily the state is also. Any control that makes J_k small will also make small the excursions of the state from the origin.

On the other hand, if (A, C) is unobservable, then if the state tends to infinity in an unobservable direction of R^n, this motion will not be sensed in J_k. Thus, the boundedness of J_k does not in this case guarantee the boundedness of the state trajectory.

We have made the hypothesis of Theorem 2.4-2 unnecessarily strong. All we really require for the theorem to hold is the *detectability* of (A, C). Thus, only the unstable modes need be observable through the performance index. In this case, however, the unique limiting solution to the Riccati equation can be guaranteed only to be positive *semidefinite*.

One result of these theorems is that we now have a way of stabilizing any multi-input plant. If Q and R are *any* positive definite matrices of appropriate dimension, then the state feedback $u_k = -K_\infty x_k$ based on the steady-state gain (2.4-13) determined from the unique positive definite solution S_∞ to the algebraic Riccati equation (2.4-12) will result in a stable closed-loop system. Different matrices Q and R will result in different closed-loop poles for $A - BK_\infty$, but these poles will always be inside the unit circle.

It would be quite useful to be able to select Q and R to yield a desired set of closed-loop poles. We present an example demonstrating the relation between Q, R, and the closed-loop poles for a simple system (Example 2.4-3), and then in Section 2.5 we discuss a technique similar to root locus for selecting Q and R to yield desired closed-loop poles.

One more result on the limiting, or *infinite-horizon* optimal control problem $((N - k) \to \infty)$ should be mentioned. Let (A, B) be stabilizable and $(A, \sqrt{Q})$ observable. Then the steady-state feedback $u_k = -K_\infty x_k$ is the *optimal* control for a particular problem, for it is the control that minimizes the cost over the infinite time interval $[0, \infty]$. That is, it minimizes

$$J_0 = \frac{1}{2} \sum_{k=0}^{\infty} (x_k^\mathrm{T} Q x_k + u_k^\mathrm{T} R u_k). \qquad (2.4\text{-}19)$$

This follows from our discussion.

The next examples illustrate these results.

Example 2.4-2.   Steady-state Control of a Scalar System

Let the plant be

$$x_{k+1} = a x_k + b u_k, \qquad (1)$$

with performance index

$$J_0 = \frac{1}{2} \sum_{k=0}^{\infty} (q x_k^2 + r u_k^2). \qquad (2)$$

The optimal control minimizing J_0 is the constant feedback

$$u_k = -k_\infty x_k, \qquad (3)$$

where the gain (2.4-13) is

$$k_\infty = \frac{abs_\infty}{b^2 s_\infty + r} \qquad (4)$$

and the steady-state kernel is the unique positive (definite) root of the algebraic Riccati equation

$$s = a^2 s - \frac{a^2 b^2 s^2}{b^2 s + r} + q. \qquad (5)$$

Under the influence of the control (3), the closed-loop system is

$$a^{\mathrm{cl}} = a - b k_\infty = \frac{a}{1 + (b^2/r) s_\infty}. \qquad (6)$$

The ARE can be written
$$b^2 s^2 + [(1 - a^2) r - b^2 q] s - qr = 0. \qquad (7)$$

If we define the auxiliary variable

$$\Lambda = \frac{b^2 q}{(1 - a^2) r}, \qquad (8)$$

this becomes

$$\frac{\Lambda}{q}s^2 + (1 - \Lambda)s - \frac{\Lambda r}{b^2} = 0, \tag{9}$$

which has two solutions, given by

$$s = \frac{q}{2\Lambda}\left[\pm\sqrt{(1 - \Lambda)^2 + \frac{4\Lambda}{(1 - a^2)}} - (1 - \Lambda)\right]. \tag{10}$$

We must now consider two cases.

a. Original System Stable

If $|a| < 1$, then $(1 - a^2) > 0$ and $\Lambda > 0$. In this case the unique non-negative solution to (7) is

$$s_\infty = \frac{q}{2\Lambda}\left[\sqrt{(1 - \Lambda)^2 + \frac{4\Lambda}{(1 - a^2)}} - (1 - \Lambda)\right], \tag{11}$$

and the steady-state gain is given by (4).

In the scalar case, observability of $(a, \sqrt{q})$ is equivalent to $q \neq 0$, and plant reachability is equivalent to $b \neq 0$. Therefore, the observability and reachability conditions imply $\Lambda > 0$ and hence $s_\infty > 0$. Then, according to (6)

$$\left|a^{\text{cl}}\right| < |a|, \tag{12}$$

so that the closed-loop system is stable.

Since $|a| < 1$, if $q = 0$, the system is still detectable (the unobservable mode is stable). If $q = 0$, then $\Lambda = 0$, but according to (6) $a^{\text{cl}} = a$ is still stable. Note that in this case, $s_\infty = 0$ (i.e., positive *semidefinite*).

b. Original System Unstable

If $|a| > 1$, then $(1 - a^2) < 0$ and $\Lambda < 0$. Then the unique non-negative solution to the ARE is

$$s_\infty = \frac{-q}{2\Lambda}\left[\sqrt{(1 - \Lambda)^2 + \frac{4\Lambda}{(1 - a^2)}} + (1 - \Lambda)\right]. \tag{13}$$

Again, reachability and observability imply that Λ is strictly negative, so that according to (13), $s_\infty > 0$.

According to (6), if $\Lambda < 0$, we still have $\left|a^{\text{cl}}\right| < |a|$, but it is not easy to prove that $\left|a^{\text{cl}}\right| < 1|$. Note, however, that

$$a^{\text{cl}} = \frac{a}{1 - \dfrac{1 - a^2}{2}\left[\sqrt{(1 - \Lambda)^2 + \dfrac{4\Lambda}{1 - a^2}} + (1 - \Lambda)\right]}, \tag{14}$$

so that if $|a| \gg 1$, then $\Lambda \simeq 0$ and

$$a^{\text{cl}} \simeq \frac{1}{a}. \tag{15}$$

This is certainly stable. Note that a^{cl} does not depend on b, q, and r individually, but only on the quantity b^2q/r. (This is also true if a is stable.) If $|a| > 1$, then detectability of $(a, \sqrt{q})$ is equivalent to $q \neq 0$, which is necessary for (14) to be stable.

c. a^{cl} as a Function of a

Figure 2.4-3 shows a plot of the closed-loop plant matrix a^{cl} as a function of the original plant matrix a for $b^2q/r = 1$. Note that a^{cl} is stable for all values of a.

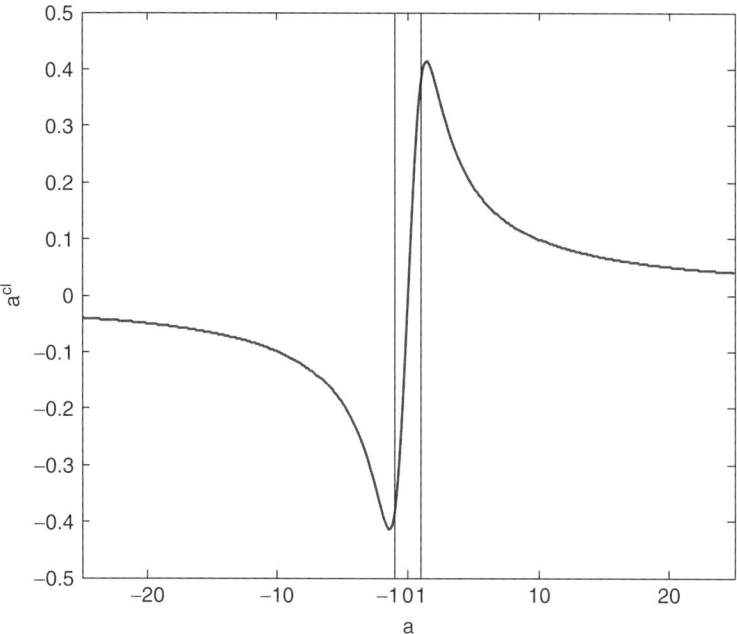

FIGURE 2.4-3   Closed-loop plant matrix as a function of an open-loop plant matrix.

■

Example 2.4-3.   Limiting Control Behavior for Systems Obeying Newton's Laws

Let us reconsider the discretized *Newton's system* of Example 2.3-2,

$$x_{k+1} = \begin{bmatrix} 1 & T \\ 0 & 1 \end{bmatrix} x_k + \begin{bmatrix} T^2/2 \\ T \end{bmatrix} u_k, \tag{1}$$

where $x_k = [d_k, v_k]^T$, with d_k and v_k the kth samples of position and velocity. The sampling period is $T = 0.5$ sec for this example, so that

$$x_{k+1} = \begin{bmatrix} 1 & 0.5 \\ 0 & 1 \end{bmatrix} x_k + \begin{bmatrix} 0.125 \\ 0.5 \end{bmatrix} u_k. \tag{2}$$

Let $Q = I$, so that the cost is

$$J_i = \frac{1}{2}x_N^T S_N x_N + \frac{1}{2}\sum_{k=i}^{N-1}(d_k^2 + v_k^2 + u_k^2). \tag{3}$$

Using the solution to the optimal control problem given in Example 2.3-2 and the associated software, we can determine the optimal cost kernel S_k, feedback gain K_k, closed-loop plant matrix A_k^{cl}, and the optimal control u^* and state trajectory x^*. By adding a polynomial root finder we can determine the poles of A_k^{cl} for each time k.

First, let us use final-state weighting of $S_N = 100I$. This is the simulation run in Example 2.3-2, and the optimal state trajectory and control are shown in Fig. 2.3-11. This behavior is quite satisfactory, and a good rendezvous is achieved. The Kalman gain K_k is shown in Fig. 2.4-4a. Note that it is defined only at integer values of k. In the graph its values are connected by lines to distinguish more easily between the two components of K_k, which are denoted by K_k^1 and K_k^2.

The poles of the closed-loop system A_k^{cl} are a function of time index k, since A_k^{cl} is a time-varying system. They are illustrated in Fig. 2.4-4b. When $k = N$, the system is marginally stable with poles at $z = 0$ and $z = 1$. As k decreases, the closed-loop poles become complex and move as shown.

For $N - i = 10$, corresponding to a run time of 5 sec, the Riccati-equation solution has reached a steady-state value that is given by S_0. It is equal to

$$S_\infty' \triangleq \begin{bmatrix} 4.035 & 2.0616 \\ 2.0616 & 4.1438 \end{bmatrix}. \tag{4}$$

The corresponding steady-state gain is

$$K_\infty = \begin{bmatrix} 0.6514 \\ 1.3142 \end{bmatrix}, \tag{5}$$

which agrees with Fig. 2.4-4a. The steady-state closed-loop plant matrix is

$$A_\infty^{cl} = (A - BK_\infty) = \begin{bmatrix} 0.9185 & 0.3357 \\ -0.3257 & 0.3429 \end{bmatrix}, \tag{6}$$

which has poles at

$$z = 0.6307 \pm j0.1628. \tag{7}$$

These steady-state closed-loop poles agree with Fig. 2.4-4b.

We might now be interested in trying to use the suboptimal feedback

$$u_k = -K_\infty x_k \tag{8}$$

with K_∞ given by (5). This control is easier to implement than the optimal control $u_k = -K_k x_k$ since the Kalman gain sequence K_k need not be stored. If we simulate this suboptimal control, the suboptimal state trajectory in Fig. 2.4-4c results. This trajectory is barely distinguishable from the optimal path in Fig. 2.3-11, except that the final values $d(5)$ and $v(5)$ are not quite equal to zero. It would probably be suitable for most purposes.

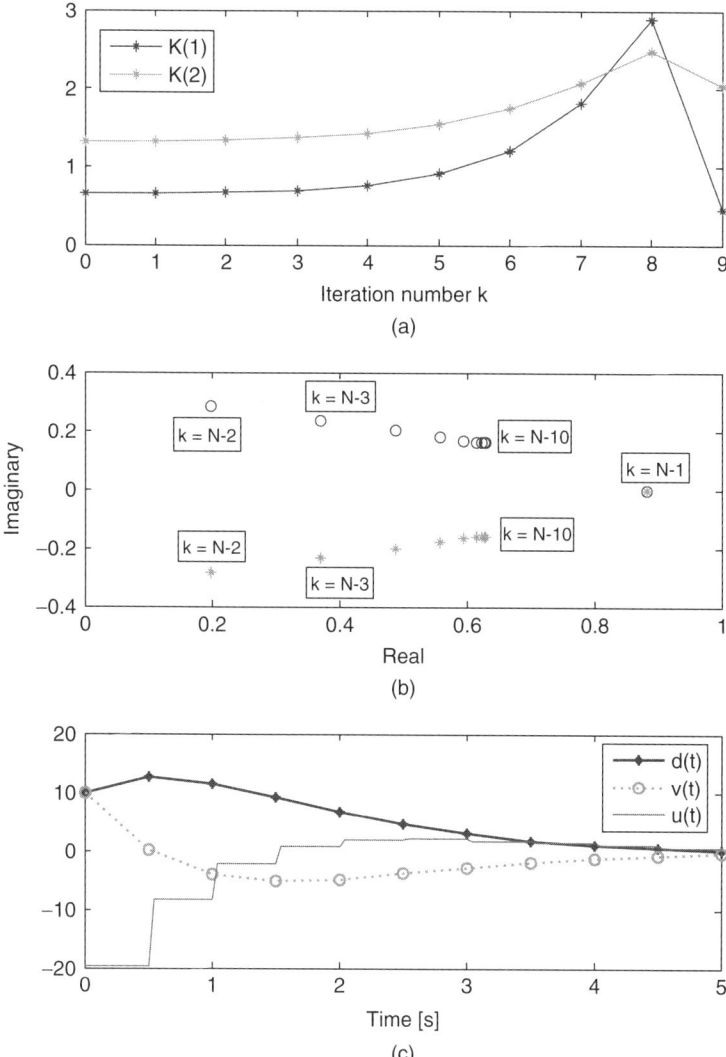

FIGURE 2.4-4 Position and velocity weighting, $S_N = 100I$. (a) Optimal feedback gain sequence. (b) Locus of the closed-loop poles. (c) Suboptimal state trajectory using steady-state feedback gain K_∞.

Let us now check our theorems. Since (A, B) is reachable, the solution S_k to the Riccati equation converges to a finite S_∞. A root of Q is given by $C = I$. Since (A, C) is certainly observable, S_∞ is positive definite. Furthermore, A_∞^{cl} is asymptotically stable.

An important consequence of these facts is that if we use control law (8) and give the system enough time, the state will always go to zero.

To show that (8) may not always be satisfactory, suppose we want to rendezvous in 2 sec, corresponding to a discrete interval of length $N - i = 4$. According to Fig. 2.4-4a,

K_k has not yet reached steady state by four iterations backward from $N = 10$ (i.e., K_6 is not yet equal to $K_0 = K_\infty$). Thus, we might anticipate problems in using the steady-state control law (8). Figure 2.4-5 shows the optimal trajectories using $u_k = -K_k x_k$ and the suboptimal trajectories using (8). For this short rendezvous time, the steady-state gains would not be satisfactory.

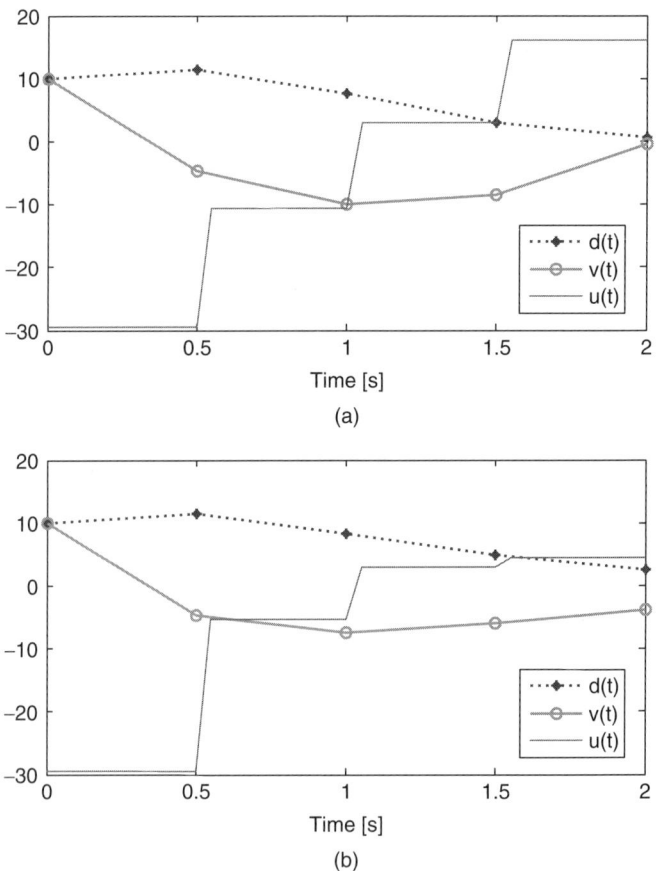

FIGURE 2.4-5 System trajectories for a short run time. (a) Trajectories using optimal control. (b) Trajectories using suboptimal steady-state feedback gains.

■

An Analytic Solution to the Riccati Equation

The optimal control is given in terms of the Riccati equation (2.4-2) with boundary condition S_N. One way to solve this equation is by iteration. We discuss here a nonrecursive solution for S_k with important theoretical uses that applies in the case of time-invariant plant and cost-weighting matrices. These results are due to Vaughan (1970). These results are for the case where A is nonsingular, as it is whenever a continuous-time system is discretized. In some cases, however, A

can be singular (e.g., system with pure delays). The solution of the discrete time ARE for the case of singular A is covered in a subsequent subsection.

We wrote the LQ regulator Hamiltonian system as the backward recursion (2.2-11), which is

$$\begin{bmatrix} x_k \\ \lambda_k \end{bmatrix} = H \begin{bmatrix} x_{k+1} \\ \lambda_{k+1} \end{bmatrix}, \tag{2.4-20}$$

with

Hamiltonian

$$H \triangleq \begin{bmatrix} A^{-1} & A^{-1}BR^{-1}B^{\mathrm{T}} \\ QA^{-1} & A^{\mathrm{T}} + QA^{-1}BR^{-1}B^{\mathrm{T}} \end{bmatrix}. \tag{2.4-21}$$

The final condition for (2.4-20) if the final state is free is $\lambda_N = S_N x_N$. The initial condition is x_0.

We assumed that

$$\lambda_k = S_k x_k \tag{2.4-22}$$

for all $k \leq N$ (cf. (2.2-50)), and based on this assumption derived the results in Table 2.2-1. Matrix S_k in (2.4-22) turned out to be given by (2.4-2). Let us now demonstrate that the Riccati-equation solution S_k can be computed in terms of the eigenvalues and eigenvectors of H. Define

$$J = \begin{bmatrix} 0 & I \\ -I & 0 \end{bmatrix}. \tag{2.4-23}$$

Then, it can be shown with only a few lines of work that

$$H^{\mathrm{T}}JH = J. \tag{2.4-24}$$

A matrix H satisfying (2.4-24) is called *symplectic*. Since H is symplectic, its inverse is very easy to find, because by (2.4-24)

$$H^{\mathrm{T}}J = JH^{-1},$$

$$J^{-1}H^{\mathrm{T}}J = H^{-1},$$

so that (since $J^{-1} = -J$)

$$H^{-1} = -JH^{\mathrm{T}}J. \tag{2.4-25}$$

Performing these multiplications yields

$$H^{-1} = \begin{bmatrix} A + BR^{-1}B^{\mathrm{T}}A^{-\mathrm{T}}Q & -BR^{-1}B^{\mathrm{T}}A^{-\mathrm{T}} \\ -A^{-\mathrm{T}}Q & A^{-\mathrm{T}} \end{bmatrix}. \tag{2.4-26}$$

(Remember that Q and R are symmetric. $A^{-\mathrm{T}}$ means $(A^{-1})^{\mathrm{T}}$.) Now it can be shown that if μ is an eigenvalue of H, then so is $1/\mu$. If μ is an eigenvalue with eigenvector $\begin{bmatrix} f \\ g \end{bmatrix}$ partitioned comfortably with H, then

$$\begin{bmatrix} A^{-1} & A^{-1}BR^{-1}B^{\mathrm{T}} \\ QA^{-1} & A^{\mathrm{T}} + QA^{-1}BR^{-1}B^{\mathrm{T}} \end{bmatrix} \begin{bmatrix} f \\ g \end{bmatrix} = \mu \begin{bmatrix} f \\ g \end{bmatrix}. \tag{2.4-27}$$

This can be rearranged to read

$$\begin{bmatrix} A^{\mathrm{T}} + QA^{-1}BR^{-1}B^{\mathrm{T}} & -QA^{-1} \\ -A^{-1}BR^{-1}B^{\mathrm{T}} & A^{-1} \end{bmatrix} \begin{bmatrix} g \\ -f \end{bmatrix} = \mu \begin{bmatrix} g \\ -f \end{bmatrix}. \tag{2.4-28}$$

Now note that the coefficient matrix on the left-hand side is just $H^{-\mathrm{T}}$. This means that μ is also an eigenvalue of $H^{-\mathrm{T}}$, and hence of H^{-1}. Therefore, $1/\mu$ is an eigenvalue of H, as we wanted to show (see Appendix A).

What this means is that the $2n$ eigenvalues of H can be arranged in a matrix

$$D = \begin{bmatrix} M & 0 \\ 0 & M^{-1} \end{bmatrix}, \tag{2.4-29}$$

where M is a diagonal matrix containing n eigenvalues outside the unit circle. (Hence, M^{-1} is stable.)

There is a nonsingular matrix W whose columns are the eigenvectors of H such that

$$W^{-1}HW = D. \tag{2.4-30}$$

Define a state space transformation W^{-1} so that for each k,

$$\begin{bmatrix} x_k \\ \lambda_k \end{bmatrix} = W \begin{bmatrix} w_k \\ z_k \end{bmatrix}$$
$$= \begin{bmatrix} W_{11} & W_{12} \\ W_{21} & W_{22} \end{bmatrix} \begin{bmatrix} w_k \\ z_k \end{bmatrix}, \tag{2.4-31}$$

where W_{ij} are partitions of W. Then the Hamiltonian system (2.4-20) takes on its Jordan normal form

$$\begin{bmatrix} w_k \\ z_k \end{bmatrix} = D \begin{bmatrix} w_{k+1} \\ z_{k+1} \end{bmatrix}. \tag{2.4-32}$$

The solution to (2.4-32) in terms of the final conditions is

$$\begin{bmatrix} w_k \\ z_k \end{bmatrix} = \begin{bmatrix} M^{N-k} & 0 \\ 0 & M^{-(N-k)} \end{bmatrix} \begin{bmatrix} w_N \\ z_N \end{bmatrix}. \tag{2.4-33}$$

The problem with this solution is that M^{N-k} does not go to zero as $(N-k) \rightarrow \infty$, since M is not stable. Therefore, rewrite (2.4-33) as

$$\begin{bmatrix} w_N \\ z_k \end{bmatrix} = \begin{bmatrix} M^{-(N-k)} & 0 \\ 0 & M^{-(N-k)} \end{bmatrix} \begin{bmatrix} w_k \\ z_N \end{bmatrix} \tag{2.4-34}$$

Now we consider the relations between x_k and λ_k (e.g., (2.4-22)) and between w_k and z_k. According to (2.4-22) and (2.4-31), at the final time N

$$\lambda_N = W_{21}W_N + W_{22}z_N = S_N x_N = S_N(W_{11}w_N + W_{12}z_N).$$

Solving for z_N in terms of w_N yields

$$z_N = Tw_N, \tag{2.4-35}$$

where

$$T = -(W_{22} - S_N W_{12})^{-1}(W_{21} - S_N W_{11}). \tag{2.4-36}$$

Now, by (2.4-34)

$$z_k = M^{-(N-k)}z_N = M^{-(N-k)}Tw_N = M^{-(N-k)}TM^{-(N-k)}w_k, \tag{2.4-37}$$

so that at each value of k

$$z_k = T_k w_k, \tag{2.4-38}$$

where

$$T_k = M^{-(N-k)}TM^{-(N-k)}. \tag{2.4-39}$$

It remains to relate S_k in (2.4-22) to T_k in (2.4-38). To do this, use (2.4-31) to write

$$\lambda_k = W_{21}w_k + W_{22}z_k = S_k x_k = S_k(W_{11}w_k) + W_{12}z_k),$$

so that by (2.4-38)

$$(W_{21} + W_{22}T_k)w_k = S_k(W_{11} + W_{12}T_k)w_k. \tag{2.4-40}$$

Since this must hold for all x_0, and hence for all trajectories w_k, it implies that

$$S_k = (W_{21} + W_{22}T_k)(W_{11} + W_{12}T_k)^{-1}. \tag{2.4-41}$$

Equations (2.4-36), (2.4-39), and (2.4-41) give a nonrecursive analytic solution to the Riccati equation for any $k \leq N$ in terms of S_N and the eigenvalues and eigenvectors of Hamiltonian matrix H. One important special case is the following. As $N \to \infty$, the Riccati equation tends to the ARE (2.4-12). We have seen that if (A, B) is reachable and $(A, \sqrt{Q})$ is observable, then the steady-state feedback gain K_∞ defined in terms of the ARE solution by (2.4-13) is often a satisfactory choice as a suboptimal and easy-to-implement control law. One way to find the positive definite solution S_∞ to the ARE is to select any S_N and iterate (2.4-2) until S_k converges. Our new results provide an alternative way to find S_∞, which is important theoretically. If $(N - k) \to \infty$, then $M^{-(N-k)}$ goes to zero since M^{-1} is stable. This means that $T_k \rightarrow 0$, so that in the steady-state limit, (2.4-41), yields

$$S_\infty = W_{21} W_{11}^{-1} \qquad (2.4\text{-}42)$$

as an expression for the positive definite ARE solution. Thus, S_∞ can be manufactured from the *unstable* eigenvectors

$$\begin{bmatrix} W_{11} \\ W_{21} \end{bmatrix}$$

of H (or the *stable* eigenvectors of H^{-1}). Hence, the optimal steady-state feedback K_∞ can be found (by (2.4-13)) *without solving* a Riccati equation.

Example 2.4-4.   Analytic Solution to a Scalar Riccati Equation

This example is from Vaughan (1970). Let the plant and cost function be

$$x_{k+1} = x_k + u_k, \qquad (1)$$

$$J_i = \frac{10}{2} x_N^2 + \frac{1}{2} \sum_{k=1}^{N-1} (x_k^2 + u_k^2). \qquad (2)$$

Then

$$H = \begin{bmatrix} 1 & 1 \\ 1 & 2 \end{bmatrix} \qquad (3)$$

has eigenvalues of 0.382, 2.618, so

$$M = 2.618. \qquad (4)$$

The matrix of eigenvectors (the unstable one first!) is

$$W = \begin{bmatrix} 1.0 & 1.0 \\ 1.618 & -0.618 \end{bmatrix}. \qquad (5)$$

Since $S_N = 10$,

$$T = -0.789 \qquad (6)$$

and

$$T_k = -0.789(0.382)^{2(N-k)}. \qquad (7)$$

Therefore, (2.4-41) yields

$$s_k = \frac{1.618 + 0.488(0.382)^{2(N-k)}}{1 - 0.789(0.382)^{2(N-k)}} \qquad (8)$$

as the analytic solution to the Riccati equation (2.4-2). In the limiting case $(N - k) \rightarrow \infty$ this yields the ARE solution

$$s_\infty = 1.618, \qquad (9)$$

so the steady-state feedback is given by (2.4-13) as

$$K_\infty = \frac{1.618}{2.618} = 0.618. \tag{10}$$

The control law
$$u_k = -0.618x_k \tag{11}$$

results in a stable closed-loop plant of

$$a_\infty^{cl} = (a - bK_\infty) = 0.382. \tag{12}$$

We point out now that the closed-loop pole of 0.382 is the stable eigenvalue of Hamiltonian matrix H. This is not a coincidence, and we shall have more to say about it in Section 2.5. ■

Analytic Solution to the Discrete Riccati Equation: System Matrix A Singular

In several digital control problems the system matrix A may either be singular or ill conditioned with respect to inversion. In that case the method described previously may either fail or produce misleading results due to the fact the A is ill conditioned with respect to inversion. This problem can be circumvented by employing the generalized eigenvalue problem for the analytic solution of the discrete Riccati equation presented by Pappas, Laub, and Sandell (1980).

In the previous section we defined the Hamiltonian matrix H as

$$H = \begin{bmatrix} A^{-1} & A^{-1}BR^{-1}B^{\mathrm{T}} \\ QA^{-1} & A^{\mathrm{T}} + QA^{-1}BR^{-1}B^{\mathrm{T}} \end{bmatrix},$$

associated with the backward recursion (2.4-20). By noting that the Hamiltonian matrix H can be decomposed as the product of two matrices as

$$H = \begin{bmatrix} A & 0 \\ -Q & I \end{bmatrix}^{-1} \begin{bmatrix} I & BR^{-1}B^{\mathrm{T}} \\ 0 & A^{\mathrm{T}} \end{bmatrix},$$

we can rewrite the backward recursion (2.4-20) as a forward generalized recursion

$$\begin{bmatrix} I & BR^{-1}B^{\mathrm{T}} \\ 0 & A^{\mathrm{T}} \end{bmatrix} \begin{bmatrix} x_{k+1} \\ \lambda_{k+1} \end{bmatrix} = \begin{bmatrix} A & 0 \\ -Q & I \end{bmatrix} \begin{bmatrix} x_k \\ \lambda_k \end{bmatrix}. \tag{2.4-43}$$

The key idea of this method is to study the generalized eigenvalue problem

$$Gz = \mu Fz \tag{2.4-44}$$

where

$$F = \begin{bmatrix} I & BR^{-1}B^{\mathrm{T}} \\ 0 & A^{\mathrm{T}} \end{bmatrix} \quad \text{and} \quad G = \begin{bmatrix} A & 0 \\ -Q & I \end{bmatrix}. \tag{2.4-45}$$

As before, we can extend the definition of a symplectic matrix H to the definition of a symplectic pair of matrices (F, G). It is easy to show that (F, G) have the following property:

$$GJG^T = FJF^T = \begin{bmatrix} 0 & A \\ -A^T & 0 \end{bmatrix}, \qquad (2.4\text{-}46)$$

where J is defined in (2.4-23). A pair that satisfies equation (2.4-46) is called a *symplectic* pair.

We now show that the solution to the discrete algebraic Riccati equation can be obtained from the eigenvectors of the Hamiltonian. Before we proceed, we show that the Hamiltonian has n stable and n unstable eigenvalues. Under the assumption of stabilizability of (A, B) and detectability of $(\sqrt{Q}, A)$, none of the eigenvalues of the Hamiltonian H given by (2.4-21), which corresponds to the discrete algebraic Riccati equation, lie on the unit circle. To see this, consider the generalized eigenvalue problem

$$Gz = \mu Fz$$

where

$$F = \begin{bmatrix} I & BR^{-1}B^T \\ 0 & A^T \end{bmatrix} \quad \text{and} \quad G = \begin{bmatrix} A & 0 \\ -Q & I \end{bmatrix}.$$

Suppose that $|\mu| = 1$, then there exists some $z \neq 0$ such that

$$\begin{bmatrix} A & 0 \\ -Q & I \end{bmatrix} \begin{bmatrix} z_1 \\ z_2 \end{bmatrix} = \mu \begin{bmatrix} I & BR^{-1}B^T \\ 0 & A^T \end{bmatrix} \begin{bmatrix} z_1 \\ z_2 \end{bmatrix}.$$

This implies

$$Az_1 = \mu z_1 + \mu BR^{-1}B^T z_2 \quad \text{and} \quad -Qz_1 + z_2 = \mu A^T z_2.$$

By premultiplying the first equation by $\mu^* z_2^H$ and postmultiplying the conjugate transpose of the second equation by z_1, one gets

$$\mu^* z_2^H A z_1 = |\mu|^2 z_2^H z_1 + |\mu|^2 z_2^H BR^{-1}B^T z_2, \quad \text{and} \quad z_2^H z_1 = \mu^* z_2^H A z_1 + z_1^H Q z_1.$$

By adding the above two equations and noting that $|\mu^2| = 1$, we get

$$z_2^H BR^{-1}B^T z_2 + z_1^H Q z_1 = 0,$$

which, in turn, implies that

$$B^T z_2 = 0 \quad \text{and} \quad \sqrt{Q} z_1 = 0.$$

Note that under these conditions

$$Az_1 = \mu z_1 \quad \text{and} \quad A^T z_2 = (1/\mu) z_2,$$

which implies that the system is unstabilizable and undetectable, which is a contradiction; so there are no eigenvalues on the unit circle.

Further, it is straightforward to show that $\det(\mu F - G) \neq 0$. Note that if the determinant were identically zero, it would also be zero for $|\mu| = 1$, which contradicts the fact that the pair (F, G) has no eigenvalues on the unit circle. Next we show, as in the case of the Hamiltonian H, that if $v \neq 0$ is an eigenvalue of $(\mu F - G)$, then $1/v$ is also an eigenvalue. To show this, assume that y is a left eigenvector corresponding to eigenvalue v. Then

$$y^H G = v y^H F,$$

$$y^H GJG^T = v y^H FJG^T,$$

$$y^H FJF^T = v y^H FJG^T,$$

$$(1/v)x^T F^T = x^H G^T,$$

$$Gx = (1/v)Fx$$

(where $x^H = y^H FJ$), which shows that if $v \neq 0$ is an eigenvalue of $(\mu F - G)$, then $1/v$ is also an eigenvalue. What this means is that there exist matrices V and W such that the $2n$ eigenvalues can be arranged as (see Appendix A.5)

$$\begin{bmatrix} \mu I - M_0 & 0 & 0 & 0 \\ 0 & \mu I - M_f & 0 & 0 \\ 0 & 0 & \mu I - M_f^{-1} & 0 \\ 0 & 0 & 0 & \mu N_0 - I \end{bmatrix} = \begin{bmatrix} \mu N_s - M_s & 0 \\ 0 & \mu N_u - M_u \end{bmatrix}$$

where the n generalized eigenvalues of $(N_s = I, M_s)$ are stable, i.e., $|\mu_i| < 1$, and the eigenvalues of (N_u, M_u) are the reciprocals of the generalized eigenvalues of (N_s, M_s), i.e., $|\mu_i| > 1$; that is, the n generalized eigenvalues of (N_u, M_u) are unstable. Note that in the case where A is singular some of the unstable eigenvalues lie at infinity.

Let V_s be the $2n \times n$ matrix corresponding to the stable eigenvalues, that is, the first n columns of the matrix V. Then

$$GV_s = FV_s M_s, \tag{2.4-47}$$

where M_s is in Jordan canonical form, corresponding to all stable eigenvalues. By substituting for F, G and conformably partitioning V_s, equation (2.4-47) can be rewritten as

$$\begin{bmatrix} A & 0 \\ -Q & I \end{bmatrix} \begin{bmatrix} V_{1s} \\ V_{2s} \end{bmatrix} = \begin{bmatrix} I & BR^{-1}B^T \\ 0 & A^T \end{bmatrix} \begin{bmatrix} V_{1s} M_s \\ V_{2s} M_s \end{bmatrix}$$

or

$$AV_{1s} = V_{1s} M_s + BR^{-1}B^T V_{2s} M_s \tag{2.4-48}$$

$$-QV_{1s} + V_{2s} = A^T V_{2s} M_s. \tag{2.4-49}$$

Since V_{1s} is nonsingular (see Appendix A), equation (2.4-48) can be rewritten as

$$A = V_{1s} M_s V_{1s}^{-1} + BR^{-1} B^{\mathrm{T}} V_{2s} M_s V_{1s}^{-1} = (V_{1s} + BR^{-1} B^{\mathrm{T}} V_{2s}) M_s V_{1s}^{-1} \quad (2.4\text{-}50)$$

and (2.4-49) can be written as

$$S = V_{2s} V_{1s}^{-1} = A^{\mathrm{T}} V_{2s} M_s V_{1s}^{-1} + Q. \quad (2.4\text{-}51)$$

In the sequel it is shown that S described in (2.4-51) is a solution to the discrete ARE. Note that using the inversion lemma, the ARE can be written as

$$S = A^{\mathrm{T}} S (I + BR^{-1} B^{\mathrm{T}} S)^{-1} A + Q. \quad (2.4\text{-}52)$$

Substituting (2.4-51), the ARE (2.4-52) becomes

$$A^{\mathrm{T}} V_{2s} M_s V_{1s}^{-1} = A^{\mathrm{T}} V_{2s} V_{1s}^{-1} (I + BR^{-1} B^{\mathrm{T}} V_{2s} V_{1s}^{-1})^{-1} A.$$

Substituting A from equation (2.4-50), the ARE becomes

$$A^{\mathrm{T}} V_{2s} M_s = A^{\mathrm{T}} V_{2s} (V_{1s} + BR^{-1} B^{\mathrm{T}} V_{2s})^{-1} (V_{1s} M_s + BR^{-1} B^{\mathrm{T}} V_{2s} M_s),$$

or

$$A^{\mathrm{T}} V_{2s} M_s = A^{\mathrm{T}} V_{2s} (V_{1s} + BR^{-1} B^{\mathrm{T}} V_{2s})^{-1} (V_{1s} + BR^{-1} B^{\mathrm{T}} V_{2s}) M_s,$$

which shows that (2.4-51) is a solution to the ARE.

It remains to show that S is the stabilizing solution to the ARE. That is, the closed-loop system matrix $(A - B(B^{\mathrm{T}} SB + R)^{-1} B^{\mathrm{T}} SA)$ has stable eigenvalues. The closed-loop matrix is

$$\begin{aligned} A - B(B^{\mathrm{T}} SB + R)^{-1} B^{\mathrm{T}} SA &= (I - B(B^{\mathrm{T}} SB + R)^{-1} B^{\mathrm{T}} S) A \\ &= (I + BR^{-1} B^{\mathrm{T}} S)^{-1} A \\ &= (I + BR^{-1} B^{\mathrm{T}} V_{2s} V_{1s}^{-1})^{-1} A \\ &= V_{1s} (V_{1s} + BR^{-1} B^{\mathrm{T}} V_{2s})^{-1} A \\ &= V_{1s} M_s V_{1s}^{-1}, \end{aligned}$$

which shows that the closed-loop system matrix $(A - B(B^{\mathrm{T}} SB + R)^{-1} B^{\mathrm{T}} SA)$ and M_s have the same spectrum. Therefore, the closed-loop spectrum corresponds to the stable eigenvalues of the symplectic pair (F, G).

The calculation of generalized eigenvectors has significant difficulties that may lead to inaccurate results; in particular, in the case of multiple eigenvalues. To avoid this problem, the solution to the ARE is computed by using the QZ algorithm and the generalized Schur form. For more details on efficient and stable computations of solutions to algebraic Riccati equations see Laub (1979), Pappas, Laub, and Sandell (1980), and Bittanti, Laub, and Willems (1991).

Example 2.4-5.   Discrete Riccati Equation: System Matrix A Singular

This example is from Åström and Wittenmark (1984). Let the plant be described by

$$x_{k+1} = \begin{bmatrix} 1 & 1 \\ 0 & 0 \end{bmatrix} x_k + \begin{bmatrix} 0 \\ 1 \end{bmatrix} u_k$$

$$y_k = \begin{bmatrix} 1 & 0 \end{bmatrix} x_k,$$

and the cost function by

$$J = \frac{1}{2} \sum_{k=0}^{\infty} (y_k^{\mathrm{T}} y_k + u_k^{\mathrm{T}} u_k).$$

Then the matrices for the generalized eigenvalue problem are (see equation (2.4-45))

$$F = \begin{bmatrix} 1 & 0 & 0 & 0 \\ 0 & 1 & 0 & 1 \\ 0 & 0 & 1 & 0 \\ 0 & 0 & 1 & 0 \end{bmatrix}, \quad G = \begin{bmatrix} 1 & 1 & 0 & 0 \\ 0 & 0 & 0 & 0 \\ -1 & 0 & 1 & 0 \\ 1 & 0 & 0 & 1 \end{bmatrix}.$$

The finite eigenvalues of (F, G) are the roots of the equation

$$\det(\mu F - G) = \mu(\mu^2 - 3\mu + 1) = 0,$$

which are $\mu_1 = 0$, $\mu_2 = 0.3820$, and $\mu_3 = 2.6180$. The fourth eigenvalue is an infinite one. The eigenvectors corresponding to the two stable eigenvalues are

$$V_s = \begin{bmatrix} V_{1s} \\ V_{2s} \end{bmatrix} = \begin{bmatrix} 0.4777 & -0.5774 \\ -0.2952 & 0.5774 \\ 0.7730 & -0.5774 \\ 0.2952 & 0.0000 \end{bmatrix}.$$

Then the solution to the ARE is

$$S = V_{1s} V_{2s}^{-1} = \begin{bmatrix} 0.7730 & -0.5774 \\ 0.2952 & 0.0000 \end{bmatrix} \begin{bmatrix} 0.4777 & -0.5774 \\ -0.2952 & 0.5774 \end{bmatrix}^{-1}$$

$$= \begin{bmatrix} 2.6180 & 1.6180 \\ 1.6180 & 1.6180 \end{bmatrix}.$$

The feedback gain is

$$K = (B^{\mathrm{T}} S B + I)^{-1} B^{\mathrm{T}} S A = \begin{bmatrix} 0.6180 & 0.6180 \end{bmatrix}.$$

■

Design of Steady-state Regulators by Eigenstructure Assignment

The results of the previous subsection gave us an alternative design procedure for the optimal steady-state LQ regulator that did not involve solving a Riccati equation. It involved finding S_∞ from the unstable eigenvectors of the Hamiltonian matrix H in (2.4-21). Let us now discuss a method for finding the optimal steady-state gain K_∞ directly from the eigenstructure of the Hamiltonian system.

We assume here that (A, B) is stabilizable and $(A, \sqrt{Q})$ is detectable. First, write the Hamiltonian system (2.4-20) as the forward recursion

$$\begin{bmatrix} x_{k+1} \\ \lambda_{k+1} \end{bmatrix} = \begin{bmatrix} A + BR^{-1}B^{\mathrm{T}}A^{-\mathrm{T}}Q & -BR^{-\mathrm{T}}B^{\mathrm{T}}A^{-\mathrm{T}} \\ -A^{-\mathrm{T}}Q & A^{-\mathrm{T}} \end{bmatrix} \begin{bmatrix} x_k \\ \lambda_k \end{bmatrix}, \qquad (2.4\text{-}53)$$

where the coefficient matrix is H^{-1} in (2.4-26). Let μ be an eigenvector of H^{-1}. Then the eigenvectors of H^{-1} corresponding to μ are the eigenvectors of H corresponding to $1/\mu$ (Appendix A). Hence, W_{11}, W_{21} in (2.4-42) can alternatively be found by partitioning the *stable* eigenvectors of H^{-1}. In terms of the state vector only, the steady-state closed-loop system with the optimal control $u_k = -K_\infty x_k$ is

$$x_{k+1} = (A - BK_\infty)x_k. \qquad (2.4\text{-}54)$$

Both (2.4-53) and (2.4-54) characterize the optimal closed-loop plant. We want to demonstrate that the eigenvalues of the optimal closed-loop system (2.4-54) are simply the stable eigenvalues of H^{-1} and that the n eigenvectors of (2.4-54) are given by the columns of W_{11}.

Suppose that μ_i is an eigenvalue of the optimal closed-loop system (Kailath 1980). Then, if only the mode corresponding to μ_i is excited, the state, control, and costate are described by

$$x_k = X_i \mu_i^k, \qquad (2.4\text{-}55a)$$

$$u_k = U_i \mu_i^k, \qquad (2.4\text{-}55b)$$

$$\lambda_k = \Lambda_i \mu_i^k, \qquad (2.4\text{-}55c)$$

for some vectors X_i, U_i, Λ_i. But $x_{k+1} = Ax_k + Bu_k$, or

$$X_i \mu_i^{k+1} = AX_i \mu_i^k + BU_i \mu_i^k,$$

so that

$$(\mu_i I - A)X_i = BU_i. \qquad (2.4\text{-}56)$$

The optimal control is $u_k = -K_\infty x_k$, so that

$$U_i = -K_\infty X_i \qquad (2.4\text{-}57)$$

and

$$(\mu_i I - A + BK_\infty)X_i = 0. \qquad (2.4\text{-}58)$$

Thus, X_i is an eigenvector of the closed-loop plant (2.4-54) for eigenvalue μ_i.

Now focus on the representation (2.4-53). Using (2.4-55a) and (2.4-55c) we have

$$\mu_i \begin{bmatrix} X_i \\ \Lambda_i \end{bmatrix} = H^{-1} \begin{bmatrix} X_i \\ \Lambda_i \end{bmatrix}. \qquad (2.4\text{-}59)$$

Hence,

$$\begin{bmatrix} X_i \\ \Lambda_i \end{bmatrix}$$

is an eigenvector of H^{-1} for eigenvalue μ_i.

According to Theorem 2.4-2, $(A - BK_\infty)$ is stable, so that $|\mu_i| < 1$. Hence, the eigenvalues of (2.4-54) are the stable eigenvalues of H^{-1} and the eigenvectors of (2.4-54) are the top halves of the stable eigenvectors of H^{-1}.

To see why this is useful, observe that H^{-1} can be written down by inspection. Its stable eigenvalues and eigenvectors can be found, and these are the desired pole locations and associated eigenvectors of the optimal closed-loop plant.

If the plant is single input, we do not need to determine the eigenvectors of H^{-1}, since given only the desired closed-loop eigenvalues we can use *Ackermann's formula* to find the required feedback K_∞. According to this formula, the state feedback K is required to assign a desired closed-loop characteristic polynomial $\Delta^d(s)$ is

$$K = e_n^T U_n^{-1} \Delta^d(A), \qquad (2.4\text{-}60)$$

where e_n is the last column of the $n \times n$ identity matrix, U_n is the reachability matrix (2.4-17), and $\Delta^d(A)$ is the desired characteristic polynomial evaluated at A (Franklin and Powell 1980).

In the multivariable case, the desired closed-loop eigenvalues are not sufficient to determine the required feedback—the closed-loop eigenvectors are also required. In general, we can compute K_∞ from the eigenstructure of H^{-1} as follows.

Suppose the closed-loop eigenvalues are distinct. Then the optimal control is $u_k = -R^{-1}B^T\lambda_{k+1}$, so that

$$U_i = -R^{-1}B^T \mu_i \Lambda_i. \qquad (2.4\text{-}61)$$

The optimal feedback satisfies (2.4-57), so that

$$K_\infty X_i = R^{-1}B^T \mu_i \Lambda_i. \qquad (2.4\text{-}62)$$

Let X be a matrix whose columns are the X_i, and Λ be a matrix whose columns are the corresponding Λ_i (where

$$\begin{bmatrix} X_i \\ \Lambda_i \end{bmatrix}$$

is an eigenvector of the stable eigenvalue μ_i of H^{-1}). Let $M = \text{diag}[\mu_1, \ldots, \mu_n]$. Then, evidently,

$$K_\infty = R^{-1}B^{\mathrm{T}}\Lambda M X^{-1}. \tag{2.4-63}$$

Compare this with (2.4-42) ($W_{11} = X$, $W_{21} = \Lambda$).

If μ_i is complex, then so are X_i and Λ_i. In this event, there is a block in (2.4-63) of the form

$$[\Lambda_i \quad \lambda_i^*]\begin{bmatrix} \mu_i & 0 \\ 0 & \mu_i^* \end{bmatrix}[X_i \quad X_i^*]^{-1}.$$

By premultiplying and postmultiplying $\text{diag}[u_i, \ \mu_i^*]$ by

$$I = \frac{1}{2}\begin{bmatrix} 1 & -j \\ 1 & j \end{bmatrix} \cdot \begin{bmatrix} 1 & 1 \\ j & -j \end{bmatrix},$$

we see that this block can be replaced by

$$[\text{Re}(\Lambda_i) \quad \text{Im}(\Lambda_i)]\begin{bmatrix} \text{Re}(\mu_i) & \text{Im}(\mu_i) \\ -\text{Im}(\mu_i) & \text{Re}(\mu_i) \end{bmatrix}[\text{Re}(X_i) \quad \text{Im}(X_i)]^{-1},$$

which results in a real feedback gain K_∞.

Note that $|X| \neq 0$ since the μ_i were assumed to be distinct. If this is not the case, then generalized eigenvectors must be used in manufacturing X.

It is worth remarking that if the state-weighting matrix Q is zero, then

$$H^{-1} = \begin{bmatrix} A & -BR^{-1}B^{\mathrm{T}}A^{-\mathrm{T}} \\ 0 & A^{-\mathrm{T}} \end{bmatrix}. \tag{2.4-64}$$

In this case the eigenvalues of H^{-1} are the eigenvalues of A plus those of A^{-1} (i.e., those of $A^{-\mathrm{T}}$). The optimal closed-loop poles are therefore found simply by taking the stable poles of A, and the reciprocals of the unstable poles of A, since this yields the set of stable eigenvalues of H^{-1}. These are also the optimal closed-loop poles in the case of infinite control weighting $R \to \infty$.

Example 2.4-6.   *Eigenstructure Design of Steady-state Regulator for Harmonic Oscillator*

Suppose our plant is the harmonic oscillator

$$\dot{x} = \begin{bmatrix} 0 & 1 \\ -\omega_n^2 & -2\delta\omega_n \end{bmatrix}x + \begin{bmatrix} 0 \\ 10 \end{bmatrix}u \tag{1}$$

with natural frequency $\omega_n = \sqrt{2}$ and damping ratio $\delta = -1/\sqrt{2}$, so that

$$\dot{x} = \begin{bmatrix} 0 & 1 \\ -2 & 2 \end{bmatrix}x + \begin{bmatrix} 0 \\ 10 \end{bmatrix}u. \tag{2}$$

The plant is unstable with poles at $1 \pm j$. Discretizing with $T = 25$ msec yields (we give only three decimal places)

$$x_{k+1} = \begin{bmatrix} 0.999 & 0.026 \\ -0.051 & 1.051 \end{bmatrix} x_k + \begin{bmatrix} 0.003 \\ 0.256 \end{bmatrix} u_k$$

$$\stackrel{\Delta}{=} A x_k + B u_k. \tag{3}$$

The open-loop poles are at

$$z = 1.025 \pm j0.026. \tag{4}$$

Let us associate with (3) the infinite-horizon performance index

$$J_0 = \frac{1}{2} \sum_{k=0}^{\infty} (q x_k^T x_k + u_k^2) \tag{5}$$

(i.e., $Q = qI$), where the infinite time interval $[0, \infty]$ means we are seeking the optimal steady-state control.

a. Locus of Optimal Closed-loop Poles versus q

Here we investigate the effect of q on the optimal closed-loop poles. To do this, we can look at the eigenstructure of the Hamiltonian matrix $(R = 1)$

$$H^{-1} = \begin{bmatrix} A + B B^T A^{-T} q & -B B^T A^{-T} \\ -A^{-T} q & A^{-T} \end{bmatrix}. \tag{6}$$

A program was written to plot the eigenvalues of H^{-1} as q varies. The resulting root-locus is shown in Fig. 2.4-6. The poles of the optimal closed-loop plant

$$x_{k+1} = (A - B K_\infty) x_k \stackrel{\Delta}{=} A_\infty^{cl} x_k \tag{7}$$

are given by the stable poles of H^{-1}. For $q = 0$, these are

$$z = 0.975 \pm j0.024, \tag{8}$$

which are simply the poles of the original plant reflected inside the unit circle (i.e., their reciprocals). As $q \to \infty$ the optimal closed-loop poles tend to

$$z = 0, 0.975. \tag{9}$$

We now discuss three ways to find the optimal steady-state feedback

$$u_k = -K_\infty x_k \tag{10}$$

for the case of $q = 0.07$.

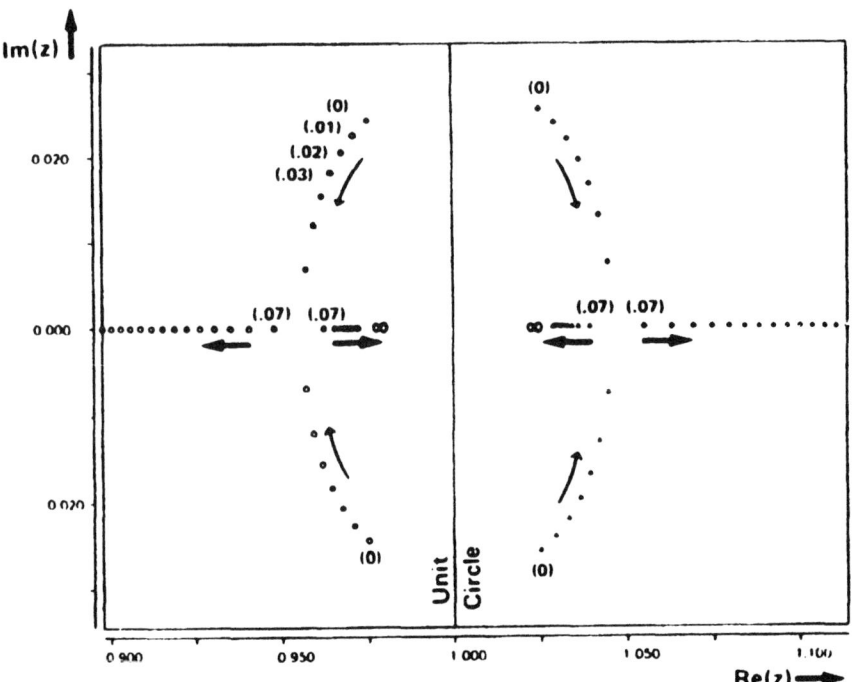

FIGURE 2.4-6 Locus of eigenvalues of H^{-1} as state weighting q varies. (Values of q are in parentheses. The increment in q is 0.01.)

b. Solution of the Algebraic Riccati Equation

To solve the ARE (2.4-12), we use a final condition of $S = I$ (any final condition will do since (A, B) is reachable and $(A, \sqrt{Q})$ is observable) in (2.4-2) and iterate until the solution converges. This occurs after 200 iterations (5 sec) and yields

$$S_\infty = \begin{bmatrix} 6.535 & 0.528 \\ 0.528 & 2.314 \end{bmatrix}. \tag{11}$$

We now use (2.4-13) to find the optimal steady-state feedback gain

$$K_\infty = [0.109 \quad 0.545]. \tag{12}$$

The resulting closed-loop plant is

$$A_\infty^{\text{cl}} = (A - BK_\infty) = \begin{bmatrix} 0.999 & 0.024 \\ -0.079 & 0.911 \end{bmatrix}, \tag{13}$$

which has stable poles of

$$z = 0.962, 0.948. \tag{14}$$

These correspond to $q = 0.07$ in Fig. 2.4-6.

c. Ackermann's Formula

We can avoid solving the ARE as follows. If $q = 0.07$, the optimal closed-loop poles are the stable eigenvalues of H^{-1}. These are found by the program that generated Fig. 2.4-6 to be 0.962, 0.948. Therefore, the desired closed-loop characteristic polynomial is

$$\Delta^{cl}(z) = (z - 0.962)(z - 0.948) = z^2 - 1.910z + 0.912. \tag{15}$$

The reachability matrix is

$$U_2 = [B \quad AB] = \begin{bmatrix} 0.003 & 0.010 \\ 0.256 & 0.269 \end{bmatrix}. \tag{16}$$

According to Ackermann's formula,

$$K_\infty = [0 \quad 1]U_2^{-1}\Delta^{cl}(A), \tag{17}$$

where $\Delta^{cl}(A) = A^2 - 1.910A + 0.912I$ is just a 2×2 real matrix. This again results in (12).

d. Eigenstructure Assignment

Another way to avoid solving the ARE is to use (2.4-63). The diagonal matrix of stable eigenvalues of H^{-1} for $q = 0.07$ is

$$M = \begin{bmatrix} 0.962 & 0 \\ 0 & 0.948 \end{bmatrix}, \tag{18}$$

and the associated eigenvectors are the columns in

$$\begin{bmatrix} 0.148 & 0.764 \\ -0.229 & -1.640 \\ \hline 0.849 & 4.130 \\ -0.452 & -3.392 \end{bmatrix} \overset{\Delta}{=} \begin{bmatrix} X \\ \hline \Lambda \end{bmatrix}. \tag{19}$$

Using these matrices in (2.4-63) yields exactly (12).

We can also check our analytic ARE solution (2.4-42). If we compute

$$S_\infty = W_{21}W_{11}^{-1} = \Lambda X^{-1}, \tag{20}$$

we get exactly (11).   ∎

Time-varying Plant

If the original plant is time varying, we need to redefine observability and reachability to discuss the asymptotic LQ regulator. Let the plant and cost function be given by

$$x_{k+1} = A_k x_k + B_k u_k, \tag{2.4-65}$$

$$J_i = \frac{1}{2}x_N^T S_N x_N + \frac{1}{2}\sum_{k=i}^{N-1}(x_k^T Q_k x_k + u_k^T R_k u_k).$$

(2.4-66)

Define the discrete state transition matrix as

$$\phi(k,i) = A_{k-1}A_{k-2}\cdots A_i \quad \text{for } k > i, \text{ with } \phi(k,k) = I.$$

(2.4-67)

Then we say the plant is *uniformly completely observable* if for every N the *observability gramian* satisfies

$$\alpha_0 I \le \sum_{k=i}^{N-1}\phi^T(k,i)Q_k\phi(k,i) \le \alpha_1 I$$

(2.4-68)

for some $I < N$, $\alpha_0 > 0$, and $\alpha_1 > 0$. This guarantees the positive definiteness of both the gramian and its inverse. Compare (2.4-68) with (2.2-24). We say the plant is *uniformly completely reachable* if for every i the *reachability gramian* satisfies

$$\alpha_0 I \le \sum_{k=i}^{N-1}\phi(N,k+1)B_k R_k^{-1} B_k^T \phi^T(N,k+1) \le a_1 I$$

(2.4-69)

for some $N > i$, $\alpha_0 > 0$, and $\alpha_0 > 0$. Compare with the time-invariant reachability gramian (2.2-36).

If the plant is time varying, then there is, in general, no constant steady-state solution to the Riccati equation. However, uniform complete observability and reachability (and boundedness of A_k, B_k, Q_k, R_k) guarantee that for large $(N - k)$ the behavior of S_k is unique, independent of S_N. They also guarantee uniform asymptotic stability of the closed-loop plant $(A - BK_k)$. See Kalman and Bucy (1961) and Kalman (1963).

2.5   FREQUENCY-DOMAIN RESULTS

In the steady-state case with time-invariant plant, the optimal closed-loop system is also time invariant, and we can work in the frequency domain to derive two important results. One of these yields further insight on the fictitious output (2.4-15), and the other gives a frequency-domain approach to the design of optimal regulators that is similar to the classical root-locus technique.

A Factorization Result

The optimal steady-state LQ regulator is given by the constant feedback

$$u_k = -Kx_k,$$

(2.5-1)

where

$$K = (B^{\mathrm{T}}S_\infty B + R)^{-1}B^{\mathrm{T}}S_\infty A. \qquad (2.5\text{-}2)$$

(For notational convenience, we write K instead of K_∞.) S_∞ is the unique positive definite solution to the ARE

$$S = A^{\mathrm{T}}[S - SB(B^{\mathrm{T}}SB + R)^{-1}B^{\mathrm{T}}S]A + Q \qquad (2.5\text{-}3)$$

(we assume (A, B) is stabilizable and $(A, \sqrt{Q})$ is observable). The resulting time-invariant closed-loop system

$$x_{k+1} = (A - BK)x_k \qquad (2.5\text{-}4)$$

is asymptotically stable.

To derive a relation between the open-loop characteristic polynomial $\Delta(z) = |zI - A|$ and the optimal closed-loop characteristic polynomial $\Delta^{\mathrm{cl}}(z)$, note that (Appendix A)

$$\begin{aligned} \Delta^{\mathrm{cl}}(z) &= |zI - A + BK| \\ &= |I + BK(zI - A)^{-1}| \cdot |zI - A| \qquad (2.5\text{-}5) \\ &= |I + K(zI - A)^{-1}B| \cdot \Delta(z). \end{aligned}$$

This identity will be useful shortly. According to Fig. 2.5-1, $-K(zI - A)^{-1}B$ can be interpreted as a loop gain matrix, so that $I + K(zI - A)^{-1}B$ is a return difference matrix.

To derive the result on which this section is based, note that

$$S - A^{\mathrm{T}}SA = (z^{-1}I - A)^{\mathrm{T}}S(zI - A) + (z^{-1}I - A)^{\mathrm{T}}SA + A^{\mathrm{T}}S(zI - A). \qquad (2.5\text{-}6)$$

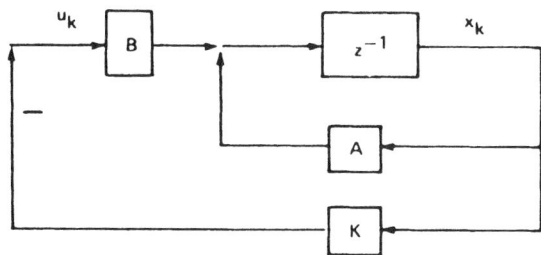

FIGURE 2.5-1   Optimal closed-loop system with control drawn as a state feedback.

Use the ARE to write

$$(z^{-1}I - A)^T S(zI - A) + (z^{-1}I - A)^T SA + A^T S(zI - A)$$

$$+ A^T SB(B^T SB + R)^{-1} B^T SA = Q.$$

Premultiply this by $B^T(z^{-1}I - A)^{-T}$ and postmultiply it by $(zI - A)^{-1}B$ to get

$$B^T SB + B^T SA(zI - A)^{-1}B + B^T(z^{-1}I - A)^{-T}A^T SB$$

$$+ B^T(z^{-1}I - A)^{-T}A^T SB(B^T SB + R)^{-1}B^T SA(zI - A)^{-1}B$$

$$= B^T(z^{-1}I - A)^{-T}Q(zI - A)^{-1}B.$$

Substitute from (2.5-2) to obtain

$$B^T SB + (B^T SB + R)K(zI - A)^{-1}B + B^T(z^{-1}I - A)^{-T}K^T(B^T SB + R)$$

$$+ B^T(z^{-1}I - A)^{-T}K^{-T}(B^T SB + R)K(zI - A)^{-1}B$$

$$= B^T(z^{-1}I - A)^{-T}Q(zI - A)^{-1}B.$$

Finally, add R to both sides and factor to see that

$$B^T(z^{-1}I - A)^{-T}Q(zI - A)^{-1}B + R$$

$$= [I + K(z^{-1}I - A)^{-1}B]^T(B^T SB + R)[I + K(zI - A)^{-1}B]. \tag{2.5-7}$$

Let us briefly discuss this result to try to get a feel for it. From (2.4-6) and (2.4-15) we can write the transfer-function relation in the plant from the control input to the fictitious output as

$$Y(z) = \begin{bmatrix} C(zI - A)^{-1}B \\ D \end{bmatrix} U(z), \tag{2.5-8}$$

where $Q = C^T C$ and $R = D^T D$. But

$$\begin{bmatrix} C(z^{-1}I - A)^{-1}B \\ D \end{bmatrix}^T \begin{bmatrix} C(zI - A)^{-1}B \\ D \end{bmatrix} = B^T(z^{-1}I - A)^{-T}Q(zI - A)^{-1}B + R; \tag{2.5-9}$$

evidently, then, (2.5-7) shows that we can factor this transfer-function product in terms of the return difference matrix.

Another point of view is obtained from Figs. 2.5-1 and 2.5-2. Define

$$H(z) = C(zI - A)^{-1}B \tag{2.5-10}$$

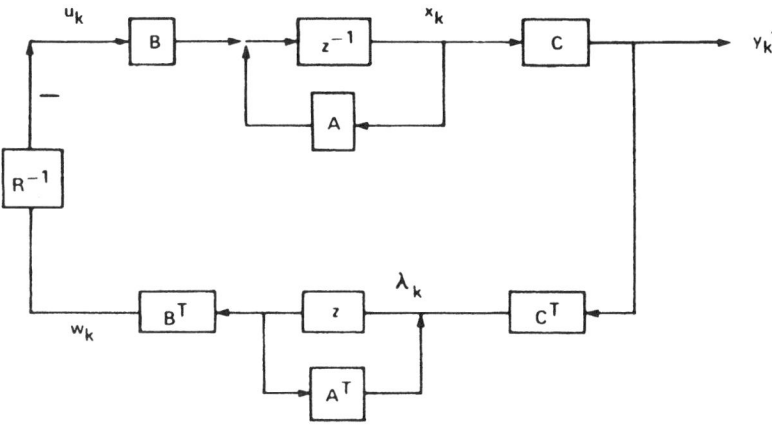

FIGURE 2.5-2   Optimal closed-loop system with control drawn as a costate feedback.

as the transfer function in the original state system from the control input to y_k^1, which is the partition of the fictitious output corresponding to Cx_k (see (2.4-15)). Then it is evident that

$$H^T(z^{-1}) = B^T(z^{-1}I - A)^{-T}C^T = B^T(z^{-1}I - A^T)^{-1}C^T \qquad (2.5\text{-}11)$$

is the transfer function of the *costate* system from y_k^1 to the intermediate signal w_k. Therefore, (2.5-7) simply expresses an equivalence between a transfer-function product in Fig. 2.5-2 and a transfer-function (i.e., return difference) product in Fig. 2.5-1. It is simply another way of expressing the equivalence of the two formulations of the LQ regulator (2.2-9) and Table 2.2-1.

We can use (2.5-5) and (2.5-7) to derive a frequency-domain method for designing steady-state regulators, as we now show.

Chang–Letov Design Procedure for the Steady-state LQ Regulator

By using the characteristic polynomial relation (2.5-5) and the factorization result (2.5-7) we can write the *Chang–Letov equation*

$$\Delta^{cl}(z^{-1})\Delta^{cl}(z) = |H^T(z^{-1})H(z) + R| \cdot \Delta(z^{-1})\Delta(z) \cdot |B^T SB + R|^{-1}, \quad (2.5\text{-}12)$$

where $H(z)$ is defined in (2.5-10) ($Q = C^T C$) (Chang 1961, Letov 1960). This is a very useful result, since it provides an alternative frequency-domain method for steady-state regulator design that is very similar to the classical root-locus technique.

To see this, note that $H(z)$ and the open-loop characteristic equation $\Delta(z)$ can be computed immediately, given the plant and performance index. The term

$|B^{T}SB + R|$ depends on the as yet unknown ARE solution S (i.e., S_{∞}), but it is irrelevant since it only provides a normalizing constant. Thus, the entire right-hand side of the Chang-Letov equation is known to within a multiplicative constant.

The term $\Delta^{cl}(z^{-1})\Delta^{cl}(z)$ on the left-hand side is a polynomial with quite interesting properties. Its roots are the roots z_i of $\Delta^{cl}(z)$ and their reciprocals $1/z_i$ (which are the roots of $\Delta^{cl}(z^{-1})$). The closed-loop plant (2.5-4) is stable, so that the optimal closed-loop poles can be determined by selecting the stable roots of the right-hand polynomial in (2.5-12)!

The importance of the Chang–Letov equation is now clear. It allows us to determine directly from A, B, Q, and R, which are all known, the optimal closed-loop poles. Then, in the single-input case, Ackermann's formula or an equivalent pole-placement technique can be used to determine the required optimal feedback K. (In the multivariable case we need to know the optimal closed-loop eigenvectors also before K can be uniquely determined. Chang–Letov cannot tell us these, but the techniques of Section 2.4 can be used.)

In the single-input case with $Q = qI$, we have

$$H(z) = \sqrt{q}\frac{[adj(zI - A)]B}{\Delta(z)} \triangleq \sqrt{q}\frac{N(z)}{\Delta(z)}, \tag{2.5-13}$$

where $N(z)$ is a column vector. Then (2.5-12) becomes

$$\Delta^{cl}(z^{-1})\Delta^{cl}(z) = \frac{(q/r)N^{T}(z^{-1})N(z) + \Delta(z^{-1})\Delta(z)}{1 + B^{T}SB/r}. \tag{2.5-14}$$

The roots of the right-hand side are the zeros of

$$1 + \frac{q}{r}\frac{N^{T}(z^{-1})N(z)}{\Delta(z^{-1})\Delta(z)}, \tag{2.5-15}$$

which is in exactly the form required for root-locus analysis. That entire body of theory therefore applies here. Evidently, then, as q/r varies from 0 (no state weighting) to ∞ (no control weighting), the optimal closed-loop poles move from the stable poles of

$$G(z) \triangleq H^{T}(z^{-1})H(z) \tag{2.5-16}$$

to its stable zeros. The ratio of cost weights q/r can therefore be selected to yield suitable closed-loop poles.

The next example illustrates these ideas. Good references are Schultz and Melsa (1967) and Kailath (1980).

Example 2.5-1.   Root-locus Design of Steady-state Regulator for Harmonic Oscillator

To compare the Chang–Letov method with the steady-state regulator design methods in Section 2.4, let us reconsider Example 2.4-6. The only thing we need to begin our design is the "fictitious" transfer function

$$H(z) = \sqrt{Q}(zI - A)^{-1}B \tag{1}$$

Using A and B from Example 2.4-6, and $Q = qI$ and $R = 1$ as in that example, this is found to be (we show only three decimal places)

$$H(z) \triangleq \sqrt{q}\frac{N(z)}{\Delta(z)} = \sqrt{q}\frac{\begin{bmatrix} 0.003z + 0.003 \\ 0.256z - 0.256 \end{bmatrix}}{z^2 - 2.050z + 1.051}. \tag{2}$$

The design is based on the rational function

$$G(z) = \frac{N^T(z^{-1})N(z)}{\Delta(z^{-1})\Delta(z)}, \tag{3}$$

since the zeros of

$$1 + \frac{q}{r}G(z) \tag{4}$$

are the roots of $\Delta^{cl}(z^{-1})\Delta^{cl}(z)$. This function is

$$G(z) = \frac{\begin{bmatrix} 0.003z^{-1} + 0.003 \\ 0.256z^{-1} - 0.256 \end{bmatrix}^T \begin{bmatrix} 0.003z + 0.003 \\ 0.256z - 0.256 \end{bmatrix}}{(z^{-2} - 2.050z^{-1} + 1.051)(z^2 - 2.050z + 1.051)} \tag{5}$$

$$= \frac{-0.066z^3 + 0.131z^2 - 0.066z}{1.051z^4 - 4.205z^3 + 6.308z^2 - 4.205z + 1.051}.$$

Note the symmetric form of the coefficients of the numerator and denominator; this means that if z is a root of either of these polynomials, then z^{-1} is also.

$G(z)$ can be factored as

$$G(z) = \frac{-0.063z(z - 0.975)(z - 1.025)}{[(z - 0.975)^2 + 0.024^2][(z - 1.025)^2 + 0.026^2]}. \tag{6}$$

If we draw the root locus of (4) as q varies from 0 to ∞, we obtain exactly Fig. 2.4-6! Thus, the eigenvalues of the Hamiltonian matrix H^{-1} in (2.4-43) are exactly the roots of $(q/r)N^T(z^{-1})N(z) + \Delta(z^{-1})\Delta(z)$.

According to the Chang-Letov equation, the optimal poles of the closed-loop plant $(A - BK)$ with LQ regulator are the stable zeros of (4) for any given q and r. When $q = 0$, they are the original plant *poles* reflected inside the unit circle (i.e., unstable plant poles z_i are replaced by z_i^{-1}), and when $q \to \infty$, they are the *zeros* of $H(z)$ reflected inside the unit circle.

Suppose we examine the root locus and decide that the closed-loop poles corresponding to $q = 0.07$ are satisfactory for our purposes. Then we can use Ackermann's formula to find the required optimal feedback K as in part c of Example 2.4-6. ∎

PROBLEMS

Section 2.1

2.1-1. Optimal control of a bilinear system. Let the scalar plant

$$x_{k+1} = x_k u_k + 1 \tag{1}$$

have performance index

$$J = \frac{1}{2} \sum_{k=0}^{N-1} u_k^2, \tag{2}$$

with final time $N = 2$. Given x_0, it is desired to make $x_2 = 0$.
a. Write state and costate equations with u_k eliminated.
b. Assume the final costate λ_2 is known. Solve for λ_0, λ_1 in terms of λ_2 and the state. Use this to express x_2 in terms of λ_2 and x_0. Hence, find a quartic equation for λ_2 in terms of initial state x_0.
c. If $x_0 = 1$, find the optimal state and costate sequences, the optimal control, and the optimal value of the performance index.

2.1-2. Optimal control of a bilinear system. Consider the bilinear system

$$x_{k+1} = A x_k + D x_k u_k + b u_k, \tag{1}$$

where $x_k \in R^n$, $u_k \in R$, with quadratic performance index

$$J = \frac{1}{2} x_N^T S_N x_N + \frac{1}{2} \sum_{k=0}^{N-1} (x_k^T Q x_k + r u_k^2), \tag{2}$$

where $S_N \geq 0$, $Q \geq 0$, $r > 0$. Show that the optimal control is the bilinear state–costate feedback,

$$u_k = -(b + D x_k)^T \lambda_{k+1} / r, \tag{3}$$

and that the state and costate equations after eliminating u_k are

$$x_{k+1} = A x_k - (b + D x_k)(b + D x_k)^T \lambda_{k+1} / r, \tag{4}$$

$$\lambda_k = Q x_k + A^T \lambda_{k+1} - (b + D x_k)^T \lambda_{k+1} D^T \lambda_{k+1} / r. \tag{5}$$

2.1-3. Optimal control of a generalized state-space system. Rederive the equations in Table 2.1-1 to find the optimal controller for the nonlinear *generalized state-space* (or *descriptor*) system

$$E x_{k+1} = f^k(x_k, u_k), \tag{6}$$

where E is singular. These systems often arise in circuit analysis, economics, and similar areas.

Section 2.2

2.2-1.   Writing the Lyapunov equation as a vector equation. Show that the Lyapunov equation (2.2-21) can be written as the vector equation

$$s(S_k) = (A_k^T \otimes A_k^T)s(S_{k+1}) + s(Q_k), \tag{1}$$

where the Kronecker product and stacking operator are defined in Appendix A.

2.2-2.   Solutions to the algebraic Lyapunov equation.
a. Find all possible solutions to (2.2-26) if

$$A = \begin{bmatrix} \frac{1}{2} & 1 \\ 0 & -\frac{1}{2} \end{bmatrix}, \qquad C = [2 \quad 0], \qquad Q = C^T C.$$

(*Hint*: Let

$$P = \begin{bmatrix} p_1 & p_2 \\ p_3 & p_4 \end{bmatrix},$$

substitute into (2.2-26), and solve for the scalars p_i. Alternatively, the results of Problem 2.2-1 can be used.)
b. Now find the symmetric solutions.

2.2-3.   Prove that (2.2-57) and the Joseph-stabilized Riccati equation (2.2-62) are equivalent to (2.2-53).

2.2-4.   Control of a scalar system. Let $x_{k+1} = 2x_k + u_k$.
a. Find the homogeneous solution x_k for $k = 0, 5$ if $x_0 = 3$.
b. Find the minimum-energy control sequence u_k required to drive $x_0 = 3$ to $x_5 = 0$. Check your answer by finding the resulting state trajectory.
c. Find the optimal feedback gain sequence K_k to minimize the performance index

$$J_0 = 5x_5^2 + \frac{1}{2} \sum_{k=0}^{4} (x_k^2 + u_k^2).$$

Find the resulting state trajectory and the costs to go J_k^* for $k = 0, 5$.

2.2-5.   Comparison of different discrete controllers

$$x_{k+1} = \begin{bmatrix} -\frac{1}{2} & \frac{1}{2} \\ 3 & -1 \end{bmatrix} x_k + \begin{bmatrix} 2 & 1 \\ 2 & 0 \end{bmatrix} u_k, \qquad x_0 = \begin{bmatrix} 8 \\ 4 \end{bmatrix}.$$

a. Find the open-loop control u_0, u_1 to drive the initial state to $x_2 = 0$ while minimizing the cost

$$J_a = \frac{1}{2} \sum_{k=0}^{1} u_k^T \begin{bmatrix} 2 & 0 \\ 0 & 1 \end{bmatrix} u_k.$$

Check your answer by "simulation" (i.e., apply your u_0, u_1 to the plant to verify that $x_2 = 0$).

b. Find a constant state-variable feedback to input component one of the form

$$u_k^1 = -Kx_k,$$

where $u_k = [u_k^1 \ u_k^2]^T$, to yield a deadbeat control (all closed-loop poles at the origin). Find the closed-loop state trajectory.

c. Let

$$J_c = 10x_2^T x_2 + J_a,$$

with J_a as in part a. Solve the Riccati equation to determine the optimal control u_0, u_1. Find the optimal cost.

d. Compare the state trajectories of parts a, b, and c.

e. Now suppose $x_0 = [1 \ 2]^T$. How must the controls of parts a, b, and c be modified?

2.2-6.   Linear performance index. Let

$$x_{k+1} = Ax_k + Bu_k,$$

$$J - S_N x_N + \sum_{K=0}^{N-1}(Qx_k + Ru_k),$$

with J a scalar. Write state and costate equations and stationarity condition. What is the problem? (We shall learn how to deal with linear cost indices in Section 5.2.)

2.2-7.   Cubic performance index. Let

$$x_{k+1} = ax_k + bu_k,$$

where x_k and u_k are scalars, and

$$J = \frac{1}{3}s_N x_N^3 + \frac{1}{3}\sum_{k=0}^{N-1}(qx_k^3 + ru_k^3).$$

a. Write state and costate equations and stationarity condition.

b. When can we solve for u_k? Under this condition, eliminate u_k from the state equation.

c. Solve the open-loop control problem (i.e., x_N fixed, $s_N = 0$, $q = 0$).

2.2-8.   Optimal control with weighting of state-input inner product

a. Redo Problem 2.2-4c if the cost index is

$$J = 5x_5^2 + \frac{1}{2}\sum_{k=0}^{4}(x_k^2 + u_k^2 + 2x_k u_k).$$

b. Redo Problem 2.2-5c if the term $2x_k^T u_k$ is added to J_a.

2.2-9.   Information formulation of the Riccati equation. If S_N is very large, then it is convenient to use the *information formulation* of the discrete Riccati equation, which propagates S_k^{-1} instead of S_k. (The name derives from the filtering application of the Riccati equation.)

Separate the Riccati equation into two parts by defining the intermediate matrix

$$\overline{S}_{k+1} \stackrel{\Delta}{=} S_{k+1} - S_{k+1}B(B^T S_{k+1} B + R)^{-1} B^T S_{k+1}. \tag{1}$$

Then

$$S_k = A^T \overline{S}_{k+1} A + Q. \tag{2}$$

The first of these equations incorporates the effect of the control input on the performance index J_k; note that $\overline{S}_{k+1} \le S_{k+1}$. The second equation shows the effects of the plant dynamics and state weighting Q on J_k; these effects generally make S_k larger than $\overline{K}_{k+1}$.

a. Show that *input update* (1) can be written

$$\overline{S}_{k+1}^{-1} = S_{k+1}^{-1} + BR^{-1}B^T. \tag{3}$$

b. Assume $|A| \ne 0$, $Q > 0$, and define

$$F_{k+1} = A^{-1}\overline{S}_k^{-1} A^{-T}. \tag{4}$$

Show that the *state update* (2) can be written

$$S_k^{-1} = F_{k+1} - F_{k+1}(F_{k+1} + Q^{-1})^{-1} F_{k+1}. \tag{5}$$

If the information Kalman gain is defined as

$$G_k = (F_{k+1} + Q^{-1})^{-1} F_{k+1}, \tag{6}$$

this becomes

$$S_k^{-1} = F_{k+1}(I - G_k). \tag{7}$$

The information update is given by (3), (4), (6), and (7).

2.2-10.   Square-root Riccati formulations. Split the Riccati equation into two parts as in Problem 2.2-9 and define the roots P_k, $\overline{P}_k$ by $S_k = P_k^T P_k$, $\overline{S}_k = \overline{P}_k^T P_k$. The following results show how to propagate $\overline{P}_k$, P_k instead of $\overline{S}_k$, S_k and yield numerically stable algorithms. See Schmidt (1967, 1970), Businger and Golub (1965), Dyer and McReynolds (1969), Bierman (1977), Morf and Kailath (1975), and Kaminski et al. (1971).

a. Show that the state update is equivalent to

$$T_1 \begin{bmatrix} \overline{P}_{k+1} A \\ C \end{bmatrix} = \begin{bmatrix} P_k \\ 0 \end{bmatrix}, \tag{1}$$

where $Q = C^T C$, and T_1 is any orthogonal transformation (i.e., $T_1^T T_1 = I$) selected so that P_k has n rows.

b. Show that the input update is equivalent to

$$
T_2 \begin{bmatrix} D & 0 \\ P_{k+1}B & P_{k+1} \end{bmatrix} = \begin{bmatrix} (B^T S_{k+1} B + R)^{1/2} & \overline{K}_{k+1} \\ 0 & \overline{P}_{k+1} \end{bmatrix}, \tag{2}
$$

where $R = D^T D$, and T_2 is any orthogonal matrix such that $\overline{P}_{k+1}$ has n rows.

c. Show that both updates can be expressed together in square-root form as

$$
T_3 \begin{bmatrix} D & 0 \\ P_{k+1}B & P_{k+1}A \\ 0 & C \end{bmatrix} = \begin{bmatrix} (B^T S_{k+1} B + R)^{1/2} & \overline{K}_{k+1}A \\ 0 & P_k \\ 0 & 0 \end{bmatrix}, \tag{3}
$$

where T_3 is orthogonal.

Section 2.3

2.3-1. Digital control of harmonic oscillator. A harmonic oscillator is described by

$$
\begin{aligned}
\dot{x}_1 &= x_2 \\
\dot{x}_2 &= -\omega_n^2 x_1 + u.
\end{aligned} \tag{1}
$$

a. Discretize the plant using a sampling period of T.

b. With the discretized plant, associate a performance index of

$$
J = \frac{1}{2}[s_1(x_N^1)^2 + s_2(x_N^2)^2] + \frac{1}{2} \sum_{k=0}^{N-1} [q_1(x_k^1)^2 + q_2(x_k^2)^2 + ru_k^2], \tag{2}
$$

where the state is $x_k = [x_k^1 \quad x_k^2]^T$. Write scalar equations for a digital optimal controller.

c. Write a MATLAB subroutine to simulate the plant dynamics, and use the time response program *lsim.m* to obtain zero-input state trajectories.

d. Write a MATLAB subroutine to compute and store the optimal control gains and to update the control u_k given the current state x_k. Write a MATLAB driver program to obtain time response plots for the optimal controller.

2.3-2. Digital control of an unstable system. Repeat the previous problem for

$$
\begin{aligned}
\dot{x}_1 &= x_2, \\
\dot{x}_2 &= a^2 x_1 + bu.
\end{aligned} \tag{3}
$$

2.3-3. Digital controller with weighting of state–input inner product. Modify your controller in Problem 2.3-1 to include terms like

$$v_1 x_k^1 u_k + v_2 x_k^2 u_k$$

in the performance index, where v_1 and v_2 are scalar weightings.

2.3-4. General digital control subroutine. For a general time-invariant plant and performance index as in Table 2.2-1, use MATLAB to write m-files to compute and store the optimal feedback gains K_k.

Section 2.4

2.4-1. Steady-state behavior. In this problem we consider a rather unrealistic discrete system because it is simple enough to allow an analytic treatment. Thus, let the plant

$$x_{k+1} = \begin{bmatrix} 0 & 1 \\ 0 & 0 \end{bmatrix} x_k + \begin{bmatrix} 0 \\ 1 \end{bmatrix} u_k \tag{1}$$

have performance index of

$$J_0 = \frac{1}{2} x_N^T x_N + \frac{1}{2} \sum_{k=0}^{N-1} \left(x_k^T \begin{bmatrix} q_1 & q_2 \\ q_2 & q_1 \end{bmatrix} x_k + r u_k^2 \right). \tag{2}$$

a. Find the optimal steady-state (i.e., $N \to \infty$) Riccati solution S_∞^* and show that it is positive definite. Find the optimal steady-state gain K_∞^* and determine when it is nonzero.

b. Find the optimal steady-state closed-loop plant and demonstrate its stability.

c. Now the suboptimal constant feedback

$$u_k = -K_\infty^* x_k \tag{3}$$

is applied to the plant. Find scalar updates for the components of the suboptimal cost kernel S_k. Find the suboptimal steady-state cost kernel S_∞ and demonstrate that $S_\infty = S_\infty^*$.

2.4-2. Analytic Riccati solution. Let

$$A = \begin{bmatrix} 1 & 1 \\ 0 & 1 \end{bmatrix}, \quad B = \begin{bmatrix} 0 \\ 1 \end{bmatrix}, \quad S_N = I, \quad Q = I.$$

a. Let $r = 0.1$. Find the Hamiltonian matrix H and its eigenvalues and eigenvectors. Find the analytic expression for Riccati solution S_k. Find the steady-state solution S_∞ using (2.4-42). Find the optimal steady-state gain K_∞ using (2.4-63) and also using Ackermann's formula.
b. Let $r = 1$. Find the Hamiltonian matrix and its eigenstructure. Find the steady-state solution S_∞ and gain K_∞. (*Hint*: See the discussion following (2.4-63).)

2.4-3. Software for plotting optimal closed-loop poles
a. For the LQ case, write a computer program to
   1. Compute the Hamiltonian matrix H.
   2. Find eigenstructure of H. You can use MATLAB routine *eig.m*.
   3. Compute the steady-state Riccati solution S_∞.
   4. Compute the steady-state gain K_∞ and closed-loop system A^{cl}.

b. Modify your program to find the optimal steady-state closed-loop poles as a function of r, the control weighting. (Note that you only need to do 1 and 2 for this.) You now have a design tool to select the cost-weighting matrices $Q = qI$ and $R = rI$ to yield desired closed-loop performance at steady state.

Section 2.5

2.5-1. Chang–Letov design.
a. For the system in Problem 2.4-2, use the Chang–Letov procedure and Ackermann's formula to find the optimal steady-state feedback gain K_∞ and closed-loop plant if $r = 0.1$.
b. Plot a root locus of the closed-loop poles as r varies from ∞ to 0.

2.5-2. Reciprocal polynomials.
Let $\phi(z) = z^2 + 2\alpha z + \omega^2$. Find and sketch the roots z_1, z_2 of $\phi(z)$. Define the *reciprocal polynomial* as $z^2\phi(z^{-1}) = \omega^2 z^2 + 2\alpha z + 1$. Show that the roots of $z^2\phi(z^{-1})$ are the roots of $\phi(z)$ reflected about the unit circle. That is, they are equal to $1/z_1$ and $1/z_2$.

2.5-3. Decomposition of polynomials using a reciprocal polynomial.
a. Show that any polynomial $P(z)$ of degree n with real coefficients can be decomposed as

$$P(z) = P_1(z) + P_2(z), \tag{1}$$

where the *mirror-image* and *anti-mirror-image* polynomials are defined by

$$P_1(z) = \tfrac{1}{2}(P(z) + z^n P(z^{-1})), \tag{2}$$
$$P_2(z) = \tfrac{1}{2}(P(z) + z^n P(z^{-1})). \tag{3}$$

b. Show that

$$P_1(z) = z^n P_1(z^{-1}),$$ (4)

$$P_2(z) = -z^n P_2(z^{-1}).$$ (5)

(Compare this with the decomposition of a real matrix into symmetric and antisymmetric matrices. Hence, note that the reciprocal polynomial $z^n P(z^{-1})$ plays a role similar to that of the transpose of a matrix.)

3

OPTIMAL CONTROL
OF CONTINUOUS-TIME SYSTEMS

We shall now discuss optimal control for systems with a continuous time index. From a glance at the table of contents, it is apparent that this chapter will follow the development of Chapter 2.

There are several distinctions between the optimal control problems for continuous and discrete systems, the most noticeable of which is that the continuous control laws are based on equations of a simpler form than their discrete counterparts. That will allow us to obtain some analytic solutions in this chapter. Another distinction arises in the initial stages of the derivation of the control law. For continuous systems, we must distinguish between *differentials* and *variations* in a quantity, which we did not need to do in Chapter 2. This means that we shall need to use the calculus of variations, which is briefly reviewed in Section 3.1.

The continuous dependence on time also makes it fairly simple to talk about minimum-time problems, which we do in Chapter 5. The derivations in this chapter are for the most part similar to those for discrete systems, and we shall attempt to set them down in a manner that makes clear what is going on without duplicating too much of our work from Chapter 2.

3.1 THE CALCULUS OF VARIATIONS

Only a few ideas from the calculus of variations will be needed, so our review will be short. For an in-depth discussion, see Athans and Falb (1966) or Kirk (1970). In Section 3.2 we shall be concerned with minimizing an augmented performance index J' exactly as we were in Chapter 2. To perform this minimization, we need to find the change induced in J' by independent changes in all of its arguments (cf. (2.1-6)). Unfortunately, we shall run into a slight problem. The change in J' will depend on the time and state differentials dt and dx. However, these quantities are not independent. The purpose of this section is to clear up this point and to derive a relation that will soon be useful.

If $x(t)$ is a continuous function of time t, then the differentials $dx(t)$ and dt are not independent. We can, however, define a small change in $x(t)$ that *is* independent of dt. Let us define the *variation* in $x(t)$, $\delta x(t)$, as the incremental change in $x(t)$ when time t is held fixed.

To find the relations among dx, δx, and dt, examine Fig. 3.1-1. Here we show the original function $x(t)$ and a neighboring function $x(t) + dx(t)$ over an interval specified by initial time t_0 and final time T (Bryson and Ho 1975). In addition to the increment $dx(t)$ at each time t, the final time has been incremented by dT. It is clear from the illustration that the overall increment in x at T, $dx(T)$, depends on dT. According to our definition, the variation $\delta x(T)$ occurs at the fixed value of $t = T$, as shown and is independent of dT. Since $x(t)$ and $x(t) + dx(t)$ have approximately the same slope $\dot{x}(T)$ at $t = T$, and since dT is small, we have

$$dx(T) = \delta x(T) + \dot{x}(T)\,dT. \qquad (3.1\text{-}1)$$

This relation is the one we shall need later.

Another relation we shall need is *Leibniz's rule* for functionals: if $x(t) \in R^n$ is a function of t and

$$J(x) = \int_{t_0}^{T} h(x(t), t)\,dt, \qquad (3.1\text{-}2)$$

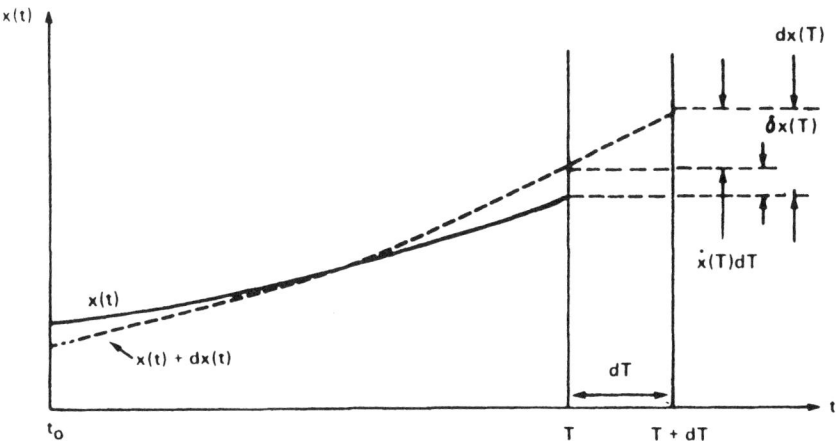

FIGURE 3.1-1   Relation between the variation δx and the differential dx.

where $J(\cdot)$ and $h(\cdot)$ are both real scalar functionals (i.e., functions of the function $x(t)$), then

$$dJ = h(x(T), T) \, dT - h(x(t_0), t_0) \, dt_0$$
$$+ \int_{t_0}^{T} \left[h_x^T(x(t), t) \delta x \right] dt. \qquad (3.1\text{-}3)$$

Our notation is

$$h_x \triangleq \frac{\partial h}{\partial x}.$$

3.2 SOLUTION OF THE GENERAL CONTINUOUS-TIME OPTIMIZATION PROBLEM

The philosophy in this chapter is to derive the solution to the continuous optimal control problem in the most general case. This is accomplished in the present section. Then, in subsequent sections, we consider various special cases of the general solution. The discussion at the beginning of Chapter 2 and the comments in Section 2.1 also apply here; they provide some insight on the formulation of the optimal control problem.

Problem Formulation and Solution

Suppose the plant is described by the nonlinear time-varying dynamical equation

$$\dot{x}(t) = f(x, u, t), \qquad (3.2\text{-}1)$$

with state $x(t) \in R^n$ and control input $u(t) \in R^m$. With this system let us associate the performance index

$$J(t_0) = \phi(x(T), T) + \int_{t_0}^{T} L(x(t), u(t), t) \, dt, \qquad (3.2\text{-}2)$$

where $[t_0, T]$ is the time interval of interest. The final weighting function $\phi(x(T), T)$ depends on the final state and final time, and the weighting function $L(x, u, t)$ depends on the state and input at intermediate times in $[t_0, T]$.

The performance index is selected to make the plant exhibit a desired type of performance. Some different possibilities for $J(t_0)$ are discussed in Example 2.1-1, which carries over to the continuous case.

The *optimal control problem* is to find the input $u^*(t)$ on the time interval $[t_0, T]$ that drives the plant (3.2-1) along a trajectory $x^*(t)$ such that the cost function (3.2-2) is minimized, *and* such that

$$\psi(x(T), T) = 0 \qquad (3.2\text{-}3)$$

for a given function $\psi \in R^p$. This corresponds to the function-of-final-state-fixed discrete problem solved in Section 4.5.

The roles of the final weighting function ϕ and the fixed final function ψ should not be confused. $\phi(x(T), T)$ is a function of the final state, which we want to make *small*. An illustration might be the energy, which is $[x^T(T)S(T)x(T)]/2$, where $S(T)$ is a given weighting matrix. On the other hand, $\psi(x(T), T)$ is a function of the final state, which we want *fixed* at exactly zero. As an illustration, consider a satellite with state $x = [r\dot{r}\theta\dot{\theta}]^T$, where r and θ are radius and angular position. If we want to place the satellite in a circular orbit with radius R, then the final state function to be zeroed would be

$$\psi(x(T), T) = \begin{bmatrix} r(T) - R \\ \dot{r}(T) \\ \dot{\theta}(T) = \sqrt{\dfrac{\mu}{R^3}} \end{bmatrix},$$

with $\mu = GM$ the gravitational constant of the attracting mass M.

To solve the continuous optimal control problem, we shall use Lagrange multipliers to adjoin the constraints (3.2-1) and (3.2-3) to the performance index (3.2-2). Since (3.2-1) holds at each $t \in [t_0, T]$, we require an associated multiplier $\lambda(t) \in R^n$, which is a function of time. Since (3.2-3) holds only at one time, we require only a constant associated multiplier $v \in R^p$. The augmented performance index is thus

$$
\begin{aligned}
J' = {} & \phi(x(T), T) + v^T\psi(x(T), T) \\
& + \int_{t_0}^{T} \left[L(x, u, t) + \lambda^T(t)(f(t)(x, u, t) - \dot{x}) \right] dt.
\end{aligned}
\tag{3.2-4}
$$

If we define the *Hamiltonian function* as

$$H(x, u, t) = L(x, u, t) + \lambda^T f(x, u, t), \tag{3.2-5}$$

then we can rewrite (3.2-4) as

$$
\begin{aligned}
J' = {} & \phi(x(T), T) + v^T\psi(x(T), T) \\
& + \int_{t_0}^{T} \left[H(x, u, t) - \lambda^T\dot{x} \right] dt.
\end{aligned}
\tag{3.2-6}
$$

Using Leibniz's rule, the increment in J' as a function of increments in x, λ, v, u, and t is

$$
\begin{aligned}
dJ' = {} & (\phi_x + \psi_x^T v)^T dx\,|_T + (\phi_t + \psi_t^T v)\,dt\,|_T + \psi^T|_T dv \\
& + (H - \lambda^T\dot{x})\,dt\,|_T - (H - \lambda^T\dot{x})\,dt\,|_{t_0} \\
& + \int_{t_0}^{T} \left[H_x^T\delta x + H_u^T\delta u - \lambda^T\delta\dot{x} + (H_\lambda - \dot{x})^T\delta\lambda \right] dt.
\end{aligned}
\tag{3.2-7}
$$

To eliminate the variation in $\dot{x}$, integrate by parts to see that

$$-\int_{t_0}^{T} \lambda^{\mathrm{T}} \delta\dot{x}\, dt = -\lambda^{\mathrm{T}} \delta x|_T + \lambda^{\mathrm{T}} \delta x|_{t_0} + \int_{t_0}^{T} \dot{\lambda}^{\mathrm{T}} \delta x\, dt. \tag{3.2-8}$$

If we substitute this into (3.2-7), there result terms at $t = T$ dependent on both $dx(t)$ and $\delta x(T)$. We can express $\delta x(T)$ in terms of $dx(t)$ and dT using (3.1-1). The result after these two substitutions is

$$dJ' = (\phi_x + \psi_x^{\mathrm{T}} v - \lambda)^{\mathrm{T}}\, dx\,|_T + (\phi_t + \psi_t^{\mathrm{T}} v + H - \lambda^{\mathrm{T}}\dot{x} + \lambda^{\mathrm{T}}\dot{x})\, dt\,|_T$$

$$+ \psi^{\mathrm{T}}|_T\, dv - (H - \lambda^{\mathrm{T}}\dot{x} + \lambda^{\mathrm{T}}\dot{x})\, dt\,|_{t_0} + \lambda^{\mathrm{T}}\, dx\,|_{t_0}$$

$$+ \int_{t_0}^{T} \left[(H_x + \dot{\lambda})^{\mathrm{T}} \delta x + H_u^{\mathrm{T}} \delta u + (H_\lambda - \dot{x})^{\mathrm{T}} \delta\lambda \right] dt. \tag{3.2-9}$$

According to the Lagrange theory, the constrained minimum of J is attained at the unconstrained minimum of J'. This is achieved when $dJ' = 0$ for all independent increments in its arguments. Setting to zero the coefficients of the independent increments dv, δx, δu, and $\delta\lambda$ yields necessary conditions for a minimum as shown in Table 3.2-1. For our applications, t_0 and $x(t_0)$ are both fixed and known, so that dt_0 and $dx(t_0)$ are both zero. The two terms evaluated at $t = t_0$ in (3.2-9) are thus automatically equal to zero.

The final condition (3.2-10) in the table needs further discussion. We have seen that $dx(T)$ and dT are not independent (Fig. 3.1-1). Therefore, we cannot simply set the coefficients of the first two terms on the right-hand side of (3.2-9) separately equal to zero. Instead, the entire expression (3.2-10) must be zero at $t = T$. Compare it with (2.7-7). The extra term in (3.2-10) arises in the present situation since we have allowed for possible variations in the final time T. This will allow us to deal with minimum-time problems, which we shall do in Chapter 5.

For convenience, we have shown the conditions in the table both in terms of H and in terms of L and f. Compare these results with Table 2.1-1 and see the associated discussion for further insight. Note that the discrete and continuous costate equations are both dynamical equations that develop *backward* in time. In the continuous case, this amounts to making the rate of change (i.e., $\dot{\lambda}$) negative. The costate equation is also called the *adjoint* to the state equation.

As in the discrete case, the optimal control in Table 3.2-1 depends on the solution to a two-point boundary-value problem, since $x(t_0)$ is given and $\lambda(T)$ is determined by (3.2-10). It is, in general, very difficult to solve these problems. We do not really care about the value of $\lambda(t)$, but it must evidently be determined as an intermediate step in finding the optimal control $u^*(t)$, which depends on $\lambda(t)$ through the stationarity condition.

An important point is worth noting. The time derivative of the Hamiltonian is

$$\dot{H} = H_t + H_x^{\mathrm{T}}\dot{x} + H_u^{\mathrm{T}}\dot{u} + \dot{\lambda}^{\mathrm{T}} f = H_t + H_u^{\mathrm{T}}\dot{u} + (H_x + \dot{\lambda})^{\mathrm{T}} f. \tag{3.2-11}$$

TABLE 3.2-1   Continuous Nonlinear Optimal Controller with Function of Final State Fixed

System model:

$$\dot{x} = f(x, u, t), \quad t \geq t_0, \quad t_0 \text{ fixed}$$

Performance index:

$$J(t_0) = \phi(x(T), T) + \int_{t_0}^{T} L(x, u, t)\, dt$$

Final-state constraint:

$$\psi(x(T), T) = 0$$

Optimal controller Hamiltonian:

$$H(x, u, t) = L(x, u, t) + \lambda^{\mathrm{T}} f(x, u, t)$$

State equation:

$$\dot{x} = \frac{\partial H}{\partial \lambda} = f, \quad t \geq t_0$$

Costate equation:

$$-\dot{\lambda} = \frac{\partial H}{\partial x} = \frac{\partial f^{\mathrm{T}}}{\partial x}\lambda + \frac{\partial L}{\partial x}, \quad t \leq T$$

Stationarity condition:

$$0 = \frac{\partial H}{\partial u} = \frac{\partial L}{\partial u} + \frac{\partial f^{\mathrm{T}}}{\partial u}\lambda$$

Boundary conditions:

$$x(t_0) \text{ given}$$

$$(\phi_x + \psi_x^{\mathrm{T}}\nu - \lambda)^{\mathrm{T}}|_T\, dx(T) + (\phi_t + \psi_t^{\mathrm{T}}\nu + H)|_T\, dT = 0 \qquad (3.2\text{-}10)$$

If $u(t)$ is an optimal control, then

$$\dot{H} = H_t. \qquad (3.2\text{-}12)$$

Now, in the time-invariant case, f and L are not explicit functions of t, and so neither is H. In this situation

$$\dot{H} = 0. \qquad (3.2\text{-}13)$$

Hence, for time-invariant systems and cost functions, the Hamiltonian is a *constant* on the optimal trajectory.

Let us begin to develop a feel for the continuous optimal controller by looking at some examples.

Examples

The first two examples make the point that the solution to the optimization problem given in Table 3.2-1 is very general; it does not only apply in system theory. The next examples illustrate the computation of the optimal controller for dynamical systems, the last example emphasizing that the optimal control equations apply for general nonlinear systems.

Example 3.2-1.   Hamilton's Principle in Classical Dynamics

Hamilton's principle for conservative systems in classical physics says that "of all possible paths along which a dynamical system may move from one point to another within a specified time interval (consistent with any constraints), the actual path followed is that which minimizes the time integral of the difference between the kinetic and potential energies" (Marion 1965).

a. Lagrange's Equations of Motion

We can derive Lagrange's equations of motion from this principle by defining (Bryson and Ho 1975)

$$q \triangleq \text{generalized coordinate vector,}$$

$$u = \dot{q} \triangleq \text{generalized velocities,}$$

$$U(q) \triangleq \text{potential energy,}$$

$$T(q, u) \triangleq \text{kinetic energy,}$$

$$L(q, u) \triangleq T(q, u) - U(q), \text{ the Lagrangian of the system.}$$

The "plant" is then described by

$$\dot{q} = u \triangleq f(q, u), \tag{1}$$

where the function f is given by the physics of the problem. To find the trajectories of the motion, Hamilton's principle says that we must minimize the performance index

$$J(0) = \int_0^T L(q, u)\, dt. \tag{2}$$

Therefore, the Hamiltonian is

$$H = L + \lambda^T u. \tag{3}$$

According to Table 3.2-1, for a minimum we require

$$-\dot{\lambda} = \frac{\partial H}{\partial q} = \frac{\partial L}{\partial q} \tag{4}$$

and

$$0 = \frac{\partial H}{\partial u} = \frac{\partial L}{\partial u} + \lambda. \tag{5}$$

Combining these equations yields Lagrange's equations of motion

$$\frac{\partial L}{\partial q} - \frac{d}{dt}\frac{\partial L}{\partial \dot{q}} = 0. \tag{6}$$

It is worth emphasizing that in this context, the costate equation and the stationarity condition are equivalent to Lagrange's equation. In the general context of variational problems, (6) is called *Euler's equation*. The costate equation and stationarity condition in Table 3.2-1 are, therefore, an alternative formulation of Euler's equation.

In the context of this example, condition (3.2-13) is nothing more than a statement of the conservation of energy!

b. Hamilton's Equations of Motion

If we define the *generalized momentum vector* by

$$\lambda = -\frac{\partial L}{\partial \dot{q}}, \tag{7}$$

then the equations of motion can be written in Hamilton's form as

$$\dot{q} = \frac{\partial H}{\partial \lambda}, \tag{8}$$

$$-\dot{\lambda} = \frac{\partial H}{\partial q}. \tag{9}$$

Hence, in the optimal control problem, the state and costate equations are a generalized formulation of Hamilton's equations of motion! ∎

Example 3.2-2.* *Shortest Distance between Two Points

The length of a curve $x(t)$ dependent on a parameter t between $t = a$ and $t = b$ is given by

$$J = \int_a^b \sqrt{1 + \dot{x}^2(t)}\, dt. \tag{1}$$

To specify that the curve join two points $(a, A)(b, B)$ in the plane, we need to impose the boundary conditions

$$x(a) = A, \tag{2}$$

$$x(b) = B. \tag{3}$$

See Shultz and Melsa (1967).

It is desired to find the curve $x(t)$ joining (a, A) and (b, B) that minimizes (1). To put this into the form of an optimal control problem, define the "input" by

$$\dot{x} = u. \tag{4}$$

This is the "plant." Then (1) becomes

$$J = \int_a^b \sqrt{1 + u^2}\, dt. \tag{5}$$

The Hamiltonian is

$$H = \sqrt{1+u^2} + \lambda u. \tag{6}$$

Now, Table 3.2-1 yields the conditions

$$\dot{x} = H_\lambda = u, \tag{7}$$
$$-\dot{\lambda} = H_x = 0, \tag{8}$$
$$0 = H_u = \lambda + \frac{u}{\sqrt{1+u^2}}. \tag{9}$$

To solve these for the optimal slope u, note that by (9)

$$u = \frac{-\lambda}{\sqrt{1-\lambda^2}}, \tag{10}$$

but according to (8), λ is constant. Hence,

$$u = \text{const} \tag{11}$$

is the optimal "control." Now use (7) to get

$$x(t) = c_1 t + c_2. \tag{12}$$

To determine c_1 and c_2, use the boundary conditions (2) and (3) to see that

$$x(t) = \frac{(A-B)t + (aB-bA)}{a-b}. \tag{13}$$

The optimal trajectory (13) between two points is thus a straight line. ∎

Example 3.2-3.   Temperature Control in a Room

It is desired to heat a room using the least possible energy. If $\theta(t)$ is the temperature in the room, θ_a the ambient air temperature outside (a constant), and $u(t)$ the rate of heat supply to the room, then the dynamics are

$$\dot{\theta} = -a(\theta - \theta_a) + bu \tag{1}$$

for some constants a and b, which depend on the room insulation and so on. By defining the state as

$$x(t) \overset{\Delta}{=} \theta(t) - \theta_a, \tag{2}$$

we can write the state equation

$$\dot{x} = -ax + bu. \tag{3}$$

See McClamroch (1980). To control the temperature on the fixed time interval $[0, T]$ with the least possible supplied energy, define the performance index as

$$J(0) = \frac{1}{2} \int_0^T u^2(t)\, dt. \tag{4}$$

We shall discuss two possible control objectives in parts a and b below. The Hamiltonian is

$$H = \frac{u^2}{2} + \lambda(-ax + bu). \tag{5}$$

According to Table 3.2-1, the optimal control $u(t)$ is determined by solving

$$\dot{x} = H_\lambda = -ax + bu, \tag{6}$$
$$\dot{\lambda} = -H_x = a\lambda, \tag{7}$$
$$0 = H_u = u + b\lambda. \tag{8}$$

The stationarity condition (8) says that the optimal control is given by

$$u(t) = -b\lambda(t), \tag{9}$$

so to determine $u^*(t)$ we need only find the optimal costate $\lambda^*(t)$.

Substituting (9) into (6) yields the state–costate equations

$$\dot{x} = -ax - b^2\lambda, \tag{10a}$$
$$\dot{\lambda} = a\lambda, \tag{10b}$$

which must now be solved for $\lambda^*(t)$ and the optimal state trajectory $x^*(t)$. We do not yet know the final costate $\lambda(T)$, but let us solve (10) as if we did. The solution to (10b) is

$$\lambda(t) = e^{-a(T-t)}\lambda(T). \tag{11}$$

Using this in (10a) yields

$$\dot{x} = -ax - b^2\lambda(T)e^{-a(T-t)}. \tag{12}$$

Using Laplace transforms to solve this gives

$$\begin{aligned} X(s) &= \frac{x(0)}{s+a} - \frac{b^2\lambda(T)e^{-aT}}{(s+a)(s-a)} \\ &= \frac{x(0)}{s+a} - \frac{b^2}{a}\lambda(T)e^{-aT}\left(\frac{-1/2}{s+a} + \frac{1/2}{s-a}\right) \end{aligned} \tag{13}$$

so that

$$x(t) = x(0)e^{-at} - \frac{b^2}{a}\lambda(T)e^{-aT}\sinh at. \tag{14}$$

Equations (11) and (14) give the optimal costate $\lambda^*(t)$ and state $x^*(t)$ in terms of the as yet unknown final costate $\lambda(T)$. The initial state $x(0)$ is given.

Now we consider two possible control objectives, which will give two ways to determine $\lambda(T)$.

a. Fixed Final State

Suppose that the initial temperature of the room is equal to $\theta_a = 60°$. Then

$$x(0) = 0°. \tag{15}$$

Let our control objective be to drive the final temperature $\theta(T)$ exactly to $70°$ at the given final time of T seconds. Then the final state is required to take on the fixed value of

$$x(T) = 10°. \tag{16}$$

Note that since the final time and final state are both fixed, dT and $dx(T)$ are both zero, so that (3.2-10) is satisfied.

Using (15) and (16), we must determine $\lambda(T)$; then we can find $\lambda(t)$ by using (11) and the optimal control by using (9). To find $\lambda(T)$, use (14) to write

$$x(T) = x(0)e^{-aT} - \frac{b^2}{2a}\lambda(T)(1 - e^{-2aT}). \tag{17}$$

Taking into account (15) and (16) shows that the final costate is

$$\lambda(T) = \frac{20a}{b^2(1 - e^{-2aT})}, \tag{18}$$

and so the optimal costate trajectory is

$$\lambda^*(t) = -\frac{10\,ae^{at}}{b^2 \sinh}aT. \tag{19}$$

Finally, the optimal rate of heat supply to the room is given by (9) or

$$u^*(t) = \frac{10\,ae^{at}}{b \sinh aT} \quad 0 \le t \le T. \tag{20}$$

To check our answer, apply $u^*(t)$ to the system (3). Solving for the state trajectory yields

$$x^*(t) = 10\frac{\sinh at}{\sinh aT}. \tag{21}$$

Indeed $x^*(T) = 10$ as desired.

b. Free Final State

Now suppose that we are not so concerned that the final state $x(T)$ be exactly $10°$. Let us demand only that the control $u(t)$ minimize

$$J(0) = \frac{1}{2}s(x(T) - 10)^2 + \frac{1}{2}\int_0^T u^2(t)\,dt \tag{22}$$

for some weighting s (i.e., some real number s) to be selected later. If s is large, then the optimal solution will have $x(T)$ near $10°$, since only then will the first term make a small contribution to the cost.

According to Table 3.2-1, the state and costate equations are still given by 10, and the optimal control by (9). Therefore, (11) and (14) are still valid.

The initial condition is still (15), but the final condition must be determined by using (3.2-10). The final time T is fixed, so $dT = 0$ and the second term of (3.2-10) is automatically equal to zero. Since $x(T)$ is not fixed, $dx(T)$ is not zero (as it was in part a). Therefore, it is required that

$$\lambda(T) = \left. \frac{\partial \phi}{\partial x} \right|_T = s(x(T) - 10). \tag{23}$$

(Note that there is no function ψ in this problem.) This is our new terminal condition, and from (15) and (23) we must determine $\lambda(T)$. To do this, note that

$$x(T) = \frac{\lambda(T)}{s} + 10, \tag{24}$$

and use this and (15) and (17). Solving for the final costate gives

$$\lambda(T) = \frac{-20\, as}{2a + b^2 s(1 - e^{-2aT})}. \tag{25}$$

Using (11) gives the optimal costate trajectory

$$\lambda^*(t) = \frac{-10\, ase^{at}}{ae^{aT} + sb^2 \sinh aT}. \tag{26}$$

Finally (9) yields the optimal control

$$u^*(t) = \frac{10\, abse^{at}}{ae^{aT} + sb^2 \sinh aT}. \tag{27}$$

To check our answer, we "simulate" the control by using $u^*(t)$ in the plant (3). Solving for the optimal state trajectory yields

$$x^*(t) = \frac{10\, sb^2 \sinh at}{ae^{aT} + sb^2 \sinh aT}. \tag{28}$$

At the final time,

$$x^*(T) = \frac{10\, sb^2 \sinh aT}{ae^{aT} + sb^2 \sinh aT}. \tag{29}$$

c. Discussion

The final value $x^*(T)$ in (29) is not equal to the desired $10°$. It is a function of the final-state weighting s in the performance index. As s becomes larger, we are making it more important relatively for $x(T)$ to equal $10°$ than for $u^2(t)$ to be small on $[0, T]$. In fact, in the limit $s \to \infty$, the costate (26), control (27), and state trajectory (28) tend to the expressions found in part a. In this limit, the final state $x^*(T)$ in (29) does indeed become exactly $10°$.

By examining (29), we can determine $x^*(T)$ for various values of s and select a value that gives a good compromise between driving $x(t)$ to the desired final value and conserving control energy. Using this value of s in (27) yields the optimal control that we would actually apply to heat the room.   ∎

Example 3.2-4.   The Intercept and Rendezvous Problems

a. Problem Formulation

In Example 2.3-2 we constructed a digital controller for the rendezvous problem. Let us now find an analytical expression for the continuous-time optimal control. In the next subsection we show an easy way to program the continuous optimal controller on a digital computer that does not require the analysis done here.

The geometry of the problem is shown in Fig. 3.2-1, where $y(t)$ and $v(t)$ are the vertical position and velocity of the pursuit aircraft A relative to the target aircraft A_t, which we can assume is at rest. Its initial horizontal distance from the pursuer is D. The horizontal velocity of the pursuit aircraft relative to A_t is V; so the final time T, at which the two aircraft will have the same horizontal distance, is fixed and known to be

$$T = t_0 + \frac{D}{V}. \tag{1}$$

The *line-of-sight angle* is $\sigma(t)$. See Bryson and Ho (1975).

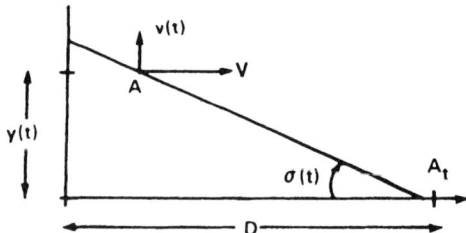

FIGURE 3.2-1   Intercept and rendezvous geometry.

In the *rendezvous* problem it is desired that final position $y(T)$ and velocity $v(T)$ *both* be zero. In the *intercept* problem we are not concerned with final velocity, and it is only desired that the final position $y(T)$ be zero.

The vertical dynamics are described by the state equations

$$\dot{y} = v, \tag{2}$$
$$\dot{v} = u, \tag{3}$$

where $u(t)$ is the vertical control input acceleration. Let the performance index be

$$J(t_0) = \frac{s_y y^2(T)}{2} + \frac{s_v v^2(T)}{2} + \frac{1}{2} \int_{t_0}^{T} u^2(t)\, dt. \tag{4}$$

For intercept, $s_v = 0$ and s_y is made large so the optimal control will make $y^2(T)$ small. For rendezvous, s_v and s_y are both selected large.

b. Problem Solution

The optimal control must now be selected to minimize (4). Each state component must have an associated scalar Lagrange multiplier; hence let $\lambda \triangleq [\lambda_y \lambda_v]^T$. Then the

Hamiltonian is

$$H = \tfrac{1}{2}u^2 + v\lambda_y + u\lambda_v. \tag{5}$$

The costate equations are therefore

$$\dot{\lambda}_y = -\frac{\partial H}{\partial y} = 0, \tag{6}$$

$$\dot{\lambda}_v = -\frac{\partial H}{\partial v} = -\lambda_y. \tag{7}$$

The stationarity condition is

$$0 = \frac{\partial H}{\partial u} = u + \lambda_v, \tag{8}$$

so the optimal control is the negative of the velocity multiplier

$$u(t) = -\lambda_v(t). \tag{9}$$

The initial conditions are

$$y(t_0), v(t_0) \text{ given.} \tag{10}$$

The final conditions are determined by (3.2-10). Since the final time is fixed, $dT = 0$ and so only the first term gives binding conditions. They are

$$\lambda_y(T) = \frac{\partial \phi}{\partial y}(T) = s_y y(T), \tag{11}$$

$$\lambda_v(T) = \frac{\partial \phi}{\partial v}(T) = s_v v(T). \tag{12}$$

We must now solve the two-point boundary-value problem defined by the state and costate equations, with u as in (9) and with boundary conditions (10)–(12). To do this, we proceed as we did in Example 3.2-3, assuming at the outset that $\lambda_y(T)$ and $\lambda_v(T)$ are known. The costate equation is first solved *backward* in time (i.e., in terms of $\lambda(T)$), and the state equation is then solved *forward* in time (i.e., in terms of $x(t_0)$).

Integrating both sides of (6) from t to T yields the constant costate component

$$\lambda_y(t) = \lambda_y(T) \overset{\Delta}{=} \lambda_y. \tag{13}$$

Integrating (7) then gives

$$\lambda_v(T) - \lambda_v(t) = -(T - t)\lambda_y$$

or

$$\lambda_v(t) = \lambda_v(T) + (T - t)\lambda_y. \tag{14}$$

Now, to simplify things, let us assume for a few moments that $t_0 = 0$. Substituting the control (9) into (3) gives

$$\dot{v} = -\lambda_v(t). \tag{15}$$

Using (14) and integrating both sides from 0 to t yields the quadratic expression

$$v(t) = v(0) - t(\lambda_v(T) + T\lambda_y) + \frac{t^2}{2}\lambda_y. \tag{16}$$

Taking this into account and integrating (2) yields the cubic expression

$$y(t) = y(0) + tv(0) - \frac{t^2}{2}(\lambda_v(T) + T\lambda_y) + \frac{t^3}{6}\lambda_y. \tag{17}$$

The state and costate equations have now been solved in terms of $\lambda(T)$ and the given $y(0)$, $v(0)$. Unfortunately, the final costate is unknown. To find it, we must use the relations (11) and (12) between the final state and costate. Thus, use these relations and (16), (17) to get

$$\lambda_y = s_y \left[y(0) + Tv(0) - \frac{T^2}{2}(\lambda_v(T) + T\lambda_y) + \frac{T^3}{6}\lambda_y \right] \tag{18}$$

and

$$\lambda_v(T) = s_v \left[v(0) - T(\lambda_v(T) + T\lambda_y) + \frac{T^2}{2}\lambda_y \right]. \tag{19}$$

These two equations can be written as

$$\begin{bmatrix} 1 + \dfrac{s_y T^3}{3} & \dfrac{s_y T^2}{2} \\ \dfrac{s_v T^2}{2} & 1 + s_v T \end{bmatrix} \begin{bmatrix} \lambda_y \\ \lambda_v(T) \end{bmatrix} = \begin{bmatrix} s_y & s_y T \\ 0 & s_v \end{bmatrix} \begin{bmatrix} y(0) \\ v(0) \end{bmatrix}. \tag{20}$$

Solving this yields the final costate

$$\begin{bmatrix} \lambda_y \\ \lambda_v(T) \end{bmatrix} = \frac{1}{\Delta(T)} \begin{bmatrix} \bar{s}_v + T & T\left(\bar{s}_v + \dfrac{T}{2}\right) \\ -T^2/2 & \bar{s}_y - \dfrac{T^3}{6} \end{bmatrix} \begin{bmatrix} y(0) \\ v(0) \end{bmatrix}, \tag{21}$$

where

$$\Delta(T) = (\bar{s}_y + T^3/3)(\bar{s}_v + T) - T^4/4 \tag{22}$$

and the reciprocal final weights are

$$\bar{s}_y \equiv \frac{1}{s_y}, \tag{23a}$$

$$\bar{s}_v \equiv \frac{1}{s_v}. \tag{23b}$$

The initial time is, in fact t_0 not 0. Since the state and costate equations are linear, all we need do to correct this is to substitute $(T - t_0)$ for T on the right-hand side of (21). Before we do this, however, consider the following. At the current time $t \leq T$ we know

$y(t)$ and $v(t)$, so we can take the current time t as the initial time. This corresponds to minimizing $J(t)$, the *remaining* cost on the interval $[t, T]$.

Substituting $(T - t)$ for T in (21) yields an expression for the final costate in terms of the *current* state:

$$\begin{bmatrix} \lambda_y \\ \lambda_v(T) \end{bmatrix} = \frac{1}{\Delta(T - t)} \begin{bmatrix} \bar{s}_v + (T - t) & (T - t)\left[\bar{s}_v + \dfrac{T - t}{2}\right] \\ -\dfrac{(T - t)^2}{2} & \bar{s}_y - \dfrac{(T - t)^3}{6} \end{bmatrix} \begin{bmatrix} y(t) \\ v(t) \end{bmatrix}. \quad (24)$$

We are finally in a position to compute the optimal control, for according to (9) and (14),

$$u(t) = -\begin{bmatrix} T - t & 1 \end{bmatrix} \begin{bmatrix} \lambda_y \\ \lambda_v(T) \end{bmatrix}. \quad (25)$$

Taking into account (24) therefore yields the optimal control

$$u(t) = -\frac{(T - t)\bar{s}_v + (T - t)^2/2}{\Delta(T - t)} y(t)$$
$$- \frac{\bar{s}_y + (T - t)^2 \bar{s}_v + (T - t)^3/3}{\Delta(T - t)} v(t). \quad (26)$$

This is a *feedback* control law since the current control is given in terms of the current state.

c. Proportional Navigation

For the intercept problem, we select $s_v = 0$ and $s_y \to \infty$. Taking the limit in (26) yields

$$u(t) = -\frac{3}{(T - t)^2} y(t) - \frac{3}{(T - t)} v(t) \quad (27)$$

as the optimal control for intercept. To make this look neater, note that for a small line-of-sight angle

$$\sigma(t) \simeq \tan \sigma(t) = \frac{y(t)}{(T - t)V}, \quad (28)$$

so that

$$\dot{\sigma} \simeq \frac{\dot{y}(t)}{(T - t)V} + \frac{y(t)}{(T - t)^2 V}. \quad (29)$$

Therefore, the optimal control is

$$u(t) = -3V\dot{\sigma}. \quad (30)$$

This is the *proportional navigation* control law. Every pilot knows that for an intercept it is necessary only to keep the angle to the target constant so that there is no relative bearing drift!

See Bryson and Ho (1975) for further discussion. ∎

Example 3.2-5.   Thrust Angle Programming

This example emphasizes that the optimal controller in Table 3.2-1 applies to general nonlinear systems.

a. The Bilinear Tangent Law

A particle of mass m is acted on by a constant thrust F applied at a variable angle of $\gamma(t)$. Its position is $(x(t), y(t))$, and its x and y velocities are $u(t)$ and $v(t)$. See Fig. 3.2-2. The nonlinear state equations $\dot{X} = f(X, \gamma, t)$ are

$$\dot{x} = u, \tag{1}$$
$$\dot{y} = v, \tag{2}$$
$$\dot{u} = a \cos \gamma, \tag{3}$$
$$\dot{v} = a \sin \gamma, \tag{4}$$

where the state is $X = [x\ y\ u\ v]^T$, $a \triangleq F/m$ is the known thrust acceleration, and thrust angle $\gamma(t)$ is the control input.

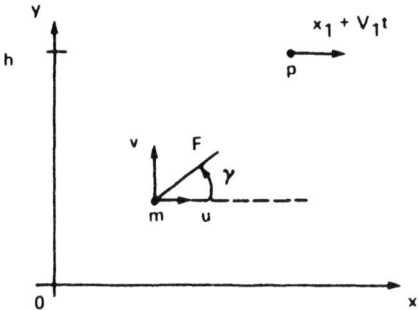

FIGURE 3.2-2   Thrust angle programming.

Let the performance index be a function only of the final time and state, so that

$$J = \phi(X(T), T) \tag{5}$$

(i.e., $L(X, \gamma, t) = 0$). Suppose that a given function ψ of the final state must be zeroed, so that

$$\psi(X(T), T) = 0. \tag{6}$$

The form of the control $\gamma(t)$ required to minimize J and satisfy (6) is easy to determine. The Hamiltonian is

$$H = L + \lambda^T f = \lambda_x u + \lambda_y v + \lambda_u a \cos \gamma + \lambda_v a \sin \gamma, \tag{7}$$

where the Lagrange multiplier $\lambda(t) = [\lambda_x\ \lambda_y\ \lambda_u\ \lambda_v]^T$ has a component associated with each state component.

According to Table 3.2-1, the costate equations are $\dot{\lambda} = -f_X^T \lambda$ or

$$\dot{\lambda}_x = -f_x^T \lambda = 0, \tag{8}$$

$$\dot{\lambda}_y = -f_y^T \lambda = 0, \tag{9}$$

$$\dot{\lambda}_u = -f_u^T \lambda = -\lambda_x, \tag{10}$$

$$\dot{\lambda}_v = -f_v^T \lambda = -\lambda_y. \tag{11}$$

(Note that subscripts on f indicate partial derivatives, while subscripts on λ indicate the costate components.) The stationarity condition is

$$0 = H\gamma = -\lambda_u a \sin\gamma + \lambda_v a \cos\gamma, \tag{12}$$

or

$$\tan\gamma(t) = \frac{\lambda_v(t)}{\lambda_u(t)}. \tag{13}$$

Integrating the costate equations backward from the final time T yields

$$\lambda_x(t) = \lambda_x(T) \overset{\Delta}{=} \lambda_x, \tag{14}$$

$$\lambda_y(t) = \lambda_y(T) \overset{\Delta}{=} \lambda_y, \tag{15}$$

$$\lambda_u(t) = (\lambda_u(T) + T\lambda_x) - t\lambda_x \overset{\Delta}{=} c_1 - t\lambda_x, \tag{16}$$

$$\lambda_v(t) = (\lambda_v(T) + T\lambda_y) - t\lambda_y \overset{\Delta}{=} c_2 - t\lambda_y. \tag{17}$$

Substituting into (13) yields the optimal control law

$$\tan\gamma(t) = \frac{t\lambda_y - c_2}{t\lambda_x - c_1}. \tag{18}$$

This is called the *bilinear tangent law* for the optimal thrust direction $\gamma(t)$.

To determine the constants λ_x, λ_y, c_1, and c_2, we may substitute for $\gamma(t)$ into the state equations using (18), solve them, and then use the boundary conditions. To determine the boundary conditions, we need to specify ϕ and ψ, which depend on the particular control objectives. There are many possible objectives we might have in mind for the behavior of the particle m. See Bryson and Ho (1975). One that leads to an interesting and fairly simple solution is discussed next.

b. Minimum-time Intercept

Suppose m represents an aircraft that wants to intercept a target P in minimum time. P has an initial position of x_1 and a constant velocity in the x direction of V_1, so that its x position at time t is $x_1 + V_1 t$. Its y position h is constant.

The minimum-time objective can be expressed by demanding that the optimal control minimize

$$J = T = \int_0^T 1\, dt. \tag{19}$$

Since $L = 1$, the Hamiltonian is now

$$H(t) = 1 + \lambda_x u + \lambda_y v + \lambda_u a \cos \gamma + \lambda_v a \sin \gamma; \tag{20}$$

however, since L is constant, the other results in part a remain valid.

If m starts out at $t_0 = 0$ at rest at the origin, the initial conditions are

$$x(0) = 0, \quad y(0) = 0, \quad u(0) = 0, \quad v(0) = 0. \tag{21}$$

The final-state function is

$$\psi(X(T), T) = \begin{bmatrix} x(T) - (x_1 + V_1 T) \\ y(T) - h \end{bmatrix} = 0, \tag{22}$$

so that

$$x(T) = x_1 + V_1 T, \tag{23}$$

$$y(T) = h. \tag{24}$$

To find the remaining terminal conditions, we need to use (3.2-10). Both the final state and the final time are free (i.e., different choices of $\gamma(t)$ will result in different values for T and the state components $u(T)$, $v(T)$). Therefore, $dx(T) \neq 0$ and $dT \neq 0$. In this problem, however, $dx(T)$ and dT are *independent* so that (3.2-10) yields the two separate conditions

$$(\phi_x + \psi_x^T v - \lambda)|_T = 0 \tag{25}$$

and

$$(\phi_t + \psi_t^T v + H)|_T = 0, \tag{26}$$

where $v = [v_x \, v_y]^T$ is a new constant Lagrange multiplier.

Taking into account (22) (note that $\phi(x(T), T) = 0$), (25) becomes

$$\lambda(T) = \begin{bmatrix} 1 & 0 \\ 0 & 1 \\ 0 & 0 \\ 0 & 0 \end{bmatrix} \begin{bmatrix} v_x \\ v_y \end{bmatrix}$$

or

$$\lambda_x(T) = v_x, \tag{27}$$

$$\lambda_y(T) = v_y, \tag{28}$$

$$\lambda_u(T) = 0, \tag{29}$$

$$\lambda_v(T) = 0. \tag{30}$$

Note that the components of $\lambda(T)$ corresponding to the *fixed* final state components $x(T)$ and $y(T)$ are unknown variables, and the components of $\lambda(T)$ corresponding to the *free* final state components $u(T)$ and $v(T)$ are fixed at zero.

Using (20) and (22), the final condition (26) becomes

$$H(T) = -\psi_t^T v|_T = -[-V_1 \quad 0] \begin{bmatrix} v_x \\ v_y \end{bmatrix}$$

or

$$1 + v_x u(T) + v_y v(T) = V_1 v_x. \tag{31}$$

We have used (27)–(30).

Now we need to solve the state equations (1)–(4), taking into account (18) and the costate solutions (14)–(17) and the boundary conditions (21), (23), (24), and (27)–(30). We also need condition (31) to allow us to solve for the unknown optimal final time T^*.

First, note that in light of (27)–(30), the costate solutions are

$$\lambda_x(t) = v_x, \tag{32}$$

$$\lambda_y(t) = v_y, \tag{33}$$

$$\lambda_u(t) = (T - t)v_x, \tag{34}$$

$$\lambda_v(t) = (T - t)v_y, \tag{35}$$

where the terminal multipliers v_x, v_y still need to be determined. The bilinear tangent law (18) therefore takes on the simple form

$$\tan \gamma = v_y/v_x. \tag{36}$$

For this minimum-time intercept problem, the optimal thrust angle is a *constant*!

To find the optimal control $\gamma^*(t)$, all that remains is to find v_x and v_y. We shall see that this still requires a little work.

Since γ is a constant, it is easy to integrate the state equations forward from $t_0 = 0$ to get

$$v(t) = at \sin \gamma, \tag{37}$$

$$u(t) = at \cos \gamma, \tag{38}$$

$$y(t) = \frac{at^2}{2} \sin \gamma, \tag{39}$$

$$x(t) = \frac{at^2}{2} \cos \gamma, \tag{40}$$

where we have used the initial conditions (21). Evaluating (39) and (40) at $t = T$ yields

$$\tan \gamma = \frac{y(T)}{x(T)}, \tag{41}$$

and final conditions (23) and (24) then give an expression for the control in terms of the final time:

$$\tan \gamma = \frac{h}{x_1 + V_1 T}. \tag{42}$$

We still need to determine the optimal final time T^* to use in (42). The role of equation (31) is to allow us to solve for T^*, but to use it we would first need to find v_x and v_y. In this particular problem, we can use a shortcut that does not require v_x, v_y.

Indeed, note that (39), (40), (23), and (24) imply that

$$\sin \gamma = \frac{2y(T)}{aT^2} = \frac{2h}{aT^2}, \tag{43}$$

$$\cos \gamma = \frac{2x(T)}{2T^2} = \frac{2(x_1 + V_1 T)}{aT^2}. \tag{44}$$

Hence, $\sin^2 \gamma + \cos^2 \gamma = 1$, or

$$4h^2 + 4(x_1 + V_1 T)^2 = a^2 T^4, \tag{45}$$

which is

$$-\frac{a^2 T^4}{4} + V_1^2 T^2 + 2V_1 x_1 T + (x_1^2 + h^2) = 0. \tag{46}$$

This is a quartic equation, which can be solved for T^* given the initial information x_1, V_1, h about the target. Only one solution to (46) will make physical sense. The optimal control is determined by simply solving (46) for T^* and then solving (42) for the optimal thrust angle γ^*.

It is not difficult to see what our solution means intuitively. See Fig. 3.2-3, where the hypotenuse can be expressed in terms of the motion of the target as

$$d^2 = h^2 + (x_1 + V_1 T^*)^2, \tag{47}$$

or in terms of the motion of the pursuit aircraft as

$$d^2 = \left(\tfrac{1}{2}a(T^*)^2\right)^2. \tag{48}$$

Equation (45) is just an expression of the requirement that the two aircraft be at the same point at the final time! Of course, if we had not gone through our rigorous derivation of (45), we could not be sure from Fig. 3.2-3 that its solution yields the *optimal* final time (courtesy of E. Verriest).

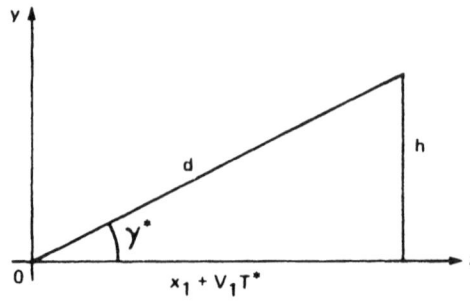

FIGURE 3.2-3   Interpretation of minimum-time intercept control law.

Solution of Two-point Boundary-value Problems

There are many computational methods for solving the optimal control problem. Our purpose is not to provide a survey of methods, for this would occupy more space than we have available. Instead, we present a few approaches that have immediate practical appeal.

Suppose it is desired to solve the optimal control problem for the nonlinear plant (3.2-1) with quadratic performance index

$$J(t_0) = \frac{1}{2}(x(T) - r(T))^T S(T)(x(T) - r(T)) + \frac{1}{2}\int_{t_0}^{T}(x^T Q x + u^T R u)\,dt,$$

(3.2-14)

where $S(T) \geq 0$, $Q \geq 0$, $R > 0$, and the desired final-state value $r(T)$ is given. Thus, we want to find the control $u(t)$ over the interval $[t_0, T]$ to minimize $J(t_0)$. The final state is constrained to satisfy (3.2-3) for some given function $\psi \in R^p$. For simplicity, let the final time T be fixed.

According to Table 3.2-1, we must solve the state equation (3.2-1) and the Euler equations

$$-\dot{\lambda} = \frac{\partial f^T}{\partial x}\lambda + Qx,$$

(3.2-15)

$$0 = Ru + \frac{\partial f^T}{\partial u}\lambda.$$

(3.2-16)

In general, the Jacobians $\partial f/\partial x$ and $\partial f/\partial u$ depend on the control $u(t)$, so that (3.2-16) is an implicit equation for $u(t)$. If $\partial f/\partial u$ is independent of $u(t)$, then we have

$$u = -R^{-1}\frac{\partial f^T}{\partial u}\lambda,$$

(3.2-17)

which we can use to eliminate $u(t)$ in the state equation and the costate equation (3.2-15), obtaining the *Hamiltonian system*

$$\dot{x} = f\left(x, -R^{-1}\frac{\partial f^T}{\partial u}\lambda, t\right),$$

(3.2-18a)

$$-\dot{\lambda} = \frac{\partial f^T}{\partial x}\lambda + Qx.$$

(3.2-18b)

The Hamiltonian system is a nonlinear ordinary differential equation in $x(t)$ and $\lambda(t)$ of order $2n$ with split boundary conditions, which are

n conditions: $x(t_0) = r(t_0)$ given,   (3.2-19)

p conditions: $\psi(x(T), T) = 0$,   (3.2-20)

$n - p$ conditions: $\lambda(T) = S(T)(x(T) - r(T)) + \left.\frac{\partial \psi^T}{\partial x}\right|_T v.$ (3.2-21)

The undetermined multipliers $v \in R^p$ allow some freedom, which means that (3.2-21) provides only the required number $n - p$ of final conditions.

There are several ways to solve the Hamiltonian system numerically. An excellent discussion is provided in Bryson and Ho (1975). One method is the following algorithm:

1. Guess the n unspecified initial conditions $\lambda(t_0)$.
2. Integrate the Hamiltonian system forward from t_0 to T.
3. Using the resulting values of $x(T)$ and $\lambda(T)$, evaluate

$$\psi(x(T), T) \tag{3.2-22}$$

and

$$\lambda(T) - S(T)(x(T) - r(T)) - \left. \frac{\partial \psi^{\mathrm{T}}}{\partial x} \right|_T v. \tag{3.2-23}$$

4. If there is no $v \in R^p$ that makes (3.2-22) and (3.2-23) equal to zero, determine changes in the final state and costate $\delta x(T)$ and $\delta \lambda(T)$ to bring these functions closer to zero.
5. Find the sensitivity matrix

$$\begin{bmatrix} \partial \mu(T) \\ \partial \lambda(t_0) \end{bmatrix}, \quad \text{where } \mu(T) = \begin{bmatrix} x(T) \\ \lambda(T) \end{bmatrix}$$

and

$$\delta \mu(T) = \begin{bmatrix} \delta x(T) \\ \delta \lambda(T) \end{bmatrix} = \frac{\partial \mu(T)}{\partial \lambda(t_0)} \delta \lambda(t_0).$$

(Several ways of doing this are given in Bryson and Ho (1975). See also our discussion of unit solutions.)
6. Calculate the change in $\lambda(t_0)$ required to produce the desired changes in the final values $x(T)$, $\lambda(T)$ by solving (3.2-24).
7. Repeat steps 2 through 6 until (3.2-22), (3.2-23) are close enough to zero for the application.

Another way to solve the two-point boundary-value problem (3.2-18)–(3.2-21) for linear systems is first to solve several initial-condition problems and then solve a system of simultaneous equations. This *unit solution method* proceeds as follows:

1. Integrate the Hamiltonian system using as initial conditions $\lambda(t_0) = 0$ and $x(t_0) = r(t_0)$, where $r(t_0)$ is the given initial state. Call the resulting solutions $x_0(t)$, $\lambda_0(t)$.

2. Suppose $\lambda \in R^n$, and let e_i represent the ith column of the $n \times n$ identity matrix. Determine n *unit solutions* by integrating the Hamiltonian system n times, using as initial conditions

$$x(t_0) = 0,$$
$$\lambda t_0 = e_i, \quad i = 1, \ldots, n. \tag{3.2-2}$$

Call the resulting unit solutions $x_i(t)$, $\lambda_i(t)$ for $i = 1, \ldots, n$.

3. General initial conditions can be expressed as

$$x(t_0) = r(t_0) \text{ given}, \tag{3.2-3}$$

$$\lambda(t_0) = \sum_{i=1}^{n} c_i e_i$$

for constants c_i. The overall solutions for these general initial conditions are

$$x(t) = x_0(t) + \sum_{i=1}^{n} c_i x_i(t),$$

$$\lambda(t) = \lambda_0(t) + \sum_{i=1}^{n} c_i \lambda_i(t).$$

Evaluate these solutions at the final time $t = T$, and then solve for the initial-costate values c_i required to ensure that the terminal conditions (3.2-20) and (3.2-21) are satisfied. The unit solutions show the effect on $x(T)$, $\lambda(T)$ of each of the n individual components of $\lambda(t_0)$, and so they can be used to find the sensitivity matrix in (3.2-24).

The next example illustrates this approach.

Example 3.2-6.   Unit Solution Method for Scalar System

The scalar plant

$$\dot{x} = ax + bu \tag{1}$$

has performance index of

$$J = \frac{s(x(T) - r(T))^2}{2} + \frac{1}{2} \int_0^T (qx^2 + ru^2)\, dt. \tag{2}$$

The desired final-state value $r(T)$ is given. The initial state $x(0) = r(0)$ is known, the final time is fixed, and the final state is free. To determine the optimal control $u(t)$ on $[0, T]$ by the method of unit solutions, we proceed as follows.

From Table 3.2-1, the Hamiltonian is

$$H = \frac{qx^2}{2} + \frac{ru^2}{2} + \lambda(ax + bu). \tag{3}$$

The Euler equations are

$$\dot{\lambda} = -\frac{\partial H}{\partial x} = -qx - a\lambda, \tag{4}$$

$$0 = \frac{\partial H}{\partial u} = ru + b\lambda. \tag{5}$$

Therefore, the optimal control is

$$u = -\frac{b}{r}\lambda. \tag{6}$$

Eliminating $u(t)$ in (1) yields the Hamiltonian system

$$\begin{bmatrix} \dot{x} \\ \dot{\lambda} \end{bmatrix} = \begin{bmatrix} a & -b^2/r \\ -q & -a \end{bmatrix} \begin{bmatrix} x \\ \lambda \end{bmatrix} \overset{\Delta}{=} A \begin{bmatrix} x \\ \lambda \end{bmatrix} \tag{7}$$

The split boundary conditions are

$$x(0) = r(0) \text{ given}, \tag{8}$$

$$\lambda(T) = s(x(T) - r(T)). \tag{9}$$

Instead of solving the split boundary-value problem (7)–(9), we solve two (i.e., $n+1$) initial-value problems, one with initial conditions $x(0) = r(0), \lambda(0) = 0$, and one with $x(0) = 0, \lambda(0) = 1$. Then we solve for the $\lambda(0)$ required to make (9) hold.

If $x(0) = r(0), \lambda(0) = 0$, then the solution can be found by Laplace transforms to be

$$\begin{bmatrix} x_0(t) \\ \lambda_0(t) \end{bmatrix} = \frac{r(0)}{2\alpha} \begin{bmatrix} \alpha - a \\ q \end{bmatrix} e^{-\alpha t} + \frac{r(0)}{2\alpha} \begin{bmatrix} \alpha + a \\ -q \end{bmatrix} e^{\alpha t}, \quad t \geq 0, \tag{10}$$

where $\alpha = \sqrt{a^2 + \frac{qb^2}{r}}$. If $x(0) = 0, \lambda(0) = 1$, the unit solution is

$$\begin{bmatrix} x_1(t) \\ \lambda_1(t) \end{bmatrix} = \frac{1}{2\alpha} \begin{bmatrix} \frac{b^2}{r} \\ \alpha + a \end{bmatrix} e^{-\alpha t} + \frac{1}{2\alpha} \begin{bmatrix} -\frac{b^2}{r} \\ \alpha - a \end{bmatrix} e^{\alpha t}, \quad t \geq 0. \tag{11}$$

Now consider the general initial condition $x(0) = r(0), \lambda(0) = c$ for some constant c. The solution with these initial conditions is

$$\begin{bmatrix} x(t) \\ \lambda(t) \end{bmatrix} = \begin{bmatrix} x_0(t) \\ \lambda_0(t) \end{bmatrix} + c \begin{bmatrix} x_1(t) \\ \lambda_1(t) \end{bmatrix}$$

$$= \frac{1}{2\alpha} \begin{bmatrix} r(0)(\alpha - a) + \frac{cb^2}{r} \\ r(0)q + c(\alpha + a) \end{bmatrix} e^{-\alpha t}$$

$$+ \frac{1}{2\alpha} \begin{bmatrix} r(0)(\alpha + a) - \frac{cb^2}{r} \\ -r(0)q + c(\alpha - a) \end{bmatrix} e^{-\alpha t}, \quad t \geq 0. \tag{12}$$

Now it remains only to determine the initial-costate value c so that boundary condition (9) holds. Evaluating (12) at $t = T$, and substituting $x(T)$ and $\lambda(T)$ into (9) yields the required initial-costate value of

$$\lambda(0) = c = \frac{r(0)[(q + sa)\sinh \alpha T + s\alpha \cosh \alpha T] - r(T)s\alpha}{(sb^2/r - a)\sinh \alpha T + \alpha \cosh \alpha T}. \tag{13}$$

Note that the initial costate is a linear combination of the initial and final states.

Using this value of c in (12) yields the optimal state and costate trajectories. Then (6) yields the optimal control. This method yields the optimal control as on open-loop control law, that is, as a function of time, not of the current state. ∎

3.3 CONTINUOUS-TIME LINEAR QUADRATIC REGULATOR

Table 3.2-1 provides the optimal controller for general nonlinear systems, but explicit expressions for the control law are hard to compute. In this section we consider the linear time-varying plant

$$\dot{x} = A(t)x + B(t)u, \tag{3.3-1}$$

where $x \in R^n, u \in R^m$ with associated quadratic performance index

$$J(t_0) = \frac{1}{2}x^{\mathrm{T}}(T)S(T)x(T) + \frac{1}{2}\int_{t_0}^{T}(x^{\mathrm{T}}Q(t)x + u^{\mathrm{T}}R(t)u)\,dt. \tag{3.3-2}$$

The time interval over which we are interested in the behavior of the plant is $[t_0, T]$. We shall determine the control $u^*(t)$ on $[t_0, T]$ that minimizes J for two cases: fixed final state and free final state. In the former case, u^* will turn out to be an open-loop control, and in the latter case a feedback control.

We assume in this section that the final time T is fixed and known, and that no function of the final state ψ is specified. The initial plant state $x(t_0)$ is given. Weighting matrices $S(T)$ and $Q(T)$ are symmetric and positive semi-definite, and $R(t)$ is symmetric and positive definite, for all $t \in [t_0, T]$.

Let us use Table 3.2-1 to write down the solution to this *linear quadratic regulator problem*.

The State and Costate Equations

The Hamiltonian is

$$H(t) = \tfrac{1}{2}(x^{\mathrm{T}}Qx + u^{\mathrm{T}}Ru) + \lambda^{\mathrm{T}}(Ax + Bu), \tag{3.3-3}$$

where $\lambda(t) \in R^n$ is an undetermined multiplier. The state and costate equations are

$$\dot{x} = \frac{\partial H}{\partial \lambda} = Ax + Bu, \tag{3.3-4}$$

$$-\dot{\lambda} = \frac{\partial H}{\partial x} = Qx + A^{\mathrm{T}}\lambda, \tag{3.3-5}$$

and the stationarity condition is

$$0 = \frac{\partial H}{\partial u} = Ru + B^{\mathrm{T}}\lambda. \tag{3.3-6}$$

Solving (3.3-6) yields the optimal control in terms of the costate

$$u(t) = -R^{-1}B^{\mathrm{T}}\lambda(t). \tag{3.3-7}$$

The control structure defined by these equations is identical to Fig. 2.2-1; however, the continuous LQ regulator cannot be implemented in this noncausal state–costate form.

Using (3.3-7) in the state equation yields the homogeneous *Hamiltonian system*.

$$\begin{bmatrix} \dot{x} \\ \dot{\lambda} \end{bmatrix} = \begin{bmatrix} A & -BR^{-1}B^{\mathrm{T}} \\ -Q & -A^{\mathrm{T}} \end{bmatrix} \begin{bmatrix} x \\ \lambda \end{bmatrix}. \tag{3.3-8}$$

The coefficient matrix is called the continuous *Hamiltonian matrix*, which we discuss further in Section 3.4.

To find the optimal control, we must take into account the boundary conditions and solve (3.3-8). We shall presently do this for two special cases: fixed and free final state. First, it is instructive to investigate the value of the performance index $J(t_0)$ when the control input $u(t)$ is zero.

Zero-input Cost and the Lyapunov Equation

We want to determine the value of the performance index J if the plant control input $u(t)$ is zero. Suppose the $n \times n$ matrix function $S(t)$ is defined as the solution to the continuous Lyapunov equation

$$-\dot{S} = A^{\mathrm{T}}S + SA + Q, \quad t \leq T, \tag{3.3-9}$$

with final condition $S(T)$ as given in (3.3-2). This equation is integrated backward in time from $t = T$. Then it is easy to show that the cost to go on any interval $[t, T]$ is given by

$$J(t) = \tfrac{1}{2}x^{\mathrm{T}}(t)S(t)x(t), \tag{3.3-10}$$

where $x(t)$ is the current state t.

To wit, note that

$$\frac{1}{2}\int_{t_0}^{T} \frac{d}{dx}(x^{\mathrm{T}}Sx)\,dt = \frac{1}{2}x^{\mathrm{T}}(T)S(T)x(T) - \frac{1}{2}x^{\mathrm{T}}(t_0)S(t_0)x(t_0). \tag{3.3-11}$$

Now add zero, in the form of the left-hand side of (3.3-11) minus its right-hand side, to $J(t_0)$ in (3.3-2) to see that $(u(t) = 0)$

$$J(t_0) = \frac{1}{2} x^{\mathrm{T}}(t_0) S(t_0) x(t_0) + \frac{1}{2} \int_{t_0}^{T} (\dot{x}^{\mathrm{T}} S x + x^{\mathrm{T}} \dot{S} x + x^{\mathrm{T}} S \dot{x} + x^{\mathrm{T}} Q x) \, dt.$$

(3.3-12)

Taking into account the state equation (3.3-1) results in

$$J(t_0) = \frac{1}{2} x^{\mathrm{T}}(t_0) S(t_0) x(t_0) + \frac{1}{2} \int_{t_0}^{T} (x^{\mathrm{T}} (A^{\mathrm{T}} S + \dot{S} + SA + Q) x \, dt; \qquad (3.3\text{-}13)$$

but $S(t)$ satisfies the Lyapunov equation, so that (3.3-10) follows since the current time t can be interpreted as the initial time of the remaining interval $[t, T]$.

This result allows us to compute, in terms of the known current state, the cost to go till time T of failing to apply any control to the plant. Note that $S(t)$ does not depend on the state, so it can be precomputed off-line and stored. Because of the form of (3.3-10), we call $S(t)$ the cost *kernel* function. The cost is just one-half the semi-norm squared of the state with respect to the weighting $S(t)$.

The solution to (3.3-9) is given by

$$S(t) = e^{A^{\mathrm{T}}(T-t)} S(T) e^{A(T-t)} + \int_{t}^{T} e^{A^{\mathrm{T}}(T-t)} Q e^{A(T-t)} \, d\tau, \qquad (3.3\text{-}14)$$

which can be verified by Leibniz's rule. According to the Lyapunov stability theory, this converges to the *steady-state* value as $(T - t) \to \infty$ of

$$S_{\infty} = \int_{0}^{\infty} e^{A^{\mathrm{T}} \tau} Q e^{A \tau} \, d\tau \qquad (3.3\text{-}15)$$

if the plant is asymptotically stable. In this event, the cost over any interval $[t, \infty]$, is given by the *steady-state* cost (3.3-10) with $S(t)$ replaced by S_{∞}, which is finite. If A is unstable and $(A, \sqrt{Q})$ is observable, where $Q = \sqrt{Q^{\mathrm{T}}} \sqrt{Q}$, then the cost tends to infinity as the time interval grows.

In the steady-state case, $\dot{S} = 0$, so that (3.3-9) becomes the *algebraic Lyapunov equation*

$$0 = A^{\mathrm{T}} S + SA + Q. \qquad (3.3\text{-}16)$$

If A is stable, then (3.3-15) is a positive semi-definite solution to (3.3-16). If $(A, \sqrt{Q})$ is observable, the steady-state solution S_{∞} is positive definite and is the unique positive definite solution to the algebraic Lyapunov equation.

See the discussion on the discrete time counterparts to these results in Section 2.2.

We shall soon see that the optimal closed-loop control depends on an equation like (3.3-9), but with an extra term to account for the effect of the input.

Example 3.3-1.   Propagation of Cost for Uncontrolled Scalar System

Let

$$\dot{x} = ax \tag{1}$$

be an uncontrolled scalar system with cost on $[t, T]$ defined by

$$J(t) = \frac{1}{2}S(T)x^2(T) + \frac{1}{2}\int_t^T qx^2(\tau)\,d\tau. \tag{2}$$

The Lyapunov equation is

$$-\dot{s} = 2as + q, \tag{3}$$

with solution

$$s(t) = e^{2a(T-t)}S(T) + \int_t^T e^{2a(T-t)}q\,d\tau \tag{4}$$

or

$$s(t) = \left(s(T) + \frac{q}{2a}\right)e^{2a(T-t)} - \frac{q}{2a}. \tag{5}$$

If $a < 0$, then as $(T - t) \to \infty$, $s(t)$ converges to the steady-state value of

$$s_\infty = -\frac{q}{2a} > 0. \tag{6}$$

Note that this is the solution to the algebraic Lyapunov equation

$$0 = 2as + q. \tag{7}$$

If a is unstable, then $s(t)$ grows without bound as the time interval of interest grows.

The steady-state cost on $[0, \infty]$ of applying no control input to the plant (1) is

$$J_\infty = \frac{1}{2}s_\infty x^2(0) = -\frac{q}{4a}x^2(0) \tag{8}$$

when $a < 0$; otherwise it is infinite. In neither case does it depend on the final-state weighting $s(T)$.   ■

Fixed-final-state and Open-loop Control

Let us now return to the problem of determining the control required in (3.3-1) to minimize the cost (3.3-2). The state and costate equations are given by (3.3-8) and the optimal control is given by (3.3-7). It remains only to solve (3.3-8) given the boundary conditions.

Suppose that the initial state is known to be $x(t_0)$ and that the control objective is to drive the state *exactly* to the given fixed reference value of $r(T)$ at the final time. Then the final condition is

$$x(T) = r(T). \tag{3.3-17}$$

Since $dx(T) = 0$ and $dT = 0$, condition (3.2-10) is automatically satisfied.

Since $x(T)$ is fixed at $r(T)$, it is redundant to include a final-state weighting in the cost index, so let $S(T) = 0$. To allow us to get an analytic solution, let $Q = 0$ also. Then the cost function is

$$J(t_0) = \frac{1}{2} \int_{t_0}^{T} u^{\mathrm{T}} R u \, dt, \qquad (3.3\text{-}18)$$

and so we are trying to find a control that drives $x(t_0)$ to $x(T) = r(T)$ using minimum control energy.

The state and costate equations are now

$$\dot{x} = Ax - BR^{-1}B^{\mathrm{T}}\lambda, \qquad (3.3\text{-}19)$$

$$\dot{\lambda} = -A^{\mathrm{T}}\lambda. \qquad (3.3\text{-}20)$$

Setting $Q = 0$ has decoupled the costate equation from the state equation, so its solution is just

$$\lambda(t) = e^{A^{\mathrm{T}}(T-t)}\lambda(T), \qquad (3.3\text{-}21)$$

where $\lambda(T)$ is still unknown. Using this expression in the state equation yields

$$\dot{x} = Ax - BR^{-1}B^{\mathrm{T}}e^{A^{\mathrm{T}}(T-t)}\lambda(T), \qquad (3.3\text{-}22)$$

whose solution is

$$x(t) = e^{A(t-t_0)}x(t_0) - \int_{t_0}^{t} e^{A(t-\tau)}BR^{-1}B^{\mathrm{T}}e^{A^{\mathrm{T}}(T-\tau)}\lambda(T) \, d\tau. \qquad (3.3\text{-}23)$$

To find $\lambda(T)$, evaluate this at $t = T$ to get

$$x(T) = e^{A(T-t_0)}x(t_0) - G(t_0, T)\lambda(T), \qquad (3.3\text{-}24)$$

where the weighted *continuous reachability gramian* is

$$G(t_0, T) = \int_{t_0}^{T} e^{A(T-\tau)}BR^{-1}B^{\mathrm{T}}e^{A^{\mathrm{T}}(T-\tau)} \, d\tau. \qquad (3.3\text{-}25)$$

According to final condition (3.3-17), then,

$$\lambda(T) = -G^{-1}(t_0, T)\left[r(T) - e^{A(T-t_0)}x(t_0)\right]. \qquad (3.3\text{-}26)$$

Finally, the optimal control can be written using (3.3-7), (3.3-21), and (3.3-26) as

$$u^*(t) = R^{-1}B^{\mathrm{T}}e^{A^{\mathrm{T}}(T-t)}G^{-1}(t_0, T)[r(T) - e^{A(T-t_0)}x(t_0)]. \qquad (3.3\text{-}27)$$

This is our result; it is the minimum-energy control that drives the given initial state $x(t_0)$ to the desired final reference value of $x(T) = r(T)$. Note that

$$x(T) = e^{A(T-t_0)}x(t_0) \qquad (3.3\text{-}28)$$

is the final state in the absence of an input, so the optimal control is proportional to the difference between this homogeneous solution and the desired final state.

Since $u^*(t)$ is found by using $G(t_0, T)$, the optimal control exists for arbitrary $x(t_0)$ and $r(T)$ if and only if $|G(t_0, T)| \neq 0$. This corresponds to *reachability* of the plant. If (A, B) is reachable, there exists a minimum-energy control that drives any $x(t_0)$ to any desired $r(T)$.

The control (3.3-27) is an *open-loop control,* since $u^*(t)$ does not depend on the current state $x(t)$. It depends only on the initial and final states, and it is precomputed and then applied for all t in $[t_0, T]$. If, for some reason, the state is perturbed off the predicted optimal trajectory, then such an open-loop control will not, in general, result in $x(T) = r(T)$ as desired.

To compute the reachability gramian in practice, we do not need to do the integration (3.3-25), which can be very messy. The solution to the Lyapunov equation is

$$\dot{P} = AP + PA^{\mathrm{T}} + BR^{-1}B^{\mathrm{T}}, \quad t > t_0, \qquad (3.3\text{-}29)$$

$$P(t) = e^{A(t-t_0)}P(t_0)e^{A^{\mathrm{T}}(t-t_0)} + \int_{t_0}^{t} e^{A(t-\tau)}BR^{-1}B^{\mathrm{T}}e^{A^{\mathrm{T}}(t-\tau)}\,d\tau.$$

Hence, if $P(t_0) = 0$, then $G(t_0, t) = P(t)$.

To determine $u^*(t)$, then, we would first solve (3.3-29) off-line to get $G(t_0, T)$. This can be done numerically using a Runge-Kutta integrator (Appendix B.1). Then for each $t \in [t_0, T]$, we would use (3.3-27) to find $u^*(t)$, which is then applied to the plant (3.3-1).

Compare the "reachability Lyapunov equation" (3.3-29) to the "observability Lyapunov equation" (3.3-9). The former describes the interaction between plant and input, and the latter describes the interaction between plant and cost function when $u(t) = 0$.

It is a simple matter to determine the value of the cost index (3.3-18) under the influence of the optimal control (3.3-27). Representing the *final-state difference* as

$$d(t_0, T) = r(T) - e^{A(T-t_0)}x(t_0), \qquad (3.3\text{-}30)$$

we have

$$J^*(t_0) = \frac{1}{2}\int_{t_0}^{T} d^{\mathrm{T}}G^{-1}e^{A(T-t)}BR^{-1}RR^{-1}B^{\mathrm{T}}e^{A^{\mathrm{T}}(T-t)}G^{-1}d\,dt,$$

where we have used the symmetry of $G^{-1}(t_0, T)$ and R^{-1}. Realizing that $d(t_0, T)$ and $G^{-1}(t_0, T)$ do not depend on t and using the definition of the gramian yields

$$J^*(t_0) = \tfrac{1}{2}d^{\mathrm{T}}(t_0, T)G^{-1}(t_0, T)d(t_0, T), \qquad (3.3\text{-}31)$$

or

$$J^*(t_0) = \tfrac{1}{2}d^T(t_0, T)P^{-1}(T)d(t_0, T),$$ (3.3-32)

where $P(t)$ satisfies (3.3-29). Compare this with (3.3-10).

Example 3.3-2.   Open-loop Control of a Scalar System

Let the scalar plant be

$$\dot{x} = ax + bu, \qquad t \geq 0,$$ (1)

with cost

$$J(0) = \frac{1}{2}\int_0^T ru^2 \, dt.$$ (2)

The Lyapunov equation (3.3-29) is

$$\dot{p} = 2ap + \frac{b^2}{r}, \quad p(0) = 0,$$ (3)

so the reachability gramian on $[0, t]$ is

$$G(0, t) = p(t) = \frac{b^2}{r}\int_0^t e^{2a(t-\tau)} \, d\tau$$

or

$$G(0, t) = \frac{b^2}{2ar}(e^{2at} - 1).$$ (4)

Compare this to the solution of the Lyapunov equation in Example 3.3-1.

According to (3.3-27), the optimal control taking $x(0)$ to $x(T) = r(T)$ for a given $r(T)$ is

$$u(t) = \frac{b}{r}e^{a(T-t)} \cdot \frac{2ar}{b^2(e^{2aT} - 1)}(r(T) - e^{aT}x(0))$$

$$= \frac{a}{b}\frac{e^{-at}}{\sinh aT}(r(T) - e^{aT}x(0)).$$ (5)

Interestingly enough, this is independent of the control weighting r. Compare (5) to the control (20) in Example 3.2-3, which was found by direct solution of the state and costate equations.   ∎

Example 3.3-3.   Open-loop Control of Motion Obeying Newton's Laws

A particle obeying Newton's laws satisfies

$$\dot{x} = \begin{bmatrix} 0 & 1 \\ 0 & 0 \end{bmatrix}x + \begin{bmatrix} 0 \\ 1 \end{bmatrix}u$$ (1)

where $x = [d\ v]^{\mathrm{T}}$ with $d(t)$ the position, $v(t)$ the velocity, and $u(t)$ an acceleration input. It is easy to find an analytic expression for the control required to drive any given $x(0)$ to any desired $x(T)$, while minimizing

$$J(0) = \frac{1}{2} \int_0^T ru^2\, dt. \tag{2}$$

To find the reachability gramian, we solve the Lyapunov equation (3.3-29). Let

$$P(t) = \begin{bmatrix} p_1(t) & p_2(t) \\ p_2(t) & p_3(t) \end{bmatrix} \tag{3}$$

Then (3.3-29) is

$$\dot{P} = \begin{bmatrix} 0 & 1 \\ 0 & 0 \end{bmatrix} P + P \begin{bmatrix} 0 & 0 \\ 1 & 0 \end{bmatrix} + \begin{bmatrix} 0 & 0 \\ 0 & 1/r \end{bmatrix}, \tag{4}$$

which yields the scalar equations

$$\dot{p}_1 = 2p_2, \tag{5}$$

$$\dot{p}_2 = p_3, \tag{6}$$

$$\dot{p}_3 = 1/r. \tag{7}$$

For the gramian, we integrate (7), (6), and then (5) with $P(0) = 0$ to get

$$p_3 = \frac{t}{r}, \tag{8}$$

$$p_2 = \frac{t^2}{2r}, \tag{9}$$

$$p_1 = \frac{t^3}{3r}, \tag{10}$$

so that

$$G(0, t) = P(t) = \begin{bmatrix} \dfrac{t^3}{3r} & \dfrac{t^2}{2r} \\ \dfrac{t^2}{2r} & \dfrac{t}{r} \end{bmatrix}. \tag{11}$$

The state transition matrix is

$$e^{At} = \begin{bmatrix} 1 & t \\ 0 & 1 \end{bmatrix}. \tag{12}$$

To find the optimal control, we use (3.3-27), which becomes

$$u(t) = \frac{1}{r}[T - t \quad 1] \begin{bmatrix} \dfrac{12r}{T^3} & -\dfrac{6r}{T^2} \\ -\dfrac{6r}{T^2} & \dfrac{4r}{T} \end{bmatrix} \left(x(T) - \begin{bmatrix} 1 & T \\ 0 & 1 \end{bmatrix} x(0) \right), \tag{13}$$

or

$$u(t) = \begin{bmatrix} \dfrac{6T - 12t}{T^3} & \dfrac{-2T + 6t}{T^2} \end{bmatrix} \left(x(T) - \begin{bmatrix} 1 & T \\ 0 & 1 \end{bmatrix} x(0) \right). \tag{14}$$

Once again, since $u(t)$ is a scalar, it is independent of r. Note also that the control magnitude decreases as the control interval $[0, T]$ increases. More control is required to move the system more quickly from one state to another. ∎

Free-final-state and Closed-loop Control

We can find an optimal control law in the form of a state feedback by changing our control objectives for the plant (3.3-1). Instead of fixing the final state at a desired final value, let us require only that the control minimize the performance index (3.3-2). Thus, the final state is *free*, and its value can be varied in the optimization process implicit in the solution presented in Table 3.2-1.

The state and costate equations (3.3-8) are reproduced here for convenience:

$$\dot{x} = Ax - BR^{-1}B^{\mathrm{T}}\lambda, \tag{3.3-33}$$

$$-\dot{\lambda} = Qx + A^{\mathrm{T}}\lambda. \tag{3.3-34}$$

The control input is

$$u(t) = -R^{-1}B^{\mathrm{T}}\lambda. \tag{3.3-35}$$

The given initial state is $x(t_0)$, and the final state $x(T)$ is free. Thus, $dx(T) \neq 0$, and $dT = 0$ (the final time is fixed and known here) in (3.2-10), so the coefficient of $dx(T)$ must be zero:

$$\lambda(T) = \left.\frac{\partial \phi}{\partial x}\right|_T = S(T)x(T). \tag{3.3-36}$$

This is the terminal condition.

To solve the two-point boundary-value problem specified by (3.3-33) and (3.3-34), given $x(t_0)$ and (3.3-36), we shall use the *sweep method* (Bryson and Ho 1975). Thus, assume that $x(t)$ and $\lambda(t)$ satisfy a linear relation like (3.3-36) for all $t \in [t_0, T]$ for some as yet unknown matrix function $S(t)$:

$$\lambda(t) = S(t)x(t). \tag{3.3-37}$$

If we can find such a $S(t)$, then this assumption is valid.

To find the intermediate function $S(t)$, differentiate the costate to get

+ play in 3.3- 33

$$\dot{\lambda} = \dot{S}x + S\dot{x} = \dot{S}x + S(Ax - BR^{-1}B^{\mathrm{T}}Sx), \tag{3.3-38}$$

where we have used the state equation. Now, taking into account the costate equation, we must have

$$-\dot{S}x = (A^T S + SA - SBR^{-1}B^T S + Q)x \qquad (3.3\text{-}39)$$

for all t. Since this holds for all state trajectories given any $x(t_0)$, it is necessary that

$$-\dot{S} = A^T S + SA - SBR^{-1}B^T S + Q, \quad t \le T. \qquad (3.3\text{-}40)$$

This is a matrix *Riccati equation*, and if $S(t)$ is its solution with final condition $S(T)$, then (3.3-37) holds for all $t \le T$. Our assumption was evidently a good one.

In terms of the Riccati-equation solution, the optimal control is given by (3.3-35) and (3.3-37) as

$$u(t) = -R^{-1}B^T Sx(t). \qquad (3.3\text{-}41)$$

Defining the *Kalman gain* as

$$K(t) = R^{-1}B^T S(t), \qquad (3.3\text{-}42)$$

we have

$$u(t) = -K(t)x(t). \qquad (3.3\text{-}43)$$

The optimal control is determined by solving the Riccati equation (3.3-40) *backward* in time for $S(t)$. This can be done offline *before* the control run since $x(t)$ is not required to find $S(t)$. The gain $K(t)$ can be computed and stored. Finally, during the control run, $u^*(t)$ is found using (3.3-43) and applied to the plant.

The continuous optimal LQ regulator is summarized in Table 3.3-1. A block diagram of this scheme has the same structure shown in Figure 2.2-3.

In terms of the Kalman gain, the Riccati equation can be written

$$-\dot{S} = A^T S + SA - K^T RK + Q. \qquad (3.3\text{-}44)$$

The control (3.3-43) is a *time-varying state feedback*, since even if A, B, Q, and R are time invariant, $K(t)$ varies with time. The closed-loop plant is

$$\dot{x} = (A - BK)x, \qquad (3.3\text{-}45)$$

and this equation can be used to find the optimal state trajectory $x^*(t)$ given any $x(t_0)$.

In terms of the closed-loop plant matrix, the Riccati equation can be written in the *Joseph-stabilized formulation*

$$-\dot{S} = (A - BK)^T S + S(A - BK) + K^T RK + Q, \quad t \le T. \qquad (3.3\text{-}46)$$

TABLE 3.3-1   Continuous Linear Quadratic Regulatory (Final State Free)

System model:

$$\dot{x} = Ax + Bu, \quad t \geq t_0$$

Performance index:

$$J(t_0) = \frac{1}{2}x^{\mathrm{T}}(T)S(T)x(T) + \frac{1}{2}\int_{t_0}^{T}(x^{\mathrm{T}}Qx + u^{\mathrm{T}}Ru)\,dt$$

Assumptions:

$$S(T) \geq 0, Q \geq 0, R > 0, \quad \text{with all three symmetric}$$

Optimal feedback control:

$$-\dot{S} = A^{\mathrm{T}}S + SA - SBR^{-1}B^{\mathrm{T}}S + Q, \quad t \leq T, \text{ given } S(T)$$

$$K = R^{-1}B^{\mathrm{T}}S$$

$$u = -Kx$$

$$J^*(t_0) = \tfrac{1}{2}x^{\mathrm{T}}(t_0)S(t_0)x(t_0)$$

By a derivation like the ones leading to (3.3-10) and (2.2-69) we can show that the coast on any interval $[t, T]$ satisfies

$$J(t) = \frac{1}{2}x^{\mathrm{T}}(t)S(t)x(t) + \frac{1}{2}\int_{t}^{T}\left\|R^{-1}B^{\mathrm{T}}Sx + u\right\|_{R}^{2} dt, \tag{3.3-47}$$

where $S(t)$ is the solution to the Riccati equation.

If we now select the optimal control (3.3-43), then the value of the performance index on $[t, T]$ is just

$$J(t) = \tfrac{1}{2}x^{\mathrm{T}}(t)S(t)x(t). \tag{3.3-48}$$

This result is important since, if we know the current state $x(t)$, then by solving the Riccati equation we can determine the optimal cost of controlling the plan on $[t, T]$ *before* we apply the control or even compute it! If this cost is too high, we should select another control scheme, or at least change the weighting matrices $S(T)$, Q, and R and find a new feedback gain $K(t)$.

If $B = 0$, then the Riccati equation reduces to the zero-input Lyapunov equation (3.3-9).

Note from (3.3-47) that

$$\frac{\partial^2 J}{\partial u^2} = R, \tag{3.3-49}$$

so that the curvature matrix in the continuous case is R. Since $R > 0$, the optimal control *minimizes* $J(t_0)$. In the discrete case, the curvature matrix is

$(B^T S_{k+1} B + R)$, where S_k satisfies the discrete Riccati equation. This accounts for the simplicity of the continuous LQ regulator as compared to the discrete LQ regulator.

By selecting $S(T)$ very large, we can guarantee that the optimal control will drive $x(T)$ very close to zero to keep $J(t_0)$ small. In the limit as $S(T) \to \infty$, it can be shown that the control scheme in Table 3.3-1 tends to the fixed-final-state scheme (3.3-27) for the case $r(T) = 0$. See the discussion in Section 2.2.

It should be clearly realized that reachability of the plant is *not* required for the free-final-state LQ regulator. Even if (A, B) is not reachable, $u^*(t)$ will do its best to keep $J(t_0)$ small. If (A, B) is reachable, it can be expected to do a better job. In fact, we shall see in Section 3.4 that reachability results in some very desirable properties as the control interval $[t_0, T]$ becomes large.

A few examples will impart some intuition on the LQ regulator. First, let us briefly discuss a software implementation of Table 3.3-1.

Software Implementation of the LQ Regulator

In the discrete case, the Riccati equation is a simple backward recursion that can easily be programmed, as we have seen. In the continuous case, however, the Riccati equation must be *integrated* backward. Most Runge-Kutta integration routines run forward in time. The best policy is therefore to convert (3.3-40) into an equation that is integrated forward. This is easy to do.

Changing variables by

$$\tau = T - t, \tag{3.3-50}$$

we have $d\tau = -dt$, so the Riccati equation becomes (in the time-invariant case)

$$\dot{S}_b = A^T S_b + S_b A - S_b B R^{-1} B^T S_b + Q, \tag{3.3-51}$$

where

$$S(t) = S_b(T - t). \tag{3.3-52}$$

All we must do to solve (3.3-40) is to integrate it *forward* from $t = 0$ *without the minus sign* on its left-hand side, then reverse the resulting solution and shift it to $t = T$.

The control scheme in Table 3.3-1 has two parts. The first is the *control law computation* by backward integration of the Riccati equation to find $S(t)$ and then $K(t)$. Only $K(t)$ must be stored. This integration we have just discussed.

The second part is the *simulation of the control law* found in part one by applying $u = -K(t)x$ to the plant. This is accomplished by a forward integration of the state equation $\dot{x} = Ax + Bu$.

The complete simulation procedure is shown in Fig. 3.3-1. The simulation portion can be compared with Fig. 2.3-1.

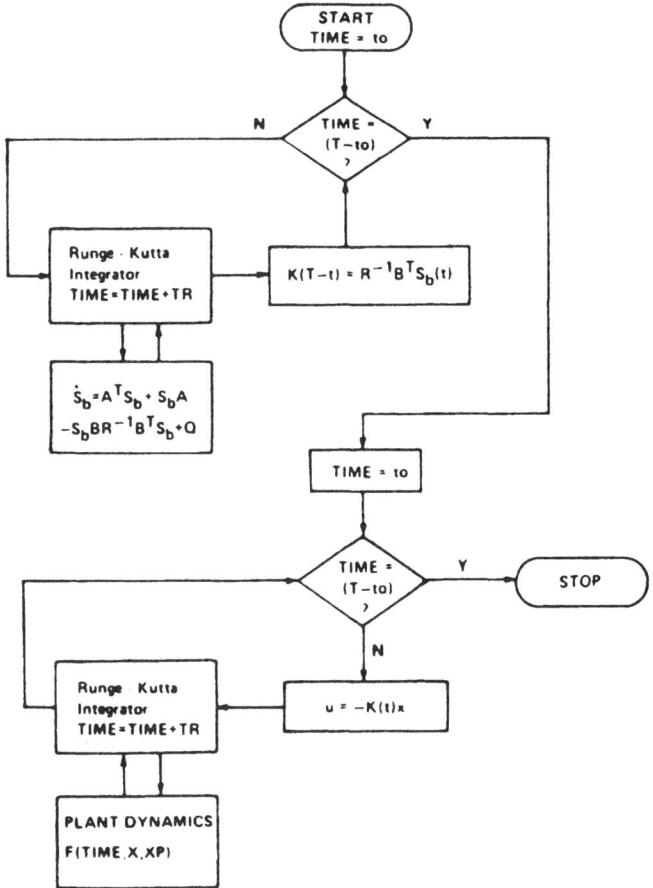

FIGURE 3.3-1   Continuous LQ regulator simulation procedure.

Examples and Exercises

Example 3.3-4.   Optimal Feedback Control of a Scalar System

This is the continuous counterpart of Example 2.2-3. Let the scalar plant be

$$\dot{x} = ax + bu \tag{1}$$

with performance index

$$J(t_0) = \frac{1}{2}s(T)x^2(T) + \frac{1}{2}\int_{t_0}^{T}(qx^2 + ru^2)\,dt. \tag{2}$$

a. Analytic Solution

The Riccati equation is

$$-\dot{s} = 2as + q - \frac{b^2 s^2}{r}, \quad t \leq T. \tag{3}$$

Using separation of variables we have

$$\int_{s(t)}^{s(T)} \frac{ds}{(b^2/r)s^2 - 2as - q} = \int_t^T dt,$$

and integrating yields

$$s(t) = s_2 + \frac{s_1 + s_2}{[(s(T) + s_1)/(s(T) - s_2)]e^{2\beta(T-t)} - 1} \tag{4}$$

where

$$\beta = \sqrt{a^2 + \frac{b^2 q}{r}}, \tag{5}$$

$$s_1 = \frac{r}{b^2}(\beta - a), \quad s_2 = \frac{r}{b^2}(\beta + a). \tag{6}$$

The steady-state value as $(T - t) \to \infty$ is given by s_2 or, if $a > 0$,

$$s_\infty = \frac{q}{\gamma}\left(1 + \sqrt{1 + \frac{\gamma}{a}}\right) \tag{7}$$

where

$$\gamma = b^2 q/ar \tag{8}$$

is a *control-effectiveness-to-plant-inertia* ratio. If $a < 0$, a similar expression holds. The steady-state value is independent of the final-state weighting $s(T)$. It is also bounded if $b \neq 0$, which corresponds to reachability, even if a is unstable. Contrast this with Example 3.3-1.

b. No Intermediate-state Weighting

Let us consider the rather interesting special case of $q = 0$. Then $\beta = |a|$ and

$$s(t) = \frac{s(T)}{b^2 s(T)/2ar + (1 - b^2 s(T)/2ar)e^{-2a(T-t)}}. \tag{9}$$

If we want to ensure that the optimal control drives $x(T)$ exactly to zero, we can let $s(T) \to \infty$ to weight $x(T)$ more heavily in $J(t_0)$. In this limit we have

$$s(t) = \frac{2ar/b^2}{1 - e^{-2a(T-t)}}, \tag{10}$$

so the optimal control is $(K(t) = bs(t)/r)$

$$u(t) = -K(t)x(t) = -\frac{2a/b}{1 - e^{-2a(T-t)}}x(t)$$

or

$$u(t) = -\frac{a}{b}\frac{e^{a(T-t)}}{\sinh a(T - t)}x(t). \tag{11}$$

Compare this with the fixed-final-state control in Example 3.3-2 for the case $r(T) = 0$. We have just discovered a *feedback* formulation of that control law.

If the plant isstable, $a < 0$, then the steady-state value $((T - t) \to \infty)$ of the cost kernel is

$$s_\infty = 0, \tag{12}$$

so that the steady-state closed-loop system

$$\dot{x} = (a - bK)x = ax \tag{13}$$

is stable. On the other hand, if $a > 0$, then

$$s_\infty = \frac{2ar}{b^2}, \tag{14}$$

and the steady-state closed-loop system is

$$\dot{x} = \left(a - \frac{b^2}{r}s_\infty\right)x = -ax. \tag{15}$$

This is still stable.

Figure 3.3-2 shows the behavior of the general solution (4) for the case of stable and unstable plant.

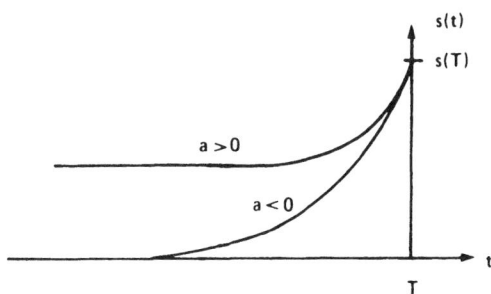

FIGURE 3.3-2   Limiting behavior of the Riccati-equation solution.

c. Simulation

To implement the LQ regulator, none of the analysis subsequent to (3) is required. The Kalman gain is

$$K(t) = \frac{b}{r}s(t), \tag{16}$$

and the complete LQ regulator is shown in Fig. 3.3-3. First, the Riccati equation (3) is integrated backward to get $s(t)$. This is accomplished by integrating

$$\dot{s}_b = 2as_b + q - b^2 s_b^2/r \tag{17}$$

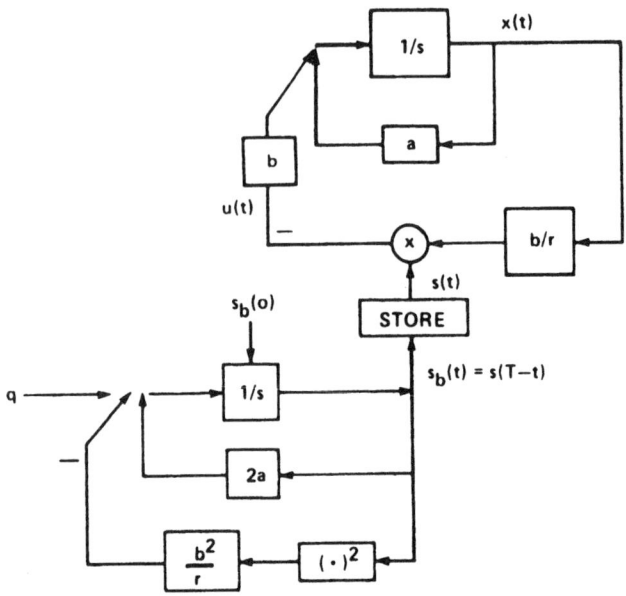

FIGURE 3.3-3   Scalar continuous LQ regulator.

forward from zero with $s_b(0) = s(T)$ and then using

$$s(t) = s_b(T - t). \tag{18}$$

The kernel $s(t)$ is stored and then used to compute $u(t)$ that takes the form

$$u = -K(t)x \tag{19}$$

as the plant dynamics are integrated in the simulation. Note that the Riccati-equation system is a sort of "doubled" or "squared" version of the plant. ∎

Example 3.3-5.   Optimal Feedback Control of a Damped Harmonic Oscillator

Let the plant be

$$\dot{x} = \begin{bmatrix} 0 & 1 \\ -\omega_n^2 & -2\delta\omega_n \end{bmatrix} x + \begin{bmatrix} 0 \\ b \end{bmatrix} u \tag{1}$$

with performance index

$$J(t_0) = \frac{1}{2} x^T(T) \begin{bmatrix} s_1(T) & 0 \\ 0 & s_3(T) \end{bmatrix} x(T)$$

$$+ \frac{1}{2} \int_{t_0}^{T} \left(x^T \begin{bmatrix} q_1 & 0 \\ 0 & q_2 \end{bmatrix} x + ru^2 \right) dt. \tag{2}$$

If we let

$$S(t) \overset{\Delta}{=} \begin{bmatrix} s_1(t) & s_2(t) \\ s_2(t) & s_3(t) \end{bmatrix} \tag{3}$$

and simplify the Riccati equation, we get the coupled scalar differential equations

$$-\dot{s}_1 = -2\omega_n^2 s_2 - \frac{b^2}{r} s_2^2 + q_1, \tag{4}$$

$$-\dot{s}_2 = s_1 - 2\delta\omega_n s_2 - \omega_n^2 s_3 - \frac{b^2}{r} s_2 s_3, \tag{5}$$

$$-\dot{s}_3 = 2s_2 - 4\delta\omega_n s_3 - \frac{b^2}{r} s_3^2 + q_2. \tag{6}$$

Writing the optimal feedback gain as

$$K(t) = [k_1(t) \quad k_2(t)], \tag{7}$$

we have $K = R^{-1}B^T S$ or

$$k_1 = \frac{b s_2}{r}, \tag{8}$$

$$k_2 = \frac{b s_3}{r}. \tag{9}$$

```
function [x, u, Sf, tf] = ex3_3_5(a, b, r, x0)
% Control of a Harmonic Oscillator
% Compute the solution to Riccati Equation
[tb, S]=ode45(@fex3_3_5,[-10:0.1:0],zeros(3,1));
% Compute Optimal Feedback Gains
Sf=flipud(S);
tf=-flipud(tb);
K=-b/r*Sf(:,2:3);
x(:,1)=x0;
u(1)=K(1,:)*x(:,1) ;
% compute Closed-loop Response
for k=1:length(tf)-1,
% Harmonic Oscillator System State Equations
x(:,k+1) =expm((a+[0; b]*K(k,:))*(tf(k+1)-tf(k)))*x(:,k);
u(k+1)=K(k+1,:)*x(:,k+1);
end

function sd=fex3_3_5(t,s)
q=1*eye(2); om=0.8; del=0.1; b=1; r=1;
sd =[-2*om^2*s(2)-b^2*s(2) ^2+q(1, 1);
      s(1)-4*del*om*s(2)-om^2*s(3)-b^2*s(2)*s(3);
      2*s(2)-4*del*om*s(3)-b^2*s(3) ^2+q(2,2)];
```

FIGURE 3.3-4 MATLAB code to use for the control of a harmonic oscillator.

The optimal control is then

$$u = -Kx = -k_1 x_1 - k_2 x_2, \tag{10}$$

where $x = [x_1 \quad x_2]^T$.

a. Software Implementation

To implement the optimal LQ controller, we need subroutines to describe the Riccati-equation dynamics (4)–(6) (without the minus signs on the left-hand side of the equalities) and to compute the gains (8)–(9). These are used in the control law computation. For the forward integration to simulate the control law, we need subroutines to compute the control (10) and provide the plant dynamics (1). The MATLAB code is shown in Fig. 3.3-4.

Using this software, the optimal state trajectories and controls for several values of $q = q_1 = q_2$ were plotted (see Fig. 3.3-5). Also shown are the cost kernel elements for two values of q.

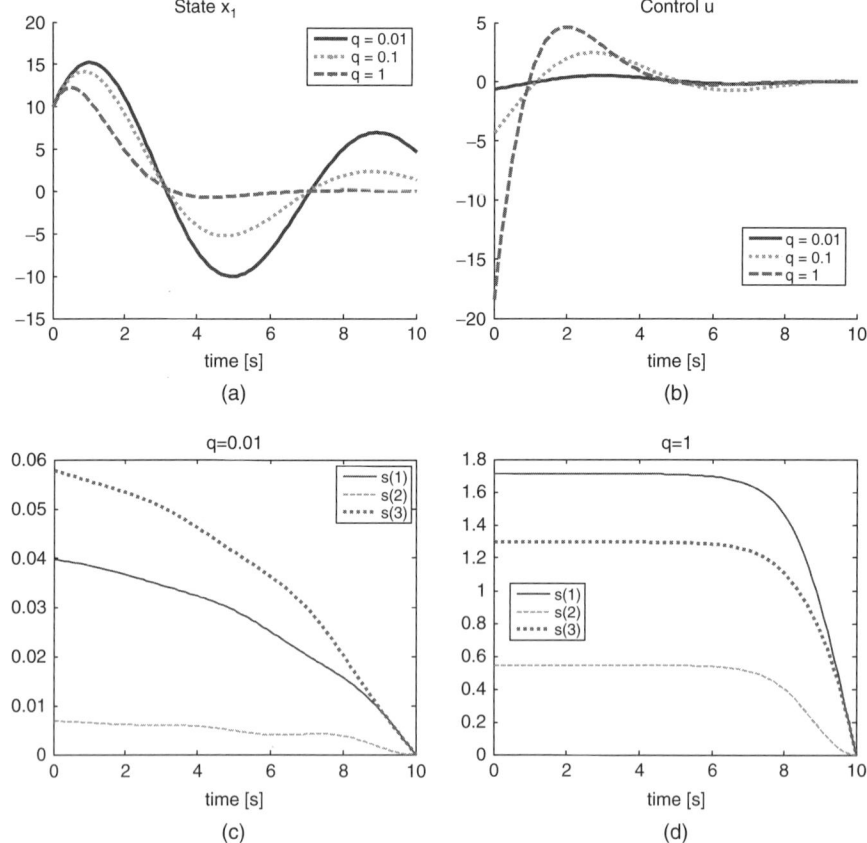

FIGURE 3.3-5 Results of a harmonic oscillator simulation using MATLAB. (a) State trajectories. (b) Optimal control inputs $u(t)$. (c) Riccati solutions $s(t)$ for $q = 0.01$. (d) Riccatti solutions for $q = 1.0$.

b. Steady-state Riccati Solution

Equations (4)–(6) are difficult to solve, but a steady-state solution is easy to obtain. Letting $\dot{s}_1 = \dot{s}_2 = \dot{s} = 0$, three algebraic equations are obtained. These can be solved to give

$$s_2(\infty) = \frac{q_1}{\gamma_1}\left(\sqrt{1 + \frac{\gamma_1}{\omega_n^2}} - 1\right), \tag{11}$$

$$s_3(\infty) = \frac{q_2}{\gamma_2}\left(\sqrt{1 + \frac{\gamma_2}{2\delta\omega_n}\left(1 + \frac{2s_2}{q_2}\right)} - 1\right), \tag{12}$$

$$s_1(\infty) = \frac{b^2}{r}\left(\frac{q_2}{\gamma_2}s_2 + \frac{q_1}{\gamma_1}s_3 + s_2s_3\right), \tag{13}$$

where

$$\gamma_1 \triangleq \frac{b^2q_1}{\omega_n^2 r}, \tag{14}$$

$$\gamma_2 \triangleq \frac{b^2q_2}{2\delta\omega_n r} \tag{15}$$

are "control effectiveness" ratios. In solving for the steady-state kernel, we select positive square roots, since $S(\infty) \geq 0$. ∎

Exercise 3.3-6.   *LQ Regulator with Weighting of State/Input Inner Product*

This is the continuous-time counterpart to Exercise 2.2-4. Let the plant

$$\dot{x} = Ax + Bu \tag{1}$$

have the modified performance index

$$J(t_0) = \frac{1}{2}x^\mathrm{T}(T)S(T)x(T) + \frac{1}{2}\int_{t_0}^{T}[x^\mathrm{T} \quad u^\mathrm{T}]\begin{bmatrix} Q & V \\ V^\mathrm{T} & R \end{bmatrix}\begin{bmatrix} x \\ u \end{bmatrix}, \tag{2}$$

where the block matrix is positive semidefinite and $R > 0$.

a. Define a modified Kalman gain as

$$K = R^{-1}(V^\mathrm{T} + B^\mathrm{T}S), \tag{3}$$

where $S(t)$ is the solution to the Riccati equation

$$-\dot{S} = A^\mathrm{T}S + SA - K^\mathrm{T}RK + Q. \tag{4}$$

(Compare (4) with (3.3-44).) Show that the optimal control is

$$u(t) = -K(t)x(t). \tag{5}$$

b. Show that the optimal remaining cost on any subinterval $[t, T]$ is

$$J(t) = \tfrac{1}{2}x^{\mathrm{T}}(t)S(t)x(t). \tag{6}$$

In summary, if the cost index contains a weighting V that picks up the state-input inner product, the only required modification to the LQ regulator is that the Kalman gain must be modified, and the Riccati-equation formulation (4) should be used.   ■

3.4   STEADY-STATE CLOSED-LOOP CONTROL AND SUBOPTIMAL FEEDBACK

In this section we present the continuous counterparts to the results in Section 2.4. It would be a very good exercise to fill in the derivations of these equations, both to become more familiar with them and to compare the continuous and discrete situations, which have subtle and interesting distinctions.

Suboptimal Feedback Gains

If the plant

$$\dot{x} = Ax + Bu \tag{3.4-1}$$

has the feedback

$$u = -Kx, \tag{3.4-2}$$

then the closed-loop plant becomes

$$\dot{x} = (A - BK)x. \tag{3.4-3}$$

The optimal feedback gain $K(t)$ is time varying and depends on the Riccati-equation solution as in Table 3.3-1.

If the gain K in (3.4-2) is arbitrary, then the resulting cost on $[t, T]$ for any t is

$$J(t) = \tfrac{1}{2}x^{\mathrm{T}}(t)S(t)x(t), \tag{3.4-4}$$

where $S(t)$ satisfies

$$-\dot{S} = (A - BK)^{\mathrm{T}}S + S(A - BK) + K^{\mathrm{T}}RK + Q, \quad t \le T, \tag{3.4-5}$$

with $S(T)$ equal to the final-state weighting.

If $K(t)$ is given, then (3.4-5) is a *Lyapunov* equation in terms of the closed-loop plant matrix. If $K(t)$ is the optimal gain in Table 3.3-1, then (3.4-5) is the Joseph stabilized Riccati equation.

If K in (3.4-2) is selected as a constant matrix, then the cost given by (3.4-4) can be examined to see if it is reasonable. If it is, and if the plant behavior is satisfactory in a simulation run, then the constant feedback can be used on the actual plant.

The Algebraic Riccati Equation

In this subsection we assume time-invariant plant and weighting matrices.

As $(T - t)$ goes to infinity, the solution to the Riccati equation can exhibit several types of behavior. It can be unbounded, or it can converge to a limiting solution $S(\infty)$, which can be zero, positive semi-definite, or positive definite.

If $S(t)$ does converge, then for $t \ll T$, $\dot{S} = 0$, so that there results in the limit the *algebraic Riccati equation* (ARE)

$$0 = A^T S + SA - SBR^{-1}B^T S + Q. \tag{3.4-6}$$

The ARE can have several solutions, and these may be real or complex, positive definite, negative definite, etc. If $S(T)$ is symmetric, then the Riccati solution $S(t)$ is symmetric and at least positive semi-definite for all $t \leq T$. $S(\infty)$ is always a solution to the ARE, but all ARE solutions are not limiting Riccati solutions for some $S(T)$.

It is worth mentioning that a real solution to

$$0 = A^T S^T + SA - SBR^{-1}B^T S^T + Q, \tag{3.4-7}$$

where S is not required to be symmetric, is given by

$$S = \sqrt{A^T(BR^{-1}B^T)^{-1}A + Q}\, L \sqrt{(BR^{-1}B^T)^{-1}} + A^T(BR^{-1}B^T)^{-1}, \tag{3.4-8}$$

where L is any orthogonal matrix (i.e., $LL^T = I$). Of course, this assumes $|BR^{-1}B^T| \neq 0$. See Schultz and Melsa (1967). In the scalar case, this reduces to the well-known formula for solving a quadratic equation! (Show that for S to be symmetric, the matrix L must satisfy a Lyapunov equation.)

If $S(\infty)$ exists, then the corresponding steady-state feedback gain is

$$K(\infty) = R^{-1}B^T S(\infty). \tag{3.4-9}$$

Under some circumstances it may be acceptable to use the time-invariant feedback law (3.4-2) with a gain of $K(\infty)$ as an alternative to the time-varying optimal feedback. The suboptimal cost associated with this control law is given by (3.4-4), where K in (3.4-5) is $K(\infty)$.

To examine the consequences of this simplified control strategy, let us discuss the limiting behavior of the closed-loop system (3.4-3) using the optimal feedback in Table 3.3-1.

Limiting Behavior of the Riccati-equation Solution

This subsection applies only for time-invariant plant and cost matrices. It is worthwhile reviewing the discrete case discussed in Section 2.4, as we shall not repeat the comments designed to impart some insight and motivation for our work.

The plant is (3.4-1) with cost index in Table 3.3-1. By defining a *fictitious output*

$$y(t) = \begin{bmatrix} C \\ 0 \end{bmatrix} x(t) + \begin{bmatrix} 0 \\ D \end{bmatrix} u(t), \tag{3.4-10}$$

where C and D are defined as any matrices such that $Q = C^T C$ and $R = D^T D$, we can write the cost as

$$J(t_0) = \frac{1}{2} x^T(T) S(T) x(T) + \frac{1}{2} \int_{t_0}^{t} y^T y \, dt. \tag{3.4-11}$$

We should first like to know when there is a finite limiting solution $S(\infty)$ to the Riccati equation. The next theorem gives us the answer.

Theorem 3.4-1. Let (A, B) be stabilizable. Then for every $S(T)$ there is a bounded limiting solution $S(\infty)$ to the Riccati equation. Furthermore, $S(\infty)$ is a positive semi-definite solution to the ARE.   ■

The proof of this theorem is similar to that of Theorem 2.4-1. Note that, in general, $S(\infty)$ is different for different $S(T)$. Remember that the free-final-state LQ regulator does not require any controllability assumptions on the plant. However, such assumptions guarantee desirable properties as the control interval $[t_0, T]$ becomes large.

If we intend to use the simplified suboptimal control law (3.4-2) with a gain of $K(\infty)$, we should certainly like for the resulting closed-loop system to be stable! The next theorem tells when we can be sure of this. It is a strengthened version of the previous result and depends on the observability of the plant by the fictitious output.

Theorem 3.4-2. Let C be any matrix so that $Q = C^T C$.
Suppose (A, C) is observable. Then (A, B) is stabilize if and only if

1. There is a unique positive definite limiting solution $S(\infty)$ to the Riccati equation. Furthermore, $S(\infty)$ is the unique positive definite solution to the ARE (3.4-6).
2. The closed-loop plant (3.4-3) is asymptotically stable, where $K = K(\infty)$ is given by (3.4-9).   ■

The comments following Theorem 2.4-2 are relevant here. The observability condition in the theorem is not really needed. If (A, C) is detectable, the result still holds, but then $S(\infty)$ can be guaranteed only to be positive semidefinite.

What these theorems tell us is that if the plant is stabilizable and if we *select* Q so that $(A, \sqrt{Q})$ is observable, then the suboptimal feedback grain (3.4-9) results in a stable closed-loop plant. Note that (3.4-9) is *the optimal* control law for the *infinite horizon* performance index

$$J_\infty = \frac{1}{2}\int_0^\infty (x^T Q x + u^T R u)\, dt. \tag{3.4-12}$$

Thus, as the control interval $[t_0, T]$ gets larger, it makes more and more sense to use a constant feedback with gain of $K(\infty)$.

A useful side result of these theorems is that we have a way of stabilizing any multivariable plant. Let Q and R be *any* positive definite matrices with the correct dimensions. Then $u = -K_\infty x$, where $K_\infty = R^{-1} B^T S$, with S the positive definite solution to (3.4-6), will result in a stable closed-loop plant. Different Q and R will result in different closed-loop poles for $(A - BK(\infty))$, but these poles will always be in the open left half-plane. Later we show examples of how the closed-loop poles move as Q and R vary.

Example 3.4-1.   *Steady-state Control of a System Obeying Newton's Law*

Let the plant

$$\dot{x} = \begin{bmatrix} 0 & 1 \\ 0 & 0 \end{bmatrix} x + \begin{bmatrix} 0 \\ 1 \end{bmatrix} u \tag{1}$$

have the infinite-horizon cost

$$J(0) = \frac{1}{2}\int_0^\infty \left(x^T \begin{bmatrix} q_d & 0 \\ 0 & q_v \end{bmatrix} x + ru^2 \right) dt. \tag{2}$$

We could do a computer simulation using MATLAB, and such a simulation would have results similar to those of Example 2.4-3. However, let us instead capitalize on the simple form of the continuous LQ regulator to get some analytic solutions.

Using A, B, Q, r in the ARE (3.4-6) yields the three coupled scalar algebraic equations:

$$0 = -\frac{s_2^2}{r} + q_d, \tag{3}$$

$$0 = s_1 - \frac{s_2 s_3}{r}, \tag{4}$$

$$0 = 2s_2 - \frac{s_3^2}{r} + q_v, \tag{5}$$

where

$$S = \begin{bmatrix} s_1 & s_2 \\ s_2 & s_3 \end{bmatrix}. \tag{6}$$

These are easily solved to yield

$$s_2 = \sqrt{q_d r}, \tag{7}$$

$$s_3 = \sqrt{q_v r + 2r\sqrt{q_d r}}, \tag{8}$$

$$s_1 = \sqrt{q_d q_v + 2 q_d \sqrt{q_d r}},\tag{9}$$

where we have selected the positive definite solution $S(\infty)$. (Show this.)

The optimal feedback gain is

$$K(\infty) = R^{-1} B^{\mathrm{T}} S(\infty) = \left[\sqrt{\frac{q_d}{r}} \quad \sqrt{\frac{q_v}{r} + 2\sqrt{\frac{q_d}{r}}} \right].\tag{10}$$

Since this depends only on the ratios q_d/r and q_v/r, let us now assume that $r = 1$. The closed-loop plant is

$$a^{\mathrm{cl}} = (A - BK(\infty)) = \begin{bmatrix} 0 & 1 \\ -\sqrt{q_d} & -\sqrt{q_v + 2\sqrt{q_d}} \end{bmatrix},\tag{11}$$

whence the optimal closed-loop characteristic equation is

$$s^2 + \sqrt{q_v + 2\sqrt{q_d}}\,s + \sqrt{q_d} = 0.\tag{12}$$

Comparing this to $s^2 + 2\delta\omega_n s + \omega_n^2$, we conclude that the optimal closed-loop poles are a complex pair with a natural frequency and damping ratio of

$$\omega_n = (q_d)^{1/4},\tag{13}$$

$$\delta = \frac{1}{\sqrt{2}}\sqrt{1 + \frac{q_v}{2\sqrt{q_d}}}.\tag{14}$$

In particular, if no velocity weighting is used ($q_v = 0$), the damping ratio is the familiar $1/\sqrt{2}$! Note that the natural frequency depends only on the position weighting, and the damping ratio only on the ratio of the velocity to the square root of the position weighting.

Knowing the relations (13) and (14), we can now pick the weights q_d and q_v that result in desirable closed-loop behavior. These values can even be used in a finite-horizon (i.e., finite final time T) performance index to design optimal time-varying feedbacks with a prescribed steady-state behavior.

It is worth remarking that if $q_d = 0$ so that $(A, \sqrt{Q})$ is not detectable, then one of the closed-loop poles is at $s = 0$ and the closed-loop plant is not stable.   ■

An Analytic Solution to the Riccati Equation

The LQ regulator Hamiltonian system is

$$\begin{bmatrix} \dot{x} \\ \dot{\lambda} \end{bmatrix} = H \begin{bmatrix} x \\ \lambda \end{bmatrix},\tag{3.4-13}$$

where the Hamiltonian matrix is

$$H = \begin{bmatrix} A & -BR^{-1}B^{\mathrm{T}} \\ -Q & -A^{\mathrm{T}} \end{bmatrix}. \tag{3.4-14}$$

By assuming

$$\lambda(t) = S(t)x(t), \tag{3.4-15}$$

we were able to derive the formulation in Table 3.3-1, wherein $S(t)$ satisfies the Riccati equation. Instead of solving the Riccati equation, $S(t)$ can be found analytically in terms of the eigenvalues and eigenvectors of H.

To find an analytic expression for $S(t)$, it is first necessary to show that if μ is an eigenvalue of H, then so is $-\mu$. Define

$$J = \begin{bmatrix} 0 & I \\ -I & 0 \end{bmatrix}. \tag{3.4-16}$$

Then by direct multiplication we see that

$$H = JH^{\mathrm{T}}J. \tag{3.4-17}$$

Therefore, if μ is an eigenvalue of H with eigenvector v,

$$Hv = \mu v,$$

so that

$$JH^{\mathrm{T}}Jv = \mu v,$$
$$H^{\mathrm{T}}Jv = -\mu Jv$$

(note $J^{-1} = -J$). Hence,

$$(Jv)^{\mathrm{T}}H = -\mu(Jv)^{\mathrm{T}}, \tag{3.4-18}$$

and (Jv) is a left eigenvector of H with eigenvalue $-\mu$.

Now we merely repeat the steps leading up to (2.4-41). The results are as follows.

Order the eigenvalues of H in a matrix

$$D = \begin{bmatrix} -M & 0 \\ 0 & M \end{bmatrix}, \tag{3.4-19}$$

where M is a diagonal matrix containing the right-half-plane eigenvalues. Let the modal matrix of eigenvectors, arranged in order to correspond to D, be

$$W \triangleq \begin{bmatrix} W_{11} & W_{12} \\ W_{21} & W_{22} \end{bmatrix}.$$ (3.4-20)

Thus,

$$\begin{bmatrix} W_{11} \\ W_{21} \end{bmatrix}$$

are the n eigenvectors of the stable eigenvalues of H.

If $S(T)$ is the Riccati-equation boundary condition, define

$$V(T) = -(W_{22} - S(T)W_{12})^{-1}(W_{21} - S(T)W_{11})$$ (3.4-21)

and

$$V(t) = e^{-M(T-t)}V(T)e^{-M(T-t)}.$$ (3.4-22)

Then an analytic solution to the Riccati equation is given by

$$S(t) = (W_{21} + W_{22}V(t))(W_{11} + W_{12}V(t))^{-1}.$$ (3.4-23)

In the limiting case $(T - t) \to \infty$, a bounded positive definite solution $S(\infty)$ exists if (A, B) is stabilizable and $(A, \sqrt{Q})$ is observable. In this limit, $V(t) \to 0$ since $-M$ is stable, so

$$S(\infty) = W_{21}W_{11}^{-1}.$$ (3.4-24)

The ARE solution is thus constructed by using the stable eigenvectors of the Hamiltonian matrix.

Design of Steady-state Regulators by Eigenstructure Assignment

We have just discovered a way to determine the optimal steady-state cost kernel $S(\infty)$ in terms of the eigenstructure of the Hamiltonian matrix H. By pursuing this line of thought a little further, we can find a way to determine the optimal steady-state feedback gain $K(\infty)$ directly from the eigenstructure of H.

We assume that (A, B) is reachable and $(A, \sqrt{Q})$ is detectable. The optimal steady-state closed-loop plant is

$$\dot{x} = (A - BK(\infty))x.$$ (3.4-25)

This and the Hamiltonian system (3.4-13) are both ways of characterizing the optimal state trajectories. We can demonstrate that if μ_i is a stable eigenvalue of H with eigenvector $[X_i^T \Lambda_i^T]^T$, where $X_i \in R^n$, then μ_i is also an eigenvalue of

$(A - BK(\infty))$ with eigenvector X_i. The argument is a straightforward modification of the discrete-time argument in Chapter 2.

H can be written down by inspection. Therefore, in the single-input case we can use this result by finding the eigenvalues of H and then realizing that the stable eigenvalues are the poles of the optimal closed-loop plant. Given these desired poles, a technique such as Ackermann's formula can be used to find the optimal feedback gain $K(\infty)$.

In the multi-input case, the optimal feedback is not uniquely specified by the closed-loop poles, so it is necessary to find the eigenvectors of H as well. A derivation virtually identical to the one for the discrete case in Chapter 2 leads to the following result.

Let the eigenvectors of the stable eigenvalues of H be placed into the $2n \times n$ matrix $[X^T \Lambda^T]^T$, where $X \in R^n$. (We called this

$$\begin{bmatrix} W_{11} \\ W_{21} \end{bmatrix}$$

in (3.4-20).) Then the optimal steady-state feedback is given by

$$K(\infty) = R^{-1} B^T \Lambda X^{-1}. \tag{3.4-26}$$

Compare this result to the corresponding result in Chapter 2, which, interestingly enough, includes in addition a matrix of stable eigenvalues M. (Why?)

Example 3.4-2. Eigenstructure Design of Steady-state Regulator for Harmonic Oscillator

Let the plant

$$\dot{x} = \begin{bmatrix} 0 & 1 \\ -\omega_n^2 & 0 \end{bmatrix} x + \begin{bmatrix} 0 \\ 1 \end{bmatrix} u \tag{1}$$

have a cost index of

$$J(0) = \frac{1}{2} \int_0^\infty \left(x^T \begin{bmatrix} q_d & 0 \\ 0 & q_v \end{bmatrix} x + ru^2 \right) dt. \tag{2}$$

a. Optimal Closed-loop Poles

The Hamiltonian matrix is

$$H = \begin{bmatrix} 0 & 1 & 0 & 0 \\ -\omega_n^2 & 0 & 0 & -1/r \\ -q_d & 0 & 0 & \omega_n^2 \\ 0 & -q_v & -1 & 0 \end{bmatrix}, \tag{3}$$

whence we can compute

$$|sI - H| = s^4 + \left(2\omega_n^2 - \frac{q_v}{r}\right)s^2 + \left(\omega_n^4 + \frac{q_d}{r}\right). \tag{4}$$

Note the even form of the characteristic polynomial of H. This means that if s is a root, then so is $-s$. Since only the ratios q_v/r and q_d/r appear, we can assume that $r = 1$. Letting $\bar{s} \overset{\Delta}{=} s^2$, (4) becomes

$$\bar{s}^2 + (2\omega_n^2 - q_v)\bar{s} + (\omega_n^4 + q_d), \tag{5}$$

which has a pair of complex roots $\bar{s}_1, \bar{s}_2$ with natural frequency of

$$\bar{\omega}_n = (\omega_n^4 + q_d)^{1/2} \tag{6}$$

and damping ratio of

$$\bar{\delta} = \frac{\omega_n^2 - q_v/2}{(\omega_n^4 + q_d)^{1/2}}. \tag{7}$$

The roots of (4) are given by $\pm\sqrt{\bar{s}_1}$ and $\pm\sqrt{\bar{s}_2}$. If the roots of (5) are represented as

$$\bar{s}_1 = \bar{\omega}_n e^{j\bar{\theta}_1}, \tag{8a}$$

$$\bar{s}_2 = \bar{\omega}_n e^{j\bar{\theta}_2}, \tag{8b}$$

where $\bar{\theta}_2 = -\bar{\theta}_1$, then the roots of (4) are

$$(\bar{\omega}_n)^{1/2} e^{\pm j\bar{\theta}_1/2}, \tag{9a}$$

$$(\bar{\omega}_n)^{1/2} e^{\pm j\bar{\theta}_2/2}, \tag{9b}$$

If $\bar{\delta} = -\cos\bar{\theta}_1$ is the damping ratio of a pole pair at angles of $\pm\bar{\theta}_1$ (i.e., $\bar{\theta}_1$ and $\bar{\theta}_2$), then

$$\pm\delta = \pm\frac{\sqrt{1-\delta}}{\sqrt{2}} \tag{10}$$

are the damping ratios of the two pole pairs at angles of $\pm\bar{\theta}_1/2$ and $\pm\bar{\theta}_2/2$. These four poles are symmetric about the imaginary axis, and the pole pair corresponding to $+\delta$ is stable, whereas the pole pair corresponding to $-\delta$ is unstable. Equation (10) follows from the trigonometric relationship

$$\cos\frac{\alpha}{2} = \frac{\sqrt{1+\cos\alpha}}{\sqrt{2}}. \tag{11}$$

In our case the stable poles of H thus have a damping ratio of

$$\delta^{cl} = \frac{1}{\sqrt{2}}\sqrt{1 - \frac{\omega_n^2 - q_v/2}{(\omega_n^4 + q_d)^{1/2}}}. \tag{12}$$

The natural frequency of the stable poles of H is

$$\omega_n^{\text{cl}} = (\omega_n^4 + q_d)^{1/4}. \tag{13}$$

Since the stable poles of H are the optimal closed-loop poles, we can write down the characteristic polynomial of $(A - BK(\infty))$:

$$\Delta^{\text{cl}}(s) = s^2 + 2\delta^{\text{cl}}\omega_n^{\text{cl}}s + (\omega_n^{\text{cl}})^2$$
$$= s^2 + \sqrt{2}\sqrt{(\omega_n^4 + q_d)^{1/2} + (q_v/2 - \omega_n^2)}s$$
$$+ (\omega_n^4 + q_d)^{1/2}. \tag{14}$$

b. Optimal Feedback Gain

According to Ackermann's formula

$$K(\infty) = \begin{bmatrix} 0 & 1 \end{bmatrix} U_2^{-1} \Delta^{\text{cl}}(A). \tag{15}$$

The reachability matrix is

$$U_2 = \begin{bmatrix} B & AB \end{bmatrix} = \begin{bmatrix} 0 & 1 \\ 1 & 0 \end{bmatrix}, \tag{16}$$

and substituting A^2, A, and I for s^2, s^1, and s^0 in (14) yields

$$\Delta^{\text{cl}}(A)$$
$$= \begin{bmatrix} -\omega_n^2 + (\omega_n^4 + q_d)^{1/2} & \sqrt{2}\sqrt{(\omega_n^4 + q_d)^{1/2} + (q_v/2 - \omega_n^2)} \\ -\omega_n^2\sqrt{2}\sqrt{(\omega_n^4 + q_d)^{1/2} + (q_v/2 - \omega_n^2)} & -\omega_n^2 + (\omega_n^4 + q_d)^{1/2} \end{bmatrix}. \tag{17}$$

The optimal feedback gain is thus

$$K(\infty) = \begin{bmatrix} -\omega_n^2 + (\omega_n^4 + q_d)^{1/2} & \sqrt{2}\sqrt{(\omega_n^4 + q_d)^{1/2} + (q_v/2 - \omega_n^2)} \end{bmatrix} \tag{18}$$

If $\omega_n = 0$, these results agree with Example 3.4-3. ∎

Time-varying Plant

If the plant is time varying, then we must redefine observability and reachability. Suppose the plant is (3.4-1) with cost index in Table 3.3-1, where A, B, Q, and R are time dependent. Let $\phi(t, t_0)$ be the state transition matrix of A.

We say the plant is *uniformly completely observable* if for every final time T the *observability gramian* satisfies

$$\alpha_0 I \leq \int_{t_0}^{T} \phi^{\text{T}}(\tau, t_0) Q(\tau)\phi(\tau, t_0) d\tau \leq \alpha_1 I \tag{3.4-27}$$

for some $t_0 < T, \alpha_0 > 0$, and $\alpha_1 > 0$. Compare this with (3.3-14). We say the plant is *uniformly completely reachable* if for every initial time t_0 the *reachability gramian* satisfies

$$\alpha_0 I \leq \int_{t_0}^{T} \phi(t, \tau) B(\tau) R^{-1}(\tau) B^{T}(\tau) \phi^{T}(t, \tau) d\tau \leq \alpha_1 I \qquad (3.4\text{-}28)$$

for some $T > t_0, \alpha_0 > 0$, and $\alpha_1 > 0$. Compare this to the gramian (3.3-25).

Uniform complete observability and reachability (and boundedness of $A(t)$, $B(t)$, $Q(t)$, $R(t)$) guarantee that for large t the behavior of $P(t)$ is unique, independent of $P(0)$. They also guarantee the uniform asymptotic stability of the closed-loop plant $(A - BK(t))$.

3.5 FREQUENCY-DOMAIN RESULTS

Several methods for designing steady-state continuous regulators have been discussed. Here we present an approach that amounts to a root-locus design method. The plant and weighting matrices are assumed time invariant, with (A, B) reachable and $(A, \sqrt{Q})$ observable.

A Factorization Result

The optimal steady-state regulator is given by a constant feedback (3.4-2), where $K = R^{-1}B^{T}S$ and S is the unique positive definite solution to the ARE (3.4-6). The resulting closed-loop system (3.4-3) is asymptotically stable.

As in Section 2.5, we can show that

$$\Delta^{cl}(s) = |I + K(sI - A)^{-1}B|\Delta(s), \qquad (3.5\text{-}1)$$

which relates the closed-loop characteristic polynomial $\Delta^{cl}(s) = |sI - A + BK|$ to the open-loop polynomial $\Delta(s) = |sI - A|$. According to Fig. 2.5-1 (with z^{-1} replaced by $1/s$), $I + K(sI - A)^{-1}B$ can be interpreted as a return-difference matrix (return difference = I-loop gain).

We can also show the factorization result

$$B^{T}(-sI - A)^{-T}Q(sI - A)^{-1}B + R$$
$$= (I + K)(-sI - A)^{-1}B)^{T}R(I + K(sI - A)^{-1}B), \qquad (3.5\text{-}2)$$

which can be interpreted as follows. Let

$$H(s) = C(sI - A)^{-1}B \qquad (3.5\text{-}3)$$

be the transfer function from $u(t)$ to $y^{1}(t) = Cx(t)$, the "top portion" of the fictitious output $y(t)$ in (3.4-10). Now examine Fig. 2.5-2 (with z^{-1}, z replaced by $1/s$, $-1/s$, respectively). It is clear that $H^{\mathrm{T}}(-s)$ is the transfer function from y^1 to the intermediate signal w. Therefore, (3.5-2) simply expresses the equivalence between a transfer-function product in the continuous version of Fig. 2.5-2 and a transfer-function product in the continuous version of Fig. 2.5-1. It is just another way of expressing the equivalence between the state–costate (3.4-13) and the closed-loop (3.4-3) formulations of the optimal LQ regulator.

Chang-Letov Design Procedure for the Steady-state LQ Regulator

According to (3.5-1) and (3.5-2) we can write

$$\Delta^{\mathrm{cl}}(-s)\Delta^{\mathrm{cl}}(s) = |H^{\mathrm{T}}(-s)H(s) + R| \cdot \Delta(-s)\Delta(s) \cdot |R|^{-1}, \qquad (3.5\text{-}4)$$

where $H(s)$ is given by (3.5-3). This is the *Chang-Letov equation* (Kailath 1980). It can be used to design optimal steady-state LQ regulators by a root-locus approach. Note that the entire right-hand side is known if the plant and weighting matrices are given, so we can use the Chang-Letov equation to determine the optimal closed-loop poles; since $(A - BK)$ is stable by Theorem 3.4-2, they are just the stable roots of the right-hand side. The roots of $\Delta^{\mathrm{cl}}(-s)\Delta^{\mathrm{cl}}(s)$ are always symmetric with respect to the imaginary axis; that is, if s_1 is a root, then so is $-s_1$.

In the single-input case with $Q = qI$, we have

$$H(s) = \frac{\sqrt{q}[\mathrm{adj}(sI - A)]B}{\Delta(s)} \triangleq \sqrt{q}\frac{N(s)}{\Delta(s)}, \qquad (3.5\text{-}5)$$

where $N(s)$ is a column vector. Then (3.5-4) becomes

$$\Delta^{\mathrm{cl}}(-s)\Delta^{\mathrm{cl}}(s) = \frac{q}{r}N^{\mathrm{T}}(-s)N(s) + \Delta(-s)\Delta(s). \qquad (3.5\text{-}6)$$

The roots of the right-hand side are the zeros of

$$1 + \left(\frac{q}{r}\right)\frac{N^{\mathrm{T}}(-s)N(s)}{\Delta(-s)\Delta(s)} = 1 + \frac{q}{r}H^{\mathrm{T}}(-s)H(s), \qquad (3.5\text{-}7)$$

which is exactly the form required for a root-locus analysis as a function of the parameter q/r. This shows that as q/r varies from zero (no state weighting) to ∞ (no control weighting), the optimal closed-loop poles move from the stable poles of

$$G(s) = H^{\mathrm{T}}(-s)H(s) \qquad (3.5\text{-}8)$$

to its stable zeros. We can therefore select the ratio of cost weights q/r to yield suitable closed-loop poles.

It is worth remarking that the stable poles of $H^{\mathrm{T}}(-s)H(s)$ are the poles of $H(s)$ with unstable poles reflected into the left half-plane (i.e., $\bar{s}_1 = -s_1$). The stable zeros of $H^{\mathrm{T}}(-s)H(s)$ are the zeros of $H(s)$ with unstable zeros reflected into the left half-plane.

The next example illustrates these ideas. For a good discussion, see Kailath (1980) or Schultz and Melsa (1967).

Example 3.5-1.   Chang-Letov Design of Aircraft Longitudinal Autopilot

The short-period longitudinal dynamics for a medium-sized jet with center of gravity unusually far aft might be described by the state equations

$$
\begin{bmatrix} \dot{\alpha} \\ \dot{\rho} \end{bmatrix} = \begin{bmatrix} -1.417 & 1.0 \\ 2.860 & -1.183 \end{bmatrix} \begin{bmatrix} \alpha \\ \rho \end{bmatrix} + \begin{bmatrix} 0 \\ -3.157 \end{bmatrix} \delta_e,
\tag{1}
$$

where α is the angle of attack, ρ the pitch rate, and δ_e the elevator deflection (Blakelock 1965). (We shall show only three decimal places.) The open-loop characteristic polynomial is

$$
\Delta(s) = |sI - A| = s^2 + 2.6s - 1.183,
\tag{2}
$$

so the open-loop poles are

$$
s = -2.995, 0.395.
\tag{3}
$$

Evidently, the center of gravity is so far aft that the short-period poles, which usually constitute a lightly damped complex pair, have become one stable and one unstable pole.

To stabilize the plant and keep the pitch rate small, we might select the performance index

$$
J(0) = \frac{1}{2} \int_0^\infty (q\rho^2 + r\delta_e^2)\, dt,
\tag{4}
$$

so that

$$
Q = \begin{bmatrix} 0 & 0 \\ 0 & q \end{bmatrix}
\tag{5}
$$

and a root of Q is

$$
C = [0 \quad \sqrt{q}].
\tag{6}
$$

Since (A, B) is reachable and (A, C) is observable (if $q \neq 0$), we know the steady-state LQ regulator results in a stable closed-loop plant.

Transfer function (3.5-3) is

$$
H(s) = \frac{-\sqrt{q}(3.157s + 4.473)}{s^2 + 2.6s - 1.183}.
\tag{7}
$$

The Chang-Letov design procedure is based on the rational function

$$
G(s) = H(-s)H(s).
\tag{8}
$$

Since we know the closed-loop characteristic polynomial $\Delta^{cl}(s)$ is stable, its roots are the stable zeros of

$$
1 + \frac{q}{r}G(s).
\tag{9}
$$

In our case

$$G(s) = \frac{(3.157s - 4.473)(-3.157s - 4.473)}{(s^2 - 2.6s - 1.183)(s^2 + 2.6s - 1.183)}$$

$$= \frac{-9.97(s - 1.417)(s + 1.417)}{(s + 0.395)(s - 2.995)(s - 0.395)(s + 2.995)}. \tag{10}$$

The poles and zeros of $G(s)$ are plotted in Fig. 3.5-1. Note that they are symmetrically placed about the imaginary axis. Also shown is the root locus as q/r varies from 0 to ∞. (Note that since the gain in (10) is *negative*, we are really plotting a root locus for negative gains $-9.97q/r$.)

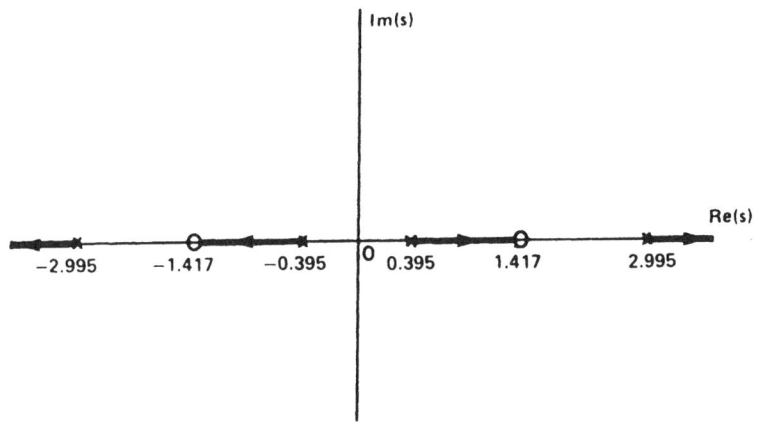

FIGURE 3.5-1   Chang-Letov root locus.

If we select $q/r = 1$, then the stable zeros of (9) are

$$s = -1.094, -4.231, \tag{11}$$

so the closed-loop characteristic polynomial is

$$\Delta^{cl}(s) = (s + 1.094)(s + 4.231)$$

$$= s^2 + 5.324s + 4.627. \tag{12}$$

Now Ackermann's formula can be used to find the optimal feedback gain K.   ∎

PROBLEMS

Section 3.2

3.2-1.   We want to minimize

$$J = \int_0^\pi \dot{x}^2 \, dt.$$

Formulate this as an optimal control problem. Find and sketch the optimal $x(t)$. Find J^*.

3.2-2.   Determining optimal control by approximation. Consider the nonlinear plant

$$\dot{x} = -x^3 + u, \quad x(0) = \tfrac{1}{2}, \tag{1}$$

with quadratic performance index

$$J = \frac{1}{2}x^2(2) + \frac{1}{2}\int_0^2 (x^2 + u^2)\, dt. \tag{2}$$

a. Write state and costate equations, stationarity condition, and boundary conditions. Eliminate $u(t)$ from the state and costate equations.
b. Prove that if $x(t)$ is small on [0, 2], then an approximate solution for the costate is

$$\lambda = x - x^3. \tag{3}$$

c. For this approximate costate, find the state solution $x(t)$.
d. Find the approximate optimal control. (Note that applying this control to the plant will tend to keep $x(t)$ small, so that approximation (3) is valid.)

3.2-3.   A model of an automobile suspension system is given by

$$m\ddot{y} + ky = u, \tag{1}$$

where m is the mass, k the spring constant, u the upward force on the frame, and y the vertical position.
a. Write the state equation if the state is $x = [y \quad \dot{y}]^T$.
   To conduct a durability test, we repetitively apply a force $u(t)$ and suddenly remove it until failure occurs. To compute the force, we can solve the following problem. Find $u(t)$ to move the automobile from $y(0) = 0$, $\dot{y}(0) = 0$ to a final position of $y(T) = h$, $\dot{y}(T) = 0$ at a given final time T. Minimize the control energy

$$J = \frac{1}{2}\int_0^T u^2\, dt. \tag{2}$$

b. Write the state and costate equations, stationarity condition, and boundary conditions. Eliminate u from the state and costate equations.
c. Solve for the costate in terms of the as yet unknown $\lambda(0)$. Solve for the state in terms of the unknown $\lambda(0)$ and the known $x(0)$.
d. Use the boundary conditions to find $\lambda(0)$. Let $m = k = 1$, $T = 2$, $h = -3$ (i.e., down) for the remainder of the problem.
e. Find optimal control and optimal state trajectory.
f. Verify that $x^*(T) = [h \quad 0]^T$ as required.

3.2-4. See Fig. P3.2-1.

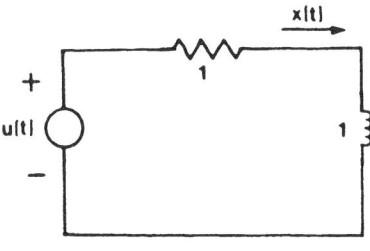

FIGURE P3.2-1

a. Find the state equation.
b. It is desired to charge up the inductance to $x(T) = 2A$ at $T = 1$ if $x(0) = 0$ while minimizing

$$J = \int_0^1 u^2 \, dt. \tag{1}$$

Find the optimal control.
c. Find the optimal state trajectory.

3.2-5.   Optimal control of a bilinear system. Let

$$\dot{x} = Ax + Dxu + bu, \tag{1}$$

where $x \in R^n$, $u \in R$, and

$$J = \frac{1}{2}x^T(T)S(T)x(T) + \frac{1}{2}\int_0^T (x^TQx + ru^2) \, dt. \tag{2}$$

Show that the optimal control consists of a state–costate inner product. Find state and costate equations after eliminating u. These cubic differential equations are very hard to solve.

3.2-6.   Numerical solution of bilinear system. Let

$$\dot{x}_1 = x_1x_2 + u,$$
$$\dot{x}_2 = x_1^2 + x_2^2,$$

with cost

$$J = \frac{s_1x_1^2(T)}{2} + \frac{s_2x_2^2(T)}{2} + \frac{1}{2}\int_0^T (q_1x_1^2 + q_2x_2^2 + ru^2) \, dt,$$

where T is fixed. Derive Euler's equations and boundary conditions. Write subroutines *fcni, fcnbf* for use with *tpoint*.

3.2-7.   Unit solution method. Consider the optimal control problem formulated by (3.2-1) and (3.2-14)–(3.2-21), where the plant is linear. We want to formalize the unit solution method by finding a system of equations to solve for the initial costate $\lambda(t_0)$.

Let $r(t_0)$ be the given initial state. Let $x_0(t)$, $\lambda_0(t)$ be the solutions of the Hamiltonian system when $x(t_0) = r(t_0)$, $\lambda(t_0) = 0$. Let $x_i(t)$, $\lambda_i(t)$ for $i = 1, \ldots, n$ be the unit solutions when $x(t_0) = 0$, $\lambda(t_0) = e_i$, where e_i is the ith column of the $n \times n$ identity matrix. Then the solutions for the general initial conditions (3.2-26) with c_i scalar constants, are given by (3.2-27).

a. *Final state fixed:* If the final condition is $x(T) = r(T)$ given, show that the initial costate $\lambda(t_0)$ that guarantees that this final condition holds is given by solving

$$[x_1(T) \cdots x_n(T)]\lambda(t_0) = r(T) - x_0(T). \tag{1}$$

b. *Final state free:* If the final state is free, show that the optimal initial costate $\lambda(t_0)$ is given by solving

$$-[\lambda_1(T) - S(T)x_1(T) \cdots \lambda_n(T) - S(T)x_n(T)]\lambda(t_0)$$
$$= -(\lambda_0(T) - S(T)x_0(T)r(T). \tag{2}$$

3.2-8.   Unit solution method. Let

$$\dot{x}_1 = x_2 + u, \tag{1a}$$

$$x_2 = x_1 - x_2, \tag{1b}$$

with

$$J = 5x_1^2(T) + \frac{1}{2} \int_0^T (x_1^2 + x_2^2 + u^2)\, dt, \tag{2}$$

where T is fixed and we require

$$x_2(T) = 1. \tag{3}$$

Solve for the optimal control by the unit solution method.

Section 3.3

3.3-1.   The scalar plant

$$\dot{x} = u \tag{1}$$

has performance index

$$J(t_0) = \frac{1}{2}sx^2(T) + \frac{1}{2} \int_{t_0}^T ru^2\, dt. \tag{2}$$

is $s(t) = $ constant?

a. First we want a closed-loop control minimizing J.
   i. Solve the Riccati equation using separation of variables.
   ii. Find the optimal control.
   iii. Sketch optimal feedback gain versus t.
b. Now we want an open-loop control that minimizes J and drives the state to
  a given $x(T)$. Initial state $x(t_0)$ is known.
   i. Find the weighted reachability gramian.
   ii. Find the optimal control as a function of $x(t_0)$ and $x(T)$. Sketch it.
   iii. Find and sketch the optimal state trajectory.
c. The open-loop control formulation of part b can be used to find a state feedback control law.
   i. Use part b.iii to solve for $x(t_0)$ in terms of $x(t)$ and $x(T)$.
   ii. Substitute this into your result in part b.ii to find $u^*(t)$ in the form of
   $u^*(t) = g(t)x(t) + h(t)$. Compare to the closed-loop control in part a.
   iii. Compare this control to the optimal control that minimizes the *cost to go*
   $J(t)$ on $[t, t + T]$.

3.3-2. See Fig. P3.2-1.
a. Write the state equation.
b. Solve Problem 3.2-4 using the reachability gramian.
c. Now let

$$J = \frac{1}{2}10x^2(T) + \frac{1}{2}\int_0^T (x^2 + u^2)\, dt.$$

   i. Solve the Riccati equation using separation of variables.
   ii. Find the optimal control.
   iii. Sketch optimal feedback gain versus t.

3.3-3. Optimal control of newton's system. Let

$$\dot{x}_1 = x_2, \tag{1a}$$

$$\dot{x}_2 = u \tag{1b}$$

have performance index

$$J = \frac{1}{2}x^T(T)x(T) + \frac{1}{2}\int_0^T (x^Tx + ru^2)\, dt, \tag{2}$$

where $x = \begin{bmatrix} x_1 & x_2 \end{bmatrix}^T$.
a. Find the Riccati equation. Write it as three scalar differential equations.
  Find the feedback gain in terms of the scalar components of $S(t)$.
b. Write subroutines to find and simulate the optimal control using MATLAB.
c. Find analytic expressions for the steady-state Riccati solution and gain.

3.3-4. Uncontrolled Newton's system. Consider the system of Problem 3.3-3. Solve the Lyapunov equation (3.3-9) to find the cost kernel $S(t)$ if $u = 0$. Sketch the scalar components of $S(t)$.

3.3-5. Uncontrolled harmonic oscillator. Repeat problem 3.3-4 for the system in Example 3.3-5. Let $S(T) = I$, $Q = I$, $\omega_n^2 = 1$, $\delta = 0.5$.

3.3-6. Cross-weighting cost terms. Repeat Problem 3.3-3 if a term like $2x_1(t)u(t)$ is added under the integral in J. Assume $r > 1$. (Why?)

3.3-7. Writing the Lyapunov equation as a vector equation. Show that Lyapunov equation (3.3-9) can be written as the vector equation

$$-\frac{d}{dt}s(S) = \left[(A^T \otimes I) + (I \otimes A^T)\right]s(S) + s(Q),$$

where the Kronecker product and stacking operator are defined in Appendix A.

3.3-8. Prove that (3.3-42) and the Joseph-stabilized equation (3.3-46) are equivalent to (3.3-40).

3.3-9. Open-loop control with function of final state fixed. Let $\dot{x} = Ax + Bu$ with

$$J = \frac{1}{2}\int_{t_0}^{T} u^T Ru\, dt.$$

It is required to drive $x(t_0)$ to a final state $x(T)$ such that

$$Cx(T) = r(T)$$

for a given $r(T)$.
Show that the optimal control is given by

$$u(t) = -R^{-1}B^T e^{A^T(T-t)}C^T\left[CG(t_0, T)C^T\right]^{-1}(Ce^{A(t-t_0)}(t_0) - r(T)).$$

Note that if $C \neq I$, there may be a solution even if the system is not reachable.

3.3-10. Hamiltonian system. Let V, W be $n \times n$ solutions to the linear equation

$$\begin{bmatrix} \dot{V} \\ \dot{W} \end{bmatrix} = \begin{bmatrix} A & -BR^{-1}B^T \\ -Q & -A^T \end{bmatrix}\begin{bmatrix} V \\ W \end{bmatrix},$$

with boundary condition $W(T) = S(T)V(T)$. Show that the solution to (3.3-40) is given by $S(t) = W(t)V^{-1}(t)$. (*Hint:* Note that $d(V^{-1})/dt = -V^{-1}\dot{V}V^{-1}$.)

3.3-11. Writing the Riccati equation as a vector equation. Show that (3.3-40) can be written as the nonlinear vector differential equation

$$-\frac{d}{dt}s(S) = \left[(A^T \otimes I) + (I \otimes A^T)\right]s(S)$$
$$+ (S \otimes S)s(BR^{-1}B^T) + s(Q).$$

3.3-12. Relation between state and costate. Show that if $u(t) = 0$, $Q = 0$, then the inner product of the state with the costate is a constant independent of time in the continuous linear quadratic regulator.

Section 3.4

3.4-1. Newton's system. Let

$$\dot{x} = x_2, \tag{1a}$$

$$\dot{x}_2 = u, \tag{1b}$$

$$J = \frac{1}{2}\int_0^\infty (x_1^2 + 2vx_1x_2 + qx_1^2 + u^2)\,dt, \tag{2}$$

where $(q - v^2) > 0$.
a. Find the solution to the ARE.
b. Find the optimal control.
c. Find the optimal closed-loop system.
d. Plot loci of closed-loop poles as q varies from 0 to ∞. For which values of q is the system stable?

3.4-2. Suboptimal control. Let

$$\dot{x} = u, \tag{1}$$

$$J = \frac{1}{2}sx^2(T) + \frac{1}{2}\int_0^T (qx^2 + ru^2)\,dt. \tag{2}$$

a. Optimal control:
   i. Set up and solve the Riccati equation. Sketch solution $s^*(t)$.
   ii. Find the optimal feedback $K(t)$. Show that the steady-state gain is

$$K_\infty = \sqrt{q/r}. \tag{3}$$

b. Suboptimal control: We shall now use the suboptimal constant feedback

$$u(t) = -K_\infty x(t). \tag{4}$$

Find the resulting kernel $s(t)$. Sketch $s^*(t)$ and $s(t)$ on the same graph. Note $s^*(t) \leq s(t)$ for all t, but $s^*(0) = s(0)$ so the two limiting values are the same.

c. Find the steady-state closed-loop plant and plot its pole as q/r goes from 0 to ∞.

3.4-3. ARE solutions. Let

$$\dot{x}_1 = x_2, \tag{1a}$$

$$\dot{x}_2 = -ax_1 - 2x_2 + u, \tag{1b}$$

and

$$J = \frac{1}{2} \int_0^\infty (2x_1^2 + x_2^2 + u^2)\, dt. \tag{2}$$

a. Plot the root locus of the open-loop system as a varies from $-\infty$ to ∞. For which values of a is (1) stable?

b. Find all symmetric ARE solutions in terms of a. How do you know there is a unique positive definite solution?

c. In the remainder of the problem, let $a = -8$. Find open-loop poles. Is (1) stable?

d. Find all (4) symmetric ARE solutions.

e. Find the unique, negative definite solution and the unique, positive definite solution.

f. Find the steady-state Kalman gain.

g. Find the closed-loop poles.

h. What can you say about the closed-loop poles for *any* value of a?

3.4-4. Suboptimal control of Newton's system. In Example 3.4-3, the steady-state gain for Newton's system was found to be of the form

$$K_\infty = [\omega_n^2 \quad 2\delta\omega_n^2], \tag{1}$$

where δ and ω_n depend on the cost weights. In a suboptimal control scheme we apply the constant feedback

$$u(t) = -K_\infty x(t). \tag{2}$$

Determine the resulting suboptimal cost kernel $S(t)$. Compare this with Problem 3.3-3.

3.4-5. Let

$$\dot{x} = x + u, \tag{1}$$

$$J = \frac{1}{2} \int_0^\infty (x^2 + u^2)\, dt. \tag{2}$$

a. Use the eigenstructure of the Hamiltonian matrix to determine the steady-state Riccati solution s_∞. Hence, find the steady-state gain K_∞.

b. Find the solution to the ARE to check part a.

c. Use the eigenstructure of the Hamiltonian matrix to determine optimal closed-loop poles. Hence, find K_∞ by Ackermann's formula.

3.4-6. Repeat Problem 3.4-5 for the plant

$$\dot{x}_1 = x_1 + u, \tag{1a}$$

$$\dot{x}_2 = -x_1 + x_2 \tag{1b}$$

if

$$J = \frac{1}{2} \int_0^\infty (x_1^2 + x_2^2 + ru^2)\, dt, \tag{2}$$

with $r = \frac{1}{10}$.

3.4-7. ARE solutions

a. Final all symmetric solutions to (3.4-6) if

$$A = \begin{bmatrix} 0 & 1 \\ 0 & 0 \end{bmatrix}, \quad B = \begin{bmatrix} 0 \\ 1 \end{bmatrix}, \quad C = [1 \ \ 0], \quad R = 1, \quad \text{and} \quad Q = C^T C.$$

b. How do you know a priori that there is a unique, positive definite solution? Check all your solutions for definiteness.

c. Some of your solutions are complex. For each real solution S, check the stability of the closed-loop system $(A - BK)$, where $K = R^{-1}B^T S$. Hence, identify the *stabilizing* and *destabilizing* solutions to the algebraic Riccati equation as the positive definite and negative definite solutions, respectively.

3.4-8. Observability and reachability for time-varying systems

a. Show that (3.4-27) holds if and only if there is a bounded positive definite solution for some $t < T$ to

$$-\dot{S} = A^T S + SA + Q, \quad S(T) = 0. \tag{1}$$

Show that (3.4-28) holds if and only if there is a bounded positive definite solution $S^{-1}(t)$ for some $t > 0$ to

$$\frac{d}{dt}(S^{-1}) = AS^{-1} + S^{-1}A^T + BR^{-1}B^T, \quad S^{-1}(0) = 0. \tag{2}$$

b. Derive discrete counterparts to these tests.

Section 3.5

3.5-1. Prove (3.5-2).

3.5-2. Change-Letov design
a. For the system of Problem 3.4-6, use the Chang-Letov equation and Ackermann's formula to find the optimal feedback gain and closed-loop plant if $r = \frac{1}{10}$.
b. Plot a root locus for the closed-loop plant as r varies from ∞ to 0.

3.5-3. Polynomial decomposition
a. Show that any polynomial $P(s)$ with real coefficients can be decomposed as

$$P(s) = P_1(s) + P_2(s) \tag{1}$$

where

$$P_1(s) = \tfrac{1}{2}(P(s) + P(-s)),$$
$$P_2(s) = \tfrac{1}{2}(P(s) - P(-s)). \tag{2}$$

Show that

$$P_1(s) = P_1(-s), \tag{3}$$
$$P_2(s) = -P_2(-s). \tag{4}$$

b. Given the roots s_i of $P(s) = 0$, how can you find the roots of $P(-s) = 0$?

3.5-4. Chang-Letov design for systems in reachable canonical form. Let

$$\dot{x} = \begin{bmatrix} 0 & 1 \\ -\omega_n^2 & -2\delta\omega_n \end{bmatrix} x + \begin{bmatrix} 0 \\ 1 \end{bmatrix} u,$$

$$J = \frac{1}{2} \int_0^\infty (q x^T x + r u^2) \, dt.$$

a. Using the Chang-Letov procedure, plot the loci of the optimal closed-loop poles as q/r goes from 0 to ∞. Show that the poles move from the "stabilized" poles of the plant (i.e., if they are unstable, reflect them into the left half-s-plane replacing s by $-s$) to values of $s = -1, \infty$.
b. Find the optimal gain by Ackermann's formula if $\omega_n = 1$, $\delta = -\frac{1}{2}$, $q/r = 2$.

4

THE TRACKING PROBLEM
AND OTHER LQR EXTENSIONS

4.1 THE TRACKING PROBLEM

In this section we present an optimal control law that forces the plant to track a desired reference trajectory $r(t)$ over a specified time interval $[t_0, T]$.

Nonlinear Systems

Let the dynamics of the plant be described by

$$\dot{x} = f(x, u). \tag{4.1-1}$$

To keep a specified linear combination of the states

$$y(t) = Cx(t) \tag{4.1-2}$$

close to the given reference track $r(t)$, let us specify the quadratic cost index

$$J(t_0) = \frac{1}{2}(Cx(T) - r(T))^{\mathrm{T}} P(Cx(T) - r(T))$$

$$+ \frac{1}{2}\int_{t_0}^{T}[(Cx - r)^{\mathrm{T}}Q(Cx - r) + u^{\mathrm{T}}Ru]\,dt, \tag{4.1-3}$$

with $P \geq 0$, $Q \geq 0$, $R > 0$.

From Table 3.2-1, the optimal control is given by solving

State system:

$$\dot{x} = f(x, u) \tag{4.1-4}$$

Costate system:

$$-\dot{\lambda} = \left(\frac{\partial f}{\partial x}\right)^T \lambda + C^T Q C x - C^T Q r \tag{4.1-5}$$

Stationarity condition:

$$0 = \left(\frac{\partial f}{\partial u}\right)^T \lambda + R u \tag{4.1-6}$$

with

Boundary conditions (note dT $= 0$):

$$x(t_0) \quad \text{given} \tag{4.1-7}$$

$$\lambda(T) = C^T P (C x(T) - r(T)). \tag{4.1-8}$$

According to the stationarity condition, the optimal control is given in terms of the costate as

$$u(t) = -R^{-1} \left(\frac{\partial f}{\partial u}\right)^T \lambda(t). \tag{4.1-9}$$

When f_u depends on x and u in a nonlinear manner this control is a nonlinear feedback of the state and costate.

The reference track $r(t)$ has added two terms to these equations that were not present in the LQ regulator. From (4.1-8) it is apparent that the final costate is no longer a linear function of the state as it was in the regulator problem. Furthermore, even if (4.1-9) can be used to eliminate $u(t)$ in the state and costate equations, the Hamiltonian system will still be nonhomogeneous, since (4.1-5) is driven by $-C^T Q r(t)$.

The LQ Tracking Problem

If the plant has linear dynamics, then the optimal controller becomes

$$\dot{x} = A x + B u, \tag{4.1-10}$$

$$-\dot{\lambda} = A^T \lambda + C^T Q C x - C^T Q r \tag{4.1-11}$$

with

$$u = -R^{-1} B^T \lambda. \tag{4.1-12}$$

Using this control policy in the state equation yields

$$\dot{x} = A x - B R^{-1} B^T \lambda. \tag{4.1-13}$$

This formulation of the optimal control law has the structure shown in Fig. 2.2-1 (with z^{-1}, z replaced by $1/s$, $-1/s$). The control law cannot be implemented in this form since the costate equation develops backward, and the boundary conditions (4.1-7) and (4.1-8) are split. Although $u(t)$ is a linear costate feedback, it cannot be expressed as a linear state feedback because of the affine form of (4.1-8).

By use of the sweep method it can be shown (see the problems) that the continuous-time optimal LQ tracker can be expressed in the causal form shown in Table 4.1-1. Note that the control input is an *affine state feedback*; it consists of a linear state feedback *plus* an additional term. The additional term depends on the output $v(t)$ of the adjoint of the closed-loop plant when driven by the reference track $r(t)$.

The closed-loop plant under the influence of this tracker control law is

$$\dot{x} = (A - BK(t))x + BR^{-1}B^{T}v. \tag{4.1-14}$$

The optimal cost on $[t, T]$ for any t using this control is

$$J(t) = \frac{1}{2}x^{T}(t)S(t)x(t) - x^{T}(t)v(t) + w(t),$$

where the new auxiliary function $w(t)$ satisfies

$$-\dot{w} = \frac{1}{2}r^{T}Qr - \frac{1}{2}v^{T}BR^{-1}B^{T}v, \quad t \leq T, \tag{4.1-15}$$

TABLE 4.1-1   Continuous Linear Quadratic Tracker

System model:
$$\dot{x} = Ax + Bu, \quad t > t_0$$

Performance index:
$$J(t_0) = \frac{1}{2}(Cx(T) - r(T))^{T}P(Cx(T) - r(T))$$
$$+ \frac{1}{2}\int_{t_0}^{T}[(Cx - r)^{T}Q(Cx - r) + u^{T}Ru]\,dt$$

Assumptions:
$$P \geq 0, \quad Q \geq 0, \quad R > 0 \quad \text{are symmetric}$$

Optimal affine control:
$$K(t) = R^{-1}B^{T}S(t)$$
$$-\dot{S} = A^{T}S + SA - SBR^{-1}B^{T}S + C^{T}QC, \quad S(T) = C^{T}PC \cdot$$
$$-\dot{v} = (A - BK)^{T}v + C^{T}Qr, \quad v(T) = C^{T}Pr(T)$$
$$u = -Kx + R^{-1}B^{T}v$$

with

$$w(T) = \frac{1}{2} r^T(T) P r(T). \tag{4.1-16}$$

For further discussion on the LQ tracker, see Athans and Falb (1966), Kirk (1970), and Bryson and Ho (1975).

Implementation and a Suboptimal Tracker

The implementation of the tracker can be simplified by doing most of the work implied by Table 4.1-1 offline. The matrix sequence $S(t)$ is independent of the state trajectory, so the Riccati equation can be solved offline, and $S(t)$ and the feedback gain $K(t)$ can be stored. If the reference track $r(t)$ is known a priori, the auxiliary function $v(t)$ can also be precomputed and stored. The only work left to do during the actual control run is then to compute $u(t) = -K(t)x(t) + R^{-1}B^T v(t)$.

Suppose, however, that $v(t)$ has been determined by integrating backward the closed-loop adjoint system with $v(T) = C^T P r(T)$. Then $v(0)$ is known. During the actual control run we use this $v(0)$ in the *forward* equation

$$-\dot{v} = (A - BK)^T v + C^T Q r, \quad t > 0, \tag{4.1-17}$$

to compute $v(t)$. This avoids storage of the auxiliary function $v(t)$.

If $t_0 = 0$ and the final time T goes to infinity, we have the infinite-horizon tracker problem, where we also let $P = 0$. Then, in the time-invariant case, if (A, B) is reachable and $(A, C\sqrt{Q})$ is observable, the Riccati-equation solution reaches a steady-state solution $S(\infty)$. The Kalman gain then reaches a corresponding steady-state value of $K(\infty)$, and the closed-loop plant is stable. Under these circumstances the optimal tracker is given by

$$-\dot{v} = (A - BK(\infty))^T v + C^T Q r, \tag{4.1-18}$$

$$u = -K(\infty)x + R^{-1}B^T v. \tag{4.1-19}$$

A *steady-state* tracker can be devised for a finite control interval $[t_0, T]$ by using the steady-state gain $K(\infty)$. The initial condition $v(0)$ can be determined offline using (4.1-18) and then during the simulation we need use only (4.1-17) and (4.1-19). Experience shows that this simplified tracker is often satisfactory if $(T - t_0)$ is large. The suitability of the suboptimal tracker can be determined in a particular application by running a computer simulation.

See Athans and Falb (1966) for a treatment of the case when $r(t)$ is a constant.

Example 4.1-1.   Scalar LQ Tracker

If the scalar plant

$$\dot{x} = ax + bu \tag{1}$$

has cost index

$$J(0) = \frac{1}{2}p(x(T) - r(T))^2 + \frac{1}{2}\int_0^T [q(x-r)^2 + Ru^2]\,dt \qquad (2)$$

for a given final time T and reference signal $r(t)$, then according to Table 4.1-1 the optimal tracker is specified by

$$-\dot{s} = 2as - \frac{b^2 s^2}{R} + q, \quad s(T) = p, \qquad (3)$$

$$-\dot{v} = \left(a - b^2\frac{s}{R}\right)v + qr, \quad v(T) = pr(T), \qquad (4)$$

$$K = bs/R, \qquad (5)$$

$$u = -Kx + \frac{bv}{R}. \qquad (6)$$

a. Computer Simulation

Equations (3) and (4) are first solved, with the minus signs to the left of the equalities omitted, by a forward integrator using the MATLAB routine *ode45.m*. The solutions are then reversed in time to obtain $s(t)$ and $v(t)$. This is accomplished by the use of a subroutine to describe the dynamics of equations (3) and (4). The sampled times of the integrator *ode45.m* are used to compute the gain K in (5) and K and v are stored.

For the simulation run of the system use the MATLAB routine *lsim.m*. An alternative method is to sample (1) to get a discrete plant, which is studied later in the chapter.

b. Steady-state Tracker

Let $T \to \infty$ and $p = 0$, so that we have the infinite-horizon tracking problem.

By Example 3.4-2, the steady-state kernel, gain, and closed-loop matrix are

$$s(\infty) = \frac{R}{b^2}(a + \alpha), \qquad (7)$$

$$K(\infty) = \frac{1}{b}(a + \alpha), \qquad (8)$$

$$a^{cl} = (a - bK(\infty)) = -\alpha, \qquad (9)$$

where

$$\alpha \overset{\Delta}{=} \sqrt{a^2 + \frac{b^2 q}{R}}. \qquad (10)$$

The auxiliary system (4) therefore becomes

$$-\dot{v} = -\alpha v + qr. \qquad (11)$$

Suppose we want to track the constant reference

$$r(t) = hu_{-1}(t),$$ (12)

where $h \in R$ and $u_{-1}(t)$ is the unit step. Then (11) can be solved to yield

$$v(t) = \int_t^T e^{-\alpha(\tau - t)} qh \, d\tau$$

$$= \frac{qh}{\alpha}(1 - e^{-\alpha(T-t)}).$$ (13)

As $T \to \infty$, this reaches the limiting value of

$$v(\infty) = \frac{qh}{\alpha}.$$ (14)

In the infinite-horizon limit, the auxiliary tracker signal is a constant.

Now the optimal control (6) is the easy-to-implement

$$u * (t) = -K(\infty)x(t) + \frac{bqh}{\alpha R}.$$ (15)

To examine the resulting plant behavior, write the closed-loop plant (4.1-14):

$$\dot{x} = -\alpha x + \frac{qhb^2}{\alpha R}.$$ (16)

Solving (16) results in the optimal state trajectory

$$x * (t) = e^{-\alpha t} x(0) + \frac{qhb^2}{\alpha^2 R}(1 - e^{-\alpha t}).$$ (17)

Since the closed-loop plant is stable ($\alpha > 0$), this reaches a steady-state value of

$$x * (\infty) = \frac{qhb^2}{\alpha^2 R} = \frac{h}{1 + \frac{a^2 R}{b^2 q}}.$$ (18)

As the ratio of state to control weighting q/R becomes large, $x*(\infty)$ comes closer to the desired constant reference value of h.   ∎

*Exercise 4.1-2.   **System with Known Disturbance***

Suppose the plant has a known disturbance $d(t)$, so that

$$\dot{x} = Ax + Bu + d.$$ (1)

Let the cost function be

$$J(t_0) = \frac{1}{2}x^T(T)S(T)x(T) + \frac{1}{2}\int_{t_0}^T (x^T Qx + u^T Ru] \, dt.$$ (2)

Show that the optimal control is given by

$$-\dot{S} = A^{\mathrm{T}}S + SA - SBR^{-1}B^{\mathrm{T}}S + Q, \tag{3}$$

$$K = R^{-1}B^{\mathrm{T}}S, \tag{4}$$

$$-\dot{v} = (A - BK)^{\mathrm{T}}v + Sd, \tag{5}$$

$$u = -Kx + R^{-1}B^{\mathrm{T}}v. \tag{6}$$

See Bryson and Ho (1975).   ■

Exercise 4.1-3.   Formulating the Tracking Problem as a Regulator Problem

Define the *tracking error*

$$e(t) = y(t) - r(t), \tag{1}$$

where $r(t)$ is the reference signal and $y(t)$ is the output of the scalar system

$$y^{(n)} + a_1 y^{(n-1)} \cdots + a_n y = u, \tag{2}$$

with $u(t)$ the control input. Define a cost index by

$$J(t_0) = \frac{1}{2} p e^2(T) + \frac{1}{2}\int_{t_0}^{T} (q e^2 + r u^2)\, dt, \tag{3}$$

and an operator by

$$\Delta(s) = s^n + a_1 s^{n-1} + \cdots + a_n, \tag{4}$$

where s represents the time derivative. Then the plant is

$$\Delta(s)y = u. \tag{5}$$

Suppose the reference track satisfies

$$\Delta(s)r = 0. \tag{6}$$

a. Reformulate (5) and (3) as a regulator problem by defining a suitable state vector.
b. Show how to find the scalar feedback coefficients $Ki(t)$ in a tracking scheme where the error and its first $n - 1$ derivatives are fed back to make $y(t)$ match $r(t)$. See Athans and Falb (1966).   ■

4.2   REGULATOR WITH FUNCTION OF FINAL STATE FIXED

To make the final state $x(T)$ take on a fixed given value $r(T)$ for any initial condition $x(t_0)$, we can use the open-loop control law (3.3-27). If the desired fixed final value is equal to zero, we can alternatively use the feedback control of Table 3.3-1 with $Q = 0$ and $S(T) \to \infty$. In this case, it is more convenient to use not the Riccati equation, but its "inverse," the Lyapunov equation

$$\frac{d}{dt}(S^{-1}) = AS^{-1} + S^{-1}A^{\mathrm{T}} - BR^{-1}B^{\mathrm{T}},$$ (4.2-1)

with $S^{-1}(T) = 0$. (Let $Q = 0$ in the Riccati equation and use $d(S^{-1})/dt = -S^{-1}\dot{S}S^{-1}$.)

Now let us consider a slightly generalized problem. Suppose the plant

$$\dot{x} = Ax + Bu$$ (4.2-2)

has performance index

$$J(t_0) = \frac{1}{2}x^{\mathrm{T}}(T)S(T)x(T) + \frac{1}{2}\int_{t_0}^{T}(x^{\mathrm{T}}Qx + u^{\mathrm{T}}Ru)\,dt$$ (4.2-3)

on a fixed interval $[t_0, T]$. It is desired to find the control that minimizes $J(t_0)$ and guarantees that

$$\psi(x(T), T) = Cx(T) - r(T) = 0$$ (4.2-4)

for a given $r(T) \in R^p$ and matrix C. This corresponds to demanding that a given linear combination of the final-state components be equal to a given vector $r(T)$.

The nonlinear version of this problem has the solution described in Table 3.2-1, with $dT = 0$ in the final condition (3.2-10).

In this linear context, the state and costate equations are (3.3-8), with the optimal control given by (3.3-7). The initial state $x(t_0)$ is given, and the final condition (3.2-10) becomes

$$\lambda(T) = S(T)x(T) + C^{\mathrm{T}}v$$ (4.2-5)

for some unknown multiplier $v \in R^p$. By using the sweep method (more details on the sweep method are given later in the discrete counterpart), the optimal control is found to be the scheme given in Table 4.2-1.

The optimal control $u(t)$ has a feedback portion $-Kx(t)$ similar to the control law in Table 3.3-1. To see what the other terms mean, note that the matrix $V \in R^{n \times p}$ is a "modified state transition matrix" for the adjoint of the time-varying closed-loop system. The auxiliary quadrature defining $P(t) \in R^{p \times p}$ has solution

$$P(t) = -\int_{t}^{T} V^{\mathrm{T}}(\tau)BR^{-1}B^{\mathrm{T}}V(\tau)\,d\tau.$$ (4.2-6)

TABLE 4.2-1   Function of Final-state-fixed LQ Regulator

System model:

$$\dot{x} = Ax + Bu, \quad t \geq t_0$$

Performance index:

$$J(t_0) = \frac{1}{2} x^T(T) S(T) x(T) + \frac{1}{2} \int_{t_0}^{T} (x^T Q x + u^T R u)\, dt$$

$$S(T) \geq 0, \quad Q \geq 0, \quad R > 0$$

Final-state constraint:

$$Cx(T) = r(T)$$

Optimal control law:

$$-\dot{S} = A^T S + SA - SBR^{-1}B^T S + Q, \quad t \leq T, \quad \text{given } S(T)$$

$$K = R^{-1}B^T S$$

$$-\dot{V} = (A - BK)^T V, \quad t \leq T \quad V(T) = C^T$$

$$-\dot{P} = V^T BR^{-1}B^T V, \quad t \leq T, \quad P(T) = 0$$

$$u = -(K - R^{-1}B^T V P^{-1} V^T)x - R^{-1}B^T V P^{-1} r(T)$$

By comparing this with (3.3-25), we see that $-P(t)$ is a sort of weighted reachability gramian. Now examine (3.3-27) to see that the extra terms in $u(t)$ in Table 4.2-1 are just a fixed-final-state type of control, which guarantees that at the final time $Cx(T)$ is equal to the desired $r(T)$.

If $C = 0$, the control in Table 4.2-1 reduces to the free-final-state control in Table 3.3-1. If $C = I$, it reduces to a feedback formulation of the fixed-final-state control (3.3-27). If $|P(t)| = 0$ for all $t \in (t_0, T)$, the problem is said to be *abnormal* and there is no solution. The nonsingularity of $P(t)$ is a reachability condition on a particular subspace of the state space R^n. It makes sense that a portion of the state space must be reachable if we want to drive $Cx(t)$ to a particular value at time $t = T$.

4.3   SECOND-ORDER VARIATIONS IN THE PERFORMANCE INDEX

All the results presented in this chapter have been deduced by finding conditions under which the first variation dJ' of an augmented performance index vanishes. Therefore, Tables 3.2-1, 3.3-1, 4.1-1, and 4.2-1 give *necessary* conditions for an optimal control. In this section we study the second-order variation $\delta^2 J'$ in the augmented performance index to obtain several results, including sufficient conditions for a minimum and an approach to the control of nonlinear systems. See Bryson and Ho (1975).

Perturbation Control

Consider the nonlinear plant with controller as in Table 3.2-1, where the final state
is free and the final time is fixed so that boundary condition (3.2-10) becomes

$$\lambda(T) = (\phi_x + \psi_x^T v)|_T. \tag{4.3-1}$$

The conditions in the table constitute *necessary* conditions for a minimum.

These equations involve the solution of a two-point boundary-value problem,
which is not easy. But suppose we have been able to solve for the optimal
control $u^*(t)$ and state trajectory $x^*(t)$ by some means. Then $dJ' = 0$. From a
Taylor series expansion, the cost J' is then equal to the optimal value plus the
second variation $\delta^2 J'$ and higher-order terms.

From (3.2-6) the second variation in J' is given by

$$\delta^2 J' = \frac{1}{2}\delta x^T(T) \left(\phi_{xx} + (\psi_x^T v)_x\right)\big|_T \delta x(T)$$

$$+ \frac{1}{2}\int_{t_0}^{T} [\delta x^T \quad \delta u^T] \begin{bmatrix} H_{xx} & H_{xu} \\ H_{ux} & H_{uu} \end{bmatrix} \begin{bmatrix} \delta x \\ \delta u \end{bmatrix} dt. \tag{4.3-2}$$

The variations in x and u about x^* and u^* must satisfy the incremental
constraints

$$\delta \dot{x} = f_x \delta x + f_u \delta u \tag{4.3-3}$$

and

$$\delta \psi = \psi_x(x(T), T)\delta x(T), \tag{4.3-4}$$

where $\delta \psi$ has a given value. The initial condition is

$$\delta x(t_0) \quad \text{given.} \tag{4.3-5}$$

Now examine Fig. 4.3-1. Beginning at $x(t_0)$, the optimal control $u^*(t)$ drives
the state along the trajectory $x^*(t)$, ensuring that $\psi(x(T), T) = 0$. Suppose, however, that the plant begins in initial state $x(t_0) + \delta x(t_0)$, and that in addition we
want the final-state function to take on a value not of zero but of a given $\delta \psi$.
Then by considering the *perturbation-state equation* (4.3-3) with performance
index (4.3-2) and final-state constraint (4.3-4), we can solve for the optimal control increment $\delta u^*(t)$ that achieves this and minimizes $\delta^2 J'$. The overall control
required to drive $x(t_0) + \delta x(t_0)$ to a final state satisfying $\psi(x(T), T) = \delta \psi$ is
then equal to

$$u(t) = u^*(t) + \delta u^*(t). \tag{4.3-6}$$

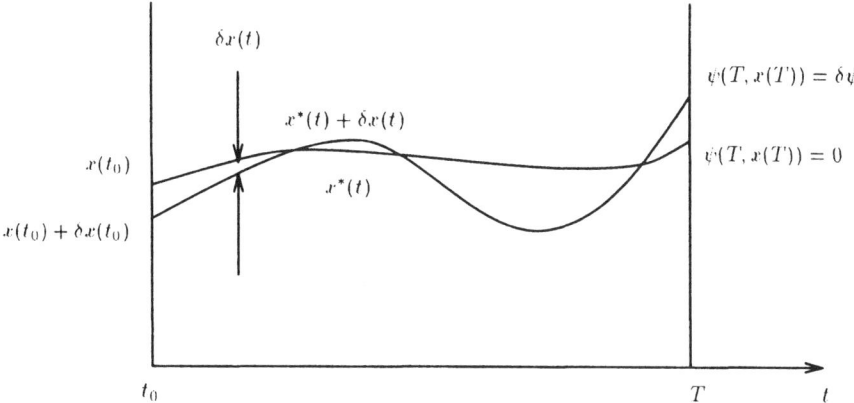

FIGURE 4.3-1 Optimal and neighboring optimal paths.

While $u^*(t)$ must be found by solving a nonlinear optimal control problem, the *perturbation control* $\delta u^*(t)$ is found by solving the *linear* quadratic perturbation problem defined by (4.3-2)–(4.3-5). Thus, once $u^*(t)$ and $x^*(t)$ have been found, desired small changes in the objectives can be attained by solving the LQ problem, which we have seen is quite tractable.

This *perturbation-control* approach amounts to linearizing the plant about the known optimal trajectory $u^*(t)$, $x^*(t)$. The state trajectory $x^*(t) + \delta x^*(t)$ that results when the overall control $u^*(t) + \delta u^*(t)$ is applied to the plant is called a *neighboring optimal* path. It satisfies $dJ' = 0$ and minimizes d^2J', so that it also minimizes J' in (3.2-6).

In Table 4.2-1 we gave the solution to the function of final-state-fixed LQ regulator. Since there are off-diagonal weighting terms in d^2J', we must use the modified Kalman gain introduced in Exercise 3.3-6 in the solution of the control problem (4.3-2)–(4.3-5). The resulting optimal control scheme is given in Table 4.3-1. See McReynolds (1966) and Bryson and Ho (1975).

If the plant is linear and the cost index is quadratic, then Table 4.3-1 reduces to Table 4.2-1 (then $H_{xu} = 0$, $H_{ux} = 0$), so that the optimal control $u^*(t)$ and the optimal increment $\delta u^*(t)$ are found by using the same equations.

Sufficient Conditions for a Minimizing Control

Our results on neighboring optimal paths can be used to find sufficient conditions for the control $u^*(t)$ determined by using Table 3.2-1 to be one that minimizes the cost index (as opposed to maximizing it, for example). This discussion should be compared with the derivation of the constrained curvature matrix L_{uu}^f in Section 1.2 and, of course, to the discrete-time discussion in Section 4.6.

First, we shall solve the simpler problem of determining when the optimal LQ regulator found as in Table 3.3-1 is a minimizing control. In this case the plant

TABLE 4.3-1   Perturbation Control

Perturbation system model:
$$\delta \dot{x} = f_x \delta x + f_u \delta u, \quad \delta x(t_0) \text{ given}$$

Cost function second variation:
$$\delta^2 J' = \frac{1}{2} \delta x^T(T)(\phi_{xx} + (\psi_x^T v)_x)|_T \delta x(T)$$
$$+ \frac{1}{2} \int_{t_0}^{T} [\delta x^T \quad \delta u^T] \begin{bmatrix} H_{xx} & H_{xu} \\ H_{ux} & H_{uu} \end{bmatrix} \begin{bmatrix} \delta x \\ \delta u \end{bmatrix} dt$$

Final-state incremental constraint:
$$\psi_x(x(T), T)\delta x(T) = \delta \psi, \ \delta \psi \text{ given}$$

Optimal control increment:
$$K = H_{uu}^{-1} \left(H_{ux} + f_u^T S \right)$$
$$-\dot{S} = f_x^T S + S f_x - K^T H_{uu} K + H_{xx}, \quad t \le T$$
$$-\dot{V} = (f_x - f_u K)^T V, \quad t \le T$$
$$\dot{P} = V^T f_u H_{uu}^{-1} f_u^T V, \quad t \le T$$
$$\delta u = - \left(K - H_{uu}^{-1} f_u^T V P^{-1} V^T \right) \delta x - H_{uu}^{-1} f_u^T V P^{-1} \delta \psi$$

Boundary conditions:
$$S(T) = (\phi_{xx} + (\psi_x^T v)_x)|_T$$
$$V(T) = \psi_x^T(x(T), T)$$
$$P(T) = 0$$

is linear with a quadratic performance index, and $\psi(x(T), T)$ is identically the zero function, since there are no final-state constraints. The Hamiltonian is

$$H = \frac{1}{2} \left(x^T Q x + u^T R u \right) + \lambda^T (A x + B u). \tag{4.3-7}$$

Suppose that the optimal control $u^*(t)$ has been found by using Table 3.3-1, and that $x^*(t)$ is the resulting optimal state trajectory. Then the first variation dJ' in the augmented cost index (3.2-6) is equal to zero, and the second variation is given by (4.3-2), or

$$\delta^2 J' = \frac{1}{2} \delta x^T(T) S(T) \delta x(T) + \frac{1}{2} \int_{t_0}^{T} [\delta x^T \quad \delta u^T] \begin{bmatrix} Q & 0 \\ 0 & R \end{bmatrix} \begin{bmatrix} \delta x \\ \delta u \end{bmatrix} dt. \tag{4.3-8}$$

This, however, is simply $J(t_0)$ in Table 3.3-1 with u and x replaced by their variations. The equations in Table 4.3-1 thus reduce to those in Table 3.3-1.

If $\delta^2 J' > 0$ for all $\delta u \neq 0$, then $u^*(t)$ is a control that locally minimizes $J(t_0)$. Equation (3.3-47) applies since J and $\delta^2 J'$ have the same form, so the second variation can be written in terms of the integral of a perfect square

$$\delta^2 J' = \frac{1}{2}\delta x^{\mathrm{T}}(t_0)S(t_0)\delta x(t_0) + \frac{1}{2}\int_{t_0}^{T} \left\| R^{-1}B^{\mathrm{T}}S\delta x + \delta u \right\|_R^2 dt, \qquad (4.3\text{-}9)$$

where $S(t)$ is the solution of the Riccati equation. The optimal control variation is the one that minimizes $\delta^2 J'$,

$$\delta u = -R^{-1}B^{\mathrm{T}}S\delta x, \qquad (4.3\text{-}10)$$

and then $\delta^2 J' = 0$ as long as the initial condition $\delta x(t_0)$ equals zero.

Compare the optimal trajectory x^* and any other trajectory with the same initial point, so that $\delta x(t_0) = 0$. The feedback law (4.3-10) then results in $\delta u = 0$ as long as

$$R > 0, \qquad (4.3\text{-}11)$$

for then $\delta x = 0$ implies $\delta u = 0$. Thus, $\delta^2 J' > 0$ if $\delta u \neq 0$ whenever $\delta x(t_0) = 0$. A sufficient condition for $u^*(t)$ to be a minimizing control is therefore (4.3-11), which is guaranteed by our assumptions. According to (3.3-47), R is the second derivative of J with respect to the control input u, so it can be interpreted as a constrained curvature matrix in the continuous-time case.

Now let us discuss sufficient conditions for the more general control found by Table 3.2-1 to be a minimizing control. We consider the fixed-final-time case so that $dT = 0$ in (3.2-10).

Let u^* be determined as in Table 3.2-1, and let the optimal state trajectory be x^*. Then in this more general case it can be shown that, if $\delta x(t_0) = 0$ and $\delta \psi = 0$,

$$\delta^2 J' = \frac{1}{2}\int_{t_0}^{T} \left\| H_{uu}^{-1}\left[H_{ux} + f_u^{\mathrm{T}}\left(S - VP^{-1}V^{\mathrm{T}} \right) \right]\delta x + \delta u \right\|_{H_{uu}}^2 dt, \qquad (4.3\text{-}12)$$

where S, V, P are found as in Table 4.3-1. The optimal control variation

$$\delta u = -H_{uu}^{-1}\left[H_{ux} + f_u^{\mathrm{T}}\left(S - VP^{-1}V^{\mathrm{T}} \right) \right]\delta x \qquad (4.3\text{-}13)$$

results in $\delta^2 J' = 0$, and is the same as the control in Table 4.3-1 (for $\delta \psi = 0$, $\delta x(t_0) = 0$). If $\delta x(t) = 0$, then $\delta u(t) = 0$ as long as, for $t_0 \leq t < T$,

$$H_{uu} > 0, \qquad (4.3\text{-}14)$$

$$P < 0, \qquad (4.3\text{-}15)$$

and

$$S - VP^{-1}V^{\mathrm{T}} \quad \text{finite.} \tag{4.3-16}$$

Arguing as above we see that these three are sufficient conditions for the control law in Table 3.2-1 to yield a minimum cost for the fixed-final-time case.

4.4   THE DISCRETE-TIME TRACKING PROBLEM

Up to this point we have discussed in detail only performance indices that keep the state small without using too much control energy. We have also mentioned the fixed-final-state controller that drives the state to a desired value at the final time, but this control scheme is essentially open loop.

In this section we want to construct a control scheme that makes the system follow (or track) a desired trajectory over the entire time interval by using a closed-loop control law. Such control strategies are important, for example, in the control of spacecraft and robot arms.

We shall deal with time-invariant systems, but the results generalize to the time-varying case.

Nonlinear Systems

Systems such as robot arms are nonlinear and cannot conveniently be linearlized. Let us therefore discuss the tracking problem for nonlinear systems first. Unfortunately, we shall not be able to present a solution for the control law, but we can at least formulate the nonlinear problem to see some of the ways in which it differs from the regulator problem.

Consider the plant dynamics

$$x_{k+1} = f(x_k, u_k). \tag{4.4-1}$$

If we are interested in making a certain linear combination of the states

$$y_k = Cx_k \tag{4.4-2}$$

follow a desired known reference signal r_k over a time interval $[0, N]$, then we can minimize the cost function

$$J_0 = \frac{1}{2}(Cx_N - r_N)^{\mathrm{T}}P(Cx_N - r_N)$$

$$+ \frac{1}{2}\sum_{k=0}^{N-1}\left[(Cx_k - r_k)^{\mathrm{T}}Q(Cx_k - r_k) + u_k^{\mathrm{T}}Ru_k\right], \tag{4.4-3}$$

where $P \geq 0$, $Q \geq 0$, $R > 0$, and the actual value of x_N is not constrained.

According to Table 2.2-1, the optimal control u_k is given by solving

State system:

$$x_{k+1} = f(x_k, u_k) \qquad (4.4\text{-}4)$$

Costate system:

$$\lambda_k = \left(\frac{\partial f}{\partial x_k}\right)^{\mathrm{T}} \lambda_{k+1} + C^{\mathrm{T}} Q C x_k - C^{\mathrm{T}} Q r_k \qquad (4.4\text{-}5)$$

Stationarity condition:

$$0 = \left(\frac{\partial f}{\partial u_k}\right)^{\mathrm{T}} \lambda_{k+1} + R u_k \qquad (4.4\text{-}6)$$

with

Boundary conditions:

$$x_0 \quad \text{given}, \qquad (4.4\text{-}7)$$

$$\lambda_N = C^{\mathrm{T}} P (C x_N - r_N). \qquad (4.4\text{-}8)$$

From the stationarity condition, the optimal control is

$$u_k = -R^{-1} \left(\frac{\partial f}{\partial u_k}\right)^{\mathrm{T}} \lambda_{k+1}. \qquad (4.4\text{-}9)$$

The input Jacobian depends, in general, on x_k and u_k in a nonlinear manner, so that the optimal control is, in general, a nonlinear feedback of the state and the costate.

The effect of the reference track r_k is to add two terms to these equations that were not present in the regulator program. From (4.4-8), it is evident that the final costate is no longer proportional to the state as it was in the regulator problem. Furthermore, although we might be able to use (4.4-9) to eliminate u_k in the state and costate equations, it is clear from (4.4-5) that the resulting Hamiltonian system is no longer homogeneous; it is now driven by a forcing function $-C^{\mathrm{T}} Q r_k$ dependent on the desired trajectory.

The Linear Quadratic Tracking Problem

If the plant is linear so that

$$x_{k+1} = A x_k + B u_k \qquad (4.4\text{-}10)$$

with $x_k \in R^n$, then a nice form can be found for the tracking control scheme. In this case, (4.4-4)–(4.4-6) become

$$x_{k+1} = A x_k + B u_k, \qquad (4.4\text{-}11)$$

$$\lambda_k = A^{\mathrm{T}} \lambda_{k+1} + C^{\mathrm{T}} Q C x_k - C^{\mathrm{T}} Q r_k,$$

$$0 = B^{\mathrm{T}} \lambda_{k+1} + R u_k. \qquad (4.4\text{-}12)$$

The boundary conditions are still given by (4.4-7) and (4.4-8).

The stationarity condition shows that

$$u_k = -R^{-1}B^T\lambda_{k+1}, \tag{4.4-13}$$

and using this in (2.6-11) yields the nonhomogeneous Hamiltonian system

$$\begin{bmatrix} x_{k+1} \\ \lambda_k \end{bmatrix} = \begin{bmatrix} A & -BR^{-1}B^T \\ C^T Q C & A^T \end{bmatrix} \begin{bmatrix} x_k \\ \lambda_{k+1} \end{bmatrix} + \begin{bmatrix} 0 \\ -C^T Q \end{bmatrix} r_k. \tag{4.4-14}$$

The optimal control scheme given by these equations is similar to Fig. 2.2-1, but with an added input $-C^T Q r_k$. This version of the control law cannot be implemented in practice, since the boundary conditions are split between times $k = 0$ and $k = N$. Let us find a more useful version.

The optimal control is a linear *costate* feedback, but unfortunately, because of the forcing term in the costate equation and (4.4-8), it is no longer possible to express it as a linear *state* feedback as we did in Table 2.2-1 for the LQ regulator. We can, however, express u_k as a combination of a linear state variable feedback *plus* a term depending on r_k. To do this, we use a sweep method like the one in Section 2.2.

From the looks of (4.4-8), it seems reasonable to assume that for all $k \leq N$, we can write

$$\lambda_k = S_k x_k - v_k \tag{4.4-15}$$

for some as yet unknown auxiliary sequences S_k and v_k (cf. (2.2-50)). Note that S_k is an $n \times n$ matrix, whereas v_k is an n vector. This will turn out to be a valid assumption if consistent equations can be found for S_k and v_k. To find these equations, use (4.4-15) in the state equation portion of (4.4-14) to get

$$x_{k+1} = Ax_k - BR^{-1}B^T S_{k+1} x_{k+1} + BR^{-1}B^T v_{k+1},$$

which can be solved for x_{k+1} to yield

$$x_{k+1} = (I + BR^{-1}B^T S_{k+1})^{-1}(Ax_k + BR^{-1}B^T v_{k+1}). \tag{4.4-16}$$

Using (4.4-15) and (4.4-16) in the costate equation gives

$$S_k x_k - v_k = C^T Q C x_k + A^T S_{k+1}(I + BR^{-1}B^T S_{k+1})^{-1}$$
$$\times (Ax_k + BR^{-1}B^T v_{k+1}) \tag{4.4-17}$$
$$- A^T v_{k+1} - C^T Q r_k,$$

or

$$[-S_k + A^{\mathrm{T}} S_{k+1}(I + BR^{-1} B^{\mathrm{T}} S_{k+1})^{-1} A + C^{\mathrm{T}} QC] x_k$$
$$+ [v_k + A^{\mathrm{T}} S_{k+1}(I + BR^{-1} B^{\mathrm{T}} S_{k+1})^{-1} BR^{-1} B^{\mathrm{T}} v_{k+1}$$
$$- A^{\mathrm{T}} v_{k+1} - C^{\mathrm{T}} Q r_k] = 0. \tag{4.4-18}$$

This equation must hold for all state sequences x_k given any x_0, so that the bracketed terms must individually vanish. Using the matrix inversion lemma therefore allows us to write

$$S_k = A^{\mathrm{T}}[S_{k+1} - S_{k+1} B(B^{\mathrm{T}} S_{k+1} B + R)^{-1} B^{\mathrm{T}} S_{k+1}] A + C^{\mathrm{T}} QC \tag{4.4-19}$$

and

$$v_k = [A^{\mathrm{T}} - A^{\mathrm{T}} S_{k+1} B(B^{\mathrm{T}} S_{k+1} B + R)^{-1} B^{\mathrm{T}}] v_{k+1} + C^{\mathrm{T}} Q r_k. \tag{4.4-20}$$

By comparing (4.4-15) and (4.4-8), the boundary conditions for these recursions are seen to be

$$S_N = C^{\mathrm{T}} PC, \tag{4.4-21}$$

$$v_N = C^{\mathrm{T}} P r_N, \tag{4.4-22}$$

Since the auxiliary sequences S_k and v_k can now be computed, assumption (4.4-15) was a valid one, and the optimal control is

$$u_k = -R^{-1} B^{\mathrm{T}} \lambda_{k+1} = -R^{-1} B^{\mathrm{T}} (S_{k+1} x_{k+1} - v_{k+1}). \tag{4.4-23}$$

We are still not quite done, though, since this control depends on x_{k+1}, which is not known at time k. Substitute the state equation (4.4-11) into (4.4-23) and write

$$u_k = -R^{-1} B^{\mathrm{T}} S_{k+1}(A x_k + B u_k) + R^{-1} B^{\mathrm{T}} v_{k+1}.$$

Now premultiply by R and solve for u_k to see that

$$u_k = (B^{\mathrm{T}} S_{k+1} B + R)^{-1} B^{\mathrm{T}} (-S_{k+1} A x_k + v_{k+1}). \tag{4.4-24}$$

This is the control law we have been seeking.

We can improve the appearance of our equations by defining a *feedback* gain

$$K_k = (B^{\mathrm{T}} S_{k+1} B + R)^{-1} B^{\mathrm{T}} S_{k+1} A \tag{4.4-25}$$

and a *feedforward* gain

$$K_k^v = (B^{\mathrm{T}} S_{k+1} B + R)^{-1} B^{\mathrm{T}}. \tag{4.4-26}$$

Then we have the formulation in Table 4.4-1.

TABLE 4.4-1   Discrete Linear Quadratic Tracker

System model:
$$x_{k+1} = Ax_k + Bu_k, \quad k > i$$

Performance index:
$$y_k = Cx_k$$

$$J_i = \frac{1}{2}(y_N - r_N)^T P(y_N - r_N) + \frac{1}{2}\sum_{k=i}^{N-1}\left[(y_k - r_k)^T Q(y_k - r_k) + u_k^T Ru_k\right]$$

Assumptions:
$$P \geq 0, \quad Q \geq 0, \quad R > 0, \qquad \text{with all three symmetric}$$

Optimal affine control:

$$K_k = \left(B^T S_{k+1} B + R\right)^{-1} B^T S_{k+1} A, \quad S_N = C^T PC$$

$$S_k = A^T S_{k+1}(A - BK_k) + C^T QC \tag{4.4-27}$$

$$v_k = (A - BK_k)^T v_{k+1} + C^T Qr_k, \quad v_N = C^T Pr_N \tag{4.4-28}$$

$$K_k^v = \left(B^T S_{k+1} B + R\right)^{-1} B^T$$

$$u_k = -K_k x_k + K_k^v v_{k+1} \tag{4.4-29}$$

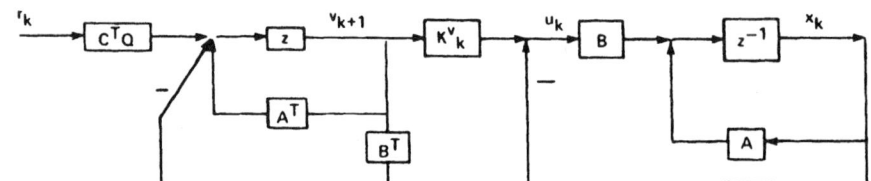

FIGURE 4.4-1   Formulation of the LQ tracker as an affine state feedback.

A schematic of the optimal LQ tracker is shown in Fig. 4.4-1. It consists of an *affine* state feedback (i.e., a term linear in x_k *plus* a term independent of x_k) whose gains are dependent on the solution to the Riccati equation (4.4-19) and whose second term is dependent on an auxiliary sequence v_k derived from reference r_k by the auxiliary difference equation (4.4-28).

The closed-loop plant under the influence of this control is the nonhomogeneous time-varying system

$$x_{k+1} = (A - BK_k)x_k + BK_k^v v_{k+1}. \tag{4.4-30}$$

Note that the auxiliary system (4.4-28) that generates v_k is simply the adjoint of this closed-loop plant!

It can be shown after two pages of tedious work that the optimal value of the performance index on the interval $[k, N]$ under the influence of the control sequence defined in Table 4.4-1 is

$$J_k = \frac{1}{2}x_k^T S_k x_k - x_k^T v_k + w_k, \qquad (4.4\text{-}31)$$

where the new auxiliary sequence w_k satisfies the backward recursion

$$w_k = w_{k+1} + \frac{1}{2}r_k^T Q r_k - \frac{1}{2}v_{k+1}B(B^T S_{k+1}B + R)^{-1}B^T v_{k+1}, \quad k < N \quad (4.4\text{-}32)$$

with boundary condition

$$w_N = \frac{1}{2}r_N^T P r_N. \qquad (4.4\text{-}33)$$

An important special case of the LQ tracker is when we select $Q = 0$ and $r_k = 0$ for $k < N$. Then we are interested only in driving the state near a desired value r_N without using too much control energy. As P in (4.4-3) gets larger, x_N approaches r_N more closely. When $Q = 0$, it is easier to solve for S_k^{-1} instead of S_k by applying the matrix inversion lemma to (4.4-19) to get the Lyapunov equation (2.2-71). If we want x_N to be exactly equal to r_N, then we would let $P \to \infty$, so that the boundary condition for the Lyapunov equation becomes $S_N^{-1} = 0$. It can be shown that under these circumstances, the optimal tracker reduces to the fixed-final-state control, which can be written in the open-loop form (2.2-38). Of course, the affine formulation in Table 4.4-1 would be used in practice as closed-loop control is more robust.

Implementation and a Suboptimal Tracker

Much of the work in Table 4.4-1 can be done off-line to simplify the implementation of the tracker. The Riccati equation does not depend on the state trajectory; so, sequences S_k, K_k, and K_k^v can be computed offline before the control is applied. Then the gains K_k and K_k^v can be stored for use during the actual control run on the plant. Presumably, the desired track r_k is known beforehand, so that the auxiliary sequence v_k can also be computed offline and stored. During the actual control run, then, the only work left to do is to compute the optimal control using (4.4-29).

An alternative to the storage of v_k is as follows. First, solve (4.4-28) off-line (backward in time) using the given values of r_k to determine the initial condition v_0. Store only this value v_0. Then, during the control run, at each step solve

$$v_{k+1} = (A - BK_k)^{-T}v_k - (A - BK_k)^{-T}C^T Q r_k \qquad (4.4\text{-}34)$$

for v_{k+1}, and use (4.4-29) to find the optimal control.

In some applications, it is satisfactory to use a suboptimal tracker constructed as follows. If (A, B) is reachable and $(A, C\sqrt{Q})$ is observable, the tracker gains K_k and K_k^v reach steady-state values K_∞ and K_∞^v as $(N - k) \to \infty$. It is then worth asking whether a suboptimal tracker defined by

$$v_{k+1} = (A - BK_\infty)^{-\mathrm{T}} v_k - (A - BK_\infty)^{-\mathrm{T}} C^\mathrm{T} Q r_k, \qquad (4.4\text{-}35)$$

$$u_k = -K_\infty x_k + K_\infty^v v_{k+1}, \qquad (4.4\text{-}36)$$

would perform adequately, where (4.4-35) is initialized using a value of v_0 computed off-line using (4.4-28). This *time-invariant tracker* has the advantage of requiring storage of no sequences other than r_k for the actual control run. It is also easier to implement than the optimal tracker.

Experience shows that the time-invariant tracker is often satisfactory. Its adequacy can be checked in a particular application by performing a computer simulation. We illustrate the time-invariant tracker in a subsequent example. See Athans and Falb (1966) for a discussion of the case where the desired track r_k is a constant. Note that the time-invariant tracker is optimal for the performance index (4.4-3) when the final time N is infinity and $P = 0$ (i.e., the *infinite-horizon tracking problem*).

Examples and Exercises

Example 4.4-1.   Tracking for a System Obeying Newton's Law

This belongs in the sequence containing Examples 2.3-2 and 2.4-3. The plant is the sampled version of "Newton's system,"

$$x_{k+1} = \begin{bmatrix} 1 & T \\ 0 & 1 \end{bmatrix} x_k + \begin{bmatrix} T^2/2 \\ T \end{bmatrix} u_k, \qquad (1)$$

where x_k contains the kth samples of position and velocity. Let $T = 0.5$, and the run time be 5 sec so that $N = 5/0.5 = 10$. The tracking cost is

$$J_0 = \frac{1}{2}(x_N - r_N)^\mathrm{T} \begin{bmatrix} p_d & 0 \\ 0 & p_v \end{bmatrix} (x_N - r_N)$$

$$+ \frac{1}{2} \sum_{k=0}^{N-1} \left[(x_k - r_k)^\mathrm{T} \begin{bmatrix} q_d & 0 \\ 0 & q_v \end{bmatrix} (x_k - r_k) + r u_k^2 \right], \qquad (2)$$

with $r_k \in R^2$ the samples of the reference track.

Suppose that we want the position $d(t)$ to be along the parabolic track

$$c_0 + c_1 t + \frac{1}{2} c_2 t^2 \qquad (3)$$

for some given constants c_i, and that we are not concerned about velocity. Then the first component of r_k is

$$r_k^d = c_0 + c_1 T k + \frac{1}{2} c_2 T^2 k^2, \tag{4}$$

the second component can be zero, and we set p_v and q_v to zero in (2). (An alternative is to reformulate (2) in terms of sampled position $d_k = C x_k$ where $C = [1 \quad 0]$.)

According to Table 4.4-1, the LQ tracker is given by the LQ regulator equations (7)–(15) in Example 2.3-2, with some equations added to compute feedforward gain K_k^v and auxiliary signal v_k. (In this example v_k does not represent velocity, but the tracker auxiliary signal.) Defining

$$K_k^v \triangleq \begin{bmatrix} k_1^v & k_2^v \end{bmatrix} \tag{5}$$

(the time dependence of the components is not shown), and

$$v_k \triangleq \begin{bmatrix} v_k^d \\ v_k^v \end{bmatrix}, \tag{6}$$

these equations are $v_k = (A_k^{cl})^T v_{k+1} + Q r_k$ or

$$v_k^d = a_{11}^{cl} v_{k+1}^d + a_{21}^{cl} v_{k+1}^v + q_d r_k^d, \tag{7}$$

$$v_k^v = a_{12}^{cl} v_{k+1}^d + a_{22}^{cl} v_{k+1}^v \tag{8}$$

and $K_k^v = (B^T S_{k+1} B + B)^{-1} B^T$ or

$$k_1^v = T^2 / 2\delta, \tag{9}$$

$$k_2^v = T / \delta, \tag{10}$$

with δ defined in Example 2.3-2. The tracker control is given by

$$u_k = -k_1 d_k - k_2 x_2(k) + k_1^v v_{k+1}^d + k_2^v v_{k+1}^v, \tag{11}$$

with $x_2(k)$ the second component of x_k (i.e., the velocity).

A software implementation is just Fig. 2.3-9 with these added equations.   ∎

Exercise 4.4-2.   System with Known Disturbance

Suppose the plant has a known disturbance d_k so that

$$x_{k+1} = A x_k + B u_k + d_k. \tag{1}$$

Let the cost function be

$$J_i = \frac{1}{2} x_N^T S_N x_N + \frac{1}{2} \sum_{k=i}^{N-1} (x_k^T Q x_k + u_k^T R u_k). \tag{2}$$

Show that the optimal control is given by

$$K_k = \left(B^T S_{k+1} B + R\right)^{-1} B^T S_{k+1} A, \quad S_N \text{ given} \tag{3}$$

$$S_k = A^T S_{k+1}(A - BK_k) + Q, \tag{4}$$

$$v_k = (A - BK_k)^T v_{k+1} - (A - BK_k)^T S_{k+1} d_k, \quad v_N = 0, \tag{5}$$

$$K_k^v = \left(B^T S_{k+1} B + R\right)^{-1} B^T \tag{6}$$

$$u_k = K_k x_k + K_k^v v_{k+1}. \tag{7}$$

■

Exercise 4.4-3. Formulating the Tracking Problem as a Regulator Problem

Under certain conditions on the reference track r_k, the LQ tracking problem can be reformulated as a LQ regulator problem (Athans and Falb 1966). Let y_k be the output of a scalar system described by the difference equation

$$y_k + a_1 y_{k-1} + \cdots + a_n y_{k-n} = u_{k-1} \tag{1}$$

for given a_i. (Note that u_k and y_k could be the input and output of a state system from which the form (1) can be determined.) We want to find a control input sequence u_k to make y_k match a known reference signal r_k. Define the *tracking error* as

$$e_k = y_k - r_k \tag{2}$$

and a performance index by

$$J_i = \frac{1}{2} p e_N^2 + \frac{1}{2} \sum_{k=i}^{N-1} \left(q e_k^2 + r u_k^2\right). \tag{3}$$

Define an operator by

$$\Delta(z^{-1}) = 1 + a_1 z^{-1} + \cdots + a_n z^{-n}, \tag{4}$$

where z^{-1} is the *delay operator* such that $z^{-1} y_k = y_{k-1}$. Then the plant can be represented as

$$\Delta(z^{-1}) y_k = u_{k-1} \tag{5}$$

Suppose r_k is selected from the class of signals satisfying

$$\Delta(z^{-1}) r_k = 0; \tag{6}$$

that is, r_k is an unforced solution to the homogeneous plant (as such it must consist of a sum of the plant's natural modes). Then the scalar tracking problem (1), (3) can be reformulated as an n-vector LQ regulator problem that can be solved using Table 2.2-1.

a. Show that

$$\Delta(z^{-1})e_k = u_{k-1},\qquad(7)$$

so that e_k is the output of a linear system driven by u_k. (*Note*: e_k is not necessarily equal to y_k. Why?)

b. Define a state vector $xk \in Rn$ by rewriting (7) in reachable canonical form (Kailath 1980); that is, find A, B, and C in

$$x_{k+1} = Ax_k + Bu_k,\qquad(8a)$$

$$e_k = Cx_k.\qquad(8b)$$

c. Rewrite (3) in terms of xk; that is, find P and Q in

$$J_i = \frac{1}{2}x_N^T P x_N + \frac{1}{2}\sum_{k=i}^{N-1}\left(x_k^T Q x_k + ru_k^2\right).\qquad(9)$$

Show that $(A, \sqrt{Q})$ is observable.

d. Hence, show how to find the scalar feedback coefficients $Ki(k)$ in the tracking scheme shown in Fig. 4.4-2.

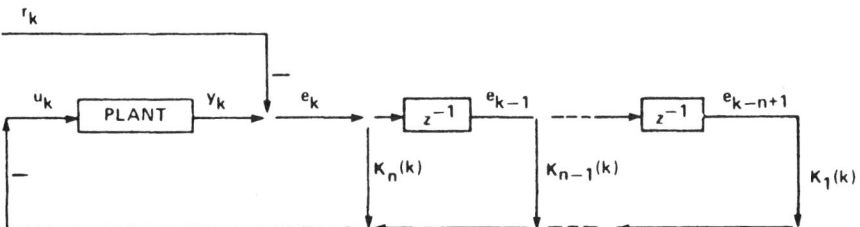

FIGURE 4.4-2   Tracking scheme using feedback of the error signal e_k and its $n - 1$ previous values.

e. Does this scheme generalize to plants described by

$$Y_k + a_1 y_{k-1} + \cdots + a_n y_{k-n} = b_1 u_{k-1} + \cdots + b_m u_{k-m}?\qquad(10)$$

∎

4.5   DISCRETE REGULATOR WITH FUNCTION OF FINAL STATE FIXED

To obtain a given fixed final value for the state x_N, we can use the open-loop control (2.2-38), which depends on the existence of an inverse to the reachability gramian. On the other hand, to make x_N take on the value of zero, we have the additional option of using the closed-loop formulation of Table 2.2-1 with $Q = 0$ and $S_N \to \infty$. Then it is convenient to use the "inverse" (2.2-71) of the Riccati equation.

In this section we are concerned with driving a function of the final state x_N to zero. An application is the case where we want to drive some of the components of x_N to zero, and we are not concerned about the other components. If $n = 4$, for example, and we want state components one and three to be zero at the final time N, then we could zero the function

$$\psi(x_N) = \begin{bmatrix} 1 & 0 & 0 & 0 \\ 0 & 0 & 1 & 0 \end{bmatrix} x_N. \tag{4.5-1}$$

Thus, we are effectively defining an "output" $\psi(x_N) = C x_N$ that we want to make vanish.

Nonlinear Systems

Let the plant be given by

$$x_{k+1} = f(x_k, u_k) \tag{4.5-2}$$

and the performance index by

$$J_i = \phi(N, x_N) + \sum_{k=i}^{N-1} L(x_k, u_k). \tag{4.5-3}$$

Our objective is to define a control sequence that will minimize J_i and *also* ensure that a specified function ψ of the final state is driven to zero. Thus, we want

$$\phi(N, x_N) = 0. \tag{4.5-4}$$

This is almost the problem we solved in Section 2.1. As a matter of fact, the solution is also almost the same. Only the terminal condition in Table 2.1-1 is different. To find the new condition, we can follow through the derivation in Section 2.1, carrying along one additional term. Since (4.5-4) is simply an additional constraint, we need to introduce an additional Lagrange multiplier v. Equation (4.5-4) only applies at one time, so v is a constant. Suppose $\psi(N, x_N) \in R^p$, then v is a p vector.

Adjoining the constraints (4.5-2) and (4.5-4) to the performance index gives an augmented performance index

$$J' = \phi(N, x_N) + v^T \psi(N, x_N)$$
$$+ \sum_{k=i}^{N-1} \left[L(x_k, u_k) + \lambda_{k+1}^T (f(x_k, u_k) - x_{k+1}) \right]. \tag{4.5-5}$$

Now we follow the work subsequent to (2.1-3). The Hamiltonian is

$$H^k = L + \lambda_{k+1}^T f, \tag{4.5-6}$$

exactly as it was there, and if we write down the steps, we obtain exactly the results of Table 2.1-1. The only difference is in the first term of (2.1-6), which now includes $y_{xN}^T v \, dx_N$. The final condition in Table 2.1-1 must therefore be replaced by

$$\left(\frac{\partial \phi}{\partial x_N} + \left(\frac{\partial \psi}{\partial x_N} \right)^T v - \lambda_N \right)^T dx_N = 0. \tag{4.5-7}$$

This seemingly minor change results in significant changes in the control law, as we now see.

The Linear Quadratic Case

What we intend to do next is follow the development of the free-final-state control as closely as we can. Exactly as in the tracking problem, we shall find it necessary to add some equations to Table 2.2-1. In fact, the LQ control with function of final state fixed looks very much like the LQ tracker in Table 4.4-1. Now, however, we need to solve a quadrature like (4.4-32) to determine the optimal control sequence (in the tracking problem, (4.4-32) was needed only to determine the optimal cost).

We shall see that if a function of the final state is fixed, the control law has a part that is similar to the free-final-state control. If $\psi_N(N, x_N)$ is identically the zero function, then our results reduce to the former type of control, and if $\psi_N(N, x_N) = x_N - r_N$ for a given r_N, then our results reduce to the latter type.

To proceed, let the plant be

$$x_{k+1} = Ax_k + Bu_k \tag{4.5-8}$$

with performance index

$$J_i = \frac{1}{2} x_N^T S_N x_N + \frac{1}{2} \sum_{k=i}^{N-1} \left(x_k^T Q x_k + u_k^T R u_k \right). \tag{4.5-9}$$

It is desired to find a control sequence to minimize J_i and also ensure that

$$\psi(N, x_N) = C x_N - r_N = 0 \tag{4.5-10}$$

for a given $r_N \in R^p$ and matrix C. Thus, we are concerned with making a specified linear function of the final state exactly match a given vector r_N.

The state and costate equations are the same as in Section 2.2, and the optimal control is still

$$u_k = -R^{-1}B^{\mathsf{T}}\lambda_{k+1}. \tag{4.5-11}$$

Hence (2.2-45) and (2.2-46) hold. We reproduce them here:

$$x_{k+1} = Ax_k - BR^{-1}B^{\mathsf{T}}\lambda_{k+1}, \tag{4.5-12}$$

$$\lambda_k = Qx_k + A^{\mathsf{T}}\lambda_{k+1}. \tag{4.5-13}$$

The new boundary condition is (4.5-7), or

$$\lambda_N = S_N x_N + C^{\mathsf{T}}v. \tag{4.5-14}$$

To use the sweep method of solution (Bryson and Ho 1975), assume that a relation like (4.5-14) holds for all k (recall that v is a constant), that is,

$$\lambda_k = S_k x_k + V_k v. \tag{4.5-15}$$

Compare this with (4.4-15). Note that the postulated V_k is an $n \times p$ *matrix*. Now we need to find S_k, V_k, and v. To do this, use (4.5-15) in the state equation (4.5-12) and solve for x_{k+1} to get the recursion

$$x_{k+1} = (I + BR^{-1}B^{\mathsf{T}}S_{k+1})^{-1}(Ax_k - BR^{-1}B^{\mathsf{T}}V_{k+1}v). \tag{4.5-16}$$

This yields the optimal state trajectory. Use this equation and (4.5-15) in (4.5-13) to obtain (cf. (4.4-18))

$$[-S_k + A^{\mathsf{T}}S_{k+1}(I + BR^{-1}B^{\mathsf{T}}S_{k+1})^{-1}A + Q]x_k$$
$$+ [-V_k - A^{\mathsf{T}}S_{k+1}(I - BR^{-1}B^{\mathsf{T}}S_{k+1})^{-1}BR^{-1}B^{\mathsf{T}}V_{k+1} + A^{\mathsf{T}}V_{k+1}]v = 0. \tag{4.5-17}$$

Since this equality holds for all trajectories x_k arising from any initial condition x_0, each term in brackets must vanish. The matrix inversion lemma therefore yields the Riccati equation

$$-S_k = A^{\mathsf{T}}\left[S_{k+1} - S_{k+1}B\left(B^{\mathsf{T}}S_{k+1}B + R\right)^{-1}B^{\mathsf{T}}S_{k+1}\right]A + Q \tag{4.5-18}$$

and the auxiliary homogeneous difference equation

$$V_k = \left[A^{\mathsf{T}} - A^{\mathsf{T}}S_{k+1}B\left(B^{\mathsf{T}}S_{k+1}B + R\right)^{-1}B^{\mathsf{T}}\right]V_{k+1}. \tag{4.5-19}$$

In terms of the Kalman gain

$$K_k = \left(B^\mathrm{T} S_{k+1} B + R\right)^{-1} B^\mathrm{T} S_{k+1} A \qquad (4.5\text{-}20)$$

these can be written

$$S_k = A^\mathrm{T} S_{k+1}(A - BK_k) + Q, \qquad (4.5\text{-}21)$$

$$V_k = (A - BK_k)^\mathrm{T} V_{k+1}, \qquad (4.5\text{-}22)$$

where $(A - BK_k)$ is the closed-loop system. Comparing (4.5-15) to (4.5-14), the boundary conditions are seen to be S_N and

$$V_N = C^\mathrm{T}. \qquad (4.5\text{-}23)$$

Now that we can find S_k and V_k, the only thing left is to determine the Lagrange multiplier ν. It is not obvious how to do this. Note that $\lambda_0 = S_0 x_0 + V_0 \nu$, but, in general, V_0 is not square, so this does not help. We are forced to make another assumption.

Accordingly, let us assume that $r_N = C x_N$ is a linear combination of x_k and ν for all k; that is,

$$r_N = C x_N = U_k x_k + P_k \nu \qquad (4.5\text{-}24)$$

for some as yet unknown matrix sequences U_k and P_k. If we can find consistent equations for these postulated variables, then the assumption is valid. Note that this relation does indeed hold for $k = N$ with $U_N = C$ and $P_N = 0$.

The left-hand side of (4.5-24) is a constant, so take the first difference to obtain

$$0 = U_{k+1} x_{k+1} + P_{k+1} \nu - U_k x_k - P_k \nu. \qquad (4.5\text{-}25)$$

Now use (4.5-16) for x_{k+1} to get, after rearrangement and application of the matrix inversion lemma,

$$\left[U_{k+1}\left(A - B\left(B^\mathrm{T} S_{k+1} B + R\right)^{-1} B^\mathrm{T} S_{k+1} A\right) - U_k \right] x_k$$
$$+ \left[P_{k+1} - P_k - U_{k+1} B \left(B^\mathrm{T} S_{k+1} B + R\right)^{-1} B^\mathrm{T} V_{k+1} \right] \nu = 0. \qquad (4.5\text{-}26)$$

The first term says that

$$U_k = U_{k+1}(A - BK_k). \qquad (4.5\text{-}27)$$

Evaluating (4.5-24) for $k = N$ yields

$$U_N = C. \qquad (4.5\text{-}28)$$

Clearly, then,

$$U_k = V_k^T. \tag{4.5-29}$$

The second term now yields a quadrature for P_k,

$$P_k = P_{k+1} - V_{k+1}^T B \left(B^T S_{k+1} B + R \right)^{-1} B^T V_{k+1}, \tag{4.5-30}$$

with $P_N = 0$. Compare this with (4.4-32)!

We are now in a position to solve for v. Suppose that $|P_i| \neq 0$ for the initial time i. Then by (4.5-24)

$$v = P_i^{-1} \left(r_N - V_i^T x_i \right). \tag{4.5-31}$$

If $|P_i| = 0$, the problem has no solution on the interval $[i, N]$, and it is said to be *abnormal*. Note that

$$v = P_k^{-1} \left(r_N - V_k^T x_k \right) \tag{4.5-32}$$

for any k where $|P_k| \neq 0$.

We can now finally compute the optimal control, for using (4.5-31) and (4.5-15) in (4.5-11) gives

$$u_k = -R^{-1} B^T \left[S_{k+1} x_{k+1} + V_{k+1} P_i^{-1} \left(r_N - V_i^T x_i \right) \right]. \tag{4.5-33}$$

To find u_k in terms of the current state x_k, use (4.5-8) in (4.5-33) and solve for u_k to get

$$u_k = - \left(B^T S_{k+1} B + R \right)^{-1} B^T \left[S_{k+1} A x_k + V_{k+1} P_i^{-1} \left(r_N - V_i^T x_i \right) \right]. \tag{4.5-34}$$

This is the optimal control law we have been seeking.

If $|P_k| \neq 0$, then we can use (4.5-32) instead of (4.5-31) to determine v. In this case, (4.5-34) can be written

$$u_k = - \left(B^T S_{k+1} B + R \right)^{-1} B^T \left[V_{k+1} P_k^{-1} r_N + \left(S_{k+1} A - V_{k+1} P_k^{-1} V_k^T \right) x_k \right], \tag{4.5-35}$$

which contains a state feedback term plus a term dependent on the desired final value r_N of Cx_N.

The function of final-state-fixed LQ regulator is summarized in Table 4.5-1. Its structure is similar to that shown in Fig. 4.4-1.

TABLE 4.5-1   Function of Final-State-Fixed LQ Regulator

System model:
$$x_{k+1} = Ax_k + Bu_k, \quad k > i$$

Performance index:
$$J_i = \frac{1}{2}x_N^T S_N x_N + \frac{1}{2}\sum_{k=i}^{N-1}\left(x_k^T Q x_k + u_k^T R u_k\right), \quad S_N \geq 0, \quad Q \geq 0, \quad R > 0$$

Final-state constraint:
$$Cx_N = r_N, \quad r_N \text{ given}$$

Optimal control law:
$$K_k = \left(B^T S_{k+1} B + R\right)^{-1} B^T S_{k+1} A, \quad S_N \text{ given}$$
$$S_k = A^T S_{k+1}(A - BK_k) + Q$$
$$V_k = (A - BK_k)^T V_{k+1}, \quad V_N = C^T$$
$$P_k = P_{k+1} - V_{k+1}^T B\left(B^T S_{k+1} B + R\right)^{-1} B^T V_{k+1}, \quad P_N = 0$$
$$K_k^u = \left(B^T S_{k+1} B + R\right)^{-1} B^T$$
$$u_k = -\left(K_k - K_k^u V_{k+1} P_k^{-1} V_k^T\right)x_k - K_k^u V_{k+1} P_k^{-1} r_N$$

use 4.5-34 !
P_k is a known !

Let us briefly examine the structure of our controller to gain further insight. The optimal control (4.5-35) consists of a feedback term like the one in the free-final-state controller (Table 2.2-1) plus another term. To see what this other term is, write the solution to (4.5-22) as

$$V_k = (A - BK_k)^T(A - BK_{k+1})^T \cdots (A - BK_{N-1})^T C^T. \tag{4.5-36}$$

Matrix V_k thus seems to be a modified-state transition matrix for the adjoint of the time-varying closed-loop system. Now, write the solution of (4.5-30):

$$P_k = -\sum_{j=k+1}^{N} V_j^T B(B^T S_j B + R)^{-1} B^T V_j. \tag{4.5-37}$$

By comparing this with (2.2-36) and identifying V_j^T with the state transition matrix A^{N-i-1} in that equation, we see that $-P_k$ is nothing more than a sort of weighted reachability gramian! We already know that the function of the final-state-fixed problem has a solution if $|P_i| \neq 0$, so this makes sense. Now compare (4.5-33) with (2.2-38) to see that the additional term in our new control law is just a fixed-final-state-type control term, which serves to guarantee that at the final time Cx_N is equal to the given r_N as desired.

It is not difficult to implement this new control law. All we need do is add a few lines to our software that implements the LQ regulator to incorporate the recursions for V_k and P_k.

4.6 DISCRETE SECOND-ORDER VARIATIONS IN THE PERFORMANCE INDEX

Up to this point in the chapter we have been concerned with the first-order differential dJ' of the augmented performance index. Our results in Tables 2.2-1, 2.2-1, 4.4-1, and 4.5-1 were all derived by finding conditions under which the differential dJ' of the appropriate performance index vanishes. These conditions are *necessary* for a control that minimizes the value of the cost index.

In this section we study the second-order variation in the augmented cost index, as we did for the static case in deriving the constrained curvature matrix (1.2-31). Our results will have several applications. They will yield *sufficient* conditions for a minimizing control. They will allow us to calculate *changes* in the optimal control sequence that cause desired changes in the final-state constraints. And, finally, they will provide a means for controlling nonlinear systems by linearizing about a nominal trajectory.

Perturbation Control

Consider the nonlinear plant (4.5-2) with performance index (4.5-3). The initial condition x_i is given, and we require that a given function $\psi(N, x_N)$ of the final state be zero. The augmented performance index J' is given by (4.5-5), and the Hamiltonian is (4.5-6).

By demanding that the first variation dJ' be equal to zero, we derive the optimal control in Table 2.1-1, but with the final condition there replaced by (4.5-7), since a function of the final state is fixed. These equations constitute *necessary* conditions for an optimal control. Assuming the final state is free, (4.5-7) becomes

$$\lambda_N = \phi_x^N + \left(\psi_x^N\right)^{\mathrm{T}} v. \tag{4.6-1}$$

Suppose that we have solved the equations in Table 2.1-1 for an optimal control sequence u_k^* and a resulting optimal state trajectory x_k^*. Thus, $dJ' = 0$. From a Taylor series expansion, we see that the cost J' is then equal to its optimal value plus the second variation d^2J' and higher-order terms. According to (4.5-5) this second variation is given by

$$d^2J' = \frac{1}{2} dx_N^{\mathrm{T}} \left(\phi_{xx}^N + \left(\psi_x^{\mathrm{T}} v\right)_x^N\right) dx_N + \frac{1}{2} \sum_{k=i}^{N-1} \begin{bmatrix} dx_k^{\mathrm{T}} & du_k^{\mathrm{T}} \end{bmatrix} \begin{bmatrix} H_{xx}^k & H_{xu}^k \\ H_{ux}^k & H_{uu}^k \end{bmatrix} \begin{bmatrix} dx_k \\ du_k \end{bmatrix},$$

$$\tag{4.6-2}$$

where

$$H_{ux}^k \triangleq \frac{\partial}{\partial x_k}\left(\frac{\partial H^k}{\partial u_k}\right),\tag{4.6-3}$$

and so on. Increments in x_k and u_k about x_k^* and u_k^* must satisfy the incremental constraints

$$dx_{k+1} = f_x^k\, dx_k + f_u^k\, du_k\tag{4.6-4}$$

and

$$d\psi(N, x_N) = \psi_x^N\, dx_N,\tag{4.6-5}$$

with initial condition

$$dx_i \quad \text{given.}\tag{4.6-6}$$

At this point we note something quite interesting. The *perturbation state equation* (4.6-4) with final constraint (4.6-5) (where $d\psi$ is specified) and performance index (4.6-2) constitute simply a linear quadratic problem with function of final state fixed! (Compare (4.5-8)–(4.5-10).) Thus, we can solve for a control increment du_k that minimizes $d^2 J'$ with constraints (4.6-4) and (4.6-5).

To see why we would be interested in doing this, examine Fig. 4.6-1. To drive the system along x_k^* beginning at x_i, we would apply the optimal control u_k^* computed by solving the equations in Table 2.1-1. Now suppose that the system begins in initial state $x_i + dx_i$, and that in addition we want to make the function of final state take on a value not equal to zero, but equal to a given desired value $d\psi$. Then we do not have to solve again the equations in the table, which may, in general, be nonlinear. Instead, we can solve the *linear* quadratic problem (4.6-2)–(4.6-6) for the optimal increment du_k given dx_i and the desired $d\psi$. The optimal control required to drive the system from $x_i + dx_i$ and satisfy $\psi(N, x_N) = d\psi$ is then given by

$$u_k = u_k^* + du_k.\tag{4.6-7}$$

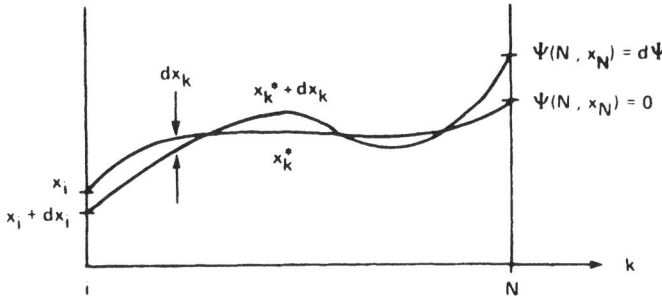

FIGURE 4.6-1   Optimal and neighboring optimal paths.

This control also minimizes the cost (4.5-3). This approach is called *perturbation control*. The state trajectory $x_k^* + dx_k$ resulting when the new control (4.6-7) is applied to the plant is called a *neighboring extremal* (or *neighboring optimal*) path.

Now that our motivation is clear, let us solve the perturbation LQ problem. In Table 4.5-1 we gave the solution to the LQ problem with a function of the final state fixed. Unfortunately, our new problem is not quite the same as the one in Section 4.5, because of the presence of the state-input cross-weighting terms H_{xu}^k and H_{ux}^k. In Exercise 2.2-4 we gave the solution to the LQ regulator problem with state-input cross-weighting terms. The only effect of the off-diagonal weighting term was that a modified Kalman gain was required.

Combining the results shown in Table 4.5-1 and those of Exercise 2.2-4, we obtain the solution to the perturbation LQ problem (4.6-2)–(4.6-6). It is presented in Table 4.6-1. (A rigorous derivation would be along the lines of the one in Section 4.5; see McReynolds (1966) and Bryson and Ho (1975).)

TABLE 4.6-1   Perturbation Control

Perturbation system model:

$$dx_{k+1} = f_x^k dx_k + f_u^k du_k, \quad dx_i \ \text{given}$$

Cost function second variation:

$$d^2 J' = \frac{1}{2} dx_N^T \left(\phi_{xx}^N + (\psi_x^T v)_x^N \right) dx_N + \frac{1}{2} \sum_{k=i}^{N-1} [dx_k^T \quad du_k^T] \begin{bmatrix} H_{xx}^k & H_{xu}^k \\ H_{ux}^k & H_{uu}^k \end{bmatrix} \begin{bmatrix} dx_k \\ du_k \end{bmatrix}$$

Final-state incremental constraint:

$$\psi_x^N dx_N = d\psi(N, x_N), \quad d\psi \ \text{given}$$

Optimal control increment:

$$K_k = \left[(f_u^k)^T S_{k+1} f_u^k + H_{uu}^k \right]^{-1} \left[(f_u^k)^T S_{k+1} f_x^k + H_{ux}^k \right]$$

$$S_k = (f_x^k)^T S_{k+1} f_x^k - K_k^T \left[(f_u^k)^T S_{k+1} f_u^k + H_{uu}^k \right] K_k + H_{xx}^k$$

$$V_k = (f_x^k - f_u^k K_k)^T V_{k+1}$$

$$P_k = P_{k+1} - V_{k+1}^T f_u^k \left[(f_u^k)^T S_{k+1} f_u^k + H_{uu}^k \right]^{-1} (f_u^k)^T V_{k+1}$$

$$K_k^u = \left[(f_u^k)^T S_{k+1} f_u^k + H_{uu}^k \right]^{-1} (f_u^k)^T$$

$$du_k = - \left(K_k - K_k^u V_{k+1} P_k^{-1} V_k^T \right) dx_k - K_k^u V_{k+1} P_k^{-1} d\psi$$

Boundary conditions:

$$S_N = \phi_{xx}^N + (\psi_x^T v)_x^N$$

$$V_N = (\psi_x^N)^T$$

$$P_N = 0$$

If the plant (4.5-2) is linear and the performance index (4.5-3) is quadratic, then Table 4.6-1 reduces to Table 4.5-1. This means that the optimal incremental control for the LQ case is computed using the same equations as the optimal control itself.

It should be clearly understood that in perturbation control the, in general, nonlinear equations in Table 2.1-1 must first be solved to obtain u_k^* in (4.6-7) and v. Then using Table 4.6-1 to find du_k corresponds to linearizing the system about the *nominal trajectory* x_k^* that results when u_k^* is applied to the plant.

Sufficient Conditions for a Minimizing Control

We can use our results on neighboring optimal paths to derive sufficient conditions for the control u_k^* found by using Table 2.1-1 to be one that minimizes the performance index. This development corresponds to the derivation of the constrained curvature matrix L_{uu}^f given by (1.2-31). Let us begin by finding sufficient conditions for the LQ regulator found in Table 2.2-1 to be a minimizing control.

Suppose that the optimal control u_k^* found by using Table 2.2-1 is used, resulting in the optimal state trajectory x_k^*. Then the differential dJ' in the augmented cost index is zero, and the second variation is given by (4.6-2), or

$$d^2J' = \frac{1}{2}dx_N^T S_n \, dx_N + \frac{1}{2}\sum_{k=1}^{N-1}[dx_k^T \quad du_k^T]\begin{bmatrix} Q & 0 \\ 0 & R \end{bmatrix}\begin{bmatrix} dx_k \\ du_k \end{bmatrix}. \qquad (4.6-8)$$

This, however, is exactly the performance index in Table 2.2-1 with u_k and x_k replaced by their increments! What this means is that the equations for determining the optimal control increment du_k in Table 4.6-1 reduce to those in Table 2.2-1.

If $d^2J' > 0$ for all $du \neq 0$, then u_k^* is a control that locally minimizes J_i (as opposed to a control that yields a local maximum, saddle point, etc.). According to (2.2-67), which applies since J_i and d^2J have the same form, we can write the second variation as the sum of perfect squares

$$d^2J' = \frac{1}{2}dx_i^T S_i \, dx_i + \frac{1}{2}\sum_{k=i}^{N-1}\left\|(B^T S_{k+1}B + R)^{-1}B^T S_{k+1}A \, dx_k + du_k\right\|^2_{(B^T S_{k+1}B+R)},$$

$$(4.6-9)$$

where S_k is determined by solution of the Riccati equation. The optimal control increment du_k is the one that makes the sum vanish,

$$du_k = -(B^T S_{k+1}B + R)^{-1}B^T S_{k+1}A \, dx_k, \qquad (4.6-10)$$

and then $d^2J' = 0$ as long as the initial condition $dx_i = 0$.

Now, compare the optimal trajectory x_k^* and any other trajectory with the same initial point so that $dx_i = 0$. Using the feedback (4.6-10) makes $du_k = 0$ as long as

$$B^T S_{k+1} B + R > 0, \tag{4.6-11}$$

for then the feedback gain is finite and $dx_k = 0$ implies $du_k = 0$. Hence, $d^2 J' > 0$ if $du_k \neq 0$ when $dx_i = 0$. The sufficient conditions for a minimizing control are therefore that u_k^* be selected according to Table 2.2-1 and that (4.6-11) hold. Examine Example 1.2-3 to see that $B^T S_{k+1} B + R$ is just a time-varying constrained curvature matrix.

Now let us discuss the more general problem of finding sufficient conditions for the control u_k^* determined using Table 2.1-1 to be a minimizing control. Extra complications arise here because of the final-state constraint and the fact that H_{xu}^k and H_{ux}^k do not necessarily vanish in (4.6-2).

Suppose that u_k^* is determined by using Table 2.1-1 and that the associated optimal state trajectory is x_k^*. Then it can be shown that in this case, if $dx_i = 0$ and $d\psi = 0$, instead of (4.6-9) we obtain

$$d^2 J' = \frac{1}{2} \sum_{k=i}^{N-1} \left\| \left(Z_{uu}^k \right)^{-1} \left[\left(f_u^k \right)^T S_{k+1} f_x^k + H_{ux}^k \right. \right.$$
$$\left. \left. - \left(f_u^k \right)^T V_{k+1} P_k^{-1} V_k^T \right] dx_k + du_k \right\|_{Z_{uu}^k}, \tag{4.6-12}$$

where S_k, V_k, P_k are determined as in Table 4.6-1 and the curvature matrix is

$$Z_{uu}^k = \left(f_u^k \right)^T S_{k+1} f_u^k + H_{uu}^k. \tag{4.6-13}$$

The proof of this is similar to the derivation of (2.2-67). See McReynolds (1966) and Bryson and Ho (1975).

The optimal control increment, which is selected to make the sum vanish, is

$$du_k = - \left(Z_{uu}^k \right)^{-1} \left[\left(f_u^k \right)^T S_{k+1} f_x^k + H_{ux}^k - \left(f_u^k \right)^T V_{k+1} P_k^{-1} V_k^T \right] dx_k, \tag{4.6-14}$$

exactly as in Table 4.6-1. Compare this with (1.2-35).

Now compare the optimal trajectory x_k^* with any other trajectory with the same initial state (so that $dx_i = 0$) and the same final constraint value (so that $d\psi = 0$). Using the control law (4.6-14) results in $du_k = 0$ as long as

$$Z_{uu}^k = \left(f_u^k \right)^T S_{k+1} f_u^k + H_{uu}^k > 0 \tag{4.6-15}$$

and

$$P_k < 0, \tag{4.6-16}$$

for then $dx_k = 0$ implies $du_k = 0$. Then, $d^2 J' > 0$ if $du_k \neq 0$ for $dx_i = 0$, and $d\psi = 0$. The sufficient conditions for guaranteeing that u_k^* is a minimizing control are therefore (4.6-15) and (4.6-16). Technically, we should also require

$$S_{k+1} f_x^k - V_{k+1} P_k^{-1} V_k^T \quad \text{finite} \quad \text{for} \quad i \leq k < N, \qquad (4.6\text{-}17)$$

but this always holds on a finite interval $[i, N]$.

PROBLEMS

Section 4.1

4.1-1. Derivation of optimal LQ tracker. Use a sweep method combining the derivations of Tables 3.3-1 and 4.4-1 to derive the LQ tracker in Table 4.1-1.

4.1-2. Design a tracker for the system in Problem 3.4-1. It is desired that $x_1(t)$ track a reference signal $r(t)$. Let weights $q = 10$, $r = 1$.
a. Find the system for generating the command input $v(t)$.
b. If $r(t) = u_{-1}(t)$, the unit step, solve for $v(t)$. Now solve for the closed-loop state trajectory if $x(0) = 0$.
c. Solution for the finite horizon control on $[0, T]$ is more difficult; however, write subroutines in MATLAB to solve and simulate the tracking problem. Simulate also the case when $x(0) = [1 \quad 0]^T$, $r(t) = u_{-1}(t - 1)$.

4.1-3. For the plant of Problem 3.4-3, it is desired to construct a tracker so that $x_1(t)$ tracks $r(t)$ on $[0, T]$. Let $P = 10I$. Set up the equations and simulate them using MATLAB. For the simulation, let $x(0) = [1 \quad 0]^T$, $r(t) = u_{-1}(t - 1)$.

Section 4.2

4.2-1. Derivation of function of final-state-fixed control law. Use the sweep method to derive the control scheme in Table 4.2-1. The approach is similar to that in Section 4.5. See also the derivation of Table 3.3-1.

4.2-2. Let $v = 0$, $q = 10$ in Problem 3.4-1. It is desired to drive $x_1(t)$ to a value of 10 at a final time of $T = 5$ if $x(0) = 0$.
a. Set up the LQ regulator equations. Write subroutines simulating it for use with MATLAB.
b. Solve analytically for the steady-state values of S, K, V, P.

Section 4.3

4.3-1. Perturbation control. Suppose that we have just solved the nonlinear Problem 3.2-2. The optimal state trajectory and control were found to be $x * (t) = \frac{1}{2}e^{-t}$, $u* = \frac{1}{8}e^{-3t} - \frac{1}{2}e^{-t}$ for the initial condition $x(0) = \frac{1}{2}$. Now the initial state varies to $\frac{1}{2}$ to $\delta x(0)$. Rather than solve the nonlinear control problem again, we want to determine the control perturbation $\delta u(t)$ needed so that $u^*(t) + \delta u(t)$ provides an optimal control to first order.

a. Set up the perturbation-control problem. Note that the linearized state in equation is time varying.

b. Write subroutines for MATLAB to solve the Riccati equation and simulate the optimal control $\delta u(t)$ on the linearized system.

c. Write subroutines for MATLAB to simulate applying $u^*(t) + \delta u(t)$ to the full nonlinear plant.

4.3-2. Perturbation control of bilinear system. Repeat Problem 4.3-1 for the bilinear system in Problem 3.2-5.

Section 4.4

4.4-1. Prove (4.4-19) and (4.4-20).

4.4-2. LQ tracker. Add a tracking capability to the plant in Problem 2.4-1. It is desired for the sum of state components 1 and 2 to track a reference signal r_k on the time interval $[0, \infty]$ (i.e., $C = [1\ 1]$). Let $Q = qI$, where q is a given scalar. Use the suboptimal feedback K_∞^* in your control law.

a. Determine the system for generating the auxiliary signal v_k.

b. Find the steady-state value for the feedforward gain K_k^v. Use this suboptimal value in your control law.

c. Suppose that $q = 10$, $r = 1$, and the reference track is $r_k = 0.9^k u_{-1}(k)$. Solve for signal v_k. Now solve for the closed-loop trajectory x_k if $x_0 = 0$. Sketch r_k and x_k. ($u_{-1}(k)$ is the unit step.)

Section 4.5

4.5-1. Derive (4.5-26).

4.5-2. Consider the system of Problem 2.4-1 with $q_1 = 10$, $q_2 = 1$, $r = 1$. It is desired to drive state component 1 to a value of 5 at final time $N = 3$ if $x_0 = 0$.

a. Determine the Riccati solution S_k and the optimal feedback gain K_k on the time interval.

b. Find auxiliary matrices V_k and P_k. Is the problem normal on the interval $[0, 3]$?

c. Find the auxiliary gain K_k^u and the control law.

d. Find the closed-loop plant.

e. Verify by simulation that the desired control objective is achieved. That is, apply your control u_k to the plant and find the resulting state trajectory.

5

FINAL-TIME-FREE
AND CONSTRAINED
INPUT CONTROL

5.1 FINAL-TIME-FREE PROBLEMS

In this section we gain more of a feel for the generalized boundary condition
(3.2-10), which is

$$(\phi_x + \psi_x^T v - \lambda)^T|_T \, dx(T) + (\phi_t + \psi_t^T v + H)|_T \, dT = 0. \qquad (5.1\text{-}1)$$

Specifically, we now let the final time T be free, so that it can be varied in
minimizing the performance index. Then $dT \neq 0$.

To solve the optimal control problem using the equations in Table 3.2-1, we can
often first eliminate the control input $u(t)$ by taking into account the stationarity
condition. Then, to solve the state and costate equations, we need the n given
components of the initial state $x(t_0)$ and n final conditions. We also need to solve
for the p components of the undetermined multiplier v and for the final time T.
The coefficient of $dx(T)$ in (5.1-1) provides n equations, the coefficient of dT
in (5.1-1) provides one equation, and the condition $\psi(x(T), T) = 0$ provides
the remaining p equations needed to specify the solution of the optimal control
problem completely.

Minimum-time Problems

One special class of final-time-free problems is defined by a performance index of

$$J(t_0) = \int_{t_0}^{T} 1 \, dt, \qquad (5.1\text{-}2)$$

which arises when we are interested in minimizing the time $(T - t_0)$ required to zero a given function of the final state $\psi(x, (T), T)$ given some initial state $x(t_0)$. We could equally well define this *minimum-time problem* by the performance index $J = (T - t_0)$, but (5.1-2) is generally more convenient. Given this performance index, the Hamiltonian is

$$H(x, u, t) = 1 + \lambda^T f(x, u, t). \qquad (5.1\text{-}3)$$

Recall that if H is not an explicit function of t, then $\dot{H} = 0$; the Hamiltonian is a constant on the optimal trajectory. (See (3.2-13).)

A special case of minimum-time problem occurs when the final state $x(T)$ is required to be fixed at a given value $r(T)$. Then $dx(T) = 0$. Since in that case

$$\psi(x(T), T) = x(T) - r(T) = 0 \qquad (5.1\text{-}4)$$

is independent of T, and since $\phi(x(T), T) = 0$ in the minimum-time problem, (5.1-1) requires that

$$H(T) = 0. \qquad (5.1\text{-}5)$$

Hence, if H is not an explicit function of t, we must have

$$H(t) = 0. \qquad (5.1\text{-}6)$$

for all $t \in [t_0, T]$. We have already discussed one minimum-time problem in Example 3.2-5b.

The Transversality Condition

Another class of final-time-free problems occurs when both $x(T)$ and T are free, but they are independent. Then (5.1-1) demands that both

$$(\phi_x + \psi_x^T v - \lambda)|_T = 0 \qquad (5.1\text{-}7)$$

and

$$(\phi_t + \psi_t^T v + H)|_T = 0. \qquad (5.1\text{-}8)$$

Yet another class of final-time-free problems occurs when both $x(T)$ and T are free, but they are dependent. An example is when the final state $x(T)$ is required to be on a specified moving point $p(t)$, but $x(T)$ and T are otherwise free. Then

$$x(T) = p(T) \qquad (5.1\text{-}9)$$

and

$$dx(T) = \frac{dp(T)}{dT} dT, \qquad (5.1\text{-}10)$$

so that (5.1-1) becomes

$$(\phi_x - \lambda)^T|_T \frac{dp(T)}{dT} dT + (\phi_t + H)|_T dT = 0. \qquad (5.1\text{-}11)$$

(note there is no $\psi(T)$ here, or $\psi(T)$ is identically the zero function). Since $dT \neq 0$, this requires

$$(\phi_x(T) - \lambda(T))^T \frac{dp(T)}{dT} + \phi_t(T) + H(T) = 0. \qquad (5.1\text{-}12)$$

The next exercise illustrates this approach, and shows that optimal control problems often have several equivalent formulations.

Exercise 5.1-1.   Alternative Formulation of the Minimum-time Intercept Problem

Example 3.2-5b can be worked another way by using conditions (5.1-9) and (5.1-12). Let us require the final state to be on the moving point

$$p(T) = \begin{bmatrix} x_i + V_1 T \\ h \end{bmatrix}, \qquad (1)$$

which is the target aircraft. Show that (5.1-9) and (5.1-12) lead to equations (23), (24), and (31) in Example 3.2-5b. In this formulation of the problem, the fixed-final-state function $\psi(x(T), T)$ is identically the zero function, and the multiplier ν is not required.   ∎

Another class of final-time-free problems occurs when the final state is required to be on a surface (or *target set*). If the surface is defined by

$$\psi(x(T), T) = 0, \qquad (5.1\text{-}13)$$

then (5.1-7) and (5.1-8) must hold independently. Let us focus on the former condition. We may write

$$\psi(T) = \begin{bmatrix} \psi_1(T) \\ \psi_2(T) \\ \vdots \\ \psi_p(T) \end{bmatrix} = 0. \qquad (5.1\text{-}14)$$

Each component $\psi_i(T) = 0$ defines a hypersurface in R^n, and the final state is required to be on the intersection of these hypersurfaces, $\psi(T) = 0$.

Writing (5.1-7) as

$$(\phi_x(T) - \lambda(T)) = - \begin{bmatrix} \left[\frac{\partial \psi_1^T(T)}{\partial x}\right]^T \\ \vdots \\ \frac{\partial \psi_p^T(T)}{\partial x} \end{bmatrix} \nu$$

$$= - \begin{bmatrix} \frac{\partial \psi_1(T)}{\partial x} & \cdots & \frac{\partial \psi_p(T)}{\partial x} \end{bmatrix} \begin{bmatrix} \nu_1 \\ \vdots \\ \nu_p \end{bmatrix}, \qquad (5.1\text{-}15)$$

it is apparent that the n vector $(\phi_x(T) - \lambda(T))$ must be a linear combination of the gradient vectors $\partial \psi_i(T)/\partial x$. This vector must therefore be *normal* or *transversal* to the surface defined by (5.1-13). As a special case, if the final-state weighting $\phi(T)$ is equal to zero, then $\lambda(T)$ itself must be normal to the surface $\psi(T) = 0$.

The requirement (5.1-7)/(5.1-15) on the final costate is known as the *transversality condition*. If $x(T)$ and dT are dependent (i.e., the surface is moving), the transversality condition looks like (5.1-12). Since (5.1-8) is also a condition on the final costate, it is often called the transversality condition. For further discussion on these concepts, see Kirk (1970), Athans and Falb (1966), and Bryson and Ho (1975). Some examples will help in understanding these ideas.

Examples

Example 5.1-2. Zermelo's Problem

This problem is taken from Bryson and Ho (1975).

A ship must travel through a region of strong currents, which depend on position. The ship has a constant speed V, and its heading $\theta(t)$ can be varied. The current is directed in the x direction with a speed of

$$u = \frac{Vy}{h}, \tag{1}$$

for a given h. See Fig. 5.1-1.

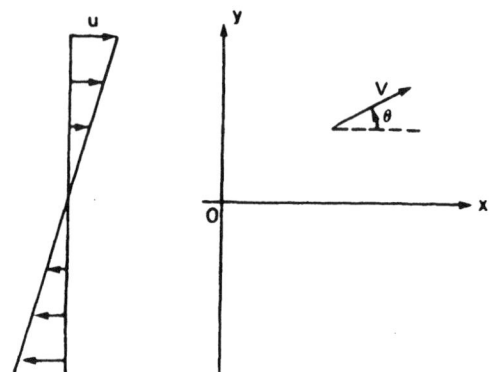

FIGURE 5.1-1   Geometry for Zermelo's problem.

It is desired to find the ship's heading $\theta(t)$ required to move from a given initial position $(x(t_0), y(t_0))$ to the origin in minimum time. The equations of motion are

$$\dot{x} = V\cos\theta + \frac{Vy}{h} \tag{2}$$

$$\dot{y} = V\sin\theta, \tag{3}$$

and the performance index is

$$J(t_0) = \int_{t_0}^{T} 1 \, dt. \tag{4}$$

The Hamiltonian is

$$H = 1 + \lambda_x(V \cos\theta + Vy/h) + \lambda_y V \sin\theta, \tag{5}$$

so that the costate equations are

$$-\dot\lambda_x = \frac{\partial H}{\partial x} = 0 \tag{6}$$

$$-\dot\lambda_y = \frac{\partial H}{\partial y} = \frac{\lambda_x V}{h}. \tag{7}$$

The control input is $\theta(t)$, and so the stationarity condition is

$$0 = \frac{\partial H}{\partial \theta} = -\lambda_x V \sin\theta + \lambda_y V \cos\theta. \tag{8}$$

According to (8), we can express the control in terms of the costate as

$$\tan\theta = \frac{\lambda_y}{\lambda_x}. \tag{9}$$

Integrating (6) and (7) and substituting into (9) yields the *linear tangent control law*

$$\tan\theta(t) = \frac{\lambda_y(T)}{\lambda_x} + \frac{V(T-t)}{h}, \tag{10}$$

where the costate component λ_x is constant. See Example 3.2-5. We could not use this to substitute for $\theta(t)$ in the state equations (2), (3) and solve for $x(t)$, $y(t)$. To find λ_x, $\lambda_y(T)$, and the final time T, we could then use the boundary conditions. Unfortunately, this approach is tedious and unfruitful. Let us try another one.

The initial conditions are $(x(t_0), y(t_0))$. The final state is fixed at $(0, 0)$, so that

$$\psi(T) \triangleq \begin{bmatrix} x(T) \\ y(T) \end{bmatrix} = 0. \tag{11}$$

Since the final state is fixed, (3.2-10) requires that

$$H(T) = 0. \tag{12}$$

The Hamiltonian is not an explicit function of t; therefore, $\dot H = 0$ and

$$H(t) = 1 + \lambda_x\left(V \cos\theta + \frac{Vy}{h}\right) + \lambda_y V \sin\theta = 0 \tag{13}$$

for all $t \in [t_0, T]$.

We can now use (8) and (13) to solve for the costate in terms of $\theta(t)$, for we have

$$\begin{bmatrix} V\cos\theta + \dfrac{Vy}{h} & V\sin\theta \\ -V\sin\theta & V\cos\theta \end{bmatrix} \begin{bmatrix} \lambda_x \\ \lambda_y \end{bmatrix} = \begin{bmatrix} -1 \\ 0 \end{bmatrix}. \tag{14}$$

which yields

$$\lambda_x = \frac{-\cos\theta}{V + V(y/h)\cos\theta}, \tag{15}$$

$$\lambda_y = \frac{-\sin\theta}{V + V(y/h)\cos\theta}. \tag{16}$$

Substitute λ_x into (6) (using λ_x and λ_y in (7) would also work) to get

$$\dot{\lambda}_x = \frac{\partial \lambda_x}{\partial \theta}\dot{\theta} + \frac{\partial \lambda_x}{\partial y}\dot{y}$$

$$= \frac{\sin\theta[V + V(y/h)\cos\theta] + \cos\theta[-V(y/h)\sin\theta]}{[V + V(y/h)\cos\theta]^2} \cdot \dot{\theta}$$

$$+ \frac{(V/h)\cos^2\theta}{[V + V(y/h)\cos\theta]^2} \cdot \dot{y} = 0. \tag{17}$$

Now use state equation (3) in this and solve for $\dot{\theta}$ to see that

$$\dot{\theta} = -\frac{V}{h}\cos^2\theta. \tag{18}$$

What we need to do at this point is to solve (2), (3), and (18) using the given boundary conditions. One approach is to integrate (18) about the final time T, use the resulting $\theta(t)$ to determine $y(t)$ by integrating (3), and then use $\theta(t)$ and $y(t)$ in (2) to determine $x(t)$. The unknowns T and $\theta(T)$ could then be determined using (11). This gives a very messy solution and no intuition.

Let us instead take θ, not the time t, as the independent variable. This will have two benefits: one is that the problem is easier to solve. We shall point out the second benefit later.

According to (6), λ_x is a constant, so we can use (15) evaluated at t and at T to write $(y(T) = 0)$

$$\frac{\cos\theta}{V + V(y/h)\cos\theta} = \frac{\cos\theta(T)}{V}. \tag{19}$$

This allows us to see that

$$\cos\theta = \frac{\cos\theta(T)}{1 - (y/h)\cos\theta(T)}, \tag{20}$$

which expresses the required ship's heading $\theta(t)$ at the current time in terms of the as yet unknown final heading $\theta(T)$ and the current y position.

It is easy to express $y(t)$ in terms of $\theta(t)$, for according to (20)

$$\frac{y(t)}{h} = \sec\theta(T) - \sec\theta(t). \tag{21}$$

To obtain $x(t)$ as a function of $\theta(t)$, use (2) to write

$$\frac{dx}{d\theta}\dot{\theta} = V\cos\theta + \frac{Vy}{h}, \tag{22}$$

or, by (18) and (21),

$$\frac{dx}{d\theta} = \frac{V\cos\theta + V\sec\theta(T) - V\sec\theta}{-(V/h)\cos^2\theta}$$

$$= -h(\sec\theta + \sec\theta(T)\sec^2\theta - \sec^3\theta). \tag{23}$$

This can be integrated to yield $(x(T) = 0)$

$$\frac{x(t)}{h} = \frac{1}{2}\left[\sec\theta(T)(\tan\theta(T) - \tan\theta) - \tan\theta(\sec\theta(T) - \sec\theta)\right.$$
$$\left. + \ln\frac{\sec\theta(T) + \tan\theta(T)}{\sec\theta + \tan\theta}\right]. \tag{24}$$

Equations (21) and (24) give the current ship's position (x, y) in terms of current heading $\theta(t)$ and final heading $\theta(T)$ along the minimum-time trajectory. Although this may not seem to be exactly what we are after, a little thought will show that our problem is now solved.

Given the current position $(x(t))$, $y(t)$, (21) and (24) can be used to solve for the current required heading $\theta(t)$ and the final heading $\theta(T)$ of the minimum-time path to the origin. Although these equations are a little messy to solve, we can express the control law in a convenient tabular or graphical form. A simple computer program can be written to compute x and y as θ and $\theta(T)$ vary, and a graph can be made like the one in Fig. 5.1-2. Then, given $x(t)$ and $y(t)$, the solution $\theta(t)$, $\theta(T)$ can be read off the graph. (Of course, for an actual application a graph with finer increments in $\theta(t)$ and $\theta(T)$ would be needed.)

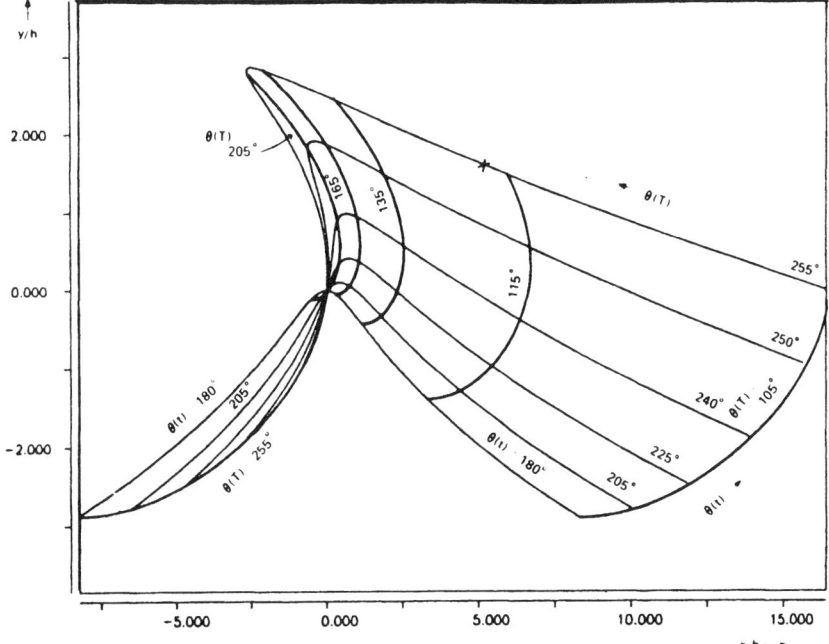

FIGURE 5.1-2   Feedback control law for Zermelo's problem.

The second benefit of using θ instead of t as the independent variable is now apparent, for the control law we have found is a closed-loop *feedback* control! Although it is highly nonlinear and of an unfamiliar form, it is a state variable feedback since the current optimal control $\theta(t)$ is given in terms of the current state $(x(t), y(t))$. Since (21) and (24) are implicit equations for $\theta(t)$, we could say they specify an *implicit feedback law*.

We have even more than this at our disposal, for according to (18)

$$\int_{\theta(t)}^{\theta(T)} \sec^2 \theta \, d\theta = -\frac{V}{h} \int_{t}^{T} dt \qquad (25)$$

$$(T - t) = \frac{h}{v}(\tan \theta(t) - \tan \theta(T)). \qquad (26)$$

Thus, once we have read $\theta(T)$ and the required control $\theta(t)$ off our graph, we can use (26) to determine the time to go to the origin; that is, the optimal value of the performance index $J(t)$!

A sample minimum-type trajectory through the region of currents is shown in Fig. 5.1-3. Since the initial position is $x(t_0) = 4.9$, $y(t_0) = 1.66$, the initial and final headings are found from Fig. 5.1-2 to be $\theta(t_0) = 255°$, $\theta(T) = 117°$. The required headings at values of $t \in [t_0, T]$ are determined in a similar manner from $(x(t), y(t))$.

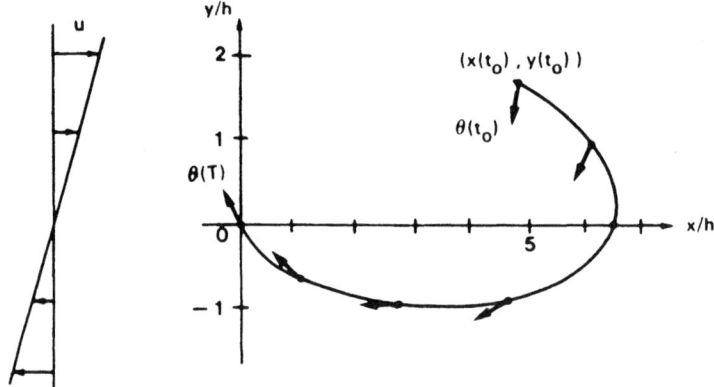

FIGURE 5.1-3   Sample minimum-time trajectory for Zermelo's problem.

Example 5.1-3.   The Brachistochrone Problem

Brachistochrone is Greek for *shortest time*. This problem was proposed and solved by Johann Bernoulli in 1696 and is one of the earliest applications of the calculus of variations.

A mass m moves in a constant force field of magnitude g starting at rest at the origin at time t_0. It is desired to find the path of minimum time to a specified final point (x_1, y_1). See Fig. 5.1-4.

If there is no friction, the field is conservative and the kinetic plus potential energy is a constant:

$$\tfrac{1}{2}mV^2(t) - mgy(t) = \tfrac{1}{2}mV^2(t_0) - mgy(t_0) = 0. \qquad (1)$$

Hence, the velocity at any time $t \geq t_0$ is given in terms of the y coordinate as

$$V = \sqrt{2gy}. \qquad (2)$$

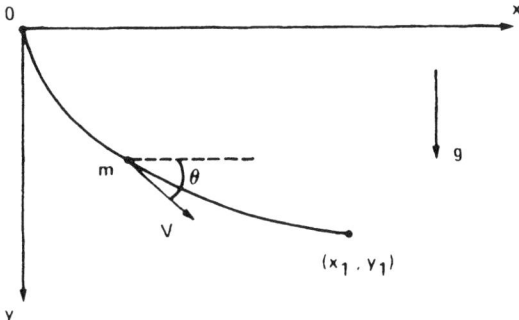

FIGURE 5.1-4   Geometry for the brachistochrone problem.

The state equations are

$$\dot{x} = V \cos \theta, \tag{3}$$

$$\dot{y} = V \sin \theta, \tag{4}$$

where the path angle $\theta(t)$ is the control input to be determined in order to minimize the minimum-time performance index

$$J(t_0) = \int_{t_0}^{T} 1 \, dt. \tag{5}$$

According to Table 3.1-1, the Hamiltonian is

$$H(t) = 1 + \lambda_x(t)V)(t) \cos \theta(t) + \lambda_y(t)V(t) \sin \theta(t), \tag{6}$$

so the Euler equations are

$$-\dot{\lambda}_x = \frac{\partial H}{\partial x} = 0, \tag{7}$$

$$-\dot{\lambda}_y = \frac{\partial H}{\partial y} = \frac{g}{V}(\lambda_x \cos \theta + \lambda_y \sin \theta), \tag{8}$$

$$0 = \frac{\partial H}{\partial \theta} = -\lambda_x V \sin \theta + \lambda_y V \cos \theta. \tag{9}$$

From the stationarity condition we can express the control in terms of the costate as

$$\tan \theta = \frac{\lambda_y}{\lambda_x} \tag{10}$$

where λ_x is constant due to (7). We shall not use (10), but will instead proceed as we did in Example 5.1-2, using θ, not t, as the independent variable. This will result in a feedback law, not simply in a description of the optimal path as a time function.

The initial conditions are $x(t_0) = 0$, $y(t_0) = 0$, and the fixed function of the final state is

$$\psi(T) = \begin{bmatrix} x(T) - x_1 \\ y(T) - y_1 \end{bmatrix} = 0. \tag{11}$$

Therefore, the terminal condition (3.2-10) requires that

$$H(T) = 0,$$ (12)

and since $H(t)$ is not explicitly dependent on t,

$$H(t) = 1 + \lambda_x V \cos\theta + \lambda_y V \sin\theta = 0$$ (13)

for all $t \in [t_0, T]$.

Using (9) and (13) we can express the costate in terms of the state and the control as

$$\lambda_x = -\frac{\cos\theta}{V},$$ (14)

$$\lambda_y = -\frac{\sin\theta}{V}.$$ (15)

Now use (14) in (7) (or (14) and (15) in (8)) to see that

$$\dot{\lambda}_x = \frac{\partial \lambda_x}{\partial\theta}\dot\theta + \frac{\partial\lambda_x}{\partial y}\dot y$$

$$= \frac{\dot\theta V \sin\theta + \dot y(g/V)\cos\theta}{V^2} = 0.$$ (16)

Taking (4) into account yields

$$\dot\theta = -\frac{g}{V}\cos\theta.$$ (17)

What we could now do is solve (3), (4), and (17) using the boundary conditions to get x, y, and θ as functions of time. This is a mess, and besides that, we want a feedback control law that gives $\theta(t)$ in terms of $x(t)$ and $y(t)$. Let us, therefore, use θ as the independent variable (Bryson and Ho 1975).

According to (7), λ_x is a constant, so evaluate (14) at both t and T to get

$$\frac{\cos\theta(t)}{V(t)} = \frac{\cos\theta(T)}{V(T)},$$ (18)

or, using (2) and $y(T) = y_1$,

$$\cos\theta = \sqrt{\frac{y(t)}{y_1}}\cos\theta(T).$$ (19)

This can be used to express $y(t)$ in terms of $\theta(t)$ and the as yet unknown $\theta(T)$ as

$$y = \frac{y_1}{\cos^2\theta(T)}\cos^2\theta.$$ (20)

To get x as a function of θ, use (3) to write

$$\frac{dx}{d\theta}\dot\theta = V\cos\theta,$$ (21)

so that, by (2) and (17),

$$\frac{dx}{d\theta} = -\frac{V^2}{g} = -2y.$$ (22)

Using (20),

$$\frac{dx}{d\theta} = \frac{-2y_1}{\cos^2 \theta(T)} \cos^2 \theta, \tag{23}$$

which can easily be integrated about the final time to give $(x(T) = x_1)$

$$x = x_1 + \frac{y_1}{2\cos^2 \theta(T)} [2(\theta(T) - \theta) + \sin 2\theta(T) - \sin 2\theta]. \tag{24}$$

We are now done. Given the current position $(x(t), y(t))$, equations (20) and (24) can be solved for the current required path angle $\theta(t)$ and the final path angle $\theta(T)$. These equations are messy to solve, but they can easily be converted to a tabular or graphical form as in the previous example by the use of a MATLAB program. It should be clearly understood that (20) and (24) specify an *implicit feedback control law*, since the control $\theta(t)$ is given once the state $(x(t), y(t))$ is known.

It is not difficult to see that given $(x(t), y(t))$ we can also calculate the time to go to (x_1, y_1), that is, the optimal value of $J(t)$. Using (17), (14), and (7) we have

$$\dot{\theta} = g\lambda_x = \text{const}, \tag{25}$$

and integrating yields

$$\theta(T) - \theta = g\lambda_x(T - t),$$

or (evaluating λ_x in (14) at the final time)

$$(T - t) = \sqrt{\frac{2y(T)}{g}} \frac{\theta - \theta(T)}{\cos \theta(T)}. \tag{26}$$

Thus, $(x(t), y(t))$ gives $(\theta(t), \theta(T))$, which then gives $(T - t)$.

Equation (25) is interesting in its own right; it shows that the rate of change of the optimal path angle is constant. It is also worth noting that according to (20), the initial path angle $\theta(t_0)$ is always $90°$; the motivation being, of course, to increase y, and hence the velocity V, as quickly as possible at the outset.

Although the optimal control problem is now solved, it is quite instructive to derive the solution to the bachistochrone problem in the form presented by Bernoulli. Let

$$\phi = \pi - 2\theta. \tag{27}$$

Then (20) becomes
$$y = a(1 + \cos(\pi - \phi))$$

or

$$y = a(1 - \cos \phi), \tag{28}$$

where

$$a \stackrel{\Delta}{=} \frac{y_1}{2\cos^2 \theta(T)} = \frac{y_1}{1 - \cos \phi(T)}. \tag{29}$$

We can also write (24) in terms of $\phi(t)$ as

$$x - x(T) + a(\phi(T) - \sin \phi(T)) = a(\phi - \sin \phi). \tag{30}$$

Equations (28) and (30) describe a *cycloid* that passes through (x_1, y_1). The final path angle $\theta(T)$ must be selected so that the cycloid passes through $(x(t_0), y(t_0))$. (A cycloid is the curve generated by a point on the circumference of a circle that rolls without slipping, in our case along the x axis.) ■

Example 5.1-4.   *Minimum-time Orbit Injection*

A spacecraft of mass m is to be placed into orbit in minimum time. See Fig. 5.1-5, where $\phi(t)$ is the thrust direction angle, F the thrust, $\gamma(t)$ the flight path angle, total velocity $V(t)$ has radial component $w(t)$ and tangential component $v(t)$, and $\mu = GM$ the gravitational constant of the attracting center. The Coriolis force

$$F_c = \frac{mVv}{r} \tag{1}$$

is directed perpendicular to V.

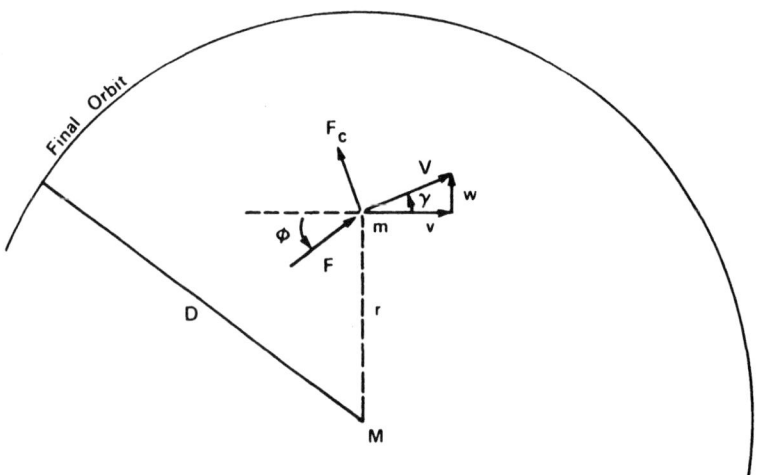

FIGURE 5.1-5   Geometry for orbit injection.

Assuming that the mass m of the spacecraft is constant, the state equations are found by summing forces to be

$$\dot{r} = w, \tag{2}$$

$$\dot{w} = \frac{v^2}{r} - \frac{\mu}{r^2} + \frac{F}{m} \sin\phi, \tag{3}$$

$$\dot{v} = \frac{-wv}{r} + \frac{F}{m} \cos\phi. \tag{4}$$

We assume that the thrust F is constant, and it is desired to select the control input $\phi(t)$ so that elapsed time

$$J(t_0) = \int_{t_0}^{T} 1\, dt \tag{5}$$

is minimized.

Using Table 3.2-1, the Hamiltonian is

$$H(t) = 1 + \lambda_r w + \lambda_w \left(\frac{v^2}{r} - \frac{\mu}{r^2} + \frac{F}{m} \sin \phi \right)$$
$$+ \lambda_v \left(\frac{-wv}{r} + \frac{F}{m} \cos \phi \right), \tag{6}$$

where a costate component is associated with each state component. The Euler equations are

$$-\dot{\lambda}_r = \frac{\partial H}{\partial r} = \left(\frac{-v^2}{r^2} + \frac{2\mu}{r^3} \right) \lambda_w + \frac{wv}{r^2} \lambda_v, \tag{7}$$

$$-\dot{\lambda}_w = \frac{\partial H}{\partial w} = \lambda_r - \frac{v}{r} \lambda_v, \tag{8}$$

$$-\dot{\lambda}_v = \frac{\partial H}{\partial v} = \frac{2v}{r} \lambda_w - \frac{w}{r} \lambda_v, \tag{9}$$

$$0 = \frac{\partial H}{\partial \phi} = \frac{F}{m} (\lambda_w \cos \phi - \lambda_v \sin \phi). \tag{10}$$

According to (10), the optimal control is expressed in terms of the costate as

$$\tan \phi = \lambda_w / \lambda_v. \tag{11}$$

Suppose the spacecraft is launched from the surface of a planet of radius R. For simplicity, we assume the planet is not rotating. Then the initial conditions are

$$r(t_0) = R, \quad w(t_0) = 0, \quad v(t_0) = 0. \tag{12}$$

If the desired orbit is circular with a radius of $r(T) = D$, then the fixed function of the final state is

$$\psi(T) = \begin{bmatrix} r(T) - D \\ w(T) \\ v(T) - \sqrt{\dfrac{\mu}{r(T)}} \end{bmatrix} = 0, \tag{13}$$

since in orbit we require the centrifugal and gravitational forces to balance:

$$\frac{mv^2}{r} = \frac{\mu m}{r^2}. \tag{14}$$

In this problem $dx(T)$ and dT are independent, so final condition (3.2-10) requires both the transversality conditions ($v = [v_r \ v_w \ v_v]^{\mathrm{T}}$)

$$\lambda_r(T) = \frac{\partial \psi^{\mathrm{T}}(T)}{\partial r} v = v_r + \frac{v_v}{2} \sqrt{\frac{\mu}{r^3(T)}}, \tag{15}$$

$$\lambda_w(T) = \frac{\partial \psi^{\mathrm{T}}(T)}{\partial w} v = v_w, \tag{16}$$

$$\lambda_v(T) = \frac{\partial \psi^{\mathrm{T}}(T)}{\partial v} v = v_v, \tag{17}$$

and

$$H(T) = 1 + \frac{F}{m}(\lambda_w(T)\sin\phi(T) + \lambda_v(T)\cos\phi(T)) = 0. \qquad (18)$$

(We have taken (13) into account to write (18).) Since $H(t)$ is not an explicit function of time, $\dot{H} = 0$ and

$$H(t) = 0 \qquad (19)$$

for all $t \in [0, T]$.

To find the optimal thrust angle $\phi(t)$, we need to solve state equations (2)–(4) and costate equations (7)–(9), taking into account (11) and the six boundary conditions (12), (15)–(17). The unknowns v_r, v_w, v_v must furthermore be selected to ensure that (13) is satisfied. Equation (18) provides an additional condition that allows the unknown final time T to be determined. This approach would yield the optimal ϕ as a function of time.

Alternatively, we could use (10) and (19) to solve for λ_w, λ_v as in previous problems. Then we could try to express r, w, and v in terms of ϕ. This approach would yield a feedback control law for ϕ.

Either approach is very complicated in this example, so we shall stop at this point. To solve the problem, a numerical approach is probably the best one.

See Bryson and Ho (1975) and Kirk (1970) for additional insight on the orbit injection problem.   ■

Example 5.1-5.   Shortest Distance from a Point to a Line

To minimize the distance from the origin to a given *target set* of admissible final states, we can define

$$\dot{x} = u, \qquad (1)$$

$$J = \int_0^T \sqrt{1 + u^2}\, dt, \qquad (2)$$

where T is in general free. See Fig. 5.1-6. In Example 3.2-2, the set of admissible final states was a single point, and T was fixed.

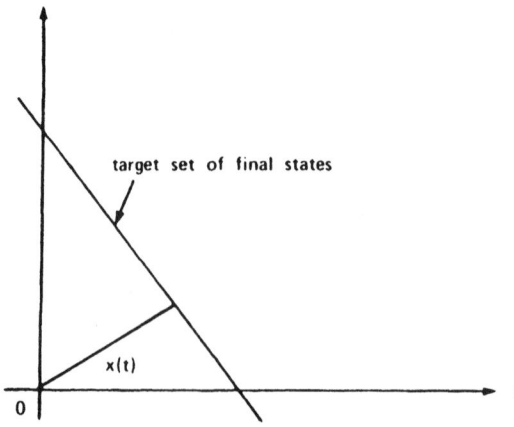

FIGURE 5.1-6   Shortest distance from the origin to a given target set.

The Hamiltonian is

$$H = \sqrt{1 + u^2} + \lambda u, \tag{3}$$

and the Euler equations are

$$-\dot{\lambda} = \frac{\partial H}{\partial x} = 0, \tag{4}$$

$$0 = \frac{\partial H}{\partial u} = \frac{u}{\sqrt{1 + u^2}} + \lambda. \tag{5}$$

The initial condition is

$$x(0) = 0. \tag{6}$$

According to the stationarity condition (5),

$$\lambda = -u/\sqrt{1 + u^2} \tag{7}$$

and

$$u = \lambda/\sqrt{1 - \lambda^2}. \tag{8}$$

The costate equation (4) shows that λ is a constant; hence, by (8), u is also. Therefore, state equation (1) yields

$$x(t) = ut, \tag{9}$$

where the initial condition (6) has been taken into account. The constant u is to be determined.

Now, suppose the target set is the line $- mt + c$. Let us solve for the optimal slope u by two methods.

a. Fixed Function of Final State

Define

$$\psi(x(T), T) = x(T) + mT - c = 0, \tag{10}$$

which ensures that $x(T)$ is in the target set. Then, by (3.2-10) the transversality condition on the costate is

$$\lambda = \frac{\partial \psi(T)}{\partial x} v = v, \tag{11}$$

where v is to be determined. The generalized boundary condition also requires

$$H(T) + \frac{\partial \psi(T)}{\partial T} v = \sqrt{1 + u^2} + \lambda u + mv = 0. \tag{12}$$

Using (7) and (11) in (12) yields

$$\sqrt{1 + u^2} - \frac{u^2}{\sqrt{1 + u^2}} - \frac{mu}{\sqrt{1 + u^2}} = 0, \tag{13}$$

which is solved to give the optimal control

$$u = \frac{1}{m}. \tag{14}$$

The optimal path from the origin to the line is, therefore,

$$x = \frac{t}{m},$$ (15)

a line perpendicular to the target set. Using (7), the costate is

$$\lambda = \nu = \frac{-1}{\sqrt{1 + m^2}},$$ (16)

and (10) can be used to solve for the final time

$$T = \frac{mc}{1 + m^2}.$$ (17)

b. Final State on Moving Point

An alternative formulation of the terminal conditions is to require that $x(t)$ be on the moving point

$$p(t) = -mt + c$$ (18)

at the final time $t = T$. Then we have no final state function (i.e., $\psi(x(T), T)$ is identically the zero function), and transversality condition (5.1-12) applies. It becomes

$$m\lambda + H(T) = 0.$$ (19)

This is identical to (12), and the remainder of the work in part a applies. Note that in this formulation the additional multiplier ν is not introduced.   ∎

Linear Quadratic Minimum-time Design

We now concern ourselves with finding an optimal control for the linear system

$$\dot{x} = Ax + Bu$$ (5.1-16)

that minimizes the performance index

$$J = \frac{1}{2}x^T(T)S(T)x(T) + \frac{1}{2}\int_{t_0}^{T} \left(\rho + x^T Q x + u^T R u\right) dt$$ (5.1-17)

with $S(T) \geq 0$, $Q \geq 0$, $R > 0$, and with the final time T free. There is no constraint on the final state; thus, the control objective is to make the final state sufficiently small. Due to the term $\rho(T - t_0)/2$ arising from the integral, there is a concern to accomplish this in a short time period.

This is a general sort of performance index that allows for a trade-off between the minimum-time objective and a desire to keep the states and the controls small. Thus, if we select smaller Q and R, the term $\rho(T - t_0)/2$ in the performance index dominates, and the control tries to make the transit time smaller. We call this the linear quadratic minimum-time (LQMT) problem. On the other hand, if we select smaller ρ the term $\frac{1}{2}\int_{t_0}^{T}(x^T Q x + u^T R u)\, dt$ in the performance index dominates and thus at the limit where ρ tends to 0 we retrieve the LQR problem.

From Table 3.2-1, the Hamiltonian is

$$H = \frac{1}{2}(\rho + x^T Q x + u^T R u) + \lambda^T (A x + B u), \qquad (5.1\text{-}18)$$

with $\lambda(t)$ the costate. The costate equation and the boundary condition are

$$-\dot{\lambda} = A^T \lambda + Q x, \qquad (5.1\text{-}19)$$

$$0 = R u + B^T \lambda, \qquad (5.1\text{-}20)$$

whence

$$u = -R^{-1} B^T \lambda. \qquad (5.1\text{-}21)$$

In (3.2-10) both $dx(T)$ and dT are nonzero, however, they are independent in this situation so that the final conditions are

$$\lambda(T) = S(T) x(T) \qquad (5.1\text{-}22)$$

$$H(T) = 0. \qquad (5.1\text{-}23)$$

Indeed, since the system and the performance index are not explicitly dependent on t, (3.2-13) shows that, for all t,

$$\dot{H}(t) = 0. \qquad (5.1\text{-}24)$$

We now remark that, with the exception of (5.1-24) this is the same boundary-value problem we solved in the closed-loop LQR problem in Section 3.3. That is, Table 3.3-1 still provides the optimal solution. The difficulty, of course, is that the final time T is unknown. To find the final time T, recall that for all times t

$$\lambda = S x, \qquad (5.1\text{-}25)$$

$$u = -R^{-1} B^T S x. \qquad (5.1\text{-}26)$$

Using these at $t = t_0$ in (5.1-18) and taking into account (5.1-24) we have

$$0 = H(t_0)$$

$$= \frac{\rho}{2} + \frac{1}{2} x^T(t_0)[S B R^{-1} B^T S + Q + (SA + A^T S) - 2 S B R^{-1} B^T S] x(t_0),$$
$$(5.1\text{-}27)$$

Therefore,

$$0 = \rho + x^T(t_0)[A^T S + S A + Q - S B R^{-1} B^T S] x(t_0), \qquad (5.1\text{-}28)$$

or, taking into account the Riccati equation from Table 3.3-1,

$$x^T(t_0) \dot{S} x(t_0) = \rho. \qquad (5.1\text{-}29)$$

We have shown that the solution procedure for the LQ minimum-time problem is to integrate the Riccati equation

$$-\dot{S} = A^T S + SA + Q - SBR^{-1}B^T S \tag{5.1-30}$$

backward from some time τ using as the final condition $S(\tau) = S(T)$. At each time t, the left-hand side of (5.1-29) is computed using the known initial state and $\dot{S}(t)$. Then, the minimum interval $(T - t_0)$ is equal to $(\tau - t)$, where t is the time for which (5.1-29) first holds. This specifies the minimum final time T. Finally, the Kalman gain and the optimal control are given using exactly the design equations from Table 3.3-1, namely

$$K = R^{-1}B^T S, \tag{5.1-31}$$

$$u = -K(t)x. \tag{5.1-32}$$

It is interesting to note that $\dot{S}$ is used to determine the optimal time interval, while S is used to determine the optimal feedback gain.

We mention that condition (5.1-29) may never hold. Then, the optimal solution is $T - t_0 = 0$; that is, the performance index is minimized by using no control. Roughly speaking, if $x(t_0)$ and/or Q and $S(T)$ are large enough, then it makes sense to apply a nonzero control $u(t)$ to make $x(t)$ decrease. On the other hand, if Q and $S(T)$ are too small for the given initial state $x(t_0)$, then it is not worthwhile to apply any control to decrease $x(t)$, for both a nonzero control and a nonzero time interval will increase the performance index.

Example 5.1-6.   *LQ minimum time for a scalar system (Verriest and Lewis 1991).*

Let the scalar plant

$$\dot{x} = ax + bu$$

$x(0) = x_0$, with performance index

$$J = \frac{1}{2}S(T)x^2(T) + \frac{1}{2}\int_0^T (\rho + qx^2 + ru^2)\,dt.$$

The Riccati equation is

$$-\dot{s} = 2as - \frac{b^2 s^2}{r} + q, \quad s_f = S(T).$$

The steady-state solution is

$$S_\infty = \frac{q}{\gamma}\left(1 + \sqrt{1 + \frac{\gamma}{a}}\right),$$

where $\gamma = b^2 q^2 / ar$ (see Example 3.3-4a).

If $x_0 = 0$ the minimum time is $T = 0$. Otherwise, the LQMT condition (5.1-29) for the minimum time T is

$$\dot{s}(0)x_0^2 = \rho. \tag{1}$$

At the steady-state, we have

$$\dot{s}_\infty x_0^2 = 0.$$

Since $s(t)$ is continuous, a necessary and sufficient condition for (1) to hold for some $t \leq T$ is

$$\dot{s}_f x_0^2 \geq \rho. \tag{2}$$

According to the Riccati equation, this is equivalent to

$$\left[\frac{b^2 s_f^2}{r} - 2as_f - q \right] x_0^2 \geq \rho$$

or

$$g(s_f) = \frac{b^2 s_f^2}{r} - 2as_f - \left(q + \frac{\rho}{x_0^2} \right) \geq 0.$$

The largest root of $g(s_f) = 0$ is given by

$$\sigma = \frac{1}{\delta} \left[1 + \sqrt{1 + \frac{\delta}{a} \left(q + \frac{\rho}{x_0^2} \right)} \right],$$

with

$$\delta = \frac{b^2}{ar}.$$

Thus, for (2) to hold it is necessary and sufficient that

$$S(T) \geq \frac{1}{\delta} \left[1 + \sqrt{1 + \frac{\delta}{a} \left(q + \frac{\rho}{x_0^2} \right)} \right]. \tag{3}$$

If this condition holds, the minimum time T is greater than zero. Otherwise, it is not worthwhile to move the state from x_0 and T is equal to 0. That is the minimum time T is nonzero if $x_0 \neq 0$ and we weight the final state $x(T)$ sufficiently in the performance index.

Some insight may be gained by noting the following points in connection with (3).

1. Parameters q, ρ, and $1/x_0^2$ have a similar effect. In fact, we could define

$$k = q + \frac{\rho}{x_0^2}.$$

Then as k increases, a larger final state weighting $S(T)$ is required for $T > 0$. As k decreases, a smaller $S(T)$ suffices for $T > 0$.

2. As r increases, δ decreases and the influence of parameter k wanes. Moreover, a larger $S(T)$ is required for $T > 0$. As the control weight r decreases, the influence of k increases, however, a smaller final weighting $S(T)$ is sufficient to make $T > 0$.

■

5.2   CONSTRAINED INPUT PROBLEMS

The optimal control law in Table 3.2-1, on which the entire chapter up to this point has been based, gives the control as a continuous implicit function of the state and costate. Under some smoothness assumptions on $f(x, u, t)$ and $L(x, u, t)$, the control is also a smooth function of time. Furthermore, it is found by solving a continuous two-point boundary-value problem.

In this section we investigate a fundamentally different sort of control law.

Pontryagin's Minimum Principle

Let the plant

$$\dot{x} = f(x, u, t) \tag{5.2-1}$$

have an associated cost index of

$$J(t_0) = \phi(x(T), T) + \int_0^T L(x, u, t)\, dt, \tag{5.2-2}$$

where the final state must satisfy

$$\psi(x(T), T) = 0 \tag{5.2-3}$$

and $x(t_0)$ is given. If the control is unconstrained, the optimal control problem is solved in Table 3.2-1, where the condition for optimality is

$$\frac{\partial H}{\partial u} = 0 \tag{5.2-4}$$

with

$$H(x, u, \lambda, t) = L(x, u, t) + \lambda^{\mathrm{T}} f(x, u, t). \tag{5.2-5}$$

Now suppose the control $u(t)$ is constrained to lie in an *admissible region*, which might be defined by a requirement that its magnitude be less than a given value. It was shown by Pontryagin et al. (1962) that in this case, Table 3.2-1 still holds, but the stationarity condition (5.2-4) must be replaced by the more general condition

$$H(x^*, u^*, \lambda^*, t) \leq H(x^*, u^* + \delta u, \lambda^*, t), \qquad \text{all admissible } \delta u,$$

where * denotes optimal quantities. That is, any variation in the optimal control occurring at time t while the state and costate maintain their optimal values at t will increase the value of the Hamiltonian. This condition can be written

$$H(x^*, u^*, \lambda^*, t) \leq H(x^*, u, \lambda^*, t), \qquad \text{all admissible } u. \tag{5.2-6}$$

The optimality requirement (5.2-6) is called *Pontryagin's minimum principle*: "the Hamiltonian must be minimized over all admissible u for optimal values of the state and costate." We shall soon see how useful the minimum principal is. Note particularly that it does *not* say $H(x^*, u^*, \lambda^*, t) \leq H(x, u, \lambda, t)$, which must certainly be true. Supplementary references are Bryson and Ho (1975), Athans and Falb (1966), and Kirk (1970).

Example 5.2-1.   Optimization with Constraints

Suppose we want to minimize

$$L = \tfrac{1}{2}u^2 - 2u + 1 \tag{1}$$

subject to

$$|u| \leq 1. \tag{2}$$

See Fig. 5.2-1. The minimum principle

$$L(u^*) \leq L(u), \qquad \text{all admissible } u, \tag{3}$$

clearly shows that the optimal value of u is

$$u^* = 1. \tag{4}$$

The optimal value of L is

$$L^* = L(1) = -\tfrac{1}{2}. \tag{5}$$

The unconstrained minimum is found by solving

$$\frac{\partial L}{\partial u} = u - 2 = 0 \tag{6}$$

to be

$$u = 2, \tag{7}$$

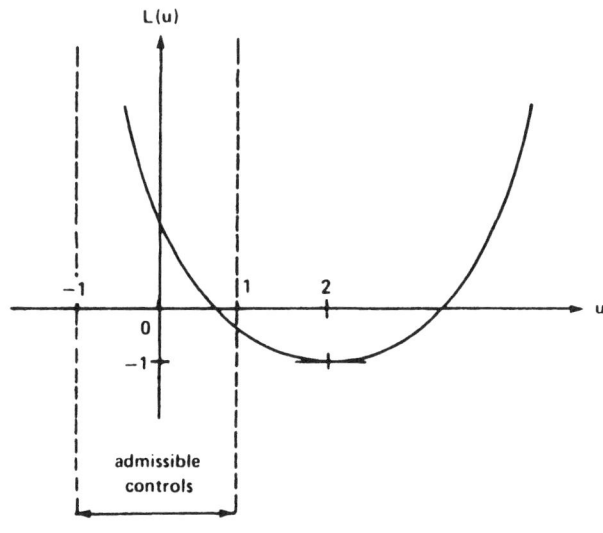

FIGURE 5.2-1   Optimization with constraints.

and

$$L(2) = -1 \qquad (8)$$

is less than (5); but $u = 2$ is inadmissible since it violates (2).   ∎

Bang-bang Control

Let us discuss the *linear minimum-time problem* with constrained input magnitude. Thus, the plant is

$$\dot{x} = Ax + Bu \qquad (5.2\text{-}7)$$

with performance index

$$J(t_0) = \int_{t_0}^{T} 1 \, dt, \qquad (5.2\text{-}8)$$

with T free. Suppose the control is required to satisfy

$$|u(t)| \leq 1 \qquad (5.2\text{-}9)$$

for all $t \in [t_0, T]$. This constraint means that *each component* of the m vector $u(t)$ must have magnitude no greater than 1.

The optimal control problem is to find a control $u(t)$ that minimizes $J(t_0)$, satisfies (5.2-9) at all times, and drives a given $x(t_0)$ to a final state $x(T)$ satisfying (5.2-3) for a given function ψ.

A requirement like (5.2-9) arises in many problems where the control magnitude is limited by physical considerations. For example, the thrust of a rocket certainly has a maximum possible value.

In Section 5.1 we talked about minimum-time problems, but only for nonlinear systems. The reason is the following. The Hamiltonian for our current problem is

$$H = L + \lambda^T f = 1 + \lambda^T (Ax + Bu). \qquad (5.2\text{-}10)$$

If we naively try to use Table 3.2-1, then the stationarity condition is found to be

$$0 = \frac{\partial H}{\partial u} = B^T \lambda. \qquad (5.2\text{-}11)$$

This does not involve u. This is because the Hamiltonian is linear in u. Clearly, to minimize H we should select $u(t)$ to make $\lambda^T(t)Bu(t)$ as small as possible. (*Small* means as far to the left as possible on the real number line; $-\infty$ is the smallest value $\lambda^T Bu$ can possibly take on!) If there is no constraint on $u(t)$, then this calls for infinite (positive or negative) values of the control variables. For this reason, we have avoided the linear minimum-time problem until we could include control input constraints.

Before we solve the linear minimum-time problem, let us consider a simple example that will give us a feel for the properties of constrained minimum-time control.

Example 5.2-2.   A One-dimensional Intercept Problem (Kirk 1970)

A pursuit aircraft starts out at rest a distance of h behind its target. See Fig. 5.2-2. The target is initially at rest also, but has an acceleration such that its motion is described by

$$y_T(t) = h + 0.1t^3. \tag{1}$$

FIGURE 5.2-2   A one-dimensional intercept problem.

The pursuer obeys Newton's laws

$$\dot{y} = v, \tag{2}$$

$$\dot{v} = u, \tag{3}$$

where $v(t)$ is its velocity and $u(t)$ its thrust per unit mass, which is constrained by

$$|u(t)| \leq 1. \tag{4}$$

For a minimum-time intercept, we want to determine $u(t)$ to minimize

$$J(0) = \int_0^T 1 \cdot dt. \tag{5}$$

This performance index is evidently minimized by selecting the *maximum admissible control* for all t. Therefore,

$$u^*(t) = 1. \tag{6}$$

Using this control and taking into account the initial state $y(0) = 0$, $v(0) = 0$, we can integrate the state equations (2) and (3) to get

$$v(t) = t, \tag{7}$$

$$y(t) = \frac{t^2}{2}. \tag{8}$$

For intercept, the target's position y_T and the pursuer's position y must be equal for some final time:

$$y_T(T) = y(T), \tag{9}$$

so T must satisfy

$$0.1T^3 - 0.5T^2 + h = 0. \tag{10}$$

A root locus for (10) as h varies from 0 to ∞ is shown in Fig. 5.2-3, where the values of h are shown in parentheses. For the intercept problem to have a solution, equation (10) must have a real positive root. From the root locus, this occurs only if $h \leq 1.85$; if the target starts out too far ahead, the pursuer cannot overtake it. For $h < 1.85$, equation (10) has two real positive roots; the smaller root is the value of T for which the pursuer first overtakes and passes the target, and the larger root is where the target again passes the pursuer.

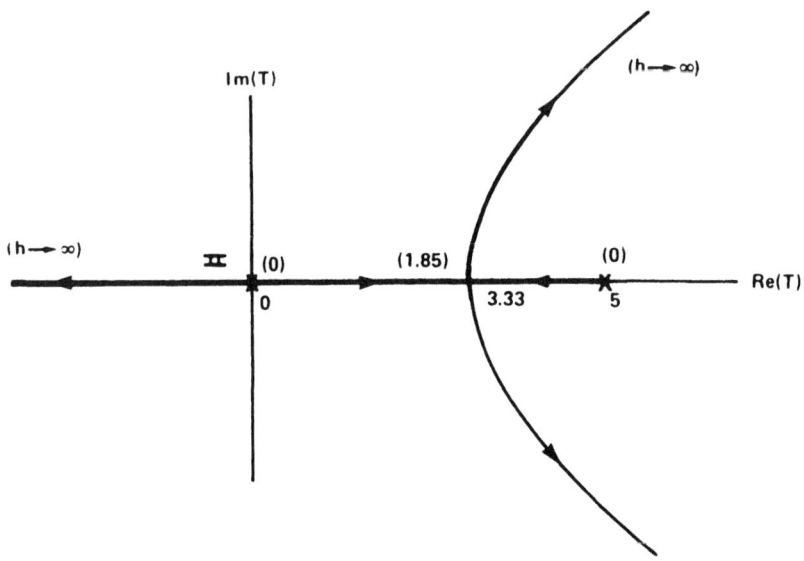

FIGURE 5.2-3   Root locus of final-time equation.

■

From this example we have learned two important properties of the linear minimum-time problem: a solution may not exist, and if one does, the optimal control strategy appears to be to apply maximum effort (i.e., plus or minus 1) over the entire control time interval. When the control takes on a value at the boundary of its admissible region, it is said to be *saturated*.

Now let us show how to find the linear minimum-time control law. For the problem formulated in (5.2-7)–(5.2-9), the Hamiltonian is given by (5.2-10). According to Pontryagin's minimum principle (5.2-6), the optimal control $u^*(t)$ must satisfy

$$1 + (\lambda^*)^{\mathrm{T}}(Ax^* + Bu^*) \le 1 + (\lambda^*)^{\mathrm{T}}(Ax^* + Bu).$$

Now we see the importance of having the *optimal* state and costate on both sides of the inequality, for this means we can say that for optimality the control $u^*(t)$ must satisfy

$$(\lambda^*)^{\mathrm{T}} Bu^* \le (\lambda^*)^{\mathrm{T}} Bu \qquad (5.2\text{-}12)$$

for all admissible $u(t)$. This condition allows us to express $u^*(t)$ in terms of the costate. To see this, let us first discuss the single-input case.

Let $u(t)$ be a scalar, and let b represent the input vector. In this case it is easy to choose $u^*(t)$ to minimize the value of $\lambda^{\mathrm{T}}(t)bu(t)$. (*Note: Minimize* means that we want $\lambda^{\mathrm{T}}(t)bu(t)$ to take on a value as close to $-\infty$ as possible.)

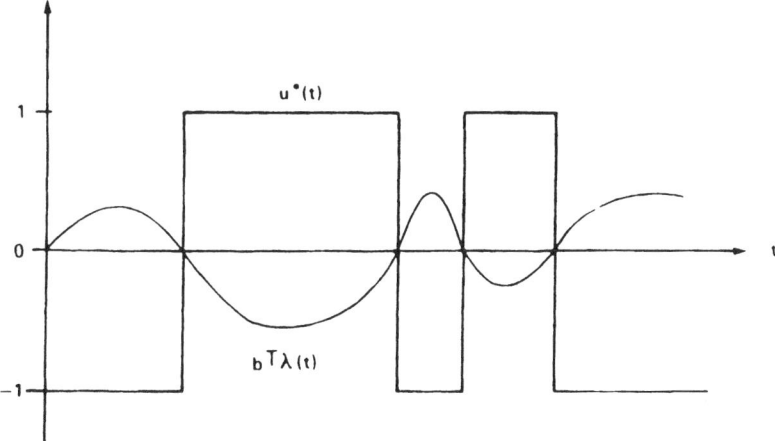

FIGURE 5.2-4   Sample switching function and associated optimal control.

If $\lambda^T(t)b$ is positive, we should select $u(t) = -1$ to get the largest possible negative value of $\lambda^T(t)bu(t)$. On the other hand, if $\lambda^T(t)b$ is negative, we should select $u(t)$ as its maximum admissible value of 1 to make $\lambda^T(t)bu(t)$ as negative as possible. If $\lambda^T(t)b$ is zero at a single point t in time, then $u(t)$ can take on any value at that time, since then $\lambda^T(t)bu(t)$ is zero for all values of $u(t)$.

This relation between the optimal control and the costate can be expressed in a neat way by defining the *signum function* $\text{sgn}(w)$ as

$$\text{sgn}(w) = \begin{cases} 1, & w > 0 \\ \text{indeterminate}, & w = 0 \\ -1, & w < 0. \end{cases} \tag{5.2-13}$$

Then the optimal control is given by

$$u^*(t) = -\text{sgn}(b^T\lambda(t)). \tag{5.2-14}$$

This expression for u^* in terms of the costate should be compared to the expression (3.3-7), which holds for linear systems with *quadratic* performance indices.

The quantity $b^T\lambda(t)$ is called the *switching function*. A sample switching function and the optimal control it determines are shown in Fig. 5.2-4. When the switching function changes sign, the control *switches* from one of its extreme values to another. The control in the figure switches four times. The optimal linear minimum-time control is always saturated since it switches back and forth between its extreme values, so it is called a *bang-bang control*.

If the control is an m vector, then according to the minimum principle (5.2-12), we need to select $u^*(t)$ to make $\lambda^T(t)Bu(t)$ take on a value as close to $-\infty$ as possible. To do this, we should select component $u_i(t)$ to be 1 if component

$b_i^T \lambda(t)$ is negative, and to be -1 if $b_i^T \lambda(t)$ is positive, where b_i is the ith column of B. This control strategy makes the quantity

$$\lambda^T(t) Bu(t) = \sum_{i=1}^{m} u_i(t) b_i^T \lambda(t) \tag{5.2-15}$$

as small as possible for all $t \in [t_0, T]$. Thus, we can write

$$u^*(t) = -\text{sgn}(B^T \lambda(t)) \tag{5.2-16}$$

if we define the signum function for a vector w as

$$v = \text{sgn } (w) \quad \text{if } v_i = \text{sgn } (w_i) \quad \text{for each } i, \tag{5.2-17}$$

where v_i, w_i are the components of v and w.

It is possible for a component $b_i^T \lambda(t)$ of the switching function $B^T \lambda(t)$ to be zero over a finite time interval. If this happens, component $u_i(t)$ of the optimal control is not well defined by (5.2-6). This is called a *singular* condition. If this does not occur, the time-optimal problem is called *normal*.

If the plant is time-invariant, then we can present some simple results on existence and uniqueness of the minimum-time control. First, we present a test for normality. See Athans and Falb (1966) and Kirk (1970) for more detail on the following results.

The time-invariant plant (5.2-7) is reachable if and only if the reachability matrix

$$U_n = [B \quad AB \quad \cdots \quad A^{n-1} B] \tag{5.2-18}$$

has full rank n. If b_i is the ith column of $B \in R^{n \times n}$, then the plant is *normal* if

$$U_i = [b_i \quad Ab_i \quad \cdots \quad A^{n-1} b_i] \tag{5.2-19}$$

has full rank n for each $i = 1, 2, \ldots, m$; that is, the plant is reachable by each separate component u_i of $u \in R^m$. Normality of the plant and normality of the minimum-time control problem are equivalent.

The next results are due to Pontryagin et al. (1962). Let the plant be normal (and hence reachable), and suppose we want to drive a given $x(t_0)$ to a desired fixed final state $x(T)$ in minimum time with a control satisfying $[u(t)] \leq 1$.

1. If the desired final state $x(T)$ is equal to zero, then a minimum-time control exists if the plant has no poles with positive real parts (i.e., no poles in the open right half-plane).
2. For any fixed $x(T)$, if a solution to the minimum-time problem exists, then it is unique.

3. Finally, if the n plant poles are all real and if the minimum-time control exists, then each component $u_i(t)$ of the time-optimal control can switch at most $n - 1$ times.

In both its computation and its final appearance, bang-bang control is fundamentally different from the smooth controls we have seen previously. The minimum principle leads to the expression (5.2-16) for $u^*(t)$, but it is difficult to solve explicitly for the optimal control. Instead, we shall see that (5.2-16) specifies several different control laws, and that we must then select which among these is the optimal control. Thus, the minimum principle keeps us from having to examine all possible control laws for optimality, giving a small subset of potentially optimal controls to be investigated.

To demonstrate these notions and show that $u^*(t)$ can still be expressed as a state feedback control law, let us consider a two-dimensional example, since the two-dimensional plane is easy to draw.

Example 5.2-3. *Bang-bang Control of Systems Obeying Newton's Laws*

Let the plant obey Newton's laws so that

$$\dot{y} = v, \tag{1}$$

$$\dot{v} = u, \tag{2}$$

with y the position and v the velocity. The state is $x = [y\,v]^{\mathrm{T}}$. Let the acceleration input u be constrained in magnitude by

$$|u(t)| \le 1. \tag{3}$$

The control objective is to bring the state from any initial point $(y(0), v(0))$ to the origin in the minimum time T. The final state is thus fixed at

$$\psi(x(T), T) = \begin{bmatrix} y(T) \\ v(T) \end{bmatrix} = 0. \tag{4}$$

a. Form of the Optimal Control

The Hamiltonian (5.2-10) is

$$H = 1 + \lambda_y v + \lambda_v u, \tag{5}$$

where $\lambda = [\lambda_y\,\lambda_v]^{\mathrm{T}}$ is the costate, so according to Table 3.2-1, the costate equations are

$$\dot{\lambda}_y = 0, \tag{6}$$

$$\dot{\lambda}_v = -\lambda_y. \tag{7}$$

The transversality condition is, from (3.2-10),

$$0 = H(T) = 1 + \lambda_y(T)v(T) + \lambda_v(T)u(T), \tag{8}$$

or, using (4),

$$\lambda_v(T)u(T) = -1. \tag{9}$$

Pontryagin's minimum principle requires (5.2-16), or

$$u(t) = -\text{sgn}(\lambda_v(t)) \tag{10}$$

so that costate component $\lambda_v(t)$ is the switching function. To determine the optimal control, we need only determine $\lambda_v(t)$.

Solving (6) and (7) with respect to the final time T yields

$$\lambda_y(t) = \text{const} \overset{\Delta}{=} \lambda_y, \tag{11}$$

$$\lambda_v(t) = \lambda_v(T) + (T - t)\lambda_y. \tag{12}$$

Using (9) and the fact that $u^*(t)$ is saturated at 1 or -1 requires either

$$u^*(T) = 1 \quad \text{and} \quad \lambda_v^*(T) = -1 \tag{13}$$

or

$$u^*(T) = -1 \quad \text{and} \quad \lambda_v^*(T) = 1. \tag{14}$$

There are several possibilities for the switching function $\lambda_v^*(t)$, depending on the values of $\lambda_v^*(T)$ and λ_y. Some possibilities are shown in Fig. 5.2-5. The actual $\lambda_v^*(t)$ depends on the initial state $(y(0), v(0))$. Note, however, that since $\lambda_v^*(t)$ is linear, it crosses the axis at most one time, so that there is at most one control switching. This agrees with our result on the maximum number of switchings when the plant poles are all real ($n - 1 = 1$). Therefore, the optimal control is one of the choices below:

1. -1 for all t
2. -1 switching to $+1$

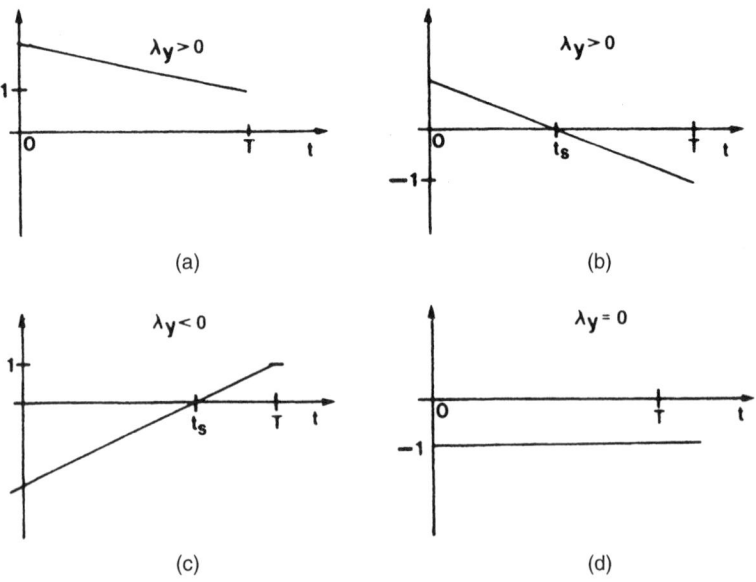

(a)

(b)

(c)

(d)

FIGURE 5.2-5   Possible switching functions $\lambda_v(t)$.

3. +1 switching to −1

4. +1 for all t

These control policies correspond to the switching functions of the forms shown in Fig. 5.2-5a, b, c, and d, respectively.

It remains to determine which of the choices 1 through 4 is the correct optimal control. We must also find the *switching time* t_s (see Fig. 5.2-5) if applicable. We can find a *state feedback* control law that tells us all this by working in the *phase plane*.

b. Phase-plane Trajectories

Let us determine the state trajectories under the influence of the two possible control inputs: $u(t) = 1$ for all t, and $u(t) = -1$ for all t. Since in either of these two cases the input u is constant, we can easily integrate state equations (2) and (1) to get

$$v(t) = v(0) + ut, \tag{15}$$

$$y(t) = y(0) + v(0)t + \tfrac{1}{2}ut^2. \tag{16}$$

To eliminate the time variable, use (15) to say $t = (v(t) - v(0))/u$, and then substitute into (16) to see that

$$u(y - y(0)) = v(0)(v - v(0)) + \tfrac{1}{2}(v - v(0))^2. \tag{17}$$

This is a parabola passing through $(y(0), v(0))$ and, as the initial state varies, a family of parabolas is defined.

The *phase plane* is a coordinate system whose axes are the state variables. The phase-plane plots of several members of the family of state trajectories (17) are shown for $u = 1$ and for $u = -1$ in Fig. 5.2-6. The arrows indicate the direction of increasing time. Hence, for example, if the initial state is $(y(0), v(0))$ as shown in Fig. 5.2-6a, then under the influence of the control $u = 1$, the state will develop along the parabola with velocity passing through zero and then increasing linearly, and position decreasing to zero and then increasing quadratically. For the particular initial condition shown, the state will be brought exactly to the origin by the control $u = 1$. For the same initial state, if the control $u = -1$ is applied, the trajectory will move along the parabola in Fig. 5.2-6b.

c. Bang-bang Feedback Control

Since the control input is saturated at 1 or −1, the parabolas in Fig. 5.2-6 are minimum-time paths in the phase plane. Unfortunately, they do not go through the origin for all initial states, so, in general, (4) is not satisfied. We can construct minimum-time paths *to the origin* by superimposing the two parts of Fig. 5.2-6. See Fig. 5.2-7. We shall now demonstrate that this figure represents a *state feedback control law*, which brings any state to the origin in minimum time.

We have argued using the minimum principle and the costate trajectories (11), (12) that at any given time t there are only two control alternatives: $u(t) = 1$ or $u(t) = -1$. Furthermore, at most one control switching is allowed.

Suppose the initial state is as shown in Fig. 5.2-6. Then the only way to arrive at the origin while satisfying these conditions is to apply $u = -1$ to move the state along a parabola to the dashed curve. At this point (labeled 1), the control is switched to $u = 1$

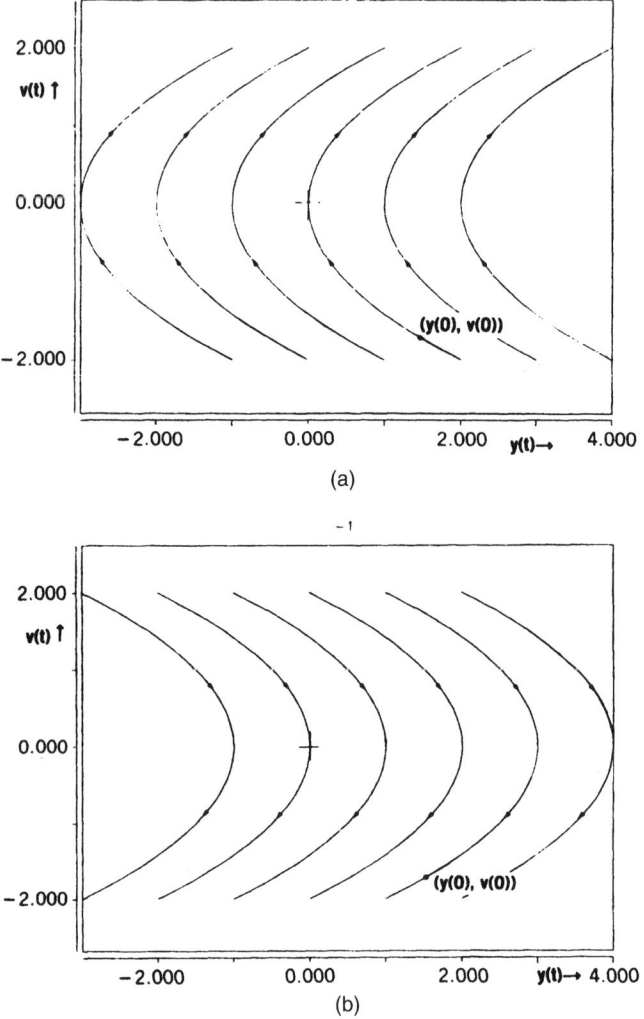

FIGURE 5.2-6 Phase plane trajectories. (a) $u = 1$. (b) $u = -1$.

to drive the state into the origin. Hence, the resultant seemingly roundabout trajectory is in fact a minimum-time path to the origin.

The dashed curve is known as the *switching curve*. For initial states on this curve, a control of $u = 1$ (if $v(0) < 0$) or $u = -1$ (if $v(0) > 0$) for the entire control interval will bring the state to zero. For initial states off this curve, the state must first be driven onto the switching curve, and then the control must be switched to its other extreme value to bring the final state to zero. By setting $v(0) = 0$, $y(0) = 0$ in (17), we can see that the equation of the switching curve is

$$y = \begin{cases} \frac{1}{2}v^2, & v < 0 \\ -\frac{1}{2}v^2, & v > 0 \end{cases}$$

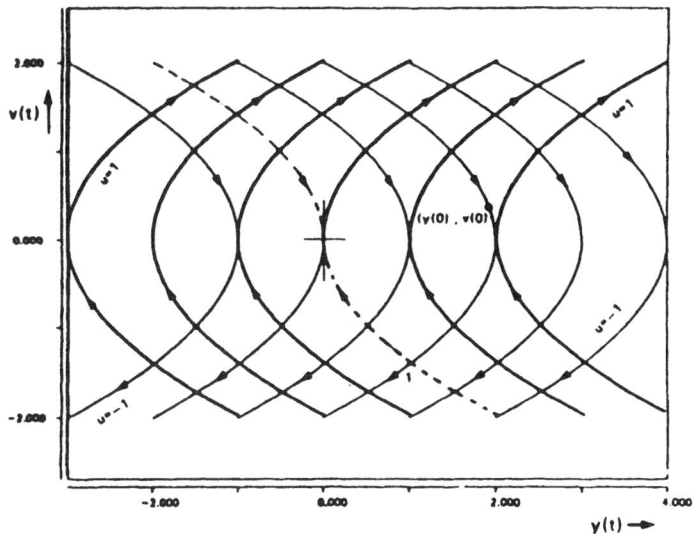

FIGURE 5.2-7   Bang-bang feedback control law.

or

$$y = -\tfrac{1}{2}v|v|. \tag{18}$$

Simply put, for initial states above the switching curve, the optimal control is $u = -1$, followed by $u = 1$, with the switching occurring when $y(t) = \tfrac{1}{2}v^2(t)$. For initial states below the switching curve, the optimal control is $u = 1$, followed by $u = -1$, with the switching occurring when $y(t) = -\tfrac{1}{2}v^2(t)$. Since the control at each time t is completely determined by the state (i.e., by the phase plane location), Fig. 5.2-7 yields a *feedback control law*.

This feedback law, which is represented graphically in the figure, can be stated as

$$u = \begin{cases} -1 & \text{if } y > -\tfrac{1}{2}v|v| \\ & \text{or if } y = -\tfrac{1}{2}v|v| \text{ and } y < 0 \\ 1 & \text{if } y < -\tfrac{1}{2}v|v| \\ & \text{or if } y = -\tfrac{1}{2}v|v| \text{ and } y > 0 \end{cases} \tag{19}$$

This control scheme should be contrasted with the control laws developed in Examples 2.3-2, 2.4-3, 3.3-3, and 3.4-3.

d. Simulation

It is easy to implement the feedback control (19) and simulate its application using MATLAB. The function *ex5_2_8* in Fig. 5.2-8 implements (19). It is based on the function

$$SW = y + \tfrac{1}{2}v|v|. \tag{20}$$

Note that SW is considered equal to zero if it is within a threshold of 10^{-4} on either side of zero. Note also that the control input must be turned off (i.e., set to zero) when the state is sufficiently close to the origin, in order to bring the plant to rest there.

```
function [x, u, t]=ex5_2_8 (a, b, x0, T, N)
% Simulation of Bang_Bang Control for Newton's System
x(:,1)=x0;
epsilon=1e-4;
t=0:T:T*N;
for k=1:N
% Compute the Switching Function
sw=x(1,k)+0.5*x(2,k)*abs(x(2,k));
if (abs (sw) <epsilon)
if (x(1, k) > 0) u(k) = 1; end
if (x(1, k) < 0) u(k) = -1; end
else
if (sw < 0) u(k) = 1; end
if (sw > 0) u(k) = -1; end
end
if (x(1,k)^2+x(2,k)^2 < epsilon) u(k) =0; end
% Harmonic Oscillator System State Equations
y=lsim (a, b, eye(2), zeros (2, 1), u(k)*ones(1, 2),...
[(k-1)*T, k*T], x(:,k));
x(:,k+1) =y(2,:)';
end
```

FIGURE 5.2-8   MATLAB simulation of bang-bang control.

Function *lsim.m* from the Control Toolbox is used to implement the plant dynamics (1), (2) at each time interval. The state trajectories resulting from the simulation when $y(0) = 10$ and $v(0) = 10$ are shown in Fig. 5.2-9. The time interval at each step of the

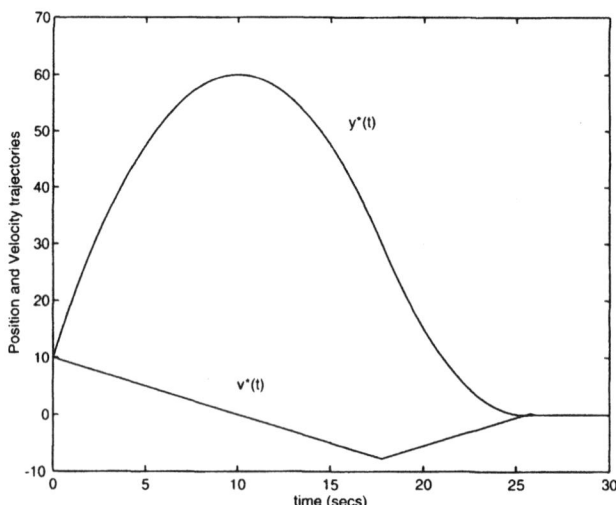

FIGURE 5.2-9   Optimal bang-bang state trajectories.

iteration for *lsim* was 25 msec. The number of iterations was $N = 1200$. It can be seen from the slope of $v(t)$ that the control input switches from $u = -1$ to $u = 1$ at $t_s = 16$ sec, and that the minimum time to the origin is 26 sec. These trajectories should be compared to the results of Examples 2.3-2 and 2.4-3.

e. Computation of Time to the Origin

It is not difficult to compute the minimum time to the origin in terms of the initial state $(y(0), v(0))$. Let us suppose that the initial state is above the switching curve, that is

$$y(0) > -\tfrac{1}{2}v(0)|v(0)|. \tag{21}$$

This situation is shown in Fig. 5.2-7. Then the initial control of $u = -1$ is applied to drive the state along the parabola passing through $(y(0), v(0))$ to the switching curve, at which time t_s the control is switched to $u = 1$ to bring the state to the origin.

The switching curve for $v < 0$ is described by $y = \tfrac{1}{2}v^2$. We can find the switching time t_s by determining when the state is on this curve. Using (15) and (16) with $u = -1$ yields

$$y(t) = y(0) + v(0)t - \frac{t^2}{2}$$

$$= \frac{v^2(t)}{2} = \frac{v^2(0)}{2} = -v(0)t + \frac{t^2}{2}$$

on the switching curve, or

$$t^2 - 2v(0)t + \frac{v^2(0)}{2} - y(0) = 0. \tag{22}$$

The switching time is therefore

$$t_s = v(0) + \sqrt{y(0) + \frac{v^2(0)}{2}}, \tag{23}$$

where the positive root is selected to make t_s positive for all $v(0)$.

At the switching time, the state is on the switching curve (at point 1 in Fig. 5.2-7), and using (15)

$$v(t_s) = v(0) - t_s. \tag{24}$$

Also using (15) for the remaining time $(T - t_s)$ yields (now $u = 1$!)

$$0 = v(T) = v(t_s) + (T - t_s). \tag{25}$$

Taking (24) into account gives the minimum time to the origin of

$$T = 2t_s - v(0)$$

or

$$T = v(0) + 2\sqrt{y(0) + \frac{v^2(0)}{2}}. \tag{26}$$

To check (23) and (26), let $y(0) = 10$, $v(0) = 10$. Then $t_s = 17.75$ and $T = 25.49$. These numbers agree with Fig. 5.2-9. A similar development holds if $(y(0), v(0))$ is below the switching curve. ∎

Bang–Off-bang Control

We now discuss the *linear minimum-fuel problem* with constrained input magnitude. Minimum-fuel control is important in aerospace applications, where fuel is limited and must be conserved. Let the plant be

$$\dot{x} = Ax + Bu. \tag{5.2-20}$$

Assuming the fuel used in each component of the input is proportional to the magnitude of that component, define the cost function

$$J(t_0) = \int_{t_0}^{T} \sum_{i=1}^{m} c_i |u_i(t)| \, dt, \tag{5.2-21}$$

where we allow the possibility of penalizing differently the fuel burned in each of the m input components $u_i(t)$ by using the scalar weights c_i. By defining the vector absolute value as

$$|u| = \begin{bmatrix} |u_1| \\ \vdots \\ |u_m| \end{bmatrix} \tag{5.2-22}$$

(this is the same definition used in (5.2-9)) and the vector $C = [c_1 c_2 \cdots c_m]^T$, we have

$$J(t_0) = \int_{t_0}^{T} C^T |u(t)| \, dt. \tag{5.2-23}$$

Suppose the control is required to satisfy

$$|u(t)| \leq 1 \tag{5.2-24}$$

for all $t \in [t_0, T]$.

We want to find a control that minimizes $J(t_0)$, satisfies (5.2-24), and drives a given $x(t_0)$ to a final state satisfying (5.2-3) for a given function ψ. The final time T can be either free or fixed; we shall discuss this further in an example. Note, however, that T must be at least as large as the *minimum* time required to drive $x(t_0)$ to a final state $x(T)$ satisfying (5.2-3).

The Hamiltonian is

$$H = C^T |u| + \lambda^T (Ax + Bu), \tag{5.2-25}$$

and according to the minimum principle (5.2-6), the optimal control must satisfy

$$C^T |u^*| + (\lambda^*)^T (Ax^* + Bu^*) \leq C^T |u| + (\lambda^*)^T (Ax^* + Bu) \tag{5.2-26}$$

for all admissible $u(t)$. Since the *optimal* state and costate appear on both sides, we require

$$C^T |u^*| + (\lambda^*)^T Bu^* \leq C^T |u| + (\lambda^*)^T Bu \tag{5.2-27}$$

for all admissible $u(t)$.

To translate (5.2-27) into a rule for determining $u^*(t)$ from the costate $\lambda(t)$, assume that the m components of the control are independent so that for each $i = 1, \ldots, m$ we require for all admissible $u_i(t)$ the scalar inequality

$$|u_i^*| + \frac{(\lambda^*)^{\mathrm{T}} b_i u_i^*}{c_i} \leq |u_i| + \frac{(\lambda^*)^{\mathrm{T}} b_i u_i}{c_i}, \tag{5.2-28}$$

where b_i denotes the ith column of B. We must now demonstrate how to select control component $u_i^*(t)$ from $\lambda^{\mathrm{T}}(t) b_i$. Since

$$|u_i| = \begin{cases} u_i, & u_i \geq 0 \\ -u_i, & u_i \leq 0 \end{cases} \tag{5.2-29}$$

we can write the quantity we are trying to maximize by selection of $u_i(t)$ as

$$q_i(t) \overset{\Delta}{=} |u_i| + \frac{b_i^{\mathrm{T}} \lambda u_i}{c_i} = \begin{cases} \left(1 + \dfrac{b_i^{\mathrm{T}} \lambda}{c_i}\right) |u_i|, & u_i \geq 0 \\ \left(1 - \dfrac{b_i^{\mathrm{T}} \lambda}{c_i}\right) |u_i|, & u_i \leq 0. \end{cases} \tag{5.2-30}$$

This quantity is plotted in Fig. 5.2-10 for $u_i = 1$, $u_i = 0$, and $u_i = -1$. In general, for values of $u_i(t)$ between -1 and 1, the quantity $q_i(t)$ will take on values inside the cross-hatched region.

To *minimize* q_i so that (5.2-28) holds, we should select values for $u_i^*(t)$ corresponding to the lower boundary of the cross-hatched region in the figure. However, if $b_i^{\mathrm{T}} \lambda / c_i$ is equal to 1, then *any* nonpositive value of $u_i(t)$ will make

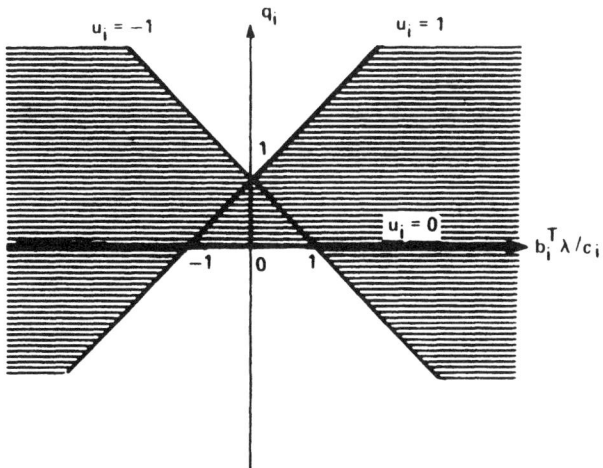

FIGURE 5.2-10   The quantity to be minimized by selecting u_i^* plotted versus $b_i^{\mathrm{T}} \lambda / c_i$.

q_i equal to zero (see (5.2-30)). On the other hand, if $\lambda^T b_i/c_i$ is equal to -1, then any non-negative value of $u_i(t)$ will make q_i equal to zero. Therefore, the minimum-fuel control law expressed as a (nonlinear!) costate feedback is, for $i = 1, \ldots, m$,

$$
u_i(t) = \begin{cases}
1, & b_i^T \lambda(t)/c_i < -1 \\
\text{non-negative}, & b_i^T \lambda(t)/c_i = -1 \\
0, & -1 < b_i^T \lambda(t)/c_i < 1 \\
\text{non-positive}, & b_i^T \lambda(t)/c_i = 1 \\
-1, & 1 < b_i^T \lambda(t)/c_i
\end{cases}
\tag{5.2-31}
$$

If we define the *dead-zone function* (Athans and Falb 1966) by

$$
\text{dez } (w) = \begin{cases}
-1 & w < -1 \\
\text{between } -1 \text{ and } 0, & w = -1 \\
0, & -1 < w < 1 , \\
\text{between } 0 \text{ and } 1, & w = 1 \\
1, & w > 1
\end{cases}
\tag{5.2-32}
$$

then we can write the minimum-fuel control as

$$
u_i^*(t) = -\text{dez} \left(\frac{b_i^T \lambda(t)}{c_i} \right), \quad i = 1, \ldots, m.
\tag{5.2-33}
$$

Since each component of $u(t)$ is either saturated or equal to zero, we call this a *bang–off-bang* control law.

If $b_i^T \lambda(t)/c_i$ is equal to 1 or -1 over a finite nonzero time interval, then the minimum principle does not define the component $u_i(t)$ there. This is called a *singular interval*. If $b_i^T \lambda(t)/c_i$ is equal to 1 or -1 only at a finite number of isolated times t, the minimum-fuel problem is said to be *normal*.

If the plant is time-invariant and the final time T is fixed, then we have the following results (Athans and Falb 1966). The minimum-fuel problem is normal if $|A| \neq 0$ and if the plant is normal, that is, if U_i defined by (5.2-19) is nonsingular for $i = 1, \ldots, m$. If the minimum-fuel problem is normal and a minimum-fuel control exists, then it is unique.

The next example will impart a feel for minimum-fuel control.

Example 5.2-4.   Bang–Off-bang Control of System Obeying Newton's Laws

Let the plant dynamics be described by

$$
\dot{y} = v,
\tag{1}
$$

$$
\dot{v} = u,
\tag{2}
$$

as in the previous example, with $x = [yv]^T$ the state. The input is constrained by

$$
|u(t)| \leq 1.
\tag{3}
$$

Suppose the objective is to determine a control input to bring any given initial state $(y(0), v(0))$ to the origin so that

$$\psi(x(T), T) = \begin{bmatrix} y(T) \\ v(T) \end{bmatrix} = 0. \tag{4}$$

The control should use minimum fuel, so let

$$J(0) = \int_0^T |u(t)| \, dt. \tag{5}$$

We do not yet care whether the final time T is free or fixed, although we shall end up covering both cases.

a. Form of the Optimal Control

The Hamiltonian is

$$H = |u| + \lambda_y v + \lambda_v u, \tag{6}$$

where the costate is $\lambda = \begin{bmatrix} \lambda_y & \lambda_v \end{bmatrix}^\mathsf{T}$. Therefore, the costate equations, as in Example 5.2-3, are

$$\dot{\lambda}_y = 0, \tag{7}$$

$$\dot{\lambda}_v = -\lambda_y. \tag{8}$$

Transversality demands

$$0 = H(T) = |u(T)| + \lambda_v(T)u(T), \tag{9}$$

where (4) has been taken into consideration.

The solution of (7), (8) is

$$\lambda_y(t) = \text{const} \overset{\Delta}{=} \lambda_y, \tag{10}$$

$$\lambda_y(t) = \lambda_v(T) + (T - t)\lambda_y. \tag{11}$$

Costate component $\lambda_v(t)$ is linear. Depending on the as yet unknown λ_y and $\lambda_v(T)$ (which depend on the initial state), $\lambda_v(t)$ can be constant ($\lambda_y = 0$), increasing ($\lambda_y < 0$), or decreasing ($\lambda_y > 0$). See Fig. 5.2-5.

Pontryagin's minimum principle requires

$$u(t) = -\text{dez}(\lambda_v(t)); \tag{12}$$

so the optimal control is

$$\begin{aligned}
u(t) &= 1, & \text{if } \lambda_v(t) &< -1, \\
0 \le u(t) &\le 1, & \text{if } \lambda_v(t) &= -1, \\
u(t) &= 0, & \text{if } -1 < \lambda_v(t) &< 1, \\
-1 \le u(t) &\le 0, & \text{if } \lambda_v(t) &= 1, \\
u(t) &= -1, & \text{if } \lambda_v(t) &> 1.
\end{aligned} \tag{13}$$

Taking into account the linearity of $\lambda_v(t)$, we therefore realize that $u = 1$ cannot switch to $u = -1$ without the intermediate value of $u = 0$, and vice versa. Furthermore, at most two switchings are allowed. The control laws that are admissible candidates for the optimal control must satisfy these requirements.

Now we must determine which is the optimal control law and find the times when the control switches to a new value.

b. Phase-plane Trajectories

Ignoring the possibility of singular intervals, there are three possible values for $u(t)$: $-1, 0$, and 1. Figure 5.2-6 shows the phase-plane trajectories for $u = 1$ and $u = -1$.

If $u(t) = 0$ for all t, then the state is given by (Example 5.2-3, equations (15), (16))

$$v(t) = v(0), \tag{14}$$

$$y(t) = y(0) + v(0)t. \tag{15}$$

These contours of constant v are horizontal lines in the phase plane as shown in Fig. 5.2-11, where the motion over a one-second interval with $y(0) = 0$ is shown by a vector on each contour. Since v is smaller nearer the origin, the system moves more slowly along paths nearer the origin.

The phase-plane paths for $u = 0$ represent paths of zero fuel consumption. To minimize the fuel used, we should take advantage of the *system drift* by using $u = 1$ or -1 to drive the state onto one of these paths letting it drift along the path toward the origin, and then applying $u = \quad 1$ or 1 to drive the state exactly to zero.

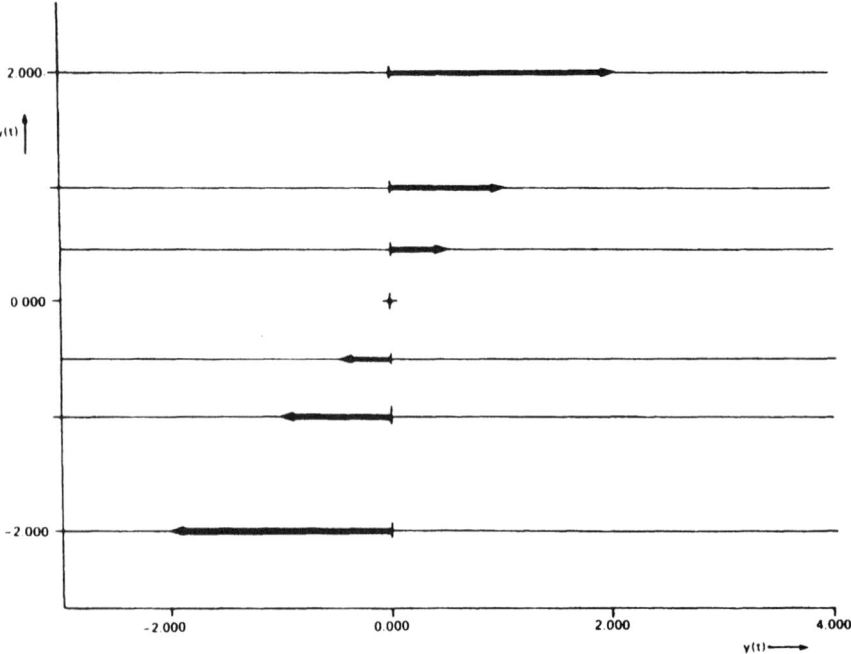

FIGURE 5.2-11   Phase-plane trajectories for $u = 0$.

To obtain a bang–Off-bang control law, we can superimpose the trajectories of Figs. 5.2-6 and 5.2-11. See Fig. 5.2-12.

Let us now discuss separately the free- and fixed-final-time situations.

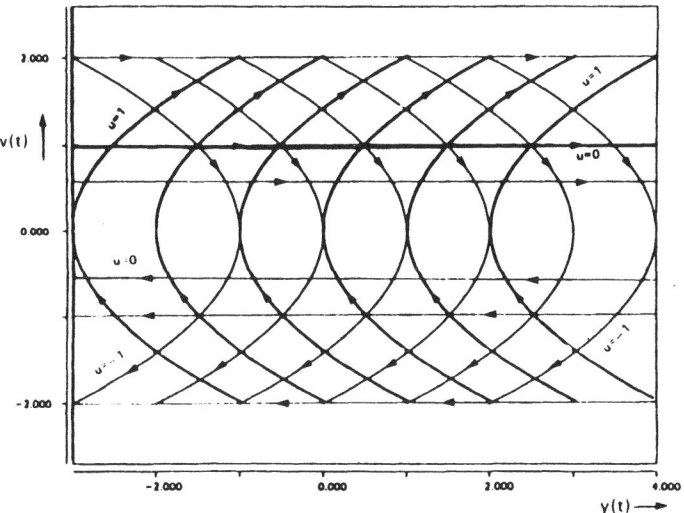

FIGURE 5.2-12   Bang–off-bang control law.

c. Free Final Time

Suppose that T is not specified. If the initial state is $(y(0), v(0))$ as shown in Fig. 5.2-12, then there is only one candidate control history that drives the plant to zero and satisfies (13): $-1, 0, 1$, with switchings at times t_1 and t_2, as shown in the figure. We shall show that if the final time T in this example is free, then a minimum-fuel control law does not generally exist!

To define the optimal control law, the switching times t_1 and t_2 must be specified. Let us postulate a candidate for the minimum-fuel control law. We need to apply $u = -1$ long enough to drive v to a negative value. Suppose we use $u = -1$ until $v = v_1 = -\epsilon$, a small quantity, and then set $u = 0$. On removing the control, the state will drift to the left along the $u = 0$ path shown according to

$$y(t) = y(t_1) - \epsilon(t - t_1), \quad t > t_1. \tag{16}$$

When the state hits the curve defined by

$$y = -\tfrac{1}{2}v|v| \tag{17}$$

at time t_2, we should apply $u = 1$ to drive the state to zero.

The shorter the interval of application of $u = -1$, the less fuel is consumed; but for any given ϵ, we can always find a control law that uses less fuel than the candidate law by turning off the control $u = -1$ when $v = v_1 = -\epsilon/2$. This will result in a leftward drift described by

$$y(t) = y(t_1) - \frac{\epsilon}{2}(t - t_1), \quad t > t_1. \tag{18}$$

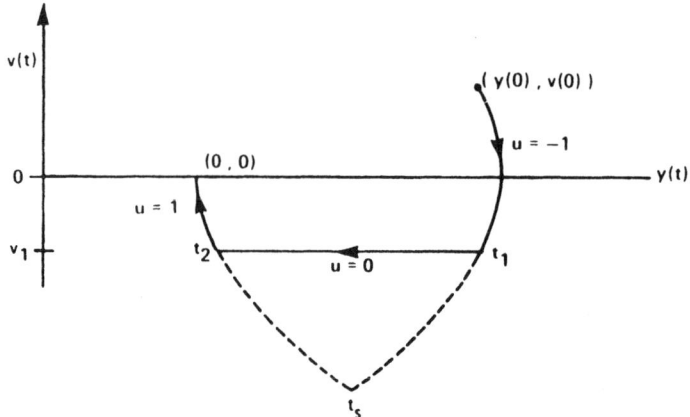

FIGURE 5.2-13   Minimum-fuel state trajectory.

When (17) is satisfied, we again apply $u = 1$. This second control law will take longer to zero the state, but it will require less fuel. Therefore, if T is not limited, no minimum-fuel control exists for the given problem. (Does one exist for any initial states other than the one we used?)

d. Fixed Final Time

Let the initial state be as shown in Fig. 5.2-13. The *minimum-time* control law is manufactured by applying $u = -1$ until $t = t_s$, and then switching to $u = 1$ (i.e., the minimum-time trajectory contains the dashed path in the figure). According to Example 5.2-3, the associated minimum final time is

$$T_{\min} = v(0) + 2\sqrt{y(0) + \frac{v^2(0)}{2}}. \tag{19}$$

Suppose that T for the minimum-fuel problem is fixed at a value

$$T > T_{\min}. \tag{20}$$

Then the minimum-fuel control law is -1, 0, 1 with switching times t_1 and t_2 to be determined. The minimum-fuel trajectory must now satisfy (4) with a fixed given T. This allows us to determine t_1, t_2, and v_1 in terms of $(y(0), v(0))$ and T.

For $0 < t < t_1$, $u(t) = -1$, so that (15) and (16) in Example 5.2-3 yield

$$v_1 = v(t_1) = v_0 - t_1, \tag{21}$$

$$y(t_1) = y(0) + v(0)t_1 - \frac{t_1^2}{2}. \tag{22}$$

For $t_1 < t < t_2$, we apply $u(t) = 0$, so that the state equation solution yields

$$v(t_2) = v_1, \tag{23}$$

$$y(t_2) = y(t_1) + v_1(t_2 - t_1). \tag{24}$$

For $t_2 < t < T$, $u(t) = 1$, so that

$$0 = v(T) = v(t_2) + (T - t_2), \tag{25}$$

$$0 = y(T) = y(t_2) + v(t_2)(T - t_2) + \frac{(T - t_2)^2}{2}, \tag{26}$$

where we have used the boundary conditions (4).

Now, use (21) and (23) in (25) to get

$$t_2 = v_0 + T - t_1. \tag{27}$$

Substituting (21), (22), (23), (24), and (27) into (26) and simplifying results in

$$t_1^2 - (v_0 + T)t_1 + \left(y_0 + v_0 T + \frac{v_0^2}{2}\right) = 0, \tag{28}$$

whence

$$t_1 = \frac{(v_0 + T) \pm \sqrt{(v_0 - T)^2 - (4y_0 + 2v_0^2)}}{2}. \tag{29}$$

Now, (27) and the fact that $t_1 < t_2$ show that

$$t_1 = \frac{(v_0 + T) - \sqrt{(v_0 - T)^2 - (4y_0 + 2v_0^2)}}{2} \tag{30}$$

and

$$t_2 = \frac{(v_0 + T) + \sqrt{(v_0 - T)^2 - (4y_0 + 2v_0^2)}}{2}. \tag{31}$$

Note that $T > T_{\min}$ guarantees the quantity under the radical sign is positive.

We can express the minimum-fuel control in open-loop form as

$$u^*(t) = \begin{cases} -1, & t < t_1 \\ 0, & t_1 \le t < t_2 \\ 1, & t_2 \le t. \end{cases} \tag{32}$$

According to (21) and (30), the minimum value of $v(t)$ attained under the influence of this control is the negative quantity

$$v_1 = \frac{(v_0 + T) + \sqrt{(v_0 - T)^2 - (4y_0 + 2v_0^2)}}{2}. \tag{33}$$

e. Computer Simulation

It is not difficult to implement (30), (31), (32) in a computer simulation. The result when $y(0) = 10$, $v(0) = 10$, and $T = 35$ sec is shown in Fig. 5.2-14. Compare to Fig. 5.2-9.

Switching times $t_1 = 12.7$ and $t_2 = 32.3$ in the figure agree with (30) and (31). The *drift velocity* during the interval when $u(t) = 0$ is $v_1 = -2.7$, agreeing with (33).   ∎

Constrained Minimum-energy Problems

To round out this section on constrained input control, let us consider the *linear minimum-energy problem*. Suppose the plant

$$\dot{x} = Ax + Bu \tag{5.2-34}$$

has associated cost index of

$$J(t_0) = \frac{1}{2} \int_{t_0}^{T} u^T(t) Ru(t) \, dt \tag{5.2-35}$$

with $R > 0$ and T either free or fixed. A control must be found to minimize $J(t_0)$, while satisfying the constraint $|u(t)| \le 1$ and driving a given $x(t_0)$ to a final state such that $\psi(x(T), T) = 0$ for a given ψ.

The Hamiltonian is

$$H = \tfrac{1}{2} u^T Ru + \lambda^T(Ax + Bu) \tag{5.2-36}$$

and so the costate equation is

$$-\dot{\lambda} = \frac{\partial H}{\partial x} = A^T \lambda \tag{5.2-37}$$

as given in Table 3.2-1.

```
function [x, u, t]=ex5_2_4 (a, b, x0, Tf, N)
% Simulation of Bang-off-bang Control for Newton's System
epsilon = 1e-4;   x(:,1)=x0;   Ti=Tf/N; t=0:Ti:Tf;
% Compute the Switching Times
ts=0;
p=[1 -(x0(2)+Tf) (x0(1)+x0(2)*Tf+x0(2)^2/2) ];
t_sw=sort(roots(p));
for k=1:N
ts=ts+Ti;
% Compute the Switching Function
u(k)=0;
if (ts<t_sw(1)) u(k)=-1;end
if (ts>t_sw(2)) u(k)=1; end
% Harmonic Oscillator System State Equations
y=lsim(a, b, eye (2), zeros(2, 1), u(k)*ones(1, 2),...
[(k-1)*Ti, k*Ti], x(:,k)); x(:,k+1) =y(2, :)';
end
```

(a)

FIGURE 5.2-14   (a) MATLAB simulation of bang–off-bang control. (b) Optimal bang–off-bang state trajectories. (c) Switching function $u^*(t)$.

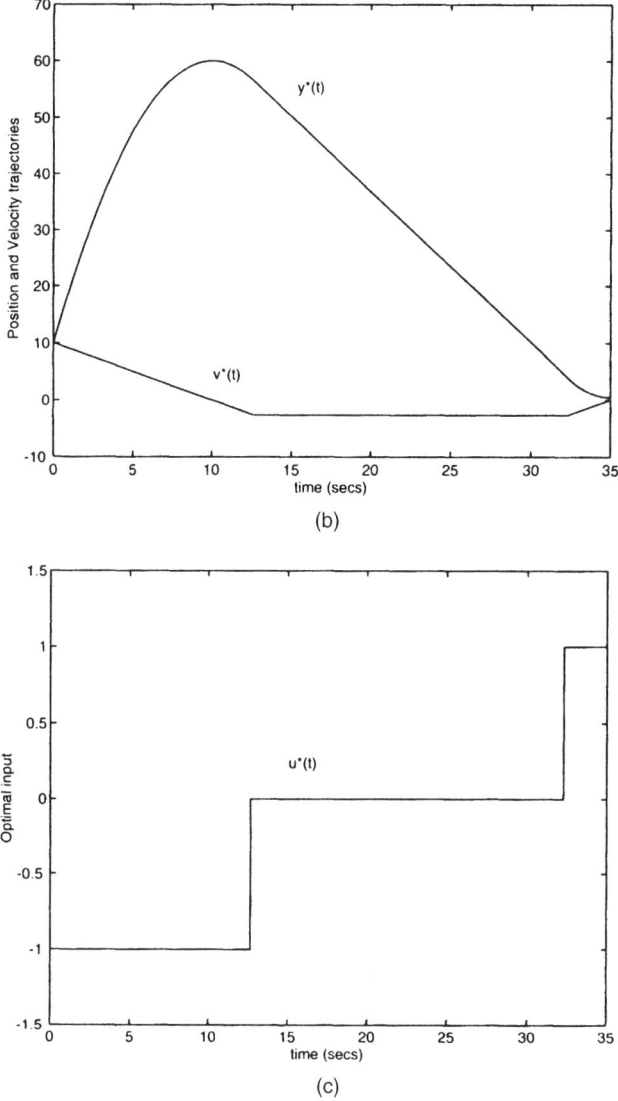

FIGURE 5.2-14   *(continued)*

Pontryagin's minimum principle (5.2-6) requires that

$$\tfrac{1}{2}(u^*)^{\mathrm{T}} R u^* + (\lambda^*)^{\mathrm{T}} B u^* \le \tfrac{1}{2} u^{\mathrm{T}} R u + \lambda^{\mathrm{T}} B u \qquad (5.2\text{-}38)$$

for all admissible $u(t)$. We must therefore select the input to minimize $\tfrac{1}{2} u^{\mathrm{T}} R u + \lambda^{\mathrm{T}} B u$. Adding to this the term $\tfrac{1}{2}\lambda^{\mathrm{T}} B R^{-1} B^{\mathrm{T}} \lambda$, which does not depend on $u(t)$, we can alternatively minimize

$$\tfrac{1}{2}(u + R^{-1}B^{T}\lambda)^{T}R(u + R^{-1}B^{T}\lambda) \overset{\Delta}{=} \tfrac{1}{2}w^{T}Rw. \qquad (5.2\text{-}39)$$

We want to work separately with each component $u_i(t)$ of the input. To do this, we must first diagonalize R. As a matter of fact, we can show by the following argument that R can be simply deleted from the right-hand side of (5.2-39). See Athans and Falb (1966).

Since $R > 0$, there is an orthogonal ($MM^{T} = I$) matrix M such that

$$D = MRM^{T} \qquad (5.2\text{-}40)$$

is diagonal. Then (5.2-39) can be written

$$\tfrac{1}{2}(Mw)^{T}D(Mw) = \tfrac{1}{2}\sum_{i=1}^{m} d_i v_i^2(t), \qquad (5.2\text{-}41)$$

where $D = \mathrm{diag}(d_i)$ and $v_i(t)$ is the ith component of $v(t) \overset{\Delta}{=} Mw(t)$.

Since $d_i > 0$, it can be seen that

$$\min_{u}\left(\tfrac{1}{2}w^{T}Rw\right) = \tfrac{1}{2}\sum_{i=1}^{m} d_i \min_{v_i}(v_i^2). \qquad (5.2\text{-}42)$$

Hence, $u(t)$ minimizes (5.2-39) if and only if $w(t)$ is such that

$$v^{T}v = (Mw)^{T}Mw \qquad (5.2\text{-}43)$$

is a minimum. The orthogonality of M, however, implies that

$$w^{T}w = (Mw)^{T}Mw \qquad (5.2\text{-}44)$$

Therefore, if $R > 0$, then selecting $u(t)$ to minimize (5.2-39) is equivalent to selecting $u(t)$ to minimize

$$\tfrac{1}{2}(u + R^{-1}B^{T}\lambda)^{T}(u + R^{-1}B^{T}\lambda) = \tfrac{1}{2}w^{T}w. \qquad (5.2\text{-}45)$$

Depending on the magnitude of the ith component of the quantity

$$q(t) = R^{-1}B^{T}\lambda(t), \qquad (5.2\text{-}46)$$

there are three cases for component $u_i(t)$ of the control. If $|q_i(t)| \leq 1$, then $u_i^*(t)$ is determined by setting to zero the derivative of (5.2-45) with respect to $u_i(t)$:

$$[u + R^{-1}B\lambda]_i = 0, \qquad (5.2\text{-}47)$$

where $[w]_i$ represents the ith component of a vector w. Hence,

$$u_i(t) = -q_i(t) \quad \text{if } |q_i(t)| \le 1. \tag{5.2-48}$$

Note that this is the same result obtained by using the stationarity condition

$$0 = \frac{\partial H}{\partial u} = Ru + B^T\lambda \tag{5.2-49}$$

from Table 3.2-1.

On the other hand, if $|q_i(t)| > 1$, then the minimum along the ith direction found by (5.2-49) corresponds to an inadmissible value of $u_i(t)$. In this case, the best we can do is select $u_i(t)$ to make $w_i(t)$ as small as possible; we cannot make it equal to zero. Hence,

$$u_i(t) = \begin{cases} 1, & q_i(t) < -1 \\ -1, & q_i(t) > 1 \end{cases}. \tag{5.2-50}$$

Defining the saturation function as

$$\text{sat}(q_i) = \begin{cases} -1, & q_i < -1 \\ q_i, & |q_i| \le 1 \\ 1, & q_i > 1 \end{cases}, \tag{5.2-51}$$

we can express (5.2-48) and (5.2-50) as

$$u_i^*(t) = -\text{sat}([R^{-1}B^T\lambda(t)]_i), \quad i = 1, \dots, m. \tag{5.2-52}$$

This is the minimum-energy constrained-input control expressed as a costate feedback. If $|R^{-1}B^T\lambda(t)| \le 1$ for all t, then (5.2-52) is identical to the control given in Table 3.2-1.

PROBLEMS

Section 5.1

5.1-1. Thrust direction programming in a gravitational field. Redo Example 3.2-5 if there is a gravitational acceleration of g in the negative y direction.

5.1-2. Minimum-time problem with control weighting. Let

$$\dot{x} = f(x, u, t), \quad x(0) \text{ given}, \tag{1}$$

have the associated cost index

$$J = \int_0^T \left(1 + \frac{1}{2}u^T R u\right) dt \tag{2}$$

with $R > 0$. We require that

$$\psi(x(T), T) = Cx(T) - r(T) = 0, \tag{3}$$

where $r(T)$ is given. The final time T is free.
Write Euler equations and boundary conditions. Show that for all $t \in [0, T]$

$$1 + \tfrac{1}{2}u^T Ru = -\lambda^T f. \tag{4}$$

5.1-3. Linear minimum-time problem with control weighting. In Problem 5.1-2, let $\dot{x} = Ax + Bu$.
a. Write Euler equations and boundary conditions.
b. Let $r(T) \in R^p$. Write $p + 1$ equations that can be solved for the undetermined multiplier $v \in R^p$ and the final time T (cf. Problem 3.3-9).
c. Show that in the scalar case $x \in R$, $u \in R$, the optimal control can be expressed as the nonlinear state feedback

$$u = -\frac{a}{b}x \pm \sqrt{\frac{a^2 x^2}{b^2} + \frac{2}{r}}.$$

How is the correct sign selected?

5.1-4. Numerical solution of Zermelo's problem. Consider the state equations for Zermelo's problem in Example 5.1-2 with the cost index

$$J = \frac{s_x(x(T) - r_x(T))^2}{2} + \frac{s_y(y(T) - r_y(T))^2}{2} + \frac{1}{2}\int_0^T (q_1 x^2 + q_2 y^2) \, dt,$$

where T is fixed and reference values $r_x(T)$ and $r_y(T)$ are given. Derive Euler's equations and the boundary conditions. Write subroutines *fcni.m, fcnbf.m* for use with *tpoint.m* in Appendix B.2 and hence solve the optimal control problem.

Section 5.2

5.2-1. Minimum-time control of a scalar system. Let

$$\dot{x} = x - u,$$

where $x \in R$. It is desired to drive any initial state $x(0)$ to zero in minimum time if $|u(t)| \le 1$.
a. Write state equation, costate equation, boundary conditions, and Pontryagin's "stationarity condition."
b. Solve the costate equation in terms of unknown $\lambda(T)$. Sketch $\lambda(t)$.

c. Express $u^*(t)$ in terms of $\lambda(T)$ for all possible cases to find the possible values for $u^*(t)$.

d. Solve the state equation for all possible values of $u^*(t)$ if $x(T) = 0$. Sketch $x(t)$ for $u^* = +1, -1$.

e. Sketch switching curve and sample trajectories in the phase plane.

f. Find the optimal *feedback* control.

g. Find the optimal cost J^* in terms of $x(0)$.

h. In terms of $x(0)$, when does this optimal control problem have a solution?

5.2-2. Minimum-time control of harmonic oscillator. The plant is

$$\dot{x} = \begin{bmatrix} 0 & 1 \\ -\omega_n^2 & 0 \end{bmatrix} x + \begin{bmatrix} 0 \\ 1 \end{bmatrix} u. \tag{5}$$

It is desired to drive any initial state to zero in minimum time if

$$|u(t)| \le 1 \tag{6}$$

for all t.

a. Find and solve the costate equations. Use Pontryagin's minimum principle to derive the form of the optimal control law.

b. Sketch phase plane trajectories for $u = 1$ and $u = -1$.

c. Find the switching curve and derive a minimum-time feedback control law.

5.2-3. Bang–Off-bang control of a scalar system. For the plant in Problem 5.2-1, find the minimum-fuel control law to drive any $x(0)$ to the origin in a given time T if $|u(t)| \le 1$.

5.2-4. Bang–Off-bang control of a harmonic oscillator. Repeate Problem 5.2-3 for the harmonic oscillator in Problem 5.2-2.

5.2-5. Minimum-energy control of a scalar system. For the system in Problem 5.2-1, find the minimum-energy control to drive $x(0)$ to zero in a given time T if $|u(t)| \le 1$.

5.2-6. Minimum-energy control of Newton's system. Repeat Problem 5.2-5 for Newton's system in Example 5.2-3.

6

DYNAMIC PROGRAMMING

The purpose of this chapter is to present a brief introduction to *dynamic programming*, which is an alternative to the variational approach to optimal control discussed in Chapters 2 and 3. We show how some of our results from those chapters can be derived by this new approach.

Dynamic programming was developed by R. E. Bellman in the later 1950s (Bellman 1957, Bellman and Dreyfus 1962, Bellman and Kalaba 1965). It can be used to solve control problems for nonlinear, time-varying systems, and it is straightforward to program. The optimal control is expressed as a *state-variable feedback* in graphical or tabular form.

Additional references for this material include Kirk (1970), Bryson and Ho (1975), Schultz and Melsa (1967), Athans and Falb (1966), and Elbert (1984).

6.1 BELLMAN'S PRINCIPLE OF OPTIMALITY

Dynamic programming is based on Bellman's *principle of optimality*:

> An optimal policy has the property that no matter what the previous
> decision (i.e., controls) have been, the remaining decisions must
> constitute an optimal policy with regard to the state resulting from
> those previous decisions. (6.1-1)

We shall see that the principle of optimality plays a role similar to that played by Pontryagin's minimum principle (5.2-6) in the variational approach to system control. It serves to *limit the number of potentially optimal control strategies that must be investigated*. It also implies that optimal control strategies must be determined by working *backward* from the final stage; the optimal control problem is inherently a *backward-in-time problem*.

A simple routing example will serve to demonstrate all the points made so far.

Example 6.1-1. An Aircraft Routing Problem

An aircraft can fly from left to right along the paths shown in Fig. 6.1-1. Intersections $a, b, c, \ldots$ represent cities, and the numbers represent the fuel required to complete each path. Let us use the principle of optimality to solve the minimum-fuel problem.

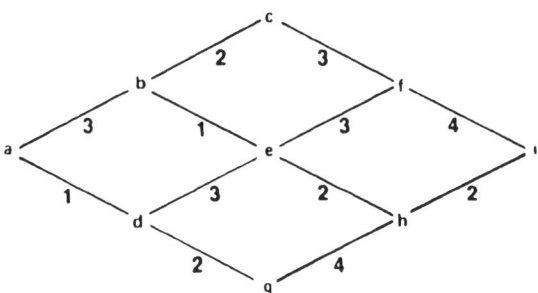

FIGURE 6.1-1   Aircraft routing network.

We can quickly construct a state-variable feedback that shows both the optimal cost and the optimal control from *any* node to i. First, however, we must define what we mean by *state* in this example.

We can label stages $k = 0$ through $k = N = 4$ of the decision-making process as shown in Fig. 6.1-2. (The arrowheads and numbers in parentheses should initially be disregarded: we shall show how to fill them in later.) At each stage $k = 0, 1, \ldots, N - 1$ a decision is required, and N is the final stage.

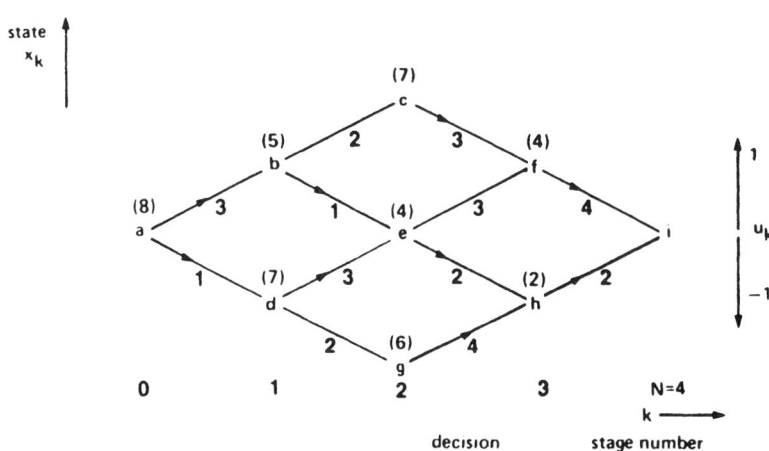

FIGURE 6.1-2   Minimum-fuel state feedback control law.

The current *state* is the node where we are making the current decision. Thus, the initial state is $x_0 = a$. At stage 1, the stage can be $x_1 = b$ or $x_1 = d$. Similarly $x_2 = c$, e, or g; $x_3 = f$ or h; and the final state is constrained to be $x_N = x_4 = i$. The control u_k at stage k can be considered to be $u_k = \pm 1$, where $u_k = 1$ results in a move up, and $u_k = -1$ results in a move down at stage $k + 1$. Now it is clear that all we have on our hands is a minimum-fuel problem with fixed final state and constrained control *and* state values.

To find a minimum-fuel feedback control law using the principle of optimality, start at $k = N = 4$. No decision is required here, so decrement k to 3. If $x_3 = f$, the optimal (only) control is $u_3 = -1$, and the cost is then 4. This is indicated by placing (4) above node f, and placing an arrowhead on path $f \rightarrow i$. If $x_3 = h$, the optimal control is $u_3 = 1$, with a cost of 2, which is now indicated on the figure.

Now decrement k to 2. If $x_2 = c$, then $u_2 = -1$ with a cost to go of $4 + 3 = 7$. This information is added to the figure. If $x_2 = e$, then we must make a decision. If we apply $u_2 = 1$ to get to f, and then go via the optimal path to i, the cost is $4 + 3 = 7$. On the other hand, if we apply $u_2 = -1$ at e and go to h, the cost is $2 + 2 = 4$. Hence, at e the optimal decision is $u_2 = -1$ with a cost to go of 4. Add this information to the figure. If $x_2 = g$, there is only one choice: $u_2 = 1$ with a cost to go of 6.

By successively decrementing k and continuing to compare the control possibilities allowed by the principle of optimality, we can fill in the remainder of the control decisions (arrowheads) and optimal costs to go shown in Fig. 6.1-2. It should be clearly realized that the only control sequences we are allowed to consider are those whose *last portions are optimal sequences*. Note that when $k = 0$, a control of either $u_0 = 1$ *or* $u_0 = -1$ yields the same cost to go of 8; the optimal control for $k = 0$ is not unique.

To examine what we have just constructed, suppose we now are told to find the minimum-fuel path from node d to the destination i. All we need to do is begin at d and follow the arrows! The optimal control u_k^* and the cost to go at each stage k are determined if we know the value of x_k. This, however, is exactly the meaning of state-variable feedback. Therefore, the grid labeled with arrows as in Fig. 6.1-2 is just a graphical *state feedback control law* for the minimum-fuel problem!

This control law formulation should be compared to those in the examples of Section 5.2. Our feedback law tells us how to get from *any* state to the fixed final state $x_4 = x_N = i$. If we change the *final* state, however (e.g., to $x_3 = x_N = f$), then the entire grid must be redone.

Several points should be noted about this example. First, note that, according to Fig. 6.1-2, there are two paths from a to i with the same cost of 8: $a \rightarrow b \rightarrow e \rightarrow h \rightarrow i$ and $a \rightarrow d \rightarrow e \rightarrow h \rightarrow i$. Evidently, the optimal solution found by dynamic programming may not be unique.

Second, suppose we had attempted, in ignorance of the optimality principle, to determine an optimal route from a to i by working *forward*. Then a near-sighted decision maker at a would compare the costs of traveling to b and d, and decide to go to d. The next myopic decision would take him to g. From there on there is no choice: he must go via h to i. The net cost of this strategy is $1 + 2 + 4 + 2 = 9$, which is nonoptimal. We can say that "any portion of an optimal path is optimal." For example, an optimal route from a to e is $a \rightarrow b \rightarrow e$, while the optimal route from e to i is $e \rightarrow h \rightarrow i$.

Finally, let us point out that Bellman's principle of optimality has reduced the number of required calculations by *restricting the number of decisions that must be made*. We could determine the optimal route from a to i by comparing *all* possible paths from a to i. This would require many more calculations than we used.   ∎

Example 6.1-2.   Computational Savings

In Example 6.1-1, determine the number of calculations we used to solve the minimum fuel problem. Now, determine the number of calculations that would be required to solve the problem by comparing *all* possible paths from *a* to *i*.   ∎

Example 6.1-3.   Maximizing a Performance Index

For the routing network in Fig. 6.1-1, find the *maximum* fuel route from *a* to *i*.   ∎

Example 6.1-4.   Target Set of Final States

In Fig. 6.1-3, the cost of traveling from one node to another is given by the numbers, and motion is allowed only from left to right.

   a. Find the minimum cost path from node *a* to the desired final state *j*.
   b. Now, define the *target set S* as $\{h, i, j\}$. That is, any value of the final state within this set is acceptable. Find the minimum cost path from *a* to *S*.
   c. Find the minimum cost path from *b* to *S*.

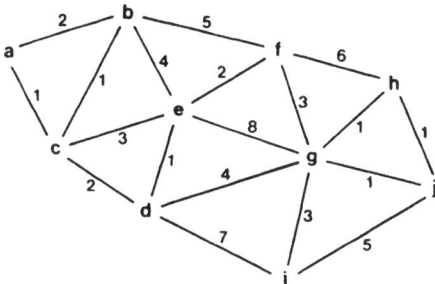

FIGURE 6.1-3   Irregular routing grid.

    ∎

   We now discuss the application of Bellman's principle of optimality to the control of dynamical systems.

6.2   DISCRETE-TIME SYSTEMS

In Chapters 2–5 we discussed the variational approach to optimal control for dynamical systems. We discovered that for nonlinear systems the state and costate equations are hard to solve, and that constraints on the control variables further complicate things. Dynamic programming, on the other hand, can easily be applied to nonlinear systems, and the more constraints there are on the control and state variables, the easier the solution!

   Let the plant be

$$x_{k+1} = f^k(x_k, u_k), \tag{6.2-1}$$

where the superscript k on f indicates that it can be time varying. Suppose we associate with this plant the performance index

$$J_i(x_i) = \phi(N, x_N) + \sum_{k=i}^{N-1} L^k(x_k, u_k), \tag{6.2-2}$$

where $[i, N]$ is the time interval of interest. We have shown the dependence of J on the initial time and state.

We want to use the principle of optimality (6.1-1) to select the control sequence u_k to minimize (6.2-2). First, we need to determine what (6.1-1) looks like in this situation. Let us express it in a more mathematical form.

Suppose we have computed the optimal cost $J_{k+1}^*(x_{k+1})$ from time $k + 1$ to the terminal time N for all possible states x_{k+1}, and that we have also found the optimal control sequences from time $k + 1$ to N for all x_{k+1}. The optimal cost results when the optimal control sequence $u_{k+1}^*, u_{k+2}^*, \ldots, u_{N-1}^*$ is applied to the plant with a state of x_{k+1}. (Note that the optimal control sequence depends on x_{k+1}.)

If we apply any arbitrary control u_k at time k and then use the known optimal control sequence from $k + 1$ on, the resulting cost will be

$$L^k(x_k, u_k) + J_{k+1}^*(x_{k+1}), \tag{6.2-3}$$

where x_k is the state at time k, and x_{k+1} is given by (6.2-1). According to Bellman, the optimal cost from time k on is equal to

$$J_k^*(x_k) = \min_{u_k}(L^k(x_k, u_k) + J_{k+1}^*(x_{k+1})), \tag{6.2-4}$$

and the optimal control u_k^* at time k is the u_k that achieves this minimum.

Equation (6.2-4) is the principle of optimality for discrete systems. Its importance lies in the fact that it allows us to optimize over *only one control vector at a time* by working backward from N. It is called the *functional equation of dynamic programming* and is the basis for computer implementation of Bellman's method. In general, we can specify additional constraints, such as the requirement that u_k belong to some admissible control set.

Without realizing it, we worked Example 6.1-1 by applying (6.2-4). Let us consider next a systems-oriented example.

Example 6.2-1. Optimal Control of a Discrete System Using Dynamic Programming

Let the plant

$$x_{k+1} = x_k + u_k \tag{1}$$

have an associated performance index of

$$J_0 = x_N^2 + \frac{1}{2}\sum_{k=0}^{N-1} u_k^2, \tag{2}$$

where the final time N is 2. The control is constrained to take on values of

$$u_k = -1, \ -0.5, \ 0, \ 0.5, \ 1, \tag{3}$$

and the state is constrained to take on only the values

$$x_k = 0, \ 0.5, \ 1.0, \ 1.5. \tag{4}$$

The control value constraint is not unreasonable, since we found in Section 5.2 that a minimum-time optimal control takes on only values of ± 1, whereas a minimum-fuel optimal control takes on only values of $0, \pm 1$. The state value constraint in this problem is also reasonable, since if the state initially takes on one of the admissible values (4), then under the influence of the allowed controls (3) subsequent states will take on integer or half-integer values. An equivalent constraint to (4) is therefore $x_0 = 0, 0.5, 1.0, 1.5$, and

$$0 \le x_k \le 1.5. \tag{5}$$

This is a positivity condition and a magnitude constraint on the state, which is often reasonable in physical situations. In an actual application N would almost certainly be greater than 2, but to handle large N we would use a computer program designed on the basis of our work in this example.

Now, the optimal control problem is to find an admissible control sequence u_0^*, u_1^* that minimizes J_0 while resulting in an admissible state trajectory x_0^*, x_1^*, x_2^*. We should like for u_k^* to be specified as a state feedback control law.

To solve this problem, we can use the principle of optimality in the form (6.2-4). First, we set up a grid of x_k versus k as shown in Fig. 6.2-1. (Disregard the arrowheads, numbers above the lines, and numbers in parentheses for now.) The lines on the figure are drawn according to the state equation (1). For each admissible x_k, the lines extend toward $x_{k+1} = x_k + u_k$, with the control value u_k written at the end of each line. Only u_k that are both admissible by (3) and result in admissible values of x_{k+1} given in (4) are considered. Thus, for example, if $x_0 = 1.0$, then the admissible u_0 are $-1, -0.5, 0, 0.5$, which result, respectively, in

$$x_1 = x_0 + u_0 = 1 - 1 = 0,$$
$$x_1 = 1 - 0.5 = 0.5,$$
$$x_1 = 1 + 0 = 1,$$
$$x_1 = 1 + 0.5 = 1.5.$$

The lines emanating from the node $x_0 = 1.0$ are directed toward these values of x_1.

Figure 6.2-1 incorporates all of the information in the state equation (1) and the constraints (3), (4). To compute the optimal control using (6.2-4), we need to work with the cost function (2). Write

$$J_k = \tfrac{1}{2}u_k^2 + J_{k+1}^* \tag{6}$$

as the *admissible costs* at time k; an admissible cost J_k is one whose last portion J_{k+1}^* is optimal. In terms of these J_k, (6.2-4) becomes

$$J_k^* = \min_{u_k}(J_k). \tag{7}$$

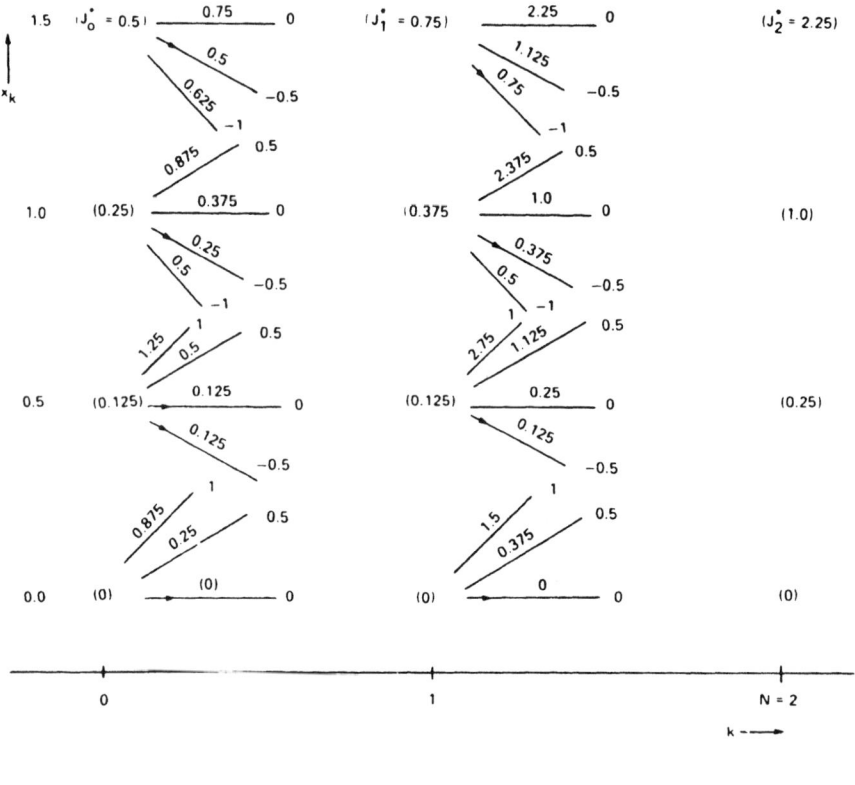

$$J_0 = \tfrac{1}{2}u_0{}^2 + J_1^*$$ $$J_1 = \tfrac{1}{2}u_1{}^2 + J_2^*$$ $$J_2^* = x_2^2$$

FIGURE 6.2-1   Decision grid for solving the optimal control problem by dynamic programming.

(The dependence of J_k and J_k^* on x_k is not explicitly shown.) Write down (6) for each k beneath the grid in Fig. 6.2-1.

To compute u_k^* and J_k^* for each x_k using (7), we need to work backward from $k = N$. First, let $k = N = 2$. The final costs

$$J_N^* = x_N^2 \tag{8}$$

must be evaluated for each admissible final state. J_N^* represents the penalty for being in the final state with value x_N. These costs are written down in parentheses in Fig. 6.2-1 at the location of each of the final states $x_N = 0$, $x_N = 0.5$, $x_N = 1.0$, and $x_N = 1.5$. They are $J_2^*(x_2 = 1.5) = 1.5^2 = 2.25$, $J_2^*(x_2 = 1.0) = 1.0$, and so on.

Now decrement to $k = 1$. For each possible state x_1 and for each admissible control u_1 we must now compute (6). Begin with $x_1 = 1.5$. If we apply $u_1 = 0$, then

$x_2 = x_1 + u_1 = 1.5$, so that $J_2^* = 2.25$. According to (6),

$$J_1 = u_1^2/2 + J_2^* = 0^2/2 + 2.25 = 2.25. \tag{9}$$

This cost is written above the line representing $u_1 = 0$ at $x_1 = 1.5$.

Now suppose we apply $u_1 = -0.5$ when $x_1 = 1.5$. Then $x_2 = x_1 + u_1 = 1.0$, and so $J_2^* = 1.0$. Thus,

$$J_1 = \frac{u_1^2}{2} + J_2^* = \frac{(-0.5)^2}{2} + 1.0 = 1.125. \tag{10}$$

Write this above the line for $u_1 = -0.5$ applied to $x_1 = 1.5$. If we apply $u_1 = -1$ when $x_1 = 1.5$, then $x_2 = x_1 + u_1 = 0.5$, so that $J_2^* = 0.25$. Thus, in this case,

$$J_1 = \frac{u_1^2}{2} + J_2^* = \frac{(-1)^2}{2} + 0.25 = 0.75. \tag{11}$$

Having computed J_1 for each control possibility with $x_1 = 1.5$, we now need to decide on the value of u_1^* if $x_1 = 1.5$. According to (7), u_1^* is simply the value of u_1 that yields the smallest J_1. Therefore,

$$u_1^* = -1, \, J_1^* = 0.75. \tag{12}$$

This can be shown in the figure by placing an arrowhead on the $u_1 = -1$ control path leading from $x_1 = 1.5$ and placing the value of J_1^* in parentheses at the location corresponding to $x_1 = 1.5$.

Now we focus our attention on $x_1 = 1.0$. Using (6) to determine J_1 for the admissible values of control $u_k = 0.5, 0, -0.5, -1$, we get the numbers shown above each line emanating from $x_1 = 1.0$ in the figure. The smallest value of J_1 is 0.375, which occurs for $u_1 = -0.5$. Hence, if $x_1 = 1.0$, then

$$u_1^* = -0.5, \quad J_1^* = 0.375. \tag{13}$$

This is indicated by placing an arrowhead on the $u_1 = -0.5$ path and $J_1^* = 0.375$ in parentheses at the location $x_1 = 1.0$. In a similar manner we obtain u_1^* and J_1^* for $x_1 = 0.5$ and $x_1 = 0.0$. See the figure for the results.

Now we decrement k to 0. Let $x_0 = 1.5$. For this value of the initial state, we use (6) with each of the possible control values $u_0 = 0, -0.5$, and -1 to compute the associated values of $J_0 = u_0^2/2 + J_1^*$, where J_1^* is the number in parentheses at the state x_1 resulting from $x_1 = 1.5 + u_0$. The values of J_0 found are shown in the figure. The smallest value of J_0 is 0.5, which occurs for $u_0 = -0.5$; so if $x_0 = 1.5$, then

$$u_0^* = -0.5, \quad J_0^* = 0.5. \tag{14}$$

In a similar fashion we compute u_0^* and J_0^* for $x_0 = 1.0, 0.5$, and 0.0. The resulting values are indicated in the figure. Note that if $x_0 = 0.5$, then a control of either $u_0 = 0$ or $u_0 = -0.5$ results in the optimal cost of $J_0^* = 0.125$; the optimal control for this initial state is not unique.

To see what we have just constructed, we can redraw Fig. 6.2-1 showing only the optimal controls u_k^* and costs to go J_k^*. See Fig. 6.2-2. Now, suppose $x_0 = 1.0$. Then to find the optimal state trajectory we need only follow the arrows. It is

$$x_0^* = 1.0, \quad x_1^* = 0.5, \quad x_0^* = 0. \tag{15}$$

Before we even apply the control, we know the optimal cost will be

$$J_0^* = 0.25. \tag{16}$$

The optimal control sequence to be applied to the plant is

$$u_0^* = -0.5, \quad u_1^* = -0.5. \tag{17}$$

Whatever the initial state x_0 is, the optimal control u_0^* and cost to go J_0^* are determined by Fig. 6.2-2 once x_0 is given. The current control u_k^* is given once the current state x_k is known; hence, the figure is just the optimal control law given in *state feedback* form.

In the variational approach to optimal control of Chapters 4 and 5, constraints make the problem solution more difficult. In the dynamic programming approach, however, more constraints mean that fewer control possibilities must be examined; constraints *simplify* the dynamic programming problem.

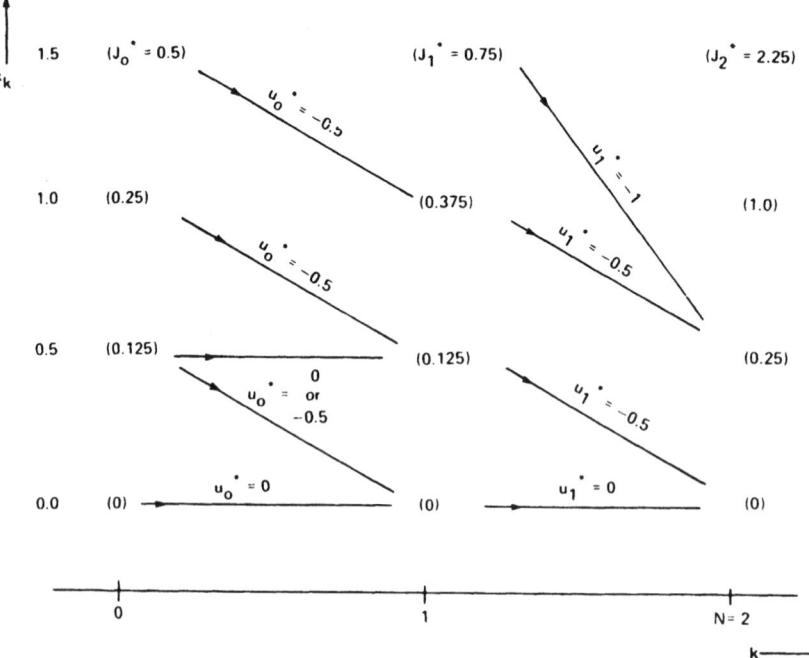

FIGURE 6.2-2   State feedback control law found by dynamic programming.

Figure 6.2-2 has the form of a *vector field* that tends to force the state down toward the origin as k increases. By changing the control weighting in J_0 in (2), the shape of the vector field changes. For instance, if we modify the cost index to

$$J_0 = x_2^2 + 2 \sum_{k=0}^{1} u_k^2 \tag{18}$$

and repeat the above procedure, the state feedback control law of Fig. 6.2-3 results. A heavier control weighting has resulted in a tendency to use less control effort. The family of optimal state trajectories shown in figures like these is called a *field of extremals*.

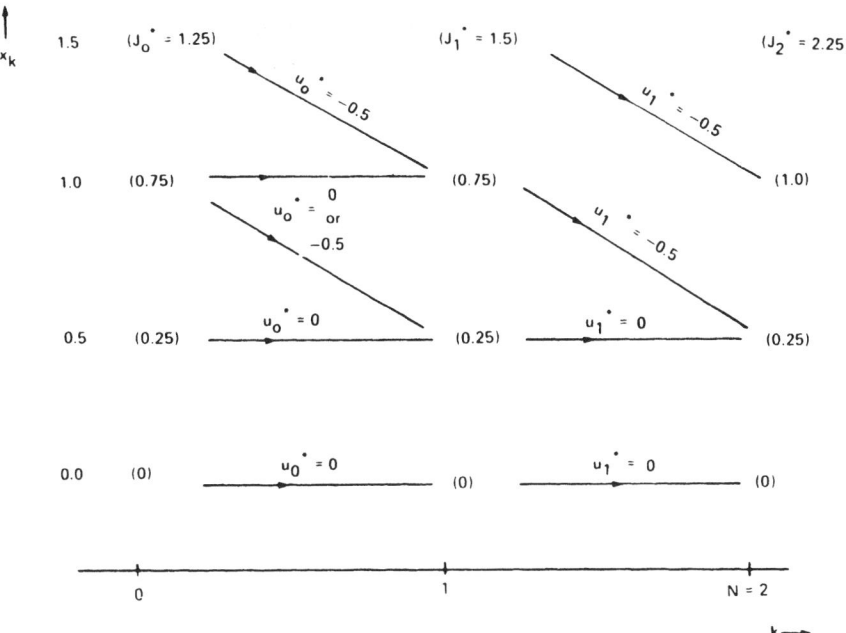

FIGURE 6.2-3   State feedback control for modified performance index.

This problem could be solved without using the optimality principle; but then *every possible* control sequence would need to be examined, and the required number of calculations would drastically increase.   ∎

In the problems we see how easy it is to extend the dynamic programming approach to nonlinear systems. It is instructive to compare the number of computations required to find the optimal control by investigating *all* possible control sequences to the number of computations required if dynamic programming is used. Let the plant be first order with a scalar control. Suppose the state has 10 admissible values and the control has 4 admissible values.

If the final time is $N = 1$, then comparing all $10 \times 4 = 40$ control possibilities requires 40 calculations. In this case, dynamic programming also requires 40 calculations. As N increases, however, the number of calculations required to investigate possible control sequences increases *exponentially* as $\Sigma_{k=1}^{N} 10 \cdot 4^k$. By contrast, the number of calculations required if we use the principle of optimality to restrict the set of potentially optimal controls increases *linearly* as $40N$ (Kirk 1970).

The next example shows that we can derive the linear quadric regulator of Table 2.2-1 by using the dynamic programming approach.

Example 6.2-2.   Discrete Linear Quadratic Regulator via Dynamic Programming

Let the plant

$$x_{k+1} = Ax_k + Bu_k \tag{1}$$

have associated performance index

$$J_i = \frac{1}{2}x_N^T S_N x_N + \frac{1}{2}\sum_{k=i}^{N-1}(x_k^T Q x_k + u_k^T R u_k), \tag{2}$$

with $S_N \geq 0$, $Q \geq 0$, and $R > 0$. (If the plant and weighting matrices are time varying, the development to follow still holds.) It is desired to find the optimal control u_k^* on the fixed time interval $[i, N]$ that minimizes J_i. The initial state x_i is given and the final state x_N is free.

The determination of u_k^* by using the principle of optimality (6.2-4) looks very much like our derivation of the zero-input cost (2.2-21), (2.2-22). To begin, let $k = N$ and write

$$J_N^* = \tfrac{1}{2}x_N^T S_N x_N, \tag{3}$$

which is the penalty for being in state x_N at time N. Now decrement k to $N-1$ and write

$$J_{N-1} = \tfrac{1}{2}x_{N-1}^T Q x_{N-1} + \tfrac{1}{2}u_{N-1}^T R u_{N-1} + \tfrac{1}{2}x_N^T S_N x_N. \tag{4}$$

According to (6.2-4), we need to find u_{N-1}^* by minimizing (4). To do this, use state equation 1 to write

$$J_{N-1} = \tfrac{1}{2}x_{N-1}^T Q x_{N-1} + \tfrac{1}{2}u_{N-1}^T R u_{N-1}$$
$$+ \tfrac{1}{2}(Ax_{N-1} + Bu_{N-1})^T S_N (Ax_{N-1} + Bu_{N-1}). \tag{5}$$

Since there are no constraints, the minimum of J_{N-1} is found by setting

$$0 = \frac{\partial J_{N-1}}{\partial u_{N-1}} = Ru_{N-1} + B^T S_N (Ax_{N-1} + Bu_{N-1}). \tag{6}$$

Solving for the optimal control yields

$$u_{N-1}^* = -(B^T S_N B + R)^{-1} B^T S_N A x_{N-1}. \tag{7}$$

Defining the Kalman gain as

$$K_{N-1} \overset{\Delta}{=} (B^\mathrm{T} S_N B + R)^{-1} B^\mathrm{T} S_N A, \tag{8}$$

we can write

$$u^*_{N-1} = -K_{N-1} x_{N-1}. \tag{9}$$

The optimal cost to go from $k = N - 1$ is found by substituting (9) into (5). If we do this and simplify, the result is

$$J^*_{N-1} = \tfrac{1}{2} x^\mathrm{T}_{N-1}[(A - BK_{N-1})^\mathrm{T} S_N (A - BK_{N-1}) + K^\mathrm{T}_{N-1} RK_{N-1} + Q]x_{N-1}. \tag{10}$$

If we define

$$S_{N-1} \overset{\Delta}{=} (A - BK_{N-1})^\mathrm{T} S_N (A - BK_{N-1}) + K^\mathrm{T}_{N-1} RK_{N-1} + Q, \tag{11}$$

this can be written

$$J^*_{N-1} = \tfrac{1}{2} x^\mathrm{T}_{N-1} S_{N-1} x_{N-1}. \tag{12}$$

Now decrement to $k = N - 2$. Then

$$J_{N-2} = \tfrac{1}{2} x^\mathrm{T}_{N-2} Q x_{N-2} + \tfrac{1}{2} u^\mathrm{T}_{N-2} R u_{N-2} + \tfrac{1}{2} x^\mathrm{T}_{N-1} S_{N-1} x_{N-1} \tag{13}$$

are the admissible costs for $N - 2$, since these are the costs that are optimal from $k = N - 1$ on. To determine u^*_{N-2}, according to (6.2-4), we must minimize (13).

Now, let us be a little tricky to save some work. Note that (13) is of the same form as (4). The optimal control and cost to go for $k = N - 2$ are therefore given by (8), (9) and (11), (12) with N there replaced by $N - 1$. If we continued to decrement k and apply the optimality principle, the result for each $k = N - 1, \ldots, 1, 0$ is

$$K_k = (B^\mathrm{T} S_{k+1} B + R)^{-1} B^\mathrm{T} S_{k+1} A, \tag{14}$$

$$u^*_k = -K_k x_k, \tag{15}$$

$$S_k = (A - BK_k)^\mathrm{T} S_{k+1}(A - BK_k) + K^\mathrm{T}_k RK_k + Q, \tag{16}$$

$$J^*_k = \tfrac{1}{2} x^\mathrm{T}_k S_k x_k, \tag{17}$$

where the final condition S_N for (16) is given in (2). Equation 16 is the Joseph stabilized Riccati equation, and so these results are identical to those found in Table 2.2-1 using the variational approach!   ∎

6.3   CONTINUOUS-TIME SYSTEMS

There are basically two approaches to the optimal control of continuous-time systems. We can discretize, solve for the optimal discrete control, and then use a zero-order hold to manufacture a digital control. Alternatively, we can solve the continuous optimal control problem to obtain a continuous input. Let us discuss both approaches.

Digital Control

Let the plant

$$\dot{x} = f(x, u, t) \tag{6.3-1}$$

have cost index of

$$J(0) = \phi(x(T), T) + \int_0^T L(x(t), u(t), t)\, dt. \tag{6.3-2}$$

To discretize the plant with a sampling period of τ sec, we can use the first order approximation

$$\dot{x}(k\tau) = (x_{k+1} - x_k)/\tau \tag{6.3-3}$$

to write (6.3-1) as

$$x_{k+1} = x_k + \tau f(x_k, u_k, k\tau), \tag{6.3-4}$$

where we have defined $x_k \overset{\Delta}{=} x(k\tau)$, $u_k \overset{\Delta}{=} u(k\tau)$ for notational convenience. Defining the discrete function

$$f^k(x_k, u_k) \overset{\Delta}{=} x_k + \tau f(x_k, u_k, k\tau), \tag{6.3-5}$$

this can be written

$$x_{k+1} = f^k(x_k, u_k), \tag{6.3-6}$$

which is exactly (6.2-1).

To discretize the cost index, we can write

$$J(0) = \phi(x(T), T) + \sum_{k=0}^{N-1} \int_{k\tau}^{(k+1)\tau} L(x(t), u(t), t)\, dt, \tag{6.3-7}$$

where

$$N = \frac{T}{\tau}. \tag{6.3-8}$$

Using a first-order approximation to each integral yields

$$J(0) = \phi(x(T), T) + \sum_{k=0}^{N-1} \tau L(x_k, u_k, k\tau). \tag{6.3-9}$$

By defining the discrete functions

$$J_0 \overset{\Delta}{=} J(0),$$

$$\phi^s(N, x_N) \overset{\Delta}{=} \phi(x(N\tau), N\tau),$$

$$L^k(x_k, u_k) \overset{\Delta}{=} \tau L(x_k, u_k, k\tau), \tag{6.3-10}$$

this becomes

$$J_0 = \phi^s(N, x_N) + \sum_{k=0}^{N-1} L^k(x_k, u_k), \qquad (6.3\text{-}11)$$

which is just (6.2-2).

In the case of a time-invariant linear plant with quadratic cost index,

$$\dot{x} = Ax + Bu, \qquad (6.3\text{-}12)$$

$$J(0) = \frac{1}{2}x^T(T)S(T)x(T) + \frac{1}{2}\int_0^T (x^T Q x + u^T R u)\, dt, \qquad (6.3\text{-}13)$$

this first-order approximation to the discretized plant becomes

$$x_{k+1} = (I + A\tau)x_k + B\tau u_k, \qquad (6.3\text{-}14)$$

$$J_0 = \frac{1}{2}x_N^T S_N x_N + \frac{1}{2}\sum_{k=0}^{N-1}(x_k^T Q^s x_k + u_k^T R^s u_k), \qquad (6.3\text{-}15)$$

where

$$S_N \overset{\Delta}{=} S(N\tau), \qquad (6.3\text{-}16)$$

$$Q^s = Q\tau, \qquad (6.3\text{-}17)$$

$$R^s = R\tau. \qquad (6.3\text{-}18)$$

In this case, however, we can do better than the Euler's approximation (6.3-14) by using the exact discrete representation of (6.3-12) including a sampler and zero-order hold

$$x_{k+1} = A^s x_k + B^s u_k, \qquad (6.3\text{-}19)$$

where

$$A^s = e^{A\tau}, \qquad (6.3\text{-}20)$$

$$B^s = \int_0^\tau e^{At}B\, dt. \qquad (6.3\text{-}21)$$

Now that the plant has been discretized, dynamic programming can be used to compute u_k^* as in Section 6.2. The digital control to be applied to the actual plant (6.3-1) is then given by

$$u(t) = u_k^*, \quad k\tau \le t < (k+1)\tau. \qquad (6.3\text{-}22)$$

To use dynamic programming, the state and control values must first be *quantized*, that is, restricted to belong to some finite set of admissible values. The finer we quantize, the more accurate our digital control will be; however, as the number of admissible values of x_k and u_k increases, so does the number of calculations required to find u_k^*. The problem can quickly become intractable even for a large digital computer. Bellman called this growth in the number of required calculations as the quantization is made finer and the number of state variables increases the *curse of dimensionality*.

An additional problem is the following. No matter how finely we quantize x_k, the relation $x_{k+1} = A^s x_k + B^s u_k$ might result in values for x_{k+1} that do not coincide with quantization levels. In this event we need to *interpolate* to find the values of $J_{k+1}^*(x_{k+1})$ to use in (6.2-4).

For further discussion on these topics, see Kirk (1970). Let us now do an example.

Example 6.3-1.   Discretization, Quantization, and Interpolation

Consider the scalar plant

$$\dot{x} = ax + bu \tag{1}$$

with cost index

$$J(0) = \frac{1}{2}x^2(T) + \frac{1}{2}\int_0^T u^2(t)dt. \tag{2}$$

and a final time of $T = 1$. The state and control are constrained by

$$0 \le x(t) \le 1 \tag{3}$$

$$|u(t)| \le 1. \tag{4}$$

To compute a control minimizing $J(0)$ by discrete dynamic programming, let us select a sampling period of $\tau = 0.5$. (In practice, a smaller τ should be selected for more accurate control, but this sampling period will serve for purposes of illustration.) Then the discretized plant is

$$x_{k+1} = e^{a\tau}x_k + \int_0^\tau e^{at}b\,dt \cdot u_k. \tag{5}$$

For simplicity, let $a = 0$ and $b = 1$ so that

$$x_{k+1} = x_k + \tau u_k = x_k + 0.5u_k. \tag{6}$$

The discretized performance index is

$$J_0 = \frac{1}{2}x_N^2 + \frac{1}{4}\sum_{k=0}^{N-1} u_k^2, \tag{7}$$

with $N = 2$.

We must now *quantize* x_k and u_k. Let us select the quantization levels

$$u_k = -1, -0.5, 0, 0.5, 1, \tag{8}$$

$$x_k = 0, 0.5, 1. \tag{9}$$

(While these levels for u_k might be satisfactory, in an actual application we would select a finer quantization for x_k.) To apply discrete dynamic programming, let us use the grid in Fig. 6.3-1, which is filled in by working backward as follows.

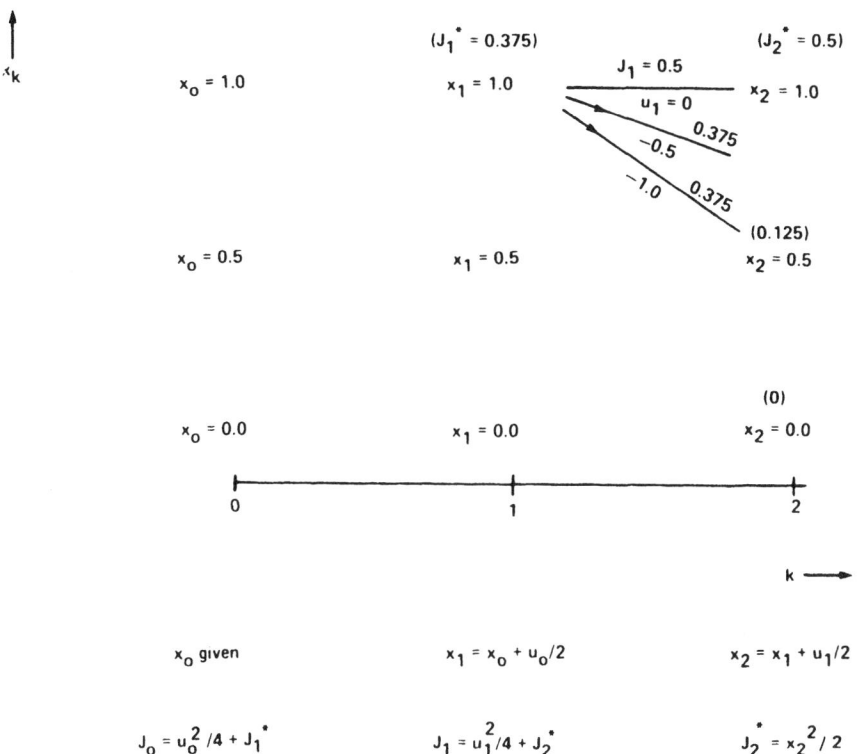

FIGURE 6.3-1 Discrete dynamic programming for continuous systems. Illustration of interpolation to find J_k^*.

Setting $k = N = 2$, we find $J_2^*(x_2)$ for admissible values of x_2 to be

$$J_2^*(x_2 = 1) = \frac{1^2}{2} = 0.5,\tag{10}$$

$$J_2^*(x_2 = 0.5) = \frac{0.5^2}{2} = 0.125,\tag{11}$$

$$J_2^*(x_2 = 0) = 0.\tag{12}$$

These values are indicated in the figure in parentheses above the respective values of x_2.

Stepping back to $k = N - 1 = 1$, we first consider the possibility $x_1 = 1$. Applying each admissible control value as u_1, the resulting values of x_2 are given by (6):

$$x_2 = x_1 + \frac{0}{2} = 1,\tag{13}$$

$$x_2 = x_1 - \frac{0.5}{2} = 0.75, \tag{14}$$

$$x_2 = x_1 - \tfrac{1}{2} = 0.5. \tag{15}$$

Lines are added to the figure between $x_1 = 1$ and these values of x_2. Below each line is indicated the corresponding value of u_1. Note that $x_2 = 0.75$ as given by (14) is not a quantization level for x_k.

We need to compute the admissible costs

$$J_1 = \frac{u_1^2}{4} + J_2^* \tag{16}$$

so that we can use (6.2-4) to find J_1^* and u_1^* when $x_1 = 1$. If $u_1 = 0$, then

$$J_1 = 0 + J_2^*(x_2 = 1) = 0.5. \tag{17}$$

This is placed above the line corresponding to $u_1 = 0$ at $x_1 = 1$.

Now we run into a problem. For $u_1 = -0.5$, we need to compute

$$J_1 = \frac{u_1^2}{4} + J_2^*(x_2 = 0.75). \tag{18}$$

Unfortunately, $J_2^*(x_2 = 0.75)$ was not calculated at stage $k = N = 2$, since 0.75 is not a value for x_k listed in (9). It can be approximated by *linear interpolation* as

$$J_2^*(x_2 = 0.75) = J_2^*(x_2 = 0.5) + \frac{J_2^*(x_2 = 1) - J_2^*(x_2 = 0.5)}{2} = 0.3125. \tag{19}$$

Then (18) yields

$$J_1 = 0.375. \tag{20}$$

Write this above the $u_1 = -0.5$ line in the figure.

For $u_1 = -1$, we obtain

$$J_1 = \frac{u_1^2}{4} + J_2^*(x_2 = 0.5) = 0.375. \tag{21}$$

Applying the principle of optimality (6.2-4), we select

$$u_1^* = -0.5 \text{ or } -1.0,$$

$$J_1^* = 0.375, \tag{22}$$

when $x_1 = 1$. (The optimal control is not unique.) To indicate this in the figure, place J_1^* above $x_1 = 1$ in parentheses, and place arrowheads on the lines corresponding to the two possible optimal controls $u_1 = -0.5$ and $u_1 = -1$.

The remainder of the optimal control law is determined exactly as in Example 6.2-1, using interpolation as necessary to approximate $J_k^*(x_k)$ for the values of x_k not listed in (9). ∎

Continuous Control and the Hamilton-Jacobi-Bellman Equation

The plant

$$\dot{x} = f(x, u, t) \qquad (6.3\text{-}23)$$

has a performance index of

$$J(x(t_0), t_0) = \phi(x(T), T) + \int_{t_0}^{T} L(x, u, t)\, dt. \qquad (6.3\text{-}24)$$

We are interested in determining a continuous optimal control $u^*(t)$ on a given interval $[t_0, T]$ that minimizes J and drives a given initial state $x(t_0)$ to a final state satisfying

$$\psi(x(T), T) = 0 \qquad (6.3\text{-}25)$$

for a given function ψ. Let us first see what form the principle of optimality (6.1-1) takes on for this problem.

Suppose t is the current time and $t + \Delta t$ is a future time close to t. Then the cost to go $J(x(t), t)$ can be written

$$J(x, t) = \phi(x(T), T) + \int_{t+\Delta t}^{T} L(x, u, \tau)\, d\tau + \int_{t}^{t+\Delta t} L(x, u, \tau)\, d\tau. \quad (6.3\text{-}26)$$

(We are using τ as a dummy variable since t is the current time.) We can therefore say

$$J(x, t) = \int_{t}^{t+\Delta t} L(x, u, \tau)\, d\tau + J(x + \Delta x, t + \Delta t), \qquad (6.3\text{-}27)$$

where $x + \Delta x$ is the state at time $t + \Delta t$ that results when the current $x(t)$ and $u(t)$ are used in (6.3-23). Note that, to first order,

$$\Delta x = f(x, u, t)\Delta t. \qquad (6.3\text{-}28)$$

Equation (6.3-27) describes all possible costs to go from time t to final time T. According to the optimality principle (6.1-1), however, the only candidates for $J^*(x, t)$ are those costs $J(x, t)$ that are optimal from $t + \Delta t$ to T. Suppose that the optimal cost $J * (x + \Delta x, t + \Delta t)$ is known for all possible $x + \Delta x$. Suppose also that the optimal control has been determined on the interval $[t + \Delta t, T]$ for each $x + \Delta x$. Then it remains only to select the current control $u(t)$ on the interval $[t, t + \Delta t]$. Hence,

$$J^*(x, t) = \min_{\substack{u(\tau) \\ t \le \tau \le t+\Delta t}} \left[\int_{t}^{t+\Delta t} L(x, u, \tau)\, d\tau + J^*(x + \Delta x, t + \Delta t) \right]. \quad (6.3\text{-}29)$$

This is the *principle of optimality for continuous-time systems*; it should be compared with (6.2-4).

Unfortunately, (6.3-29) does not provide a straightforward means of finding the optimal control and cost analytically, although some numerical solution methods use it. Let us therefore perform some further manipulations to find a way to compute these quantities.

Perform a Taylor series expansion of $J^*(x + \Delta x, t + \Delta t)$ about (x, t) and take an approximation to the integral in (6.3-29) to write, to first order,

$$
J^*(x, t) = \min_{\substack{u(\tau) \\ t \le \tau \le t + \Delta t}} \left(L\Delta t + J^*(x, t) + \left(\frac{\partial J^*}{\partial x}\right)^{\mathrm{T}} \Delta x + \frac{\partial J^*}{\partial t} \Delta t \right). \quad (6.3\text{-}30)
$$

Now use (6.3-28) and note that J^* and $J_t^* \Delta t$ are independent of $u(\tau), t \le \tau \le t + \Delta t$, to see that

$$
J^*(x, t) = J^*(x, t) + J_t^* \Delta t + \min_{\substack{u(\tau) \\ t \le \tau \le t + \Delta t}} (L\Delta t + (J_x^*)^{\mathrm{T}} f \Delta t), \quad (6.3\text{-}31)
$$

or

$$
-J_t^* \Delta t = \min_{\substack{u(\tau) \\ t \le \tau \le t + \Delta t}} (L\Delta t + (J_x^*)^{\mathrm{T}} f \Delta t). \quad (6.3\text{-}32)
$$

Letting $\Delta t \to 0$ yields finally

$$
-\frac{\partial J^*}{\partial t} = \min_{u(t)} \left(L + \left(\frac{\partial J^*}{\partial x}\right)^{\mathrm{T}} f \right). \quad (6.3\text{-}33)
$$

This is a partial differential equation for the optimal cost $J^*(x, t)$. It is called the *Hamilton-Jacobi-Bellman* (HJB) equation. It is solved backward in time from $t = T$, and by setting $t_0 = T$ in (6.3-24), its boundary condition is seen to be

$$
J^*(x(T), T) = \Phi(x(T), T) \quad \text{on the hypersurface } \psi(x(T), T) = 0. \quad (6.3\text{-}34)
$$

If we define the Hamiltonian function as

$$
H(x, u, \lambda, t) = L(x, u, t) + \lambda^{\mathrm{T}} f(x, u, t), \quad (6.3\text{-}35)
$$

then the HJB equation can be written

$$
-\frac{\partial J^*}{\partial t} = \min_u (H(x, u, J_x^*, t)). \quad (6.3\text{-}36)
$$

The HJB equation provides the solution to the optimal control problem for general nonlinear systems; however, it is in most cases impossible to solve analytically. When it can be solved it provides an optimal control in state-variable feedback (i.e., closed-loop) form.

The next examples show how to use the HJB equation to find the optimal cost and control. They also demonstrate that the nonlinear optimal control scheme of Table 3.2-1 and the linear quadratic regulator of Table 3.3-1 can both be retrieved from the HJB equation.

Example 6.3-2.   Optimal Control by Solution of HJB Equation

The scalar plant

$$\dot{x} = x + u \tag{1}$$

has a cost index of

$$J(t_0) = \frac{1}{2}x^2(T) + \frac{1}{2}\int_{t_0}^{T} ru^2 dt. \tag{2}$$

The initial time is $t_0 = 0$, the final time T is fixed, and the final state $x(T)$ is free. It is desired to find the control $u^*(t)$ minimizing $J(0)$.

The Hamiltonian is

$$H(x, u, \lambda, t) = \frac{ru^2}{2} + \lambda(x + u) \tag{3}$$

and the HJB equation is

$$-J_t^* = \min_u \left(\frac{ru^2}{2} + J_x^*(x + u) \right). \tag{4}$$

Since there are no constraints on the control u, we can determine the value of u minimizing the right-hand side of (4) by setting

$$0 = \frac{\partial H(x, u, J_x^*, t)}{\partial u} = ru + J_x^*. \tag{5}$$

The optimal control is therefore given in terms of J_x^*, the derivative of the optimal cost with respect to x, by

$$u^*(t) = -\frac{1}{r}\frac{\partial J^*(t)}{\partial x}. \tag{6}$$

Since

$$\frac{\partial^2 H}{\partial u^2} = r > 0, \tag{7}$$

u^* in (6) does indeed minimize $H(x, u, J_x^*, t)$. To find J_x^*, use (6) to evaluate

$$H^* \stackrel{\Delta}{=} H(x, u^*, J_x^*, t) = \frac{r}{2}\left(\frac{J_x^*}{r}\right)^2 + J_x^*\left(x - \frac{J_x^*}{r}\right) = xJ_x^* - \frac{(J_x^*)^2}{2r}. \tag{8}$$

Using H^* as the right-hand side of (4) yields

$$-J_t^* = xJ_x^* - \frac{(J_x^*)^2}{2r}. \tag{9}$$

We must now solve the HJB equation (9) for the optimal cost J^*. Evaluating (2) at $t_0 = T$, we see that the boundary condition for (9) is

$$J^*(T) = \frac{x^2(T)}{2}, \tag{10}$$

a quadratic function of the state. Let us assume that $J^*(t)$ is a quadratic function of the state for all $t \leq T$. Then

$$J^*(t) = \tfrac{1}{2}s(t)x^2(t) \tag{11}$$

for some function $s(t)$ that is yet to be determined.

Taking partial derivatives of (11) gives

$$\frac{\partial J^*}{\partial x} = sx, \tag{12}$$

$$\frac{\partial J^*}{\partial t} = \frac{\dot{s}x^2}{2}, \tag{13}$$

which we substitute into (9) to obtain

$$\frac{-\dot{s}x^2}{2} = sx^2 - \frac{s^2x^2}{2r}. \tag{14}$$

This must hold for all $x(t)$, so we obtain the Riccati equation in $s(t)$

$$\dot{s} = \frac{s^2}{r} - 2s. \tag{15}$$

Comparing (10) and (11), the boundary condition for (14) is found to be

$$s(T) = 1. \tag{16}$$

Equation (15) can be integrated using separation of variables:

$$\int_t^T \frac{ds}{s^2/r - 2s} = \int_t^T dt. \tag{17}$$

The result is

$$s(t) = \frac{2re^{2(T-t)}}{e^{2(T-t)} + (2r - 1)}. \tag{18}$$

This consistent solution for $s(t)$ shows that assumption (11) was valid.

Collecting our results, we see that the optimal cost is given by (11) with $s(t)$ given by (18). The optimal control (6) is

$$u^*(t) = -\frac{1}{r}J_x^*(t) = -\frac{s(t)}{r}x(t). \tag{19}$$

This is in the form of a time-varying state-variable feedback. ∎

Example 6.3-3. Deriving the Euler Equations from the HJB Equation

We can easily show that the function

$$\lambda(t) \triangleq \frac{\partial J^*}{\partial x} \tag{1}$$

is a costate for the plant (6.3-23) if $J^*(x, t)$ satisfies the HJB equation

$$-\frac{\partial J^*}{\partial t} = \min_u (L + \lambda^T f). \tag{2}$$

See Athans and Falb (1966) and Bryson and Ho (1975).

Since $u(t)$ is unconstrained, the minimization in (2) can be carried out by setting

$$0 = \frac{\partial L}{\partial u} + \frac{\partial f^T}{\partial u}\lambda. \tag{3}$$

This is just the stationarity condition from Table 3.2-1.

To examine the dynamics of $\lambda(t)$, write

$$-\frac{d\lambda}{dt} = -\frac{\partial^2 J^*}{\partial t \partial x} - \frac{\partial^2 J^*}{\partial x^2}\dot{x}. \tag{4}$$

Next, take the partial derivative of (2) with respect to x, realizing that $u = u(x, t)$ to see that

$$-\frac{\partial^2 J^*}{\partial x \partial t} = \min_u \left(\frac{\partial L}{\partial x} + \frac{\partial u^T}{\partial x}\frac{\partial L}{\partial u} + \frac{\partial \lambda^T}{\partial x}f + \frac{\partial f^T}{\partial x}\lambda + \left(\frac{\partial f}{\partial u}\frac{\partial u}{\partial x}\right)^T \lambda \right). \tag{5}$$

At the minimum, (3) holds so that

$$-\frac{\partial^2 J^*}{\partial x \partial t} = \frac{\partial L}{\partial x} + \frac{\partial \lambda^T}{\partial x}f + \frac{\partial f^T}{\partial x}\lambda. \tag{6}$$

Use (1), (6), and $\dot{x} = f(x, u, t)$ in (4) to obtain

$$-\frac{d\lambda}{dt} = \frac{\partial f^T}{\partial x}\lambda + \frac{\partial L}{\partial x}, \tag{7}$$

which is exactly the costate equation in Table 3.2-1.

If $u(t)$ is constrained, then we must select the optimal control $u^*(t)$ from the set of admissible controls to minimize $H(x, u, \lambda^*, t)$. It is apparent that the right-hand side of the HJB equation is just Pontryagin's minimum principle from Section 5.2. ■

Example 6.3-4. Deriving the Continuous Linear Quadratic Regulator from the HJB Equation

Let the plant be

$$\dot{x} = Ax + Bu \tag{1}$$

with cost index

$$J(t_0) = \frac{1}{2}x^T(T)S(T)x(T) + \frac{1}{2}\int_{t_0}^{T}(x^TQx + u^TRu)dt. \tag{2}$$

It is desired to find the optimal control to drive the state from a given $x(t_0)$ so that $J(t_0)$ is minimized. The final time T is fixed and the final state $x(T)$ is free. We can derive the Riccati-equation-based linear quadratic regulator of Table 3.3-1 from the HJB equation as follows.

Form the Hamiltonian

$$H = \frac{1}{2}(x^TQx + u^TRu) + \lambda^T(Ax + Bu). \tag{3}$$

To minimize H, set

$$0 = \frac{\partial H}{\partial u} = Ru + B^T\lambda, \tag{4}$$

so the optimal control is

$$u^* = -R^{-1}B^T\lambda. \tag{5}$$

Since

$$\frac{\partial^2 H}{\partial u^2} = R > 0, \tag{6}$$

$u^*(t)$ is a *minimizing* control. Use (5) in (3) to find the optimal Hamiltonian

$$H^* = \frac{1}{2}x^TQx + \lambda^TAx - \frac{1}{2}\lambda^TBR^{-1}B^T\lambda. \tag{7}$$

Setting $\lambda = J_x^*$, the HJB equation is

$$-J_t^* = \frac{1}{2}x^TQx + (J_x^*)^TAx - \frac{1}{2}(J_x^*)^TBR^{-1}B^TJ_x^*. \tag{8}$$

The boundary condition is

$$J^*(T) = \frac{1}{2}x^T(T)S(T)x(T). \tag{9}$$

Since $J^*(T)$ is quadratic in the state, let us assume that this is true for $J*(t), t \leq T$. That is, assume there is a symmetric matrix $S(t)$ such that

$$J^*(t) = \frac{1}{2}x^T(t)S(t)x(t) \tag{10}$$

for all $t \leq T$. Substituting (10) into (8) yields

$$0 = \frac{1}{2}x^T\dot{S}x + \frac{1}{2}x^TQx + x^TSAx - \frac{1}{2}x^TSBR^{-1}B^TSx, \tag{11}$$

or

$$x^T(\dot{S} + 2SA - SBR^{-1}B^TS + Q)x = 0. \tag{12}$$

The matrix in parentheses, which we shall call M, is not symmetric, since SA is not symmetric (the other terms are). Write

$$M = M_S + M_a \tag{13}$$

where M_S is symmetric, $M_S = M_S^T$, and M_a is antisymmetric, $M_a = -M_a^T$. Then

$$x^T M x = x^T M_S x + x^T M_a x. \tag{14}$$

But, since (14) is a scalar,

$$x^T M_a x = x^T M_a^T x = -x^T M_a x = 0, \tag{15}$$

hence,

$$x^T M x = x^T M_S x. \tag{16}$$

The symmetric part of a matrix is given by

$$M_S = (M + M^T)/2. \tag{17}$$

Therefore, (12) is equivalent to

$$x^T(\dot{S} + A^T S + SA - SBR^{-1}B^T S + Q)x = 0 \tag{18}$$

Since (18) holds for all $x(t_0)$ and hence for all state trajectories $x(t)$, we require

$$-\dot{S} = A^T S + SA - SBR^{-1}B^T S + Q. \tag{19}$$

This is the Riccati equation in Table 3.3-1. The optimal control is given by (5) as the state feedback

$$u^*(t) = -R^{-1}B^T Sx(t), \tag{20}$$

and the optimal cost is given by (10).   ∎

PROBLEMS

Section 6.1

6.1-1. A routing network is shown in Fig. P6.1-1. Find the optimal path from x_0 to x_6 if only movement from left to right is permitted. Now find the optimal path from any node as a state-variable feedback.

Section 6.2

6.2-1. Dynamic programming for bilinear system. Consider the scalar bilinear system

$$x_{k+1} = x_k u_k + u_k^2, \tag{1}$$

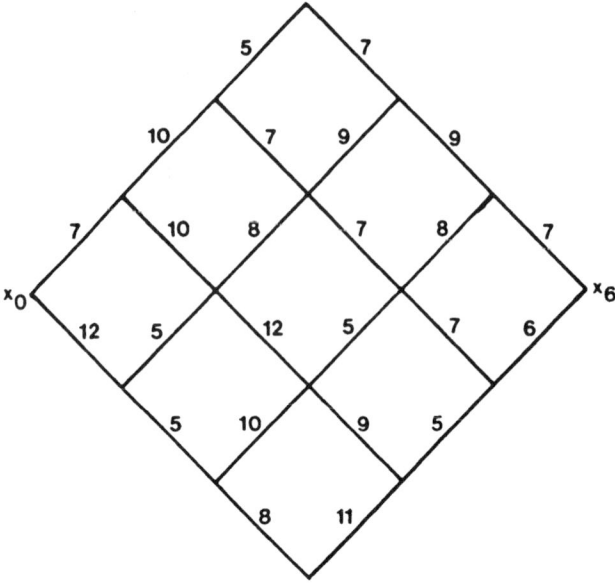

FIGURE P6.1-1   A routing network.

with cost index

$$J_0 = x_N^2 + \sum_{k=0}^{N-1} x_k u_k. \tag{2}$$

Let $N = 2$. The control is constrained to take on values of

$$u_k = -1 \text{ or } 1 \tag{3}$$

and the state to take on values of

$$x_k = -1, \ 0, \ 1, \ \text{or } 2. \tag{4}$$

a. Use dynamic programming to find an optimal state feedback control law.
b. Let $x_0 = 2$. Find the optimal cost, control sequence, and state trajectory.

6.2-2. Software for dynamic programming. Write a MATLAB program to generate Fig. 6.2.2.

6.2-3. Effect of control weighting on field of extremals. Redo Example 6.2-1 if

$$J_0 = x_N^2 + \frac{1}{2} \sum_{k=0}^{N-1} r u_k^2$$

for $r = \frac{1}{4}$, 4, 8 and compare the resulting fields of extremals.

6.2-4. Cargo loading. A vessel is to be loaded with stocks of N items. Each item $k = 1, \ldots, N$ has a weight of w_k and a known value of v_k. The maximum allowed weight of the cargo is W. We want to determine the most valuable cargo load without overloading the vessel.

Define

$$u_k = \text{number of units of item loaded,}$$

$$x_k = \text{weight capacity available for items } k, \ k+1, \ldots, N.$$

Then the problem becomes the following: Maximize

$$J_1 = \sum_{k=1}^{N} v_k u_k$$

subject to the constraint

$$\sum_{k=1}^{N} w_k u_k \leq W.$$

a. Write the state equation.
b. List the constraints for x_k and u_k.

6.2-5. Cargo loading. A ship is to be loaded with stocks of three items, whose weights and values are given in the following table:

Item	Weight (lb)	Value ($)
Dishwashers	100	360
Washing machines	125	475
Refrigerators	250	1000

a. Let

$$u_1 = \text{number of refrigerators loaded,}$$

$$u_2 = \text{number of washing machines loaded,}$$

$$u_3 = \text{number of dishwashers loaded.}$$

Use the results of Problem 6.2-4 to find the optimal loading policy if the maximum weight allowed is 730 lb.

b. By ordering the decision stages in order of increasing value per weight, show how to solve the same problem in a trivial manner.

Section 6.3

6.3-1. Software for dynamic programming
a. Write a MATLAB program to generate a feedback law in the tabular form of Fig. 6.2-2 for the scalar system

$$x_{k+1} = ax_k + bu_k, \tag{1}$$

with cost index

$$J_0 = \frac{1}{2}s_N x_N^2 + \frac{1}{2}\sum_{k=0}^{N-1}(qx_k^2 + ru_k^2).$$ (2)

The state x_k can take on values of

$$XA(1), \quad XA(2), \quad \ldots, \quad XA(NS)$$ (3)

and the control u_k can take on values

$$UA(1), \quad UA(2), \quad \ldots, \quad UA(NU),$$ (4)

all of which are specified. Assume that x_{k+1} can take on only the values (3).
b. Add to your program an interpolation scheme to find $J_{k+1}^*(x_{k+1})$ in the event that x_{k+1} takes on a value not listed in (3).

6.3-2. Optimal control by solution of the HJB equation. Let

$$\ddot{x} = u,$$ (1)

$$J(t_0) = \frac{1}{2}x^2(T) + \frac{1}{2}\int_{t_0}^{T} ru^2 dt,$$ (2)

with $t_0 = 0, T = 2$.
a. Write the HJB equation, eliminating $u(t)$.
b. Assume that

$$J^*(t) = \frac{1}{2}s_1(t)x^2(t) + s_2(t)x(t)\dot{x}(t) + \frac{1}{2}s_3(t)\dot{x}^2(t)$$ (3)

for some $s_1(t), s_2(t), s_3(t)$. Use the HJB equation to determine coupled differential equations for these $s_i(t)$. Compare with Example 3.3-5. Find boundary conditions for the equations. Express the optimal control as a linear state feedback in terms of the $s_i(t)$.

6.3-3. Continuous-time tracker via the HJB equation. Use the HJB equation to solve the linear quadratic tracking problem of Section 4.1. The procedure should basically follow Example 6.3-4. (Assume $J(x,t) = \frac{1}{2}x^T sx - x^T v + w$ and find equations for $s(t), v(t),$ and $w(t)$.)

7

OPTIMAL CONTROL
FOR POLYNOMIAL SYSTEMS

All of our discussion of optimal control has been based on a state-variable-system description. In this chapter some results are presented for optimal control of systems described by transfer functions. We shall present these results for discrete time systems, showing how to use them for digital control of continuous systems. They can also be used to design continuous controllers.

The results of this chapter are important in the area of adaptive control, and although we cover only the single-input–single-output case, they generalize to multivariable systems without too much trouble. Additional references are Clarke and Gawthrop (1975), Gawthrop (1977), and Koivo (1980). See Wolovich (1974) for further insight into the multivariable case.

7.1   DISCRETE LINEAR QUADRATIC REGULATOR

If a plant with input u_k and output y_k is described by the discrete proper (i.e., $n \geq m$) transfer function

$$H(z) = \frac{b_0 z^m + b_1 z^{m-1} + \cdots + b_m}{z^n + a_1 z^{n-1} + \cdots + a_n} = \frac{Y(z)}{U(z)}, \qquad (7.1\text{-}1)$$

we can multiply by z^{-n} to obtain

$$H(z) = \frac{z^{-d}(b_0 + b_1 z^{-1} + \cdots + b_m z^{-m})}{1 + a_1 z^{-1} + \cdots + a_n z^{-n}}, \qquad (7.1\text{-}2)$$

where

$$d = n - m \geq 0 \qquad (7.1\text{-}3)$$

is the relative degree or *control delay*. Let us therefore write the system description in the time domain as

$$A(z^{-1})y_k = z^{-d}B(z^{-1})u_k, \tag{7.1-4}$$

where

$$A(z^{-1}) = 1 + a_1 z^{-1} + \cdots + a_n z^{-n}$$
$$B(z^{-1}) = b_0 + b_1 z^{-1} + \cdots + b_m z^{-m} \tag{7.1-5}$$

are polynomials in the delay operator z^{-1} (i.e., $z^{-1}y_k = y_{k-1}$). We assume $b_0 \neq 0$. In this chapter all polynomials are written in terms of z^{-1}, not z, so some attention is required to prevent confusion.

With the plant (7.1-4) we associate the quadratic performance index

$$J_k = \left(\sum_{i=0}^{n_P} p_i y_{k+d-i} - \sum_{i=0}^{n_Q} q_i w_{k-i} \right)^2 + \left(\sum_{i=0}^{n_R} r_i u_{k-i} \right)^2 \tag{7.1-6}$$

for given integers n_P, n_Q, n_R and weighting coefficients p_i, q_i, r_i. The signal w_k is a reference or command input. For optimal control of state systems, we used quadratic cost indices that had the form of a sum of squares of state and control components. Our new J_k contains squares of sums. At time k, it depends on the output values $y_{k+d}, y_{k+d-1}, \ldots, y_{k+d-n_P}$, the control values $u_k, u_{k-1}, \ldots, u_{k-n_R}$, and the reference input values $w_k, w_{k-1}, \ldots, w_{k-n_Q}$. See Fig. 7.1-1.

We can make J_k look neater by defining *weighting polynomials*

$$P(z^{-1}) = 1 + p_1 z^{-1} + \cdots + p_{n_P} z^{-n_P},$$
$$Q(z^{-1}) = q_0 + q_1 z^{-1} + \cdots + q_{n_Q} z^{-n_Q},$$
$$R(z^{-1}) = r_0 + r_1 z^{-1} + \cdots + r_{n_R} z^{-n_R}, \tag{7.1-7}$$

where we have selected $p_0 = 1$ with no loss in generality. Then

$$J_k = (Py_{k+d} - Qw_k)^2 + (Ru_k)^2. \tag{7.1-8}$$

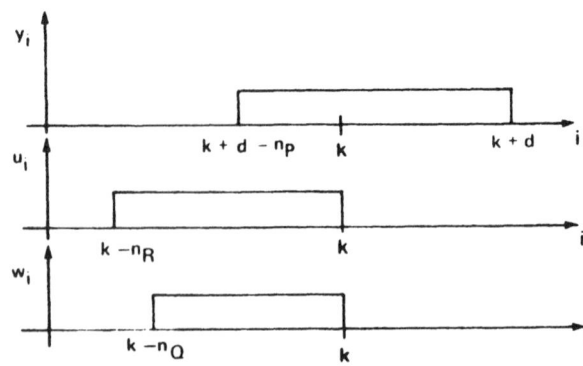

FIGURE 7.1-1   Illustration of time dependence of cost index J_k.

We have chosen this cost index for two reasons: it results in a very nice solution, and it can be selected to specify a wide range of designs that result in desirable closed-loop behavior for different applications.

To keep the output small without using too much control energy, we might simply select $P = 1$, $Q = 0$, $R = r_0$, so that

$$J_k = y_{k+d}^2 + r_0^2 u_k^2. \tag{7.1-9}$$

(In this case the reference input is not needed.) If we are more concerned about keeping small the *changes* in the control input, we could select $P = 1$, $Q = 0$, $R = r_0(1 - z^{-1})$ so that

$$J_k = y_{k+d}^2 + r_0^2 (u_k - u_{k-1})^2. \tag{7.1-10}$$

By choosing $P = Q = 1$, $R = r_0$, we get the cost index

$$J_k = (y_{k+d} - w_k)^2 + r_0^2 u_k^2, \tag{7.1-11}$$

which is a polynomial tracker; the controlled output will follow a (delayed) reference signal w_k.

We can also design a regulator that makes the plant Bz^{-d}/A behave like any designed model, for if R is chosen as zero, then the optimal controller will attempt to ensure that

$$Py_{k+d} = Qw_k. \tag{7.1-12}$$

This results in *model-following behavior*, where the plant output tracks the output of a desired model $Q(z^{-1})z^{-d}/P(z^{-1})$ that has input w_k. To see this, write (7.1-12) as

$$y_k = \frac{Q(z^{-1})z^{-d}}{P(z^{-1})} w_k. \tag{7.1-13}$$

Therefore, $Q(z^{-1})$, $P(z^{-1})$ in (7.1-8) need only be selected to be the specified model numerator and denominator dynamics.

Our objective now is to select control input sequence u_k to minimize J_k for given polynomials P, Q, R and reference w_k. Suppose $u_{k-1}, u_{k-2}, \ldots$ are known. Then we want a way of finding u_k. We cannot simply differentiate J_k as it is with respect to u_k, since y_{k+d} also depends on u_k. To make this dependence explicit, we can find a *d-step-ahead predictor* for the output.

Divide $A(z^{-1})$ into 1 to define quotient $F(z^{-1})$ and remainder $z^{-d}G(z^{-1})$:

$$A(z^{-1}) \overline{\left) 1 \right.} \quad \genfrac{}{}{0pt}{}{F(z^{-1})}{} \tag{7.1-14}$$

$$\overline{z_{-d}G(z_{-1})}$$

The division is carried out for d steps, that is, until the remainder has z^{-d} as a factor. Then

$$1 = AF + z^{-d}G, \tag{7.1-15}$$

where

$$F(z^{-1}) = 1 + f_1 z^{-1} + \cdots + f_{d-1} z^{-(d-1)},$$
$$G(z^{-1}) = g_0 + g_1 z^{-1} + \cdots + g_{n-1} z^{-(n-1)}. \tag{7.1-16}$$

Equation (7.1-15) is a simple form of *Diophantine equation* (Kučera 1979).

Now use (7.1-4) and (7.1-15) to write

$$y_{k+d} = \frac{B}{A} u_k = B\left(F + z^{-d}\frac{G}{A}\right) u_k = BF u_k + G\frac{B}{A} u_{k-d},$$

or

$$y_{k+d} = BF u_k + G y_k. \tag{7.1-17}$$

This is a *predictive formulation* of the plant (7.1-4), which expresses y_{k+d} in terms only of quantities that occur at time k and before. Using (7.1-17) in (7.1-8) yields

$$J_k = (PBF u_k + PG y_k - Q w_k)^2 + (R u_k)^2, \tag{7.1-18}$$

which explicitly reveals all dependence on u_k. To find the optimal u_k, we can now differentiate J_k with respect to u_k. Note that

$$(R u_k)^2 = (r_0 u_k + r_1 u_{k-1} + \cdots + r_{N_R} u_{k-N_R})^2$$

so that

$$\frac{\partial}{\partial u_k}(R u_k)^2 = 2 r_0 R u_k. \tag{7.1-19}$$

Note also that polynomial $PBF(z^{-1})$ has a leading coefficient of b_0. Therefore,

$$\frac{\partial J_k}{\partial u_k} = 2b_0(PBF u_k + PG y_k - Q w_k) + 2 r_0 R u_k = 0. \tag{7.1-20}$$

Solving for the optimal control at time k yields

$$\left(PFB + \frac{r_0}{b_0} R\right) u_k = -PG y_k + Q w_k. \tag{7.1-21}$$

Equation (7.1-21) provides a recursive equation for u_k in terms of $u_{k-1}, u_{k-2}, \ldots, y_k, y_{k-1}, \ldots, w_k, w_{k-1}, \ldots$, all of which are known. It is a *two-degrees-of-freedom* regulator with a feedback term depending on y_k and a feedforward term depending on w_k. The optimal linear quadratic regulator defined by this recursion is shown in Fig. 7.1-2.

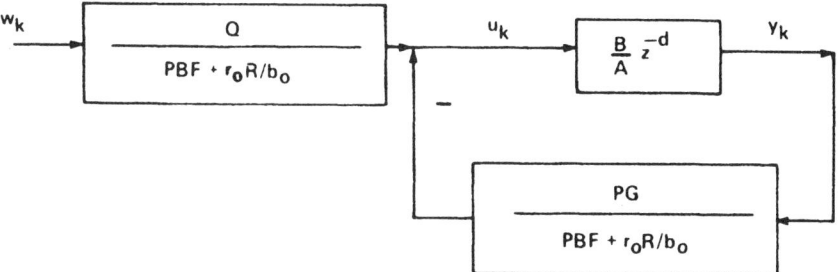

FIGURE 7.1-2   Optimal polynomial LQ regulator drawn as a two-degree-of-freedom regulator.

Several points are worth mentioning. First, note that the optimal u_k depends only on the output, the reference, and the previous control values. Unlike the optimal controls for state systems, it does not depend on the internal state of the plant. It is given in the form of a *dynamic output feedback*. We surmise that what state information is required by the controller is somehow provided by the dynamics (7.1-21): that is, the two-degrees-of-freedom regulator contains an *implicit observer for the state*.

Second, the polynomial linear quadratic regulator does not depend on first solving a recursive Riccati equation, as is the case in Table 2.2-1. The role of the Riccati equation in the design phase is taken over by the Diophantine equation (7.1-15).

Under the influence of control (7.1-21) the closed-loop transfer function (Fig. 7.1-2) is

$$H^{cl}(z) = \frac{Q}{P'} \frac{(B/A)z^{-d}}{1 + (B/A)(PG/P')z^{-d}}, \qquad (7.1\text{-}22)$$

where

$$P'(z^{-1}) \overset{\Delta}{=} PBF + r_0 R/b_0. \qquad (7.1\text{-}23)$$

Simplifying yields (use (7.1-15))

$$H^{cl}(z) = \frac{Q}{P'} \frac{BP'z^{-d}}{AP' + z^{-d}BPG} = \frac{QBz^{-d}}{PB + r_0 AR/b_0} \qquad (7.1\text{-}24)$$

so the closed-loop characteristic equation is

$$\Delta^{cl}(z) = (PB + r_0 AR/b_0)z^h, \qquad (7.1\text{-}25)$$

where h is the highest power of z^{-1} in (7.1-24) (after canceling P').

7.2   DIGITAL CONTROL OF CONTINUOUS-TIME SYSTEMS

If the continuous plant is

$$H(s) = \frac{b_0 s^m + b_1 s^{m-1} + \cdots + b_m}{s^n + a_1 s^{n-1} + \cdots + a_n}, \qquad (7.2\text{-}1)$$

then we can discretize it and design regulator (7.1-21) for the sampled plant. Based on the optimal u_k, we can determine a digital control for (7.2-1).

To discretize $H(s)$, we can write it in reachable canonical form

$$\dot{x} = \begin{bmatrix} 0 & 1 & 0 & \cdots & 0 \\ 0 & 0 & 1 & \cdots & 0 \\ \vdots & \vdots & \vdots & & \vdots \\ 0 & 0 & 0 & \cdots & 1 \\ -a_n & -a_{n-1} & -a_{n-2} & \cdots & -a_1 \end{bmatrix} x + \begin{bmatrix} 0 \\ \vdots \\ 0 \\ 1 \end{bmatrix} u,$$

$$y = \begin{bmatrix} b_m & b_{m-1} & \cdots & b_0 & 0 & \cdots & 0 \end{bmatrix} x, \qquad (7.2\text{-}2)$$

or

$$\dot{x} = Ax + Bu,$$

$$y = Cx. \qquad (7.2\text{-}3)$$

Next we discretize this system to get

$$x_{k+1} = A^s x_k + B^s u_k,$$

$$y_k = C x_k, \qquad (7.2\text{-}4)$$

where

$$A^s = e^{AT},$$

$$B^s = \int_0^T e^{A\tau} B \, d\tau, \qquad (7.2\text{-}5)$$

with sample period T. Finally, the discretized transfer function is given by

$$H^s(z) = C(zI - A^s)^{-1} B^s, \qquad (7.2\text{-}6)$$

which has the form of (7.1-1). The relative degree of $H^s(z)$ is always 1, since it is obtained by sampling, but in some cases it may be useful, or necessary, to add an integer delay to the plant to make d in (7.1-4) greater than 1 (e.g., to account for computation time).

The linear quadratic regulator (7.1-21) can be designed for (7.2-6), yielding u_k, and hence giving a digital control that can be applied to (7.2-1) as described in Section 2.3.

Example 7.2-1. Helicopter Longitudinal Autopilot

Suppose a helicopter has longitudinal dynamics described by the short-period approximation

$$H(s) = \frac{s + 0.1}{(s - 0.1)^2 + 0.25} = \frac{q(s)}{\delta_c(s)}, \tag{1}$$

with $q(t)$ the pitch rate and $\delta_c(t)$ the pilot pitch command. It is desired to design a regulator to keep $q(t)$ near zero without using abrupt changes in the command $\delta_c(t)$. The transfer function is unstable with poles at

$$s = 0.1 \pm j0.5. \tag{2}$$

a. Discretization

Since the characteristic polynomial is $\Delta(s) = s^2 - 0.2s + 0.26$, the reachable canonical form is

$$\dot{x} = \begin{bmatrix} 0 & 1 \\ -0.26 & 0.2 \end{bmatrix} x + \begin{bmatrix} 0 \\ 1 \end{bmatrix} u,$$
$$y = \begin{bmatrix} 0.1 & 1 \end{bmatrix} x. \tag{3}$$

Using a sampling period of $T = 0.001$ yields the discretized system (this can be easily computed using the subroutine *c2dm.m*, from the Control System Toolbox)

$$x_{k+1} = \begin{bmatrix} 1 & 0.01 \\ -0.003 & 1.002 \end{bmatrix} x_k + \begin{bmatrix} 0 \\ 0.01 \end{bmatrix} u_k,$$
$$y_k = \begin{bmatrix} 0.1 & 1 \end{bmatrix} x_k. \tag{4}$$

(In this example we only show three decimal places.) Computing the transfer function of (4) yields

$$H^s(z) = \frac{0.01z - 0.01z}{z^2 - 2.002z + 1.002}, \tag{5}$$

which has poles outside the unit circle at

$$z = 1.001 \pm j0.005. \tag{6}$$

b. Design of Discrete Polynomial Regulator

The relative degree of $H^s(z)$ is 1, but let us set delay $d = 2$ to allow an additional interval of $T = 10$ msec for computation of the control u_k. Therefore, the discrete plant is

$$Ay_k = z^{-2}Bu_k \tag{7}$$

with

$$A(z^{-1}) = 1 - 2.002z^{-1} + 1.002z^{-2}, \tag{8}$$
$$B(z^{-1}) = 0.01 - 0.01z^{-1}. \tag{9}$$

Long division of A into 1 yields

$$AF + z^{-2}G = 1 \tag{10}$$

where

$$F(z^{-1}) = 1 + 2.002z^{-1}, \tag{11}$$

$$G(z^{-1}) = 3.006 - 2.006z^{-1}. \tag{12}$$

Hence, the predictive system is

$$y_{k+2} = BFu_k + Gy_k$$

$$= (0.01 + 0.01\ z^{-1} - 0.02\ z^{-2})u_k + (3.006 - 2.006\ z^{-1})y_k, \tag{13}$$

or

$$y_{k+2} = 0.01\ u_k + 0.01\ u_{k-1} - 0.02\ u_{k-2} + 3.006\ y_k - 2.006\ y_{k-1}. \tag{14}$$

We must now select the performance index. To make y_{k+d} small and prevent large changes in u_k, we could use

$$J_k = y_{k+d}^2 + [0.1(u_k - u_{k-1})]^2, \tag{15}$$

which penalizes *changes* in the control input. Then the weighting polynomials are

$$P(z^{-1}) = 1, \tag{16}$$

$$Q(z^{-1}) = 0, \tag{17}$$

$$R(z^{-1}) = 0.1 - 0.1z^{-1}. \tag{18}$$

Computing the regulator polynomials in (7.1-21) we obtain

$$P' \triangleq PBF + \frac{r_0 R}{b_0}$$

$$= 1.009 - 0.988z^{-1} - 0.02z^{-2}, \tag{19}$$

$$PG = 3.006 - 2.006z^{-1}; \tag{20}$$

so the regulator is

$$P'u_k = -PGy_k, \tag{21}$$

or

$$1.009u_k = 0.988u_{k-1} + 0.02u_{k-2} - 3.006y_k + 2.006y_{k-1}. \tag{22}$$

This recursion for u_k is easily implemented. To simulate applying it to the continuous plant (1), we use the method of Section 2.3.

c. Closed-loop Dynamics

According to (7.1-25) the closed-loop characteristic equation is

$$\Delta^{cl}(z) = (1.0085 - 3.0075z^{-1} + 2.9995z^{-2} - 1.0005z^{-3} + 0.z^{-4})z^4, \qquad (23)$$

which yields closed-loop poles of

$$z = 0, 0.999, 0.992 \pm j0.099. \qquad (24)$$

Note that the closed-loop plant is stable.   ■

PROBLEMS

Section 7.1

7.1-1. Predictive formulation. The plant

$$y_k - 0.75 \, y_{k-1} = u_{k-3} - u_{k-4}$$

has zero initial conditions and an input of $u_k = 2(0.9)^k u_{-1}(k)$.
a. Find y_k for $k = 0, \dots, 10$.
b. Find the predictive formulation. Using this formulation, again find y_k for $k = 0, \dots, 10$. Your answer will agree with part a, but now y_k is found for each k using data occurring at times $k - 3$ and before.

7.1-2. Discrete polynomial regulator. It is desired for the plant

$$y_k - 2y_{k-1} + \tfrac{3}{4}y_{k-2} = u_{k-d} - \tfrac{1}{2}u_{k-d-1} \qquad (1)$$

to follow a reference signal w_k without using too many changes in the control. Hence, select the cost index

$$J_k = (y_{k+d} - w_k)^2 + r^2(u_k - u_{k-1})^2. \qquad (2)$$

a. Design a two-degrees-of-freedom regulator for $d = 1$. Is the closed-loop plant stable? Draw a block diagram of your regulator in terms of unit delays z^{-1}. Implement your regulator using MATLAB/SIMULINK.
b. Repeat part a for a control delay of $d = 2$.

7.1-3. Continuous polynomial regulator. The continuous plant is

$$\dot{x}_1 = x_2,$$
$$\dot{x}_2 = -\omega_n^2 x_1 - 2\delta w_n x_2 + bu. \qquad (1)$$

If $\delta < 0$, the plant is unstable. Let $y = x_1$. To keep $y(t)$ small without using too much control, we can select the cost index

$$J = \ddot{y}^2 + r^2 u^2. \tag{2}$$

We want to work directly with (1) without sampling.
a. Derive the continuous analog to (7.1-21) to find the transfer function $H(s)$ of the optimal feedback regulator

$$u = -H(s)y. \tag{3}$$

b. Find the closed-loop characteristic polynomial.

7.1-4. Model-following regulator. It is desired for the plant in Problem 7.1-2 to behave like the model

$$\frac{z - \frac{1}{2}}{z^2 - z + \frac{1}{4}}. \tag{1}$$

Let $d = 1$.
a. Find the required controller (set $R = 0$).
b. Check your result by finding the closed-loop transfer function.

7.1-5. Continuous model-following regulator. In Problem 7.1-3, let $\omega_n = 1$, $\delta = -0.5$, $b = 1$. It is desired for the plant to behave like the model

$$\frac{1}{(s + \frac{1}{2})^2 + \frac{3}{4}}. \tag{1}$$

a. Determine the original plant poles.
b. Find the required P, Q, R in (7.1-8).
c. Design the optimal continuous controller $u = -H(s)y + K(s)w$.
d. Find the resulting closed-loop system.

7.1-6. Model follower with control weighting. To follow a model $z^{-d}QB/D$ without using too much control, we could solve the Diophantine equation

$$PB + r_0 AR/b_0 = D \tag{1}$$

for the required weights P and R (see (7.1-24)) and then design the controller (7.1-21). (Note that the plant zeros are not moved.)
Redo problem 7.1-4 by this approach using the model

$$\frac{z - \frac{1}{2}}{z^3 - \frac{5}{4}z^2 + \frac{1}{2}z - \frac{1}{16}}. \tag{1}$$

(The nonuniqueness of the solution to (1) can be exploited to meet other design specifications.)

8

OUTPUT FEEDBACK
AND STRUCTURED CONTROL

8.1 LINEAR QUADRATIC REGULATOR WITH OUTPUT FEEDBACK

Our objective in this section is to show how to use modern control techniques to design stability augmentation systems (SAS). This is accomplished by regulating certain states of the system to zero while obtaining desirable closed-loop response characteristics. It involves the problem of stabilizing the system by placing the closed-loop poles at desirable locations.

Using classical control theory, we were forced to take a one-loop-at-a-time approach to designing multivariable SAS. In this section we select a performance criterion that reflects our concern with closed-loop stability and good time responses, and then derive matrix equations that may be solved for all the control gains simultaneously. These matrix equations are solved using digital computer programs. This approach thus closes all the loops simultaneously and results in a simplified design strategy for multi-input–multi-output (MIMO) systems or single-input–single-output (SISO) systems with multiple feedback loops.

Once the performance criterion has been selected, the control gains are explicitly computed by matrix design equations, and closed-loop stability will generally be guaranteed. This means that the engineering judgment in modern control enters in the selection of the performance criterion. Different criteria will result in different closed-loop time responses and robustness properties.

We assume the plant is given by the linear time-invariant state-variable model

$$\dot{x} = Ax + Bu \tag{8.1-1}$$

$$y = Cx, \tag{8.1-2}$$

with $x(t) \in R^n$ the state, $u(t) \in R^m$ the control input, and $y(t) \in R^p$ the measured output. The controls will be output feedbacks of the form

$$u = -Ky, \tag{8.1-3}$$

where K is an $m \times p$ matrix of constant feedback coefficients to be determined by the design procedure. Since the regulator problem only involves stabilizing the aircraft and inducing good closed-loop time responses, $u(t)$ will be taken as a pure feedback with no auxiliary input.

Output feedback will allow us to design plant controllers of any desired structure. This is another reason for preferring it over full-state feedback.

In the regulator problem, we are interested in obtaining good time responses as well as in the stability of the closed-loop system. Therefore, we shall select a performance criterion in the time domain. Let us now present this criterion.

Quadratic Performance Index

The objective of state regulation for the system is to drive any initial condition error to zero, thus guaranteeing stability. This may be achieved by selecting the control input $u(t)$ to minimize a quadratic cost or performance index (PI) of the type

$$J = \frac{1}{2} \int_0^\infty (x^T Q x + u^T R u)dt, \tag{8.1-4}$$

where Q and R are symmetric positive semidefinite weighting matrices. Positive semidefiniteness of a square matrix M (denoted $M \geq 0$) is equivalent to all its eigenvalues being nonnegative, and also to the requirement that the quadratic form $x^T M x$ be nonnegative for all vectors x. Therefore, the definiteness assumptions on Q and R guarantee that J is nonnegative and lead to a sensible minimization problem. This quadratic PI is a vector version of an integral-squared PI of the sort used in classical control.

To emphasize the motivation for the choice of (8.1-4), consider the following. If the square root $\sqrt{M}$ of a positive semidefinite matrix M is defined by

$$M = \sqrt{M}^T \sqrt{M}, \tag{8.1-5}$$

then we may write (8.1-4) as

$$J = \frac{1}{2} \int_0^\infty \left(\left\| \sqrt{Q} x \right\|^2 + \left\| \sqrt{R} u \right\|^2 \right) dt, \tag{8.1-6}$$

with $\|w\|$ the Euclidean norm of a vector w (i.e., $\|w\| = w^T w$). If we are able to select the control input $u(t)$ so that J takes on a minimum finite value, then certainly the integrand must become zero for large time. This means that both the linear combination $\sqrt{Q} x(t)$ of the states and the linear combination $\sqrt{R} u(t)$ of the controls must go to zero. In different designs we may select Q and R for different performance requirements corresponding to specified functions of the state and input. In particular, if Q and R are both chosen nonsingular, then the entire state vector $x(t)$ and all the controls $u(t)$ will go to zero with time if J has a finite value.

Since a bounded value for J will guarantee that $\sqrt{Q}x(t)$ and $\sqrt{R}u(t)$ go to zero with time, this formulation for the PI is appropriate for the regulator problem, as any initial condition errors will be driven to zero. If the state vector $x(t)$ consists of capacitor voltages $v(t)$ and inductor currents $i(t)$, then $\|x\|^2$ will contain terms like $v^2(t)$ and $i^2(t)$. Likewise, if velocity $s(t)$ is a state component, then $\|x\|^2$ will contain terms like $s^2(t)$. Therefore, the minimization of the PI (8.1-4) is a generalized minimum energy problem. We are concerned with minimizing the energy in the states without using too much control energy.

The relative magnitudes of Q and R may be selected to trade off requirements on the smallness of the state against requirements on the smallness of the input. For instance, a larger control-weighting matrix R will make it necessary for $u(t)$ to be smaller to ensure that $\sqrt{R}u(t)$ is near zero. We say that a larger R *penalizes* the controls more, so that they will be smaller in norm relative to the state vector. On the other hand, to make $x(t)$ go to zero more quickly with time, we may select a larger Q.

As a final remark on the PI, we shall see that the positions of the closed-loop poles depend on the choices for the weighting matrices Q and R. That is, Q and R may be chosen to yield good time responses in the closed-loop system. Let us now derive matrix design equations that may be used to solve for the control gain K that minimizes the PI. The result will be the design equations in Table 8.1-1.

Solution of the LQR Problem

The LQR problem with output feedback is the following. Given the linear system (8.1-1)–(8.1-2), find the feedback coefficient matrix K in the control input (8.1-3) that minimizes the value of the quadratic PI (8.1-4). In contrast with most of the classical control techniques given in previous chapters, this is a time-domain design technique. By substituting the control (8.1-3) into (8.1-1), the closed-loop system equations are found to be

$$\dot{x} = (A - BKC)x \equiv A_c x. \tag{8.1-7}$$

The PI may be expressed in terms of K as

$$J = \frac{1}{2}\int_0^\infty \left(x^\mathrm{T}(Q + C^\mathrm{T}K^\mathrm{T}RKC)\,x\right)dt. \tag{8.1-8}$$

The design problem is now to select the gain K so that J is minimized subject to the dynamical constraint (8.1-7).

This dynamical optimization problem may be converted into an equivalent static one that is easier to solve as follows. Suppose we can find a constant, symmetric, positive semi-definite matrix P so that

$$\frac{d}{dt}\left(x^{\mathrm{T}}Px\right) = -x^{\mathrm{T}}\left(Q + C^{\mathrm{T}}K^{\mathrm{T}}RKC\right)x. \tag{8.1-9}$$

Then, J may be written as (Chapter 3)

$$J = \tfrac{1}{2}x^{\mathrm{T}}(0)Px(0) - \tfrac{1}{2}\lim_{t\to\infty} x^{\mathrm{T}}(t)Px(t). \tag{8.1-10}$$

Assuming that the closed-loop system is asymptotically stable so that $x(t)$ vanishes with time, this becomes

$$J = \tfrac{1}{2}x^{\mathrm{T}}(0)Px(0). \tag{8.1-11}$$

If P satisfies (8.1-9), then we may use (8.1-7) to see that

$$-x^{\mathrm{T}}(Q + C^{\mathrm{T}}K^{\mathrm{T}}RKC)x = \frac{d}{dt}(x^{\mathrm{T}}Px) = \dot{x}^{\mathrm{T}}Px + x^{\mathrm{T}}P\dot{x}$$
$$= x^{\mathrm{T}}(A_c^{\mathrm{T}}P + PA_c)x. \tag{8.1-12}$$

Since this must hold for all initial conditions, and hence for all state trajectories $x(t)$, we may write

$$g \equiv A_c^{\mathrm{T}}P + PA_c + C^{\mathrm{T}}K^{\mathrm{T}}RKC + Q = 0. \tag{8.1-13}$$

If K and Q are given and P is to be solved for, then this is called a Lyapunov equation. (A Lyapunov equation is a symmetric linear matrix equation. Note that the equation does not change if its transpose is taken.) In summary, for any fixed feedback matrix K if there exists a constant, symmetric, positive semi-definite matrix P that satisfies (8.1-13), and if the closed-loop system is stable, then the cost J is given in terms of P by (8.1-11). This is an important result in that the $n \times n$ auxiliary matrix P is independent of the state. Given a feedback matrix K, P may be computed from the Lyapunov equation (8.1-13). Then, only the initial condition $x(0)$ is required to compute the closed-loop cost under the influence of the feedback control (8.1-3). That is, we may compute the cost of applying the feedback control $u = -Ky$ before we actually apply it.

It is now necessary to use this result to compute the gain K that minimizes the PI. By using the trace identity

$$\mathrm{tr}(AB) = \mathrm{tr}(BA) \tag{8.1-14}$$

for any compatibly dimensioned matrices A and B (with the trace of a matrix the sum of its diagonal elements) we may write (8.1-11) as

$$J = \tfrac{1}{2}\mathrm{tr}(PX), \tag{8.1-15}$$

where the $n \times n$ symmetric matrix X is defined by

$$X \equiv x(0)x^\mathrm{T}(0). \tag{8.1-16}$$

It is now clear that the problem of selecting K to minimize (8.1-8) subject to the dynamical constraint (8.1-7) on the states is equivalent to the algebraic problem of selecting K to minimize (8.1-15) subject to the constraint (8.1-13) on the auxiliary matrix P. To solve this modified problem, we use the Lagrange multiplier approach to modify the problem yet again. Thus, adjoin the constraint to the PI by defining the Hamiltonian

$$H = \mathrm{tr}(PX) + \mathrm{tr}(gS), \tag{8.1-17}$$

with S a symmetric $n \times n$ matrix of Lagrange multipliers that still needs to be determined. Then, our constrained optimization problem is equivalent to the simpler problem of minimizing (8.1-17) without constraints. To accomplish this, we need to set the partial derivatives of H with respect to all the independent variables P, S, and K equal to zero. Using the facts that for any compatibly dimensioned matrices A, B, and C and any scalar y

$$\frac{\partial}{\partial B}\mathrm{tr}(ABC) = A^\mathrm{T}C^\mathrm{T} \tag{8.1-18}$$

and

$$\frac{\partial y}{\partial B^T} = \left[\frac{\partial y}{\partial B}\right]^\mathrm{T}, \tag{8.1-19}$$

the necessary conditions for the solution of the LQR problem with output feedback are given by

$$0 = \frac{\partial H}{\partial S} = g = A_c^\mathrm{T}P + PA_c + C^\mathrm{T}K^\mathrm{T}RKC + Q \tag{8.1-20}$$

$$0 = \frac{\partial H}{\partial P} = A_c S + SA_c^\mathrm{T} + X \tag{8.1-21}$$

$$0 = \frac{1}{2}\frac{\partial H}{\partial K} = RKCSC^\mathrm{T} - B^\mathrm{T}PSC^\mathrm{T}. \tag{8.1-22}$$

The first two of these are Lyapunov equations and the third is an equation for the gain K. If R is positive definite (i.e., all eigenvalues greater than zero, which implies nonsingularity; denoted $R > 0$) and CSC^T is nonsingular, then (8.1-22) may be solved for K to obtain

$$K = R^{-1}B^\mathrm{T}PSC^\mathrm{T}\left(CSC^\mathrm{T}\right)^{-1}. \tag{8.1-23}$$

To obtain the output feedback gain K minimizing the PI (8.1-4), we need to solve the three coupled equations (8.1-20), (8.1-21), and (8.1-23). This situation is quite strange, for to find K we must determine along the way the values of two auxiliary and apparently unnecessary $n \times n$ matrices, P and S. These auxiliary quantities may, however, not be as unnecessary as it appears, for note that the optimal cost may be determined directly from P and the initial state by using (8.1-11).

The Initial Condition Problem

Unfortunately, the dependence of X in (8.1-16) on the initial state $x(0)$ is undesirable, since it makes the optimal gain dependent on the initial state through equation (8.1-21). In many applications $x(0)$ may not be known. This dependence is typical of output-feedback design. We saw in Chapter 3 that in the case of state feedback it does not occur. Meanwhile, it is usual (Levine and Athans 1970) to sidestep this problem by minimizing not the PI (8.1-4) but its expected value, that is $E\{J\}$. Then, (8.1-11) and (8.1-16) are replaced by

$$E\{J\} = \tfrac{1}{2}E\{x^T(0)Px(0)\} = \tfrac{1}{2}\text{tr}(PX), \tag{8.1-24}$$

where the symmetric $n \times n$ matrix

$$X \equiv E\{x(0)x^T(0)\} \tag{8.1-25}$$

is the initial autocorrelation of the state. It is usual to assume that nothing is known of $x(0)$ except that it is uniformly distributed on a surface described by X. That is, we assume the actual initial state is unknown, but that it is nonzero with a certain expected Euclidean norm. For instance, if the initial states are assumed to be uniformly distributed on the unit sphere, then $X = I$, the identity. This is a sensible assumption for the regulator problem, where we are trying to drive arbitrary nonzero initial states to zero.

The design equations for the LQR with output feedback are collected in Table 8.1-1 for convenient reference. We now discuss their solution for K.

Determining the Optimal Feedback Gain

The importance of this modern LQ approach to controls design is that the matrix equations in Table 8.1-1 are used to solve for all the $m \times p$ elements of K at once. This corresponds to closing all the feedback loops simultaneously. Moreover, as long as certain reasonable conditions (to be discussed) on the plant and PI weighting matrices hold, the closed-loop system is generally guaranteed to be stable. In view of the trial-and-error successive-loop-closure approach used in stabilizing multivariable systems using classical approaches, this is quite important.

The equations for P, S, and K are coupled nonlinear matrix equations in three unknowns. It is important to discuss some aspects of their solution for the optimal feedback gain matrix K.

TABLE 8.1-1   LQR with Output Feedback

System model:

$$\dot{x} = Ax + Bu$$
$$y = Cx$$

Control:

$$u = -Ky$$

Performance index:

$$J = E \int_0^\infty (x^T Q x + u^T R u)\, dt$$

With

$$Q \geq 0, \quad R > 0$$

Optimal gain design equations:

$$0 = A_c^T P + P A_c + C^T K^T R K C + Q \tag{8.1-26}$$
$$0 = A_c S + S A_c^T + X \tag{8.1-27}$$
$$K = R^{-1} B^T P S C^T (C S C^T)^{-1}, \tag{8.1-28}$$

Where

$$A_c = A - BKC, \quad X = E\{x(0)x^T(0)\}.$$

Optimal cost:

$$J = \mathrm{tr}(PX) \tag{8.1-29}$$

Numerical Solution Techniques

There are three basic numerical techniques for determining the optimal output-feedback gain K. First, we may use a numerical optimization routine like the Simplex algorithm in Nelder and Mead (1964), found in MATLAB (Optimization Toolbox). This algorithm would use only equations (8.1-26) and (8.1-29). For a given value of K, it would solve the Lyapunov equation for P, and then use P in the second equation to determine $E\{J\}$. Based on this, it would vary the elements of K to minimize $E\{J\}$. The Lyapunov equation may be solved using, for instance, subroutine *lyap.m* in MATLAB (Control System Toolbox) based on the Bartels-Stewart algorithm.

A second approach for computing K is to use a gradient-based routine found in MATLAB (Optimization Toolbox). This routine would use all of the design equations in Table 8.1-1. For a given value of K, it would solve the two Lyapunov equation in the form (8.1-22). Note that if P satisfies the first Lyapunov equation, then $g = 0$ so that (see (8.1-17)) $E\{J\} = E\{H\}$ and $\partial E\{J\}/\partial K = \partial E\{H\}/\partial K$. Thus, the third design equation gives the gradient of $E\{J\}$ with respect to K, which would be used by the routine to update the value of K.

Finally, an iterative solution algorithm was presented in Moerder and Calise (1985). It is given in Table 8.1-2. It was shown in Moerder and Calise (1985) that the algorithm converges to a local minimum for J if the following conditions hold.

TABLE 8.1-2   Optimal Output Feedback Solution Algorithm

1. Initialize:
   Set $k = 0$
   Determine a gain K_0 so that $A - BK_0C$ is asymptotically stable
2. k th iteration:
   Set $A_k = A - BK_kC$
   Solve for P_k and S_k in

$$0 = A_k^T P_k + P_k A_k + C^T K_k^T R K_k C + Q$$

$$0 = A_k S_k + S_k A_k^T + X$$

Set $J_k = \text{tr}\,(P_k X)$
   Evaluate the gain update direction

$$\Delta K = R^{-1} B^T P S C^T (C S C^T)^{-1} - K_k$$

Update the gain by

$$K_{k+1} = K_k + \alpha \Delta K$$

where α is chosen so that
$A - BK_{k+1}C$ is asymptotically stable

$$J_{k+1} \equiv \tfrac{1}{2}\text{tr}(P_{k+1} X) \le J_k$$

If J_{k+1} and J_k are close enough to each other, go to 3 Otherwise, set $k = k + 1$ and
   go to 2
3. Terminate:
   Set $K = K_{k+1}, J = J_{k+1}$ Stop

Conditions for Convergence of the LQ Solution Algorithm

1. There exists a gain K such that A_c is stable. If this is true, we call system
   (8.1-1)–(8.1-2) output stabilizable.
2. The output matrix C has full row rank p.
3. Control weighting matrix R is positive definite. This means that all the
   control inputs should be weighted in the PI.
4. Q is positive semidefinite and $(\sqrt{Q}, A)$ is detectable. That is, the observability matrix polynomial

$$O(s) \equiv \begin{bmatrix} sI - A \\ -\sqrt{Q} \end{bmatrix} \tag{8.1-30}$$

has full rank n for all values of the complex variable s not contained in the left
half plane (Kailath 1980).

If these conditions hold, then the algorithm finds an output-feedback gain
that stabilizes the plant and minimizes the PI. The detectability condition means
that any unstable system modes must be observable in the PI. Then, if the PI is

bounded, which it is if the optimization algorithm is successful, signals associated with the unstable modes must go to zero as t becomes large; that is, they are stabilized in the closed-loop system.

Initial Stabilizing Gain

Since all three algorithms for solving the matrix equations in Table 8.1-1 for K are iterative in nature, a basic issue for all of them is the selection of an initial stabilizing output-feedback gain K_0. That is, to start the algorithms, it is necessary to provide a K_0 such that $(A - BK_0C)$ is stable. See, for instance, Table 8.1-2.

One technique for finding such a gain is given in Broussard and Halyo (1983). Another possibility is to use the eigenstructure assignment techniques of the previous section to determine an initial gain for the LQ solution algorithm. We could even select a stabilizing gain using the classical techniques, and then use modern design technique to tune the control gains for optimal performance. A quite convenient technique for finding an initial stabilizing gain K_0 is discussed in Section 8.2. This involves finding a full $m \times n$ state-variable feedback matrix and then zeroing the entries that are not needed in the $n \times p$ output-feedback matrix for the given measured outputs. Note that there are many techniques for finding a full state feedback that stabilizes a system given A and B (see Chapter 3).

Iterative Design

Software that solves for the optimal output-feedback gain K can be found in the MATLAB Optimization Toolbox. Given good software, design using the LQ approach is straightforward. A design procedure would involve selecting the *design parameters* Q and R, determining the optimal gain K, and simulating the closed-loop response and frequency-domain characteristics. If the results are not suitable, different matrices Q and R are chosen and the design is repeated.

This approach introduces the notion of *tuning the design parameters* Q and R for *good performance*. In the next two sections we present sensible techniques for obtaining suitable PI weighting matrices Q and R that do not depend on individually selecting all of their entries. Example 8.1-1 will illustrate these notions.

Selection of the PI Weighting Matrices

Once the PI weighting matrices Q and R have been selected, the determination of the optimal feedback gain K is a formal procedure relying on the solution of nonlinear coupled matrix equations. Therefore, the engineering judgment in modern LQ design appears in the selection of Q and R. There are some guidelines for this, which we now discuss.

Observability in the Choice of Q

For stabilizing solutions to the output-feedback problem, it is necessary for $(\sqrt{Q}, A)$ to be detectable. The detectability condition basically means that Q

should be chosen so that all unstable states are weighted in the PI. Then, if J is bounded so that $\sqrt{Q}x(t)$ vanishes for large t, the open-loop unstable states will be forced to zero through the action of the control. This means exactly that the unstable poles must have been stabilized by the feedback control gain. A stronger condition than detectability is observability, which amounts to the full rank of $O(s)$ for all values of s. Observability is easier to check than detectability since it is equivalent to the full rank n of the observability matrix

$$O \equiv \begin{bmatrix} \sqrt{Q} \\ \sqrt{Q}A \\ \vdots \\ \sqrt{Q}A^{n-1} \end{bmatrix}, \tag{8.1-31}$$

which is a constant matrix and so easier to deal with than $O(s)$. In fact, O has full rank n if and only if the observability gramian $O^T O$ is nonsingular. Since the gramian is an $n \times n$ matrix, its determinant is easily examined using available software (e.g., singular value decomposition/condition number [MATLAB]). The observability of $(\sqrt{Q}, A)$ means basically that all states are weighted in the PI. From a numerical point of view, if $(\sqrt{Q}, A)$ is observable, then a positive-definite solution P to (8.1-26) results; otherwise, P may be singular. Since P helps determine K through (8.1-28), it is found that if P is singular it may result in some zero gain elements in K. That is, if $(\sqrt{Q}, A)$ is not observable, the LQ algorithm can refuse to close some of the feedback loops. This observability condition amounts to a restriction on the selection of Q, and is a drawback of modern control (see Example 8.1-1).

The Structure of Q

The choice of Q can be confronted more easily by considering the performance objectives of the LQR. Suppose that a performance output

$$z = Hx \tag{8.1-32}$$

is required to be small in the closed-loop system. For instance, in an aircraft lateral regulator it is desired for the sideslip angle, yaw rate, roll angle, and roll rate to be small (see Example 8.1-1). Therefore, we might select $z = |\alpha r p|^T$. Once $z(t)$ has been chosen, the performance output matrix H may be formally written down.

The signal $z(t)$ may be made small by LQR design by selecting the PI

$$J = \frac{1}{2} \int_0^\infty (x^T Q x + u^T R u)\, dt, \tag{8.1-33}$$

which amounts to using the PI in Table 8.1-1 with $Q = H^T H$, so that Q may be computed from H. That is, by weighting performance outputs in the PI, Q is directly given.

Maximum Desired Values of z(t) and u(t)

A convenient guideline for selecting Q and R is given in Bryson and Ho (1975). Suppose the performance output (8.1-32) has been defined so that H is given. Consider the PI

$$J = \frac{1}{2} \int_0^\infty (z^T \overline{Q} z + u^T R u)\, dt. \tag{8.1-34}$$

Then, in Table 8.1-1 we have $Q = H^T \overline{Q} H$. To select $\overline{Q}$ and R, one might proceed as follows, using the maximum allowable deviations in $z(t)$ and $u(t)$. Define the maximum allowable deviation in component $z_i(t)$ of $z(t)$ as z_{iM} and the maximum allowable deviation in component $u_i(t)$ of the control input $u(t)$ as u_{iM}. Then, $\overline{Q}$ and R may be selected as $\overline{Q} = \mathrm{diag}\{q_i\}$, $R = \mathrm{diag}\{r_i\}$, with

$$q_i = 1/z_{iM}^2, \quad r_i = 1/r_{iM}^2. \tag{8.1-35}$$

The rationale for this choice is easy to understand. For instance, as the allowed limits z_{iM} on $z_i(t)$ decrease, the weighting in the PI placed on $z_i(t)$ increases, which requires smaller excursions in $z_i(t)$ in the closed-loop system.

Asymptotic Properties of the LQR

Consider the PI

$$J = \frac{1}{2} \int_0^\infty (x^T Q x + \rho u^T R u)\, dt, \tag{8.1-36}$$

where ρ is a scalar design parameter. There are some quite nice results that describe the asymptotic performance of the LQR as ρ becomes small and as ρ becomes large (Kwakernaak and Sivan 1972, Harvey and Stein 1978, Grimble and Johnson 1988).

These results detail the asymptotic closed-loop eigenstructure of the LQR, and are of some assistance in selecting Q and R. Unfortunately, they are well developed only for the case of full state-variable feedback, where $C = I$ and all the states are allowed for feedback.

Example 8.1-1.   LQR Design for F-16 Lateral Regulator

In this example we should like to demonstrate the power of the LQ design equations in Table 8.1-1 by designing a lateral regulator. In our approach, we shall select the design parameters Q and R and then use the design equations there to close all the feedback loops simultaneously by computing K. The objective is to design a closed-loop controller to provide for the function of a lateral stability augmentation system as well as the closure of the roll-attitude loop. This objective involves the design of two feedback channels with multiple loops, but it is straightforward to deal with using modern control techniques. The simplicity of MIMO design using the LQR will be evident.

a. Aircraft State Equations

We use the F-16 linearized lateral dynamics at the nominal flight condition retaining the lateral states sideslip β, bank angle ϕ, roll rate p, and yaw rate r. Additional states δ_a and δ_r are introduced by the aileron and rudder actuators

$$\delta_a = \frac{20.2}{s + 20.2} u_a, \quad \delta_r = \frac{20.2}{s + 20.2} u_r . \tag{1}$$

A washout filter

$$r_w = \frac{s}{s + 1} r \tag{2}$$

is used, with r the yaw rate and r_w the washed-out yaw rate. The washout filter state is denoted x_w. Thus, the entire state vector is

$$x = [\beta \quad \phi \quad p \quad r \quad \delta_a \quad \delta_r \quad x_w]^\mathsf{T}. \tag{3}$$

The full state-variable model of the aircraft plus actuators, washout filter, and control dynamics is of the form

$$\dot{x} = Ax + Bu, \tag{4}$$

with

$$A = \begin{bmatrix} -0.3220 & 0.0640 & 0.0364 & -0.9917 & 0.0003 & 0.0008 & 0 \\ 0 & 0 & 1 & 0.0037 & 0 & 0 & 0 \\ -30.6492 & 0 & -3.6784 & 0.6646 & -0.7333 & 0.1315 & 0 \\ 8.5396 & 0 & -0.0254 & -0.4764 & -0.0319 & -0.0620 & 0 \\ 0 & 0 & 0 & 0 & -20.2 & 0 & 0 \\ 0 & 0 & 0 & 0 & 0 & -20.2 & 0 \\ 0 & 0 & 0 & 57.2958 & 0 & 0 & -1 \end{bmatrix}$$

$$B = \begin{bmatrix} 0 & 0 \\ 0 & 0 \\ 0 & 0 \\ 0 & 0 \\ 20.2 & 0 \\ 0 & 20.2 \\ 0 & 0 \end{bmatrix} . \tag{5}$$

The control inputs are the rudder and aileron servo inputs so that

$$u = \begin{bmatrix} u_a \\ u_r \end{bmatrix} \tag{6}$$

and the output is

$$y = \begin{bmatrix} r_w \\ p \\ \beta \\ \phi \end{bmatrix}. \tag{7}$$

Thus, $y = Cx$, with

$$C = \begin{bmatrix} 0 & 0 & 0 & 57.2958 & 0 & 0 & -1 \\ 0 & 0 & 57.2958 & 0 & 0 & 0 & 0 \\ 57.2958 & 0 & 0 & 0 & 0 & 0 & 0 \\ 0 & 57.2958 & 0 & 0 & 0 & 0 & 0 \end{bmatrix}. \tag{8}$$

The factor of 57.2958 converts radians to degrees. The feedback control will be output feedback of the form $u = -Ky$, so that K is a 2×4 matrix. That is, we shall select eight feedback gains. For this system the open-loop Dutch-roll mode has poles at $-0.4226 \pm j3.064$, and so has insufficient damping. The spiral mode has a pole at -0.0163.

b. LQR Output Feedback Design

For the computation of the feedback gain K, it is necessary to select PI weighting matrices Q and R in Table 8.1-1. Then, software from the Optimization Toolbox is used to compute the optimal gain K using the design equations in the table. Our philosophy for selecting Q and R follows. First, let us discuss the choice of Q. It is desired to obtain good stability of the Dutch-roll mode, so that β^2 and r^2 should be weighted in the PI by factors of q_{dr}. To obtain stability of the roll mode, which in closed-loop will consist primarily of p and ϕ, we may weight p^2 and ϕ^2 in the PI by factors of q_r. We do not care about δ_a and δ_r, so it is not necessary to weight them in the PI; the control weighting matrix R will prevent unreasonably large control inputs. Thus, so far we have

$$Q = \text{diag}\{q_{dr}, q_r, q_r, q_{dr}, 0, 0, 0\}. \tag{9}$$

We do not care directly about x_w; however, it is necessary to weight it in the PI. This is because omitting it would cause problems with the observability condition. A square root of Q in (9) is

$$\sqrt{Q} = \begin{bmatrix} \sqrt{q_{dr}} & \sqrt{q_r} & \sqrt{q_r} & \sqrt{q_{dr}} & 0 & 0 & 0 \end{bmatrix}. \tag{10}$$

Consequently, the observability matrix (8.1-31) has a right-hand column of zero; hence, the system is unobservable. This may be noted in simpler fashion by examining the A matrix in (5), where the seventh state x_w is seen to have no influence on the states that are weighted in (9). To correct this potential problem, we chose

$$Q = \text{diag}\{q_{dr}, q_r, q_r, q_{dr}, 0, 0, 1\}. \tag{11}$$

As far as the R matrix goes, it is generally satisfactory to select it as

$$R = \rho I, \tag{12}$$

with I the identity matrix and ρ a scalar design parameter.

Now, the design equations in Table 8.1-1 were solved for several choices of ρ, q_{dr}, q_r. After some trial and error, we obtained a good result using $\rho = 0.1$, $q_{dr} = 50$, $q_r = 100$. For this selection, the optimal feedback gain was

$$K = \begin{bmatrix} -0.56 & -0.44 & 0.11 & -0.35 \\ -1.19 & -0.21 & -0.44 & 0.26 \end{bmatrix}. \tag{13}$$

The resulting closed-loop poles were at

$$
\begin{aligned}
s = -3.13 &\pm j0.83 \quad \text{Dutch-roll mode } (r, \beta) \\
- 0.82 &\pm j0.11 \quad \text{roll mode } (p, \phi) \\
- 11.47 &\pm j17.18, \quad -15.02.
\end{aligned} \tag{14}
$$

To verify the design a simulation was performed. The initial state was selected as $x(0) = [1\ 0\ 0\ 0\ 0\ 0\ 0]^{\mathrm{T}}$; that is, we chose $\beta(0) = 1$. Figure 8.1-1 shows the results. Part (a) shows the Dutch-roll mode and part (b) the roll mode. Note that the responses correspond to the poles in (14), where the Dutch roll is the faster mode.

This design has two deficiencies. First, it uses eight feedback gains in (13). This is undesirable for two reasons. (1) It requires the gain scheduling of all eight gains, and (2) the control system has no structure. That is, all outputs are fed back to both inputs; zeroing some of the gains would give the controller more structure in terms of feeding back certain outputs to only one or the other of the inputs. The second deficiency is that it was necessary to juggle the entries of Q to obtain a good solution. Actually, due to our weighting of β^2 and r^2 by q_{dr}, and ϕ^2 and p_2 by q_r, the design was fairly straightforward and took about half an hour in all. It was, however, necessary to weight the washout filter state x_w, which is not obvious without considering the observability question.

c. Effect of Weighting Parameters

It is interesting to examine more closely the effects of the design parameters, namely, the entries of the PI weighting matrices Q and R. Using the same Q as above, we show the sideslip response in Fig. 8.1-2a for control weightings of $\rho = 0.1, 0.5$, and 1. Increased control weighting in the PI generally suppresses the control signals in the closed-loop system; that is, less control effort is allowed. As less control effort is allowed, the control is less effective in controlling the modes. Indeed, according to the figure, as ρ increases, the undershoot in β increases. Moreover, with increasing ρ the control is also less effective in suppressing the undesirable oscillations in the Dutch-roll mode that were noted in the open-loop system.

As far as the effect of the Dutch-roll weighting q_{dr} goes, examine Fig. 8.1-2b, where $\rho = 0.1$ and $q_r = 100$ as in part b, but the sideslip response is shown for $q_r = 0, 50$, and 100. As q_{dr} increases, the undershoot decreases, reflecting the fact that increased weighting on β^2 in the PI will result in smaller excursions in β in closed loop. One last point is worth noting. The open-loop system is stable; therefore, it is clear that it is detectable, since all the unstable modes are observable for any choice of Q (there are no unstable modes). Thus, the design would work if we omitted the weighting on x_w^2 in the Q matrix (although, it turns out, the closed-loop poles are not as good). In general, however, the detectability condition is difficult to check in large systems that are open-loop unstable; thus, the observability condition is used instead. Failing to weight an undetectable state can lead to some zero elements of K, meaning that some feedback loops are not closed.

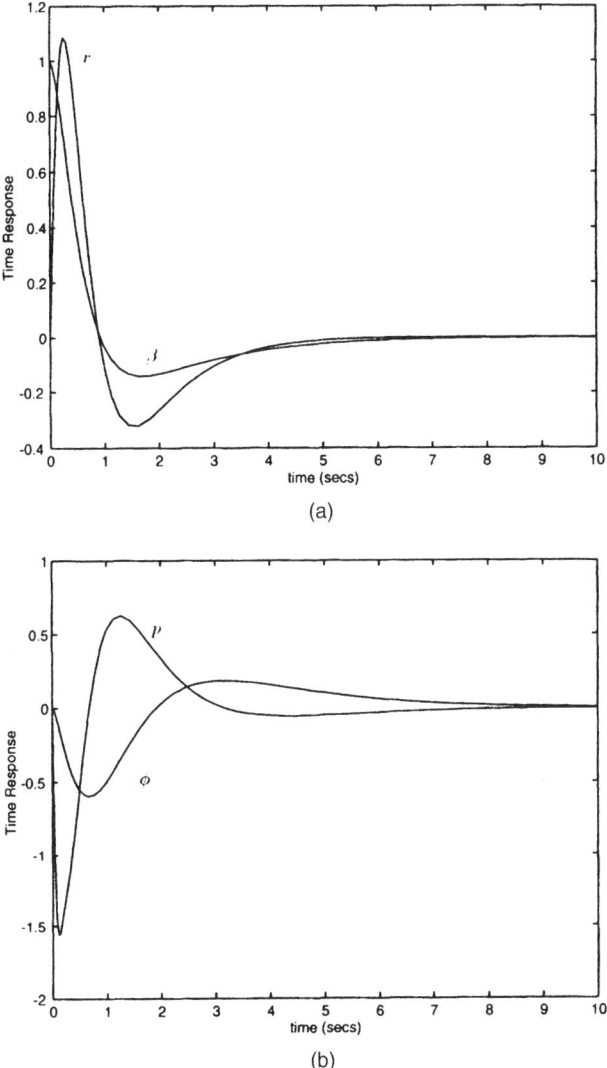

FIGURE 8.1-1   Closed-loop lateral response. (a) Dutch-roll states β and r. (b) Roll mode states ϕ and p.

Thus, to guarantee that this does not occur, Q should be selected so that $(\sqrt{Q}, A)$ is observable.

d. Gain Scheduling

For implementation on an aircraft, the control gains in (13) should be gain scheduled. To accomplish this, the nonlinear aircraft equations are linearized at several equilibrium flight conditions over the desired flight envelope to obtain state-variable models like

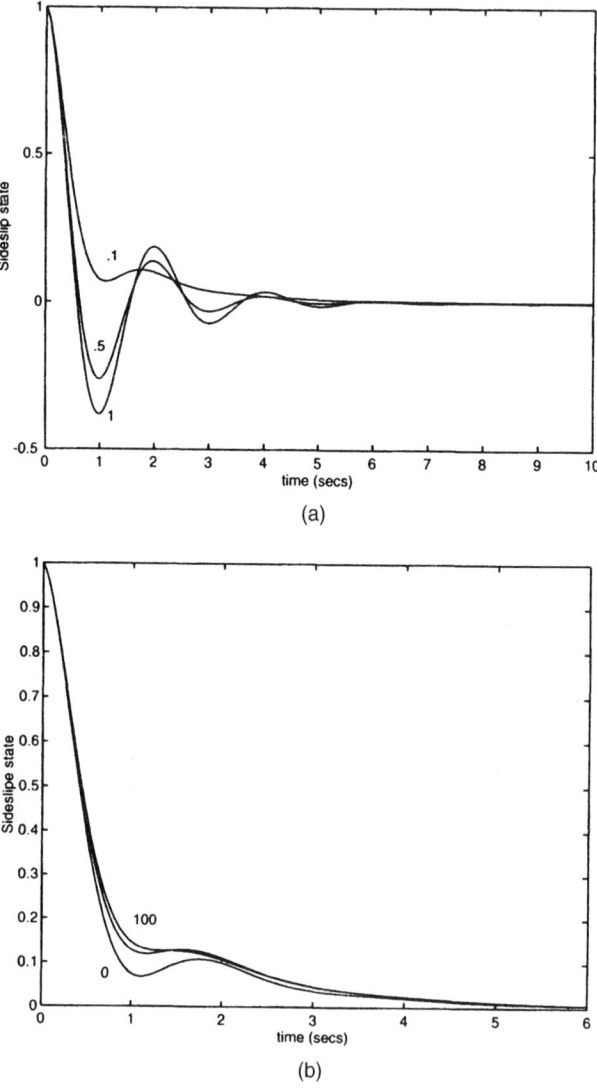

FIGURE 8.1-2 Effect of PI weighting. (a) Sideslip as a function of $\rho(\rho = 0.1, 0.5, 1)$. (b) Sideslip as a function of $q_{dr}(q_{dr} = 0, 50, 100)$.

(4) with different A and B matrices. Then, the LQR design is repeated for those different systems.

A major advantage of LQR design will quickly be apparent, for once the control structure has been selected, it takes only a minute or two to run the software to find the *optimal* gains for a new A and B using the design equations in Table 8.1-1. Note that the optimal gains for one point in the gain schedule can be used as initial stabilizing gains in the LQ solution algorithm for the next point.

It is important, however, to be aware of an additional consideration. The optimal gains at each gain scheduling point should guarantee robust stability and performance; that is, they should guarantee stability and good performance at points near the design equilibrium point. Such robust stability can be verified after the LQ design by using multivariable frequency-domain techniques. These techniques are developed in Section 9.2, where the remarks on robustness to plant parameter variations are particularly relevant to gain scheduling. ∎

8.2 TRACKING A REFERENCE INPUT

In control design we are often interested not in regulating the state near zero, which we discussed in the previous section, but in *following a nonzero reference command signal*. For example, we may be interested in designing a control system for optimal step-response shaping. This reference-input tracking or servodesign problem is important in the design of command augmentation systems (CAS). In this section and the next we cover tracker design.

It should be mentioned that the optimal linear quadratic (LQ) tracker of modern control is not a causal system (see Chapter 4). It depends on solving an "adjoint" system of differential equations backward in time, and so is impossible to implement. A suboptimal "steady-state" tracker using full state-variable feedback is available, but it offers no convenient structure for the control system in terms of desired dynamics like PI control, washout filters, and so on.

Modified versions of the LQ tracker have been presented in Davison and Ferguson (1981) and Gangsaas et al. (1986). There, controllers of desired structure can be designed since the approaches are output-feedback based. The optimal gains are determined numerically to minimize a PI with, possibly, some constraints.

It is possible to design a tracker by first designing a regulator using, for instance, Table 8.1-1. Then, some feedforward terms are added to guarantee perfect tracking (Kwakernaak and Sivan 1972). The problem with this technique is that the resulting tracker has no convenient structure and often requires derivatives of the reference command input. Moreover, servosystems designed using this approach depend on knowing the DC gain exactly. If the DC gain is not exactly known, the performance deteriorates. That is, the design is *not robust* to uncertainties in the model.

Here we discuss an approach to the design of tracking control systems that is very useful in several control applications. This approach will allow us to design a servo control system that has any structure desired. This structure will include a unity-gain outer loop that feeds the performance output back and subtracts it from the reference command, thus defining a tracking error $e(t)$ that should be kept small. See Fig. 8.2-1. It can also include compensator dynamics, such as a washout filter or an integral controller. The control gains are chosen to minimize a quadratic performance index (PI). We are able to give explicit design equations for the control gains (see Table 8.2-1), which may be solved using software available in the MATLAB Optimization Toolbox.

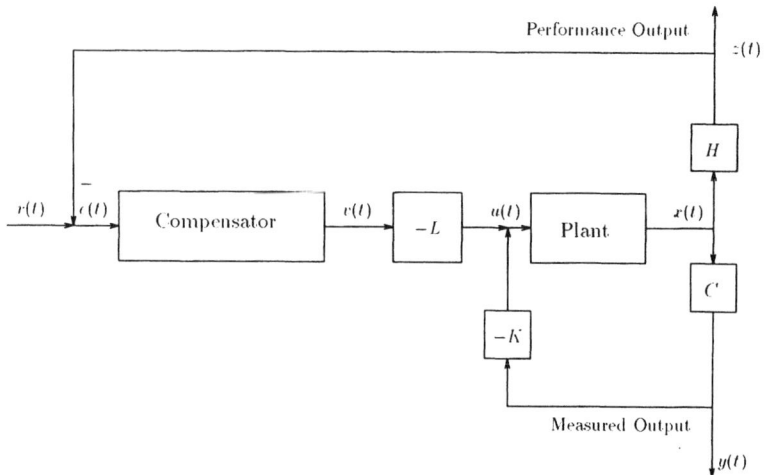

FIGURE 8.2-1   Plant with compensator of desired structure.

A problem with the tracker developed in this section is the need to select the design parameters Q and R in the PI in Table 8.2-1. Later, we show how modified PIs may be used to make the selection of Q and R almost transparent, yielding tracker design techniques that are very convenient for use in aircraft control systems design. We show, in fact, that *the key to achieving required performance using modern design strategies is in selecting an appropriate PI*.

Tracker with Desired Structure

In several control designs there is a wealth of experience and knowledge that dictates in many situations what sort of compensator dynamics yield good performance from the point of view of both the controls engineer and the pilot. For example, a washout circuit may be required, or it may be necessary to augment some feedforward channels with integrators to obtain a steady-state error of exactly zero. The control system structures used in classical aircraft design also give good robustness properties. That is, they perform well even if there are disturbances or uncertainties in the system. Thus, the multivariable approach developed here usually affords this robustness. Formal techniques for verifying closed-loop robustness for multivariable control systems are given in Chapter 9.

Our approach to tracker design allows controller dynamics of any desired structure and then determines the control gains that minimize a quadratic PI over that structure. Before discussing the tracker design, let us examine how the compensator dynamics may be incorporated into the system state equations.

A dynamic compensator of prescribed structure may be incorporated into the system description as follows. Consider the situation in Fig. 8.2-1 where the plant is described by

$$\dot{x} = Ax + Bu \tag{8.2-1}$$

$$y = Cx \tag{8.2-2}$$

with state $x(t)$, control input $u(t)$, and $y(t)$ the measured output available for feedback purposes. In addition,

$$z = Hx \tag{8.2-3}$$

is a performance output, which must track the given reference input $r(t)$. The performance output $z(t)$ is not generally equal to $y(t)$. It is important to realize that for perfect tracking it is necessary to have as many control inputs in vector $u(t)$ as there are command signals to track in $r(t)$ (Kwakernaak and Sivan 1972).

The dynamic compensator has the form

$$\dot{w} = Fw + Ge$$

$$v = Dw + Je \tag{8.2-4}$$

with state $w(t)$, output $v(t)$, and input equal to the tracking error

$$e(t) = r(t) - z(t). \tag{8.2-5}$$

F, G, D, and J are known matrices chosen to include the desired structure in the compensator. The allowed form for the plant control input is

$$u = -Ky - Lv, \tag{8.2-6}$$

where the constant gain matrices K and L are to be chosen in the controls design step to result in satisfactory tracking of $r(t)$. This formulation allows for both feedback and feedforward compensator dynamics.

These dynamics and output equations may be written in augmented form as

$$\frac{d}{dt}\begin{bmatrix} x \\ w \end{bmatrix} = \begin{bmatrix} A & 0 \\ -GH & F \end{bmatrix}\begin{bmatrix} x \\ w \end{bmatrix} + \begin{bmatrix} 0 \\ G \end{bmatrix}r \tag{8.2-7}$$

$$\begin{bmatrix} y \\ v \end{bmatrix} = \begin{bmatrix} C & 0 \\ -JH & D \end{bmatrix}\begin{bmatrix} x \\ w \end{bmatrix} + \begin{bmatrix} 0 \\ H \end{bmatrix}r \tag{8.2-8}$$

$$z = \begin{bmatrix} 0 & H \end{bmatrix}\begin{bmatrix} x \\ w \end{bmatrix}, \tag{8.2-9}$$

and the control input may be expressed as

$$u = -[K \quad L] \begin{bmatrix} y \\ v \end{bmatrix}. \tag{8.2-10}$$

Note that, in terms of the augmented plant/compensator state description, the admissible controls are represented as a *constant output feedback [K L]*. In the augmented description, all matrices are known except the gains K and L, which need to be selected to yield acceptable closed-loop performance.

A comment on the compensator matrices F, G, D, J is in order. Often, these matrices are completely specified by the structure of the compensator. Such is the case, for instance, if the compensator contains integrators. However, if it is desired to include a washout or a lead-lag, it may not be clear exactly how to select the time constants. In such cases, engineering judgment will usually give some insight. However, it may sometimes be necessary to go through the design to be proposed, and then if required return to readjust F, G, D, J and reperform the design.

LQ Formulation of the Tracker Problem

By redefining the state, the output, and the matrix variables to streamline the notation, we see that the augmented equations (8.2-7)–(8.2-9) that contain the dynamics of both the plant and the compensator are of the form

$$\dot{x} = Ax + Bu + Gr \tag{8.2-11}$$
$$y = Cx + Fr \tag{8.2-12}$$
$$z = Hx. \tag{8.2-13}$$

In this description, let us take the state $x(t) \in R^n$, control input $u(t) \in R^m$, reference input $r(t) \in R^q$, performance output $z(t) \in R^q$, and measured output $y(t) \in R^p$. The admissible controls (8.2-10) are proportional output feedbacks of the form

$$u = -Ky = -KCx - KFr, \tag{8.2-14}$$

with constant gain K to be determined. This situation corresponds to the block diagram in Fig. 8.2-2. Since K is an $m \times p$ matrix, we intend to close all the feedback loops simultaneously by computing K.

Using these equations the closed-loop system is found to be

$$\dot{x} = (A - BKC)x + (G - BKF)r$$
$$\equiv A_c x + B_c r. \tag{8.2-15}$$

In the remainder of this subsection, we shall use the formulation (8.2-11)–(8.2-14), assuming that the compensator, if required, has already been included

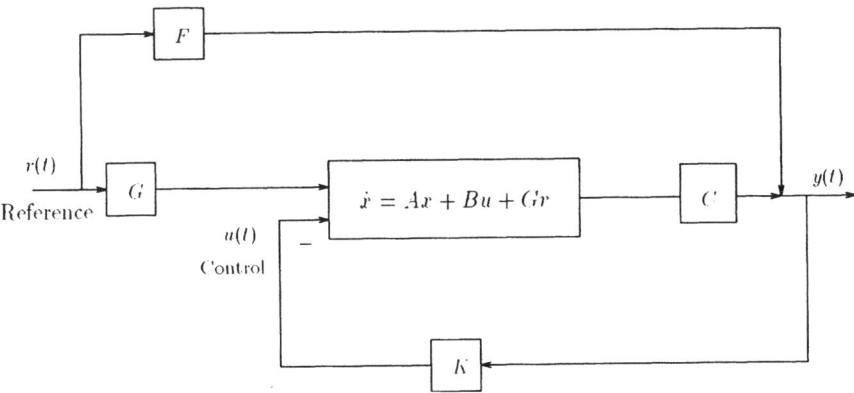

FIGURE 8.2-2   Plant/feedback structure.

in the system dynamics and demonstrating how to select the constant output feedback gain matrix K using LQ techniques.

Our formulation differs sharply from the traditional formulations of the optimal tracker problem studied in Chapter 4. Note that (8.2-14) includes both feedback and feedforward terms, so that both the closed-loop poles and compensator zeros may be affected by varying the gain K (see Example 8.2-1). Thus, we should expect better success in shaping the step response than by placing only the poles. We shall assume henceforth that the reference input $r(t)$ is a step command with magnitude r_0. Designing for such a command will yield suitable time-response characteristics. Although our design is based on step-response shaping, it should be clearly realized that the resulting control system, if properly designed, will give good time responses for any arbitrary reference command signal $r(t)$.

Let us now formulate an optimal control problem for selecting the control gain K to guarantee tracking of $r(t)$. Then, we shall derive the design equations in Table 8.2-1, which are used to determine the optimal K. These equations are solved using software like that found in the Optimization Toolbox.

The Deviation System

Denote steady-state values by overbars and deviations from the steady-state values by tildes. Then, the state, output, and control deviations are given by

$$\tilde{x}(t) = x(t) - \bar{x} \tag{8.2-16}$$

$$\tilde{y}(t) = y(t) - \bar{y} = K\tilde{x} \tag{8.2-17}$$

$$\tilde{z}(t) = z(t) - \bar{z} = H\tilde{x} \tag{8.2-18}$$

$$\tilde{u}(t) = u(t) - \bar{u} = -KCx - KFr_0 - (-KC\bar{x} - KFr_0)$$

$$= -KC\tilde{x}(t) \tag{8.2-19}$$

or

$$\tilde{u} = -K\tilde{y}. \tag{8.2-20}$$

The tracking error $e(t) = r(t) - z(t)$ is given by

$$e(t) = \tilde{e}(t) + \bar{e} \tag{8.2-21}$$

with the error deviation given by

$$\tilde{e}(t) = e(t) - \bar{e} = (r_0 - Hx) - (r_0 - H\bar{x}) = -H\tilde{x} \tag{8.2-22}$$

or

$$\tilde{e} = -\tilde{z}. \tag{8.2-23}$$

Since in any acceptable design the closed-loop plant will be asymptotically stable, A_c is nonsingular. According to (8.2-15), at steady state

$$0 = A_c\bar{x} + B_c r_0, \tag{8.2-24}$$

so that the steady-state state response $\bar{x}$ is

$$\bar{x} = -A_c^{-1} B_c r_0 \tag{8.2-25}$$

and the steady-state error is

$$\bar{e} = r_0 - H\bar{x} = (I + HA_c^{-1}B_c)r_0. \tag{8.2-26}$$

To understand this expression, note that the closed-loop transfer function from r_0 to z (see (8.2-15) and (8.2-13)) is

$$H(s) = H(sI - A_c)^{-1}B_c. \tag{8.2-27}$$

The steady-state behavior may be investigated by considering the DC value of $H(s)$ (i.e., $s = 0$); this is just $-HA_c^{-1}B_c$, the term appearing in (8.2-24).

Using (8.2-16), (8.2-19), and (8.2-23) in (8.2-15) the closed-loop dynamics of the state deviation are seen to be

$$\begin{aligned} \dot{\tilde{x}} &= A_c\tilde{x} \\ \tilde{y} &= C\tilde{x} \\ \tilde{z} &= H\tilde{x} = -\tilde{e} \end{aligned} \tag{8.2-28}$$

and the control input to the deviation system (8.2-26) is (8.2-19). Thus, the step-response shaping problem has been converted to a regulator problem for the deviation system

$$\dot{\tilde{x}} = A\tilde{x} + B\tilde{u}. \tag{8.2-29}$$

Again, we emphasize the difference between our approach and the traditional one described in Chapter 4. Once the gain K in (8.2-19) has been found, the control for the plant is given by (8.2-14), which inherently has both feedback and feedforward terms. Thus, no extra feedforward term need be added to make e zero.

Performance Index

To make the tracking error $e(t)$ in (8.2-20) small, we propose to attack two equivalent problems: the problem of regulating the error deviation $\tilde{e}(t) = -\tilde{z}(t)$ to zero, and the problem of making small the steady-state error e. Note that we do not assume a type 1 system, which would force e to be equal to zero. This can be important in aircraft controls, where it may not be desirable to force the system to be of type 1 by augmenting all control channels with integrators. This augmentation complicates the servo structure. Moreover, it is well known from classical control theory that suitable step responses may often be obtained without resorting to inserting integrators in all the feedforward channels.

To make small both the error deviation $\tilde{e}(t) = -H\tilde{x}(t)$ and the steady-state error e, we propose selecting K to minimize the performance index (PI)

$$J = \int_0^\infty (\tilde{e}^T\tilde{e} + \tilde{u}^T R\tilde{u})\, dt + \frac{1}{2}\bar{e}^T V\bar{e}, \tag{8.2-30}$$

with $R > 0$, $V \geq 0$ design parameters. The integrand is the standard quadratic PI with, however, a weighting V included on the steady-state error. Note that the PI weights the control deviations and not the controls themselves. If the system is of type 1, containing integrators in all the feedforward paths, then V may be set to zero since the steady-state error is automatically zero.

Making small the error deviation $\tilde{e}(t)$ improves the transient response, while making small the steady-state error $e(t)$ improves the steady-state response. If the system is of type 0, these effects involve a trade-off, so that then there is a design trade-off involved in selecting the size of V. We can generally select $R = rI$ and $V = vI$, with r and v scalars. This simplifies the design since now only a few parameters must be tuned during the interactive design process. According to (8.2-21), $\tilde{e}^T\tilde{e} = \tilde{x}^T H^T H\tilde{x}$. Referring to Table 8.1-1, therefore, it follows that the matrix Q there is equal to $H^T H$, where H is known. That is, weighting the error deviation in the PI has already shown us how to select the design parameter Q, affording a considerable simplification.

The problem we now have to solve is how to select the control gains K to minimize the PI J for the deviation system (8.2-29). Then, the tracker control for the original system is given by (8.2-14).

We should point out that the proposed approach is suboptimal in the sense that minimizing the PI does not necessarily minimize a quadratic function of the total error $e(t) = \bar{e} + \tilde{e}(t)$. It does, however, guarantee that both $\tilde{e}(t)$ and e are small in the closed-loop system, which is a design goal.

Solution of the LQ Tracker Problem

It is now necessary to solve for the optimal feedback gain K that minimizes the PI. The design equations needed are now derived. They appear in Table 8.2-1. By using (8.2-26) and a technique like the one used in Section 8.3 (see problems), the optimal cost is found to satisfy

$$J = \tfrac{1}{2}\tilde{x}^{\mathrm{T}}(0)P\tilde{x}(0) + \tfrac{1}{2}\bar{e}^{\mathrm{T}}V\bar{e}, \qquad (8.2\text{-}31)$$

with $P \geq 0$ the solution to

$$0 = g \equiv A_c^{\mathrm{T}}P + PA_c + Q + C^{\mathrm{T}}K^{\mathrm{T}}RKC, \qquad (8.2\text{-}32)$$

with $Q = H^{\mathrm{T}}H$ and e given by (8.2-24).

In our discussion of the linear quadratic regulator we assumed that the initial conditions were uniformly distributed on a surface with known characteristics. While this is satisfactory for the regulator problem, it is an unsatisfactory assumption for the tracker problem. In the latter situation the system starts at rest and must achieve a given final state that is dependent on the reference input, namely (8.2-23). To find the correct value of $\tilde{x}(0)$, we note that, since the plant starts at rest (i.e., $x(0) = 0$), according to (8.2-25)

$$\tilde{x}(0) = -\bar{x}, \qquad (8.2\text{-}33)$$

so that the optimal cost (8.2-31) becomes

$$J = \tfrac{1}{2}\bar{x}^{\mathrm{T}}P\bar{x} + \tfrac{1}{2}\bar{e}^{\mathrm{T}}V\bar{e} = \tfrac{1}{2}\mathrm{tr}(PX) + \tfrac{1}{2}\bar{e}^{\mathrm{T}}V\bar{e}, \qquad (8.2\text{-}34)$$

with P given by (8.2-32), e given by (8.2-24), and

$$X \equiv \bar{x}\bar{x}^{\mathrm{T}} = A_c^{-1}B_c r_0 r_0^{\mathrm{T}} B_c^{\mathrm{T}} A_c^{-\mathrm{T}}, \qquad (8.2\text{-}35)$$

with $A_c^{-\mathrm{T}} = (A_c^{-1})^{\mathrm{T}}$. The optimal solution to the unit-step tracking problem, with (8.2-11) initially at rest, may now be determined by minimizing J in (8.2-34) over the gains K, subject to the constraint (8.2-32) and equations (8.2-24), (8.2-35).

This algebraic optimization problem can be solved by any well-known numerical method (cf Press et al. 1986, Söderström 1978). A good approach for a fairly small number ($mp \leq 10$) of gain elements in K is the SIMPLEX minimization routine (Nelder and Mead 1964). To evaluate the PI for each fixed value of K in the iterative solution procedure, one may solve (8.2-32) for P using the *lyap.m* subroutine in MATLAB (Control System Toolbox) and then employ (8.2-34). Software for determining the optimal control gains K can be found in the Optimization Toolbox.

Design Equations for a Gradient-based Solution

As an alternative solution procedure one may use gradient-based techniques (e.g., the Davidson-Fletcher-Powell algorithm [Press et al. 1986]), which are generally faster than non-gradient-based approaches.

To find the gradient of the PI with respect to the gains, define the Hamiltonian

$$H = \text{tr}(PX) + \text{tr}(gS) + \tfrac{1}{2}\bar{e}^{\mathsf{T}} V \bar{e}, \tag{8.2-36}$$

with S a Lagrange multiplier. Now, using the basic matrix calculus identities

$$\frac{\partial Y^{-1}}{\partial x} = -Y^{-1}\frac{\partial Y}{\partial x}Y^{-1} \tag{8.2-37}$$

$$\frac{\partial UV}{\partial x} = \frac{\partial U}{\partial x}V + U\frac{\partial V}{\partial x} \tag{8.2-38}$$

$$\frac{\partial y}{\partial x} = \text{tr}\left[\frac{\partial y}{\partial z} \cdot \frac{\partial z^{\mathsf{T}}}{\partial x}\right], \tag{8.2-39}$$

we may proceed as in the previous section, with, however, a little more patience due to the extra terms (see the problems at the end of the chapter), to obtain the necessary conditions for a solution given in Table 8.2-1.

To find K by a gradient minimization algorithm, it is necessary to provide the algorithm with the values of J and $\partial J/\partial K$ for a given K. The value of J is given by the expression in Table 8.2-1 for the optimal cost. To find $\partial J/\partial K$ given K, solve (8.2-40), (8.2-41) for P and S. Then, since these equations hold, $\partial J/\partial K = \partial H/\partial K$, which may be found using (8.2-42).

These equations should be compared to those in Table 8.1-1. Note that the dependence of X on the gain K (see (8.2-45)) and the presence of e in the PI have resulted in extra terms being added in (8.2-42).

Determining the Optimal Feedback Gain

The issues in finding the optimal output-feedback gain K in the tracker problem of Table 8.2-1 are the same as those discussed in connection with the regulator problem of Table 8.1-1: choice of Q to satisfy detectability, choice of solution technique, finding an initial stabilizing gain, and iterative design by tuning Q and R. We emphasize that there are only a few design parameters in our approach, namely r and v (since we can generally select $R = rI$, $V = vI$). Thus, it is not difficult or time-consuming to come up with good designs. Much of the simplicity of our approach derives from the fact that Q in the PI is equal to $H^{\mathsf{T}}H$, which is known. Let us now illustrate the servodesign procedure by an example.

TABLE 8.2-1   LQ Tracker with Output Feedback

System model:

$$\dot{x} = Ax + Bu + Gr$$
$$y = Cx + Fr$$
$$z = Hx$$

Control:

$$u = -Ky$$

Performance index:

$$J = \frac{1}{2} \int_0^\infty \left(\tilde{x}^T Q \tilde{x} + \tilde{u}^T R \tilde{u} \right) dt + \frac{1}{2} \bar{e}^T V \bar{e}, \ \text{with} \ Q = H^T H$$

Optimal output feedback gain:

$$0 = \frac{\partial H}{\partial S} = A_c^T P + P A_c + Q + C^T K^T R K C \tag{8.2-40}$$

$$0 = \frac{\partial H}{\partial P} = A_c S + S A_c^T + X \tag{8.2-41}$$

$$0 = \frac{1}{2} \frac{\partial H}{\partial K} = RKCSC^T - B^T PSC^T + B^T A_c^{-T}(P + H^T VH)\bar{x}\bar{y}^T \tag{8.2-42}$$

with r a unit step of magnitude r_0 and

$$\bar{x} = -A_c^{-1} B_c r_0 \tag{8.2-43}$$

$$\bar{y} = C\bar{x} + F r_0 \tag{8.2-44}$$

$$X = \bar{x}\bar{x}^T = A_c^{-1} B_c r_0 r_0^T B_c^T A_c^{-T}, \tag{8.2-45}$$

where

$$A_c = A - BKC, \ B_c = G - BKF$$

Optimal cost:

$$J = \tfrac{1}{2}\text{tr}(PX) + \tfrac{1}{2}\bar{e}^T V \bar{e}$$

Example 8.2-1.   Normal Acceleration CAS

In this example, we show that, using the LQ design equations in Table 8.2-1, we can close all the loops simultaneously. Thus, the design procedure is more straightforward. We also demonstrate that using LQ design, the *algorithm automatically selects the zero of the compensator for optimal performance*.

a. Control System Structure

The normal acceleration control system is shown in Fig. 8.2-3, where r is a reference step input in g's and $u(t)$ is the elevator actuator voltage. An integrator has been added in the feedforward path to achieve zero steady-state error. The performance output that should track the reference command r is $z = n_z$, so that the tracking error is $e = r - n_z$. The

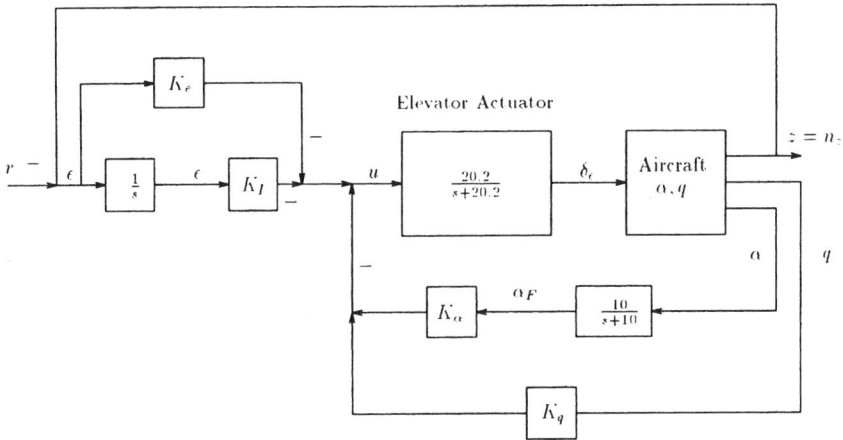

FIGURE 8.2-3   G-command system.

state and measured output are

$$
x = \begin{bmatrix} \alpha \\ q \\ \delta_e \\ \alpha_F \\ \varepsilon \end{bmatrix}, \quad y = \begin{bmatrix} \alpha_F \\ q \\ e \\ \varepsilon \end{bmatrix}, \tag{1}
$$

with $\varepsilon(t)$ the integrator output and α_F the filtered measurements of angle of attack.

The linearized F-16 dynamics about the nominal flight condition are augmented to include the elevator actuator, angle-of-attack filter, and compensator dynamics. The result is

$$
\dot{x} = Ax + Bu + Gr \tag{2}
$$
$$
y = Cx + Fr \tag{3}
$$
$$
z = Hx, \tag{4}
$$

with

$$
A = \begin{bmatrix} -1.01887 & 0.90506 & -0.00215 & 0 & 0 \\ 0.82225 & -1.07741 & -0.17555 & 0 & 0 \\ 0 & 0 & -20.2 & 0 & 0 \\ 10 & 0 & 0 & -10 & 0 \\ -16.26 & -0.9788 & 0.04852 & 0 & 0 \end{bmatrix}, \tag{5a}
$$

$$
B = \begin{bmatrix} 0 \\ 0 \\ 20.2 \\ 0 \\ 0 \end{bmatrix}, \quad G = \begin{bmatrix} 0 \\ 0 \\ 0 \\ 0 \\ 1 \end{bmatrix} \tag{5b}
$$

$$
C = \begin{bmatrix} 0 & 0 & 57.2958 & 0 & 0 \\ 0 & 57.2958 & 0 & 0 & 0 \\ -16.26 & -0.9788 & 0.04852 & 0 & 0 \\ 0 & 0 & 0 & 0 & 1 \end{bmatrix}, \quad F = \begin{bmatrix} 0 \\ 0 \\ 0 \\ 1 \\ 0 \end{bmatrix} \tag{5c}
$$

$$
H = [16.26 \quad 0.9788 \quad -0.04852 \quad 0 \quad 0]. \tag{5d}
$$

The factor of 57.2958 is added to convert angles from radians to degrees. The control input is

$$
u = -Ky = -[k_\alpha \quad k_q \quad k_e \quad k_I]y = -k_\alpha \alpha_F - k_q q - k_e e - k_I \varepsilon. \tag{6}
$$

It is desired to select the four control gains to guarantee a good response to a step command r. Note that k_α and k_q are feedback gains, while k_e and k_I are feedforward gains.

The proportional-plus-integral compensator is given by

$$
k_e + \frac{k_I}{s} = k_e \frac{s + k_I/k_e}{s}, \tag{7}
$$

which has a zero at $s = k_I/k_e$. Since the LQ design algorithm will select all four control gains, it will automatically select the optimal location for the compensator zero.

b. Performance Index and Determination of the Control Gains

Due to the integrator, the system is of type 1. Therefore, the steady-state error $\bar{e}$ is automatically equal to zero. A natural PI thus seems to be

$$
J = \frac{1}{2} \int_0^\infty (\tilde{e}^2 + \rho \tilde{u}^2)\, dt, \tag{8}
$$

with ρ a scalar weighting parameter. Since $\tilde{e} = H\tilde{x}$, this corresponds to the PI in Table 8.2-1 with

$$
Q = H^T H = \begin{bmatrix} 264.3876 & 15.9153 & -0.7889 & 0 & 0 \\ 15.9153 & 0.9580 & -0.0475 & 0 & 0 \\ -0.7889 & -0.0475 & 0.0024 & 0 & 0 \\ 0 & 0 & 0 & 0 & 0 \\ 0 & 0 & 0 & 0 & 0 \end{bmatrix}. \tag{9}
$$

This is, unfortunately, not a suitable Q matrix since (H, A) is not observable in open loop. Indeed, according to Fig. 8.2-3 observing the first two states α and q can never give information about ε in the open-loop configuration (where the control gains are zero). Thus, the integrator state is unobservable in the PI. Since the integrator pole is at $s = 0$, (H, A) is undetectable (unstable unobservable pole), so that any design based on (9) would, in fact, yield a value for the integral gain of $k_I = 0$. To correct the observability problem here let us select

$$Q = H^T H = \begin{bmatrix} 264.3876 & 15.9153 & -0.7889 & 0 & 0 \\ 15.9153 & 0.9580 & -0.0475 & 0 & 0 \\ -0.7889 & -0.0475 & 0.0024 & 0 & 0 \\ 0 & 0 & 0 & 0 & 0 \\ 0 & 0 & 0 & 0 & 1 \end{bmatrix}, \qquad (10)$$

where we include a weighting on $\varepsilon(t)$ to make it observable in the PI.

Now, we selected $\rho = 1$ and solved the design equations in Table 8.2-1 for the optimal control gain K. For this Q and ρ the feedback matrix was

$$K = [0.0005 \quad -0.1455 \quad 1.1945 \quad 1.0000] \qquad (11)$$

and the closed-loop poles were

$$s = -1.24 \pm j0.79$$
$$-1.28, \quad -10.00 \quad -20.28. \qquad (12)$$

These yield a system that is not fast enough; the complex pair is also unsuitable in terms of flying qualities requirements.

After repeating the design using several different Q and ρ, we decided on

$$Q = H^T H = \begin{bmatrix} 264.3876 & 15.9153 & -0.7889 & 0 & 0 \\ 15.9153 & 0.9580 & -0.0475 & 0 & 0 \\ -0.7889 & -0.0475 & 0.0024 & 0 & 0 \\ 0 & 0 & 0 & 0 & 0 \\ 0 & 0 & 0 & 0 & 100 \end{bmatrix}, \qquad (13)$$

$\rho = 0.01$. The decreased control weighting ρ has the effect of allowing larger control effort and so speeding up the response. The increased weighting on the integrator output $\varepsilon(t)$ has the effect of forcing n_z to its final value of r more quickly, hence also speeding up the response. The increased weighting on the second state component q has the effect of regulating excursions in $\tilde{q}(t)$ closer to zero, and hence of providing increased damping.

With this Q and ρ the control matrix was

$$K = [-0.0075 \quad -1.0504 \quad 25.6504 \quad 100.0000] \qquad (14)$$

and the closed-loop poles were at

$$s = -2.89 \pm j3.76$$
$$-16.47 \pm j3.76$$
$$-10. \qquad (15)$$

The closed-loop step response is shown in Fig. 8.2-4; it is fairly fast with an overshoot of 6%. Note the slight delay due to the nonminimum-phase zero. Further tuning of the

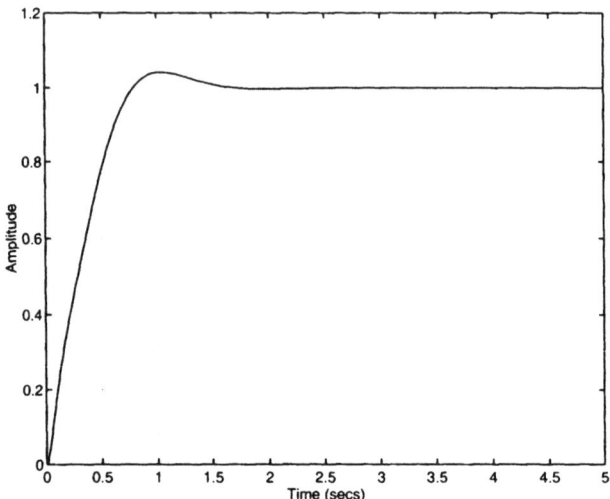

FIGURE 8.2-4   Normal acceleration step response.

elements of Q and R could provide less overshoot, a faster response, and a smaller gain for the angle-of-attack feedback.

According to (7), the compensator zero has been placed by the LQ algorithm at

$$s = -k_I/k_e = -4.06. \tag{16}$$

c. Discussion

We can now emphasize an important aspect of modern LQ design. As long as $Q \geq 0$, $R > 0$, and $(\sqrt{Q}, A)$ is observable, the closed-loop system designed using Table 8.2-1 is generally stable. Thus, the LQ theory has allowed us to tie the control system design to some design parameters that may be tuned to obtain acceptable behavior–namely, the elements of weighting matrices Q and R. If the optimal control gain K does not result in suitable performance in terms of time responses and closed-loop poles, the elements of Q and R can be changed and the design repeated. The importance of this is that for admissible Q and R, closed-loop stability is guaranteed. A disadvantage of the design equations in Table 8.2-1 is the need to try different Q and R until suitable performance is obtained, as well as the need for (H, A) to be observable.

Another point needs to be made. Using the control (6) in (2) and using (3), yields the closed-loop plant

$$\dot{x} = (A - BKC)x + (G - BKF)r, \tag{17}$$

whence the closed-loop transfer function from $r(t)$ to $z(t)$ is

$$H(s) = H(sI - (A - BKC))^{-1}(G - BKF). \tag{18}$$

Note that the transfer function numerator depends on the optimal gain K. That is, this scheme uses optimal positioning of both the poles and zeros to attain step-response shaping.

d. Selection of Initial Stabilizing Gain

To initialize the algorithm that determines the optimal K by solving the design equations in Table 8.2-1, it is necessary to find an initial gain that stabilizes the system. In this example, we simply selected gains with signs corresponding to the static loop sensitivity of the individual transfer functions, since this corresponds to negative feedback. The static loop sensitivities from u to α and from u to q are negative, so positive gains were chosen for these loops (note $(A - BKC)$). The initial gain used was

$$K = [1 \quad 1 \quad -1 \quad 1].\tag{19}$$

■

8.3   TRACKING BY REGULATOR REDESIGN

In this section we discuss an alternative tracker design technique that amounts to first designing a regulator and then adding some feedforward terms to guarantee tracking behavior. This technique does not have the advantages of the direct design approach of the previous section. There, we were able to

1. Select the form of the compensator, including a unity outer loop to allow feedforward of the error.
2. Simplify the design stage by using only a few design parameters in the PI.

However, the approach to be presented here is simple to understand and may be quite useful in some applications. It will also give us some more insight on the tracking problem.

Let us suppose that the plant-plus-compensator in Fig. 8.2-1 is described, using the technique in Section 8.2, as

$$\dot{x} = Ax + Bu + Er \tag{8.3-1}$$

$$y = Cx + Fr \tag{8.3-2}$$

$$z = Hx, \tag{8.3-3}$$

where $y(t)$ is the measurable output available for feedback and the performance output $z(t)$ is required to track the reference input $r(t)$. The tracking error is

$$e = r - z. \tag{8.3-4}$$

Thus, this augmented description contains the dynamics of both the plant and the compensator. It is desired to select the control input $u(t)$ so that the tracking error goes to zero.

Deviation System

For perfect tracking, there must exist an ideal plant state x^* and an ideal plant input u^* such that

$$\dot{x}^* = Ax^* + Bu^* + Er \tag{8.3-5}$$

$$y^* = Cx^* + Fr \qquad (8.3\text{-}6)$$

$$z^* = Hx^* = r. \qquad (8.3\text{-}7)$$

If this is not so, then we cannot have tracking with zero error. See O'Brien and Broussard (1978). What this assumption means is that there is indeed a control input $u^*(t)$ that results in a performance output $z^*(t)$ equal to the desired $r(t)$.

Defining the state, control, and output deviations as

$$\tilde{x} = x - x^*, \quad \tilde{u} = u - u^*, \quad \tilde{y} = y - y^*, \quad \tilde{z} = z - z^*, \qquad (8.3\text{-}8)$$

we may subtract (8.3-5)–(8.3-7) from (8.3-1)–(8.3-3) to obtain the dynamics of the "deviation system" given by

$$\dot{\tilde{x}} = A\tilde{x} + B\tilde{u} \qquad (8.3\text{-}9)$$

$$\tilde{y} = C\tilde{x} \qquad (8.3\text{-}10)$$

$$\tilde{z} = H\tilde{x} = -e. \qquad (8.3\text{-}11)$$

Thus, to regulate the tracking error to zero we may simply design a regulator to control the state of the deviation system to zero. To do this, it is only necessary to select a reasonable PI that weights $\tilde{x}$ and $\tilde{u}$, and then use the design equations in Table 8.1-1, not the more complicated development relating to Table 8.2-1.

Let us note, however, that now the usual restrictions on the PI weights Q and R in Table 8.1-1 apply. That is, $(\sqrt{Q}, A)$ should be observable. What this means is that we will generally be faced with selecting too many design parameters (i.e., the elements of Q and R). Thus, an important advantage of using the approach in Section 8.2 is lost.

Regulator Redesign Adding Feedforward Terms

Suppose we have obtained output feedback gains K that are optimal with respect to (8.3-9)–(8.3-11). Then

$$\tilde{u} = -K\tilde{y}, \qquad (8.3\text{-}12)$$

so that the required control for the plant is

$$u = u^* + \tilde{u} = u^* + K\tilde{y} - Ky. \qquad (8.3\text{-}13)$$

That is, the resulting control for the servo or tracker problem is the optimal regulator feedback control $-Ky$ plus some feedforward terms that are required to guarantee perfect tracking.

We should emphasize that, while the control (8.3-12), designed for the deviation system, is optimal, the control (8.3-13) is not an optimal solution to the tracker problem for the original plant. However, the closed-loop response resulting from (8.3-13) may be satisfactory for many applications; hence, this is not too severe a drawback.

It is now necessary to determine the ideal plant control u^* and output y^* in order to complete the design of the servo control law (8.3-13). To accomplish this, take the Laplace transform of the ideal plant to obtain $R = Z^* = H(sI - A)^{-1}[BU^* + ER]$, or

$$H(sI - A)^{-1}BU^*(s) = [I - H(sI - A)^{-1}E]R(s). \qquad (8.3\text{-}14)$$

Define the transfer function from $u(t)$ to $z(t)$ as

$$H(s) = H(sI - A)^{-1}B. \qquad (8.3\text{-}15)$$

There exists a solution $U^*(s)$ to (8.3-14) for all $R(s)$ if and only if $H(s)$ has full row rank. Thus, the number of control inputs should be at least equal to the number of performance outputs. This is an important and fundamental restriction on the tracking problem.

Let us assume that the number of control inputs is equal to the number of performance outputs so that $H(s)$ is square. If in addition $H(s)$ is nonsingular, then

$$U^*(s) = H^{-1}(s)[I - H(sI - A)^{-1}E]R(s) \qquad (8.3\text{-}16)$$

$$Y^*(s) = C(sI - A)^{-1}[BU*(s) + ER(s)] + FR(s). \qquad (8.3\text{-}17)$$

Using these values of $u^*(t)$ and $y^*(t)$ in (8.3-13) yields the tracker control law.

Tracking a Unit Step

If $r(t)$ is a unit step, then the feedforward terms simplify and we can gain more intuition on the servo control problem. In this case, the ideal responses are nothing but the steady-state responses since $\dot{x}^* = 0$. See Kwakernaak and Sivan (1972). Substituting the control (8.3-13) into (8.3-1) yields

$$\dot{x} = A_c x + Er + B(u^* + Ky^*) \qquad (8.3\text{-}18)$$

$$z = Hx, \qquad (8.3\text{-}19)$$

where

$$Ac = A - BKC. \qquad (8.3\text{-}20)$$

Noting that x^* is constant, we see that, at steady state

$$0 = A_c x^* + Er + B(u^* + Ky^*). \qquad (8.3\text{-}21)$$

Thus, if $r = z^*$ as desired,

$$r = Hx^* = -HA_c^{-1}[B(u^* + Ky^*) + Er], \qquad (8.3\text{-}22)$$

since A_c is stable, and, hence, invertible, in any useful design. Therefore,

$$-HA_c^{-1}B(u^* + Ky^*) = [I + HA_c^{-1}E]r. \tag{8.3-23}$$

Let us define the closed-loop transfer function from $u(t)$ to $z(t)$ as

$$H_c(s) = H(sI - (A - BKC))^{-1}B. \tag{8.3-24}$$

Then, (8.3-23) may be solved for the feedforward terms $(u^* + Ky^*)$ if and only if the closed-loop DC gain $H_c(0)$ is invertible. (Note that $H_c(0) = -HA_c^{-1}B$.) In that event

$$(u^* + Ky^*) = H_c^{-1}(0)[I + HA_c^{-1}E]r, \tag{8.3-25}$$

and the servo control (8.3-13) is given by

$$u = -Ky + H_c^{-1}(0)[I + HA_c^{-1}E]r. \tag{8.3-26}$$

The second term is a feedforward term added to achieve the correct steady-state value of $z(t)$. Thus, the servo control is equal to the optimal regulator feedback control $-Ky$ plus a term involving the inverse of the closed-loop DC gain. Equation (8.3-26) is the fundamental design equation of this section. The feedback does not change the system zeros. Consequently, the zeros of $H(s)$ and those of $H_c(s)$ are the same. Thus, $H_c(0)$ is invertible if and only if the plant has no system zeros at $s = 0$. This is the condition for perfect tracking of a unit step command, which has a pole at $s = 0$. It makes sense, since the DC behavior is related to the steady-state response.

Example 8.3-1.   Tracking by Regulator Design

Consider the plant

$$\dot{x} = \begin{bmatrix} -\alpha & 1 \\ 0 & 0 \end{bmatrix} \tag{1}$$

$$z = [1 \quad 0]x = Hx. \tag{2}$$

It is desired to use only x_1 for feedback purposes so that

$$y = [1 \quad 0]x = Cx. \tag{3}$$

The control objective is to select $u(t)$ to make the tracking error

$$e = r - z \tag{4}$$

go to zero. We assume that $r(t)$ is an arbitrary command input, not necessarily the unit step. To achieve this goal, we can select any reasonable PI that weights $\tilde{x}$ and $\tilde{u}$, and then use the equations in Table 8.1-1 to obtain the optimal LQ regulator gain K in the deviation system control law

$$\tilde{u} = -K\tilde{y}. \tag{5}$$

To convert this regulator to a tracker, we can use equations (8.3-16) and (8.3-17) to write (verify!)

$$u^*(t) = \ddot{r}(t) + \alpha \dot{r}(t) \tag{6}$$

$$y^*(t) = r(t). \tag{7}$$

Therefore, the control that ensures tracking of $r(t)$ by $z(t)$ is

$$u = \ddot{r} + \alpha \dot{r} + Kr - Ky. \tag{8}$$

An implementation of this control scheme appears in Fig. 8.3-1. Note that, generally, the feedforward term that is added to the feedback control is a linear combination of $r(t)$ and its derivatives. This may not be satisfactory in some applications, since differentiation generally makes noisy signals noisier; consequently, it is usually advisable to avoid it.

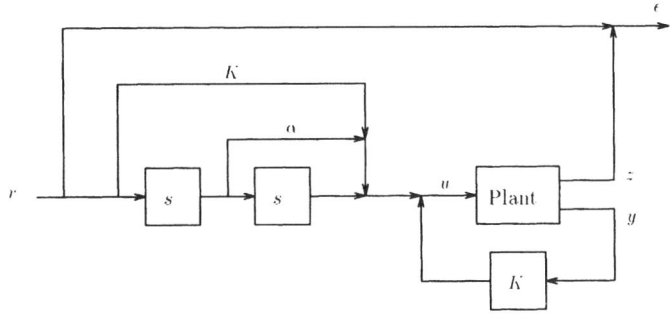

FIGURE 8.3-1   Tracker obtained using regulator redesign.

In our applications, we should like to have control schemes of the form of Fig. 8.2-1, which have a unity feedback outer loop from $z(t)$ and feedforward of the tracking error $e(t)$. This is a structure that takes into account the intuition of classical control theory. As it turns out, in this example $y = z$, so that the control law (8) may be reformulated as

$$u = \ddot{r} + \alpha \dot{r} + Ke. \tag{9}$$

However, in general, $y = z$, so it is not usually possible to formulate the control schemes resulting from the approach of this section using a unity feedback outer loop. ■

8.4   COMMAND-GENERATOR TRACKER

In Section 8.2 we selected a PI compensator to make the loop gain of type 1 in order to obtain zero steady-state tracking error in response to a step command input (i.e., a position command). If the reference input is not a unit step, then a single integrator will no longer guarantee a steady-state error of zero. In Section 8.3 we saw how to convert a regulator into a tracker by adding additional feedforward terms. However, if the reference input $r(t)$ is not constant, the feedforward terms generally contain derivatives of $r(t)$.

In this section we demonstrate the command-generator tracker (CGT) design, which is a powerful design technique that automatically gives the precompensator required to obtain zero steady-state error for a large class of command inputs $r(t)$. See Franklin et al. (1986). In this approach, we shall incorporate a model of the dynamics of $r(t)$ into the control system.

As we shall see, CGT design may be used for tracking and also for disturbance rejection.

Tracking

The plant

$$\dot{x} = Ax + Bu \tag{8.4-1}$$

has measured outputs available for control purposes given by

$$y = Cx, \tag{8.4-2}$$

and its performance output

$$z = Hx \tag{8.4-3}$$

is required to track the reference input $r(t)$.

Command-generator System

Let us suppose that for some initial conditions the reference command satisfies the differential equation

$$r^{(d)} + a_1 r^{(d-1)} + \cdots + a_d r = 0 \tag{8.4-4}$$

for a given degree d and set of coefficients a_i. Most command signals of interest satisfy such an equation. For instance, the unit step of magnitude r_0 satisfies

$$\dot{r} = 0 \tag{8.4-5}$$

with $r(0) = r_0$, while the ramp (velocity command) with slope v_0 satisfies

$$\ddot{r} = 0, \tag{8.4-6}$$

with $r(0) = 0$, $\dot{r}(0) = v_0$.

We may express (8.4-4) in state-variable (observability canonical) form (Kailath 1980). Illustrating for the case of scalar $r(t)$ and $d = 3$, this is

$$\dot{\rho} = \begin{bmatrix} 0 & 1 & 0 \\ 0 & 0 & 1 \\ -a_3 & -a_2 & -a_1 \end{bmatrix} \rho \equiv F\rho$$

$$r = [1 \quad 0 \quad 0]\rho. \tag{8.4-7}$$

Note that in this form the plant matrix is zero except for a superdiagonal of 1's and the bottom row of coefficients. We call (8.4-4)–(8.4-7) the command-generator system.

Let us define the command-generator characteristic polynomial as

$$\Delta(s) = s^d + a_1 s^{d-1} + \cdots + a_d. \tag{8.4-8}$$

Then, denoting d/dt in the time domain by s, we may write (8.4-4) as

$$\Delta(s)r = 0. \tag{8.4-9}$$

To make $z(t)$ follow $r(t)$, define the tracking error

$$e = r - z = r - Hx. \tag{8.4-10}$$

We should like to convert the servo or tracking problem into a regulator problem where the error must be regulated to zero.

Modified System

To accomplish this, write

$$\Delta(s)e = \Delta(s)r - \Delta(s)Hx = -H\xi, \tag{8.4-11}$$

where we have used (8.4-9) and defined the modified plant state vector

$$\xi = \Delta(s)x = x^{(d)} + a_1 x^{(d-1)} + \cdots + a_d x. \tag{8.4-12}$$

Note that (8.4-11) may be written in the observability canonical form

$$\dot{\varepsilon} = F\varepsilon + \begin{bmatrix} 0 \\ -H \end{bmatrix} \xi, \tag{8.4-13}$$

where $\varepsilon(t) = [e \dot{e} \cdots e^{(d-1)}]^{\mathrm{T}}$ is the vector of the error and its first $d-1$ derivatives.

To determine the dynamics of $\xi(t)$, operate on (8.4-1) with $\Delta(s)$ to obtain

$$\dot{\xi} = A\xi + B\mu, \tag{8.4-14}$$

where the modified control input is

$$\mu = \Delta(s)u = u^{(d)} + a_1 u^{(d-1)} + \cdots + a_d u. \tag{8.4-15}$$

Now we may put all the dynamics (8.4-13), (8.4-14) into a single augmented state representation by writing

$$\frac{d}{dt}\begin{bmatrix} \varepsilon \\ \xi \end{bmatrix} = \left[\begin{array}{c|cc} & 0 & \\ \hline F & -H \\ 0 & \Lambda \end{array}\right]\begin{bmatrix} \varepsilon \\ \xi \end{bmatrix} + \begin{bmatrix} 0 \\ B \end{bmatrix}\mu. \tag{8.4-16}$$

Using this system, we may now perform a LQ regulator design, since if its state goes to zero, then the tracking error $e(t)$ vanishes. For this design, we take the outputs available for feedback as

$$v = \begin{bmatrix} I & 0 \\ 0 & C \end{bmatrix}\begin{bmatrix} \varepsilon \\ \xi \end{bmatrix}. \tag{8.4-17}$$

We can select any reasonable PI that weights $[\varepsilon^{\mathrm{T}} \quad \xi^{\mathrm{T}}]^{\mathrm{T}}$ and μ and use the design equations in Table 8.1-1 to obtain optimal feedback gains so that

$$\mu = -[K_\varepsilon \quad K_y]\begin{bmatrix} \varepsilon \\ C\xi \end{bmatrix} \tag{8.4-18}$$

or

$$\Delta(s)u = -K_\varepsilon\varepsilon - K_yC\Delta(s)x. \tag{8.4-19}$$

Servo Compensator

To determine the control input $u(t)$ for the original system, write this as

$$\Delta(s)(u + K_yy) = -K_\varepsilon\varepsilon - [K_d \cdots K_2K_1]\begin{bmatrix} e \\ \dot{e} \\ \vdots \\ e^{d-1} \end{bmatrix}. \tag{8.4-20}$$

Thus, we obtain the transfer function

$$\frac{u + K_vy}{e} = \frac{K_1s^{d-1} + \cdots + K_{d-1}s + K_yd}{s^d + a_1s^{d-1} + \cdots + a_d}, \tag{8.4-21}$$

which may be implemented in reachability canonical form to obtain the servo control structure shown in Fig. 8.4-1. If $d = 1$ so that $r(t)$ is a ramp, it yields two integrators in the feedforward compensator, which results in a type 2 system and gives zero steady-state error.

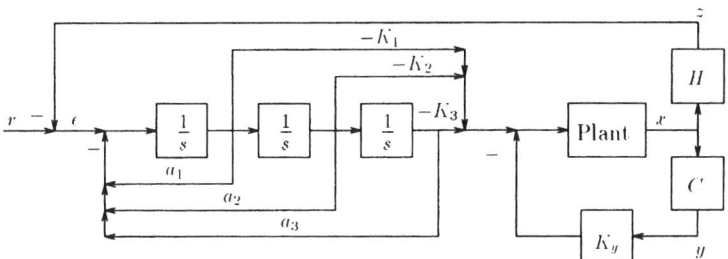

FIGURE 8.4-1 Command-generator tracker for $d = 3$.

Note that the CGT is a servo controller that has a sensible structure with no derivatives of $r(t)$. Note further that its dynamics reflect the dynamics of the reference input. In fact, the controller is said to contain an internal model of the reference input generator.

It should be emphasized that this technique is extremely direct to apply. Indeed, given the command-generator polynomial $\Delta(s)$, the system (8.4-16), (8.4-17) may be written down immediately, and the equations in Table 8.1-1 used to select the feedback gains.

It remains to say that we may place all the poles of the modified system (8.4-16) using full state feedback (i.e., $C = I$) if and only if the system is reachable. That is, the reachability matrix should have full rank. This should logically be a necessary condition if we want to accomplish the more difficult problem of pole placement using reduced-state, that is output, feedback with $C \neq I$. It is not too difficult to show by a straightforward determination of the reachability matrix of (8.4-16) that the modified system is reachable if

1. The original system (8.4-1) is reachable, and
2. The open-loop transfer function from $u(t)$ to $z(t)$ given by

$$H(s) = H(sI - A)^{-1}B \tag{8.4-22}$$

has no zeros at the roots of $\Delta(s) = 0$.

The second condition represents a restriction on the sorts of command inputs that a given plant may follow with zero steady-state error. For instance, to track a unit step the plant can have no system zeros at $s = 0$.

A word on the command-generator assumption (8.4-4) is in order. For control systems design it is not necessary to determine the coefficients a_i that describe the actual reference command, which may be a complicated function of time (e.g., the pilot's command in an aircraft design example). Instead, the performance objectives should be taken into account to select $\Delta(s)$. For instance, if it is desired for the plant to follow a position command, then for design purposes we may select the command generator $\dot{r} = 0$. On the other hand, if the plant should follow a rate (velocity) command, we may select $\ddot{r} = 0$. Then, when the actual command input is applied (which may be neither a unit step nor a unit ramp) the system will exhibit the appropriate closed-loop behavior.

Example 8.4-1.   Track Following for a Disk Drive Head-positioning System

The head-positioning mechanism for a disk drive can be either rotary or of the linear voice coil actuator type. The latter is described by

$$\ddot{y} = \frac{k}{m}u, \tag{1}$$

where m is the mass of the coil and carriage assembly, $k(nt/V)$ is the product of the motor force constant and power amplifier gain, $y(t)$ is the head position, and control input $u(t)$ is motor voltage (Bell et al. 1984). By absorbing the constant k/m into the definition of the input, we may write the state-variable description

$$\dot{x} = \begin{bmatrix} 0 & 1 \\ 0 & 0 \end{bmatrix} x + \begin{bmatrix} 0 \\ 1 \end{bmatrix} u = Ax + Bu, \tag{2}$$

with $x = [y \ v]^T$ and $v(t)$ the head velocity, where $u(t)$ is now in units of acceleration. This is just the Newton's law system we have already discussed in several examples.

There are two control problems associated with the head-positioning mechanism. One is the track access or seek problem. Here, the head must be driven from an initial position to a certain track in minimum time. The second problem is that of track following, where a servo control system must be designed to hold the head above a specified track while reading from or writing to the disk. In this example we should like to deal with the track-following problem (Franklin et al. 1986).

Let the rotational velocity of the disk be ω_0, which is usually between 2500–4000 rpm depending on the application. Because the disk is not quite centered, the tracks defined as circles on the disk actually trace out ellipses with respect to a stationary head position. The spindle design keeps the once-around runout component of the disk motion within 300 microinches. However, for reading and writing it is desired that the head follow the track with an error of less than 75 microinches.

To design a servo control system under these circumstances, we must account for the fact that the reference input $r(t)$ (i.e., the desired radius from the center of the spindle) satisfies the equation

$$\ddot{r} = \omega_0^2 r = 0. \tag{3}$$

Therefore, the command-generator polynomial is

$$\Delta(s) = s^2 + \omega_0^2. \tag{4}$$

It is desired for the head position $y(t)$ to follow $r(t)$, so that the performance output is

$$z = [1 \ 0]x = Hx. \tag{5}$$

The augmented system (8.4-16) may now be written down as

$$\frac{d}{dt}\begin{bmatrix} e \\ \dot{e} \\ \overline{\xi} \end{bmatrix} = \begin{bmatrix} 0 & 1 & 0 & 0 \\ -\omega_0^2 & 0 & -1 & 0 \\ 0 & 0 & 0 & 1 \\ 0 & 0 & 0 & 0 \end{bmatrix} \begin{bmatrix} e \\ \dot{e} \\ \overline{\xi} \end{bmatrix} + \begin{bmatrix} 0 \\ 0 \\ 0 \\ 1 \end{bmatrix} \mu. \tag{6}$$

Now, the LQ regulator design equations from Table 8.1-1 may be used to obtain the optimal feedback gains for this system so that

$$
\mu = -[k_i \quad k_{\dot{e}} \quad k_y \quad k_v] \begin{bmatrix} e \\ \dot{e} \\ \bar{\xi} \end{bmatrix}. \tag{7}
$$

For this stage of the controls design we can select the PI in Table 8.1-1 with $Q = qI$, $R = I$, where q is a design parameter, which may be chosen large for good regulation (e.g., $q = 100$).

Finally, the track-following servo controller is given by (8.4-21), which has the form of Fig. 8.4-1 with $d = 2$. The command-generator controller has an oscillator with a frequency of ω_0 as a precompensator. It is called the internal model of the reference generator (Franklin et al. 1986).

It is worth noting that in this situation (7) is a full state-variable feedback. Thus, if we prefer we may use the Riccati-equation approach in Section 3.4 to select the control gains.   ∎

Tracking with Disturbance Rejection

The CGT approach may also be used in the disturbance rejection problem.

Disturbance Generator System

If the system is driven by an unknown disturbance $d(t)$, then (8.4-1) must be modified to read

$$
\dot{x} = Ax + Bu + Dd. \tag{8.4-23}
$$

Suppose $d(t)$ satisfies the differential equation

$$
d^{(q)} + p_1 d^{(q-1)} + \cdots + p_q d = 0 \tag{8.4-24}
$$

for some degree q and known coefficients p_i. In illustration, d could be a constant unknown disturbance so that $d = 0$, or a sinusoidal disturbance that satisfies a differential equation of order two.

Define

$$
\Delta_d(s) = s^q + p_1 s^{q-1} + \cdots + p_q, \tag{8.4-25}
$$

so that

$$
\Delta_d(s)d = 0, \tag{8.4-26}
$$

where s represents d/dt in the time domain. With

$$
\xi \equiv \Delta_d(s)x, \quad \mu \equiv \Delta_d(s)u, \tag{8.4-27}
$$

it follows that (8.4-23) may be written as

$$
\dot{\xi} = A\xi + B\mu, \tag{8.4-28}
$$

which does not involve the disturbance.

Tracking and Disturbance Rejection

Now, suppose that the reference input $r(t)$ satisfies

$$\Delta_r(s)r = 0 \qquad (8.4\text{-}29)$$

for some given $\Delta_r(s)$. Defining

$$\Delta(s) = \Delta_d(s)\Delta_r(s), \qquad (8.4\text{-}30)$$

we may use the CGT technique to derive a controller that results in tracking of $r(t)$ by the performance output

$$z = Hx \qquad (8.4\text{-}31)$$

in the presence of the disturbance $d(t)$. Indeed, if the measured output is $y = Cx$, the required controller is given exactly by (8.4-21) with the modified $\Delta(s)$ of (8.4-30). Referring to the discussion following (8.4-21), we see that perfect disturbance rejection may be achieved only if the system has no zeros at the poles of the disturbance.

In point of fact, we need not select the polynomial $\Delta(s)$ given by (8.4-30) if $\Delta_d(s)$ and $\Delta_r(s)$ have common factors. Instead, we should select the least common multiple of $\Delta_d(s)$ and $\Lambda_r(s)$.

8.5 EXPLICIT MODEL-FOLLOWING DESIGN

Here, we consider the problem of controlling a plant so that it has a response like that of a prescribed model with desirable behavior. The model has desirable qualities in terms of speed of response, percent overshoot, robustness, and so on. In aircraft control, for instance, a series of performance models for different situations is tabulated in Mil. Spec. (1797). Thus, aircraft controls design often has the objective of making the aircraft behave like the specified model.

There are two fundamentally different sorts of model-following control, "explicit" and "implicit," which result in controllers of different structure (Armstrong 1980, Kreindler and Rothschild 1976). The latter, however, yields an inconvenient form of controller for the servo design or tracking problem; specifically, it usually requires derivatives of the performance output (Stevens and Lewis 1992). Therefore, in this section we consider only explicit model following. Implicit model following is important for a different application, namely, selecting the performance index weighting matrices.

Regulator with Model Following

First, we consider the regulator problem, where the objective is to drive the plant state to zero. Then, we treat the more difficult tracker or servo problem, where the plant is to follow a reference command with behavior like the prescribed model.

Let the plant be described in state-variable form by

$$\dot{x} = Ax + Bu \tag{8.5-1}$$

$$y = Cx \tag{8.5-2}$$

$$z = Hx \tag{8.5-3}$$

with state $x(t) \in R^n$ and control input $u(t) \in R^m$. The measured output $y(t)$ is available for feedback purposes. A model is prescribed with dynamics

$$\dot{\tilde{x}} = \tilde{A}\tilde{x} \tag{8.5-4}$$

$$\tilde{Z} = \tilde{H}\tilde{x}, \tag{8.5-5}$$

where the model matrix $\tilde{A}$ reflects a system with desirable handling qualities, such as speed of response, overshoot, and so on. The model states available for feedback purposes are given by

$$\tilde{y} = \tilde{C}\tilde{x}. \tag{8.5-6}$$

Model quantities shall be denoted by underbars or the subscript "m."

It is desired for the plant performance output $z(t)$ to match the model output $z(t)$, for then the plant will exhibit the desirable time response of the model. That is, we should like to make small the model mismatch error

$$e = \tilde{z} - z = \tilde{H}\tilde{x} - Hx. \tag{8.5-7}$$

To achieve this control objective, let us select the performance index

$$J = \frac{1}{2} \int_0^\infty (e^T Q e + u^T R u) \, dt, \tag{8.5-8}$$

with $Q \geq 0$ and $R > 0$.

We can cast this model-matching problem into the form of the regulator problem whose solution appears in Table 8.1-1 as follows. Define the augmented state $x' = [x^T \ \tilde{x}^T]^T$ and the augmented system

$$\dot{x}' = \begin{bmatrix} A & 0 \\ 0 & \tilde{A} \end{bmatrix} x' + \begin{bmatrix} B \\ 0 \end{bmatrix} u \equiv A'x' + B'u \tag{8.5-9}$$

$$y' = \begin{bmatrix} C & 0 \\ 0 & \tilde{C} \end{bmatrix} x' \equiv C'x' \tag{8.5-10}$$

so that

$$e = [-H \quad \tilde{H}]x' \equiv H'x'. \tag{8.5-11}$$

Then, the PI (8.5-8) may be written

$$J = \frac{1}{2} \int_0^\infty ((x')^{\mathrm{T}} Q' x' + u^{\mathrm{T}} R u) \, dt, \tag{8.5-12}$$

with

$$Q' = \begin{bmatrix} H^{\mathrm{T}} Q H & -H^{\mathrm{T}} Q \tilde{H} \\ -\tilde{H}^{\mathrm{T}} Q H & \tilde{H}^{\mathrm{T}} Q \tilde{H} \end{bmatrix}. \tag{8.5-13}$$

At this point, it is clear that the design equations of Table 8.1-1 apply if the primed quantities A', B', C', Q', are used. The conditions for convergence of the algorithm in Table 8.1-2 require that (A', B', C') be output stabilizable and $(\sqrt{Q'}, A')$ be detectable. Since the model matrix A is certainly stable, the block diagonal form of A' and C' shows that output stabilizability of the plant (A, B, C) is required. The second condition requires detectability of $(\sqrt{Q}H, A)$.

It should be noted that the detectability condition on $(\sqrt{Q}H, A)$ may be avoided by including time weighting of the form $t^k (x')^{\mathrm{T}} Q' x'$ in the PI. Then, the control gains may be computed by using a simplified version of the equations in Table 8.2-1. Specifically, since the equations there deal with the tracker problem, we can take $X = I$ and $\bar{y} = 0$ to solve the regulator design problem. This corresponds to minimizing not J but its expected value. See the discussion in Sections 8.1 and 8.2.

The form of the resulting control law is quite interesting. Indeed, the optimal feedback is of the form

$$u = -K' y' \equiv -[K_p \quad K_m] y' = -K_p y - K_m \tilde{y}. \tag{8.5-14}$$

Thus, not only the plant output but also the model output is required. That is, the model acts as a compensator to drive the plant states to zero in such a fashion that the performance output $z(t)$ follows the model output $z(t)$.

Tracker with Model Following

Unfortunately, while the model-following regulator problem has a direct solution that is easy to obtain, the model-following tracker problem is not easy. In this situation, we should like the plant (8.5-1)–(8.5-3) to behave like the model

$$\dot{\tilde{x}} = \tilde{A} \tilde{x} + \tilde{B} r, \tag{8.5-15}$$

$$\tilde{z} = \tilde{H} \tilde{x}, \tag{8.5-16}$$

which is driven by the reference input $r(t)$. The approach above yields

$$\dot{x}' = \begin{bmatrix} A & 0 \\ 0 & \tilde{A} \end{bmatrix} x' + \begin{bmatrix} B \\ 0 \end{bmatrix} u + \begin{bmatrix} 0 \\ \tilde{B} \end{bmatrix} r \equiv A' x' + B' u + G' r, \tag{8.5-17}$$

which contains a term in $r(t)$. For such systems, the determination of the optimal feedback gains is not straightforward (see Chapter 4).

Therefore, let us approach the problem by using the command-generator technique of Section 8.4. Thus, suppose for some initial conditions the reference command satisfies the differential equation

$$r^{(d)} + a_1 r^{(d-1)} + \cdots + a_d r = 0 \qquad (8.5\text{-}18)$$

for a given degree d and set of coefficients a_i. Define the command-generator characteristic polynomial as

$$\Delta(s) = s^d + a_1 s^{d-1} + \cdots + a_d. \qquad (8.5\text{-}19)$$

Then, denoting d/dt in the time domain by s, we may write

$$\Delta(s)r = 0. \qquad (8.5\text{-}20)$$

Multiplying the augmented dynamics (8.5-17) by $\Delta(s)$ results in

$$\dot{\xi} = A'\xi + B'\mu, \qquad (8.5\text{-}21)$$

where the modified state and control input are

$$\xi = \Delta(s)x' = (x')^{(d)} + a_1(x')^{(d-1)} + \cdots + a_d x', \qquad (8.5\text{-}22)$$

$$\mu = \Delta(s)u = u^{(d)} + a_1 u^{(d-1)} + \cdots + a_d u. \qquad (8.5\text{-}23)$$

We note the important point that $r(t)$ has vanished by virtue of (8.5-20). Let us denote

$$\xi = \begin{bmatrix} \xi_p \\ \xi_m \end{bmatrix}, \qquad (8.5\text{-}24)$$

with ξ_p the modified plant state and ξ_m the modified model state. Applying $\Delta(s)$ to the model mismatch error (8.5-7) results in

$$\Delta(s)e = [-H \quad \tilde{H}] = H'\xi. \qquad (8.5\text{-}25)$$

This may be expressed in terms of state-variables using the observability canonical form (Kailath 1980), which for scalar $e(t)$ and $d = 3$ is

$$\dot{\varepsilon} = \begin{bmatrix} 0 & 1 & 0 \\ 0 & 0 & 1 \\ -a_3 & -a_2 & -a_1 \end{bmatrix} \varepsilon + \begin{bmatrix} 0 \\ H' \end{bmatrix} \xi \equiv F\varepsilon + \begin{bmatrix} 0 \\ H' \end{bmatrix} \xi \qquad (8.5\text{-}26)$$

$$e = \begin{bmatrix} 1 & 0 & 0 \end{bmatrix} \varepsilon, \qquad (8.5\text{-}27)$$

where $\varepsilon(t) = [e \quad \dot{e} \cdots e^{(d-1)}]^{\mathrm{T}}$ is the vector of the error and its first $d - 1$ derivatives.

Collecting all the dynamics (8.5-21), (8.5-26) into one system yields

$$\frac{d}{dt}\begin{bmatrix}\varepsilon\\\xi\end{bmatrix}=\left[\begin{array}{c|c}F & H'\\\hline 0 & A'\end{array}\right]\begin{bmatrix}\varepsilon\\\xi\end{bmatrix}+\begin{bmatrix}0\\B'\end{bmatrix}\mu. \qquad (8.5\text{-}28)$$

Using this system, we may now perform a LQ regulator design, since if its state goes to zero, then the tracking error $e(t)$ vanishes. For this design, we shall take the outputs available for feedback as

$$v=\begin{bmatrix}I & 0 & 0\\0 & C & 0\\0 & 0 & C\end{bmatrix}. \qquad (8.5\text{-}29)$$

To achieve small error without using too much control energy, we may select the PI (8.5-8) (with $u(t)$ replaced by $\mu(t)$). According to (8.5-27), the error is given in terms of the state of (8.5-28) by

$$e=h\begin{bmatrix}\varepsilon\\\xi\end{bmatrix} \qquad (8.5\text{-}30)$$

with $h=[1\ 0\ \cdots\ 0]$ the first row of the identity matrix. Therefore, in the PI we should weight the state of (8.5-28) using

$$Q'=h^{\mathrm{T}}Qh. \qquad (8.5\text{-}31)$$

Since the observability canonical form is always observable, the augmented system (8.5-28) is detectable if the plant (H, A) and the model (H, A) are both detectable.

Now, applying the equations of Table 8.1-1 to the system (8.5-28) with outputs (8.5-29) and PI weights Q' and R yields the optimal control law

$$\mu=-[K_\varepsilon\ \ K_p\ \ K_m]\begin{bmatrix}\varepsilon\\C\xi_p\\\tilde{C}\xi_m\end{bmatrix} \qquad (8.5\text{-}32)$$

or

$$\Delta(s)u=-K_\varepsilon\varepsilon-K_pC\Delta(s)x-K_m\tilde{C}\Delta(s)x. \qquad (8.5\text{-}33)$$

To determine the optimal control input $u(t)$, write this as

$$\Delta(s)(u+K_py+K_m\tilde{y})=-K_\varepsilon\varepsilon\equiv-[K_d\cdots K_2\ \ K_1]\begin{bmatrix}e\\\dot{e}\\\vdots\\e^{(d-1)}\end{bmatrix}. \qquad (8.5\text{-}34)$$

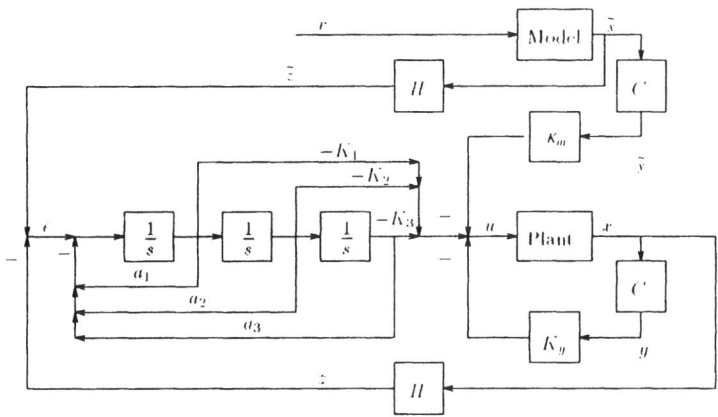

FIGURE 8.5-1   Command-generator tracker for $d = 3$.

Thus, we obtain the transfer function

$$\frac{u + K_p y + K_m \tilde{y}}{e} = \frac{K_1 s^{d-1} + \cdots + K_{d-1}s + K_d}{s^d + a_1 s^{d-1} + \cdots + a_d}, \tag{8.5-35}$$

which may be implemented in reachability canonical form to obtain the control structure shown in Fig. 8.5-1.

The structure of this model-following command-generator tracker is very interesting. It consists of an output feedback Kp, a feedforward compensator that is nothing but the reference model, and an additional feedforward filter in the error channel that guarantees perfect tracking. Note that if $d = 1$ so that $r(t)$ is a unit step, the error filter is a PI controller like that shown in Fig. 8.2-4. If $d = 2$ so that $r(t)$ is a ramp, the error filter consists of two integrators, resulting in a type 2 system that gives zero steady-state error.

It is interesting to note that the augmented state description (8.5-28) is nothing but the state description of Fig. 8.5-1 (see the problems at the end of the chapter). It should be emphasized that this technique is extremely direct to apply. Indeed, given the prescribed model and the command generator polynomial $\Delta(s)$, the system (8.5-28), (8.5-29) may be written down immediately, and the design equations in Table 8.1-1 used to select the feedback gains. See Section 8.4 for some discussion on the reachability of the augmented system.

8.6   OUTPUT FEEDBACK IN GAME THEORY AND DECENTRALIZED CONTROL

This section presents a unified formulation of the decentralized control, Nash game, and Stackelberg game problems in the linear-quadratic cost case. An interesting development is the occurrence in the Stackelberg game problem with output feedback of a "fictitious follower" set of equations that must be solved in the leader's problems. The necessary conditions for solution of each of these problems are put in a form that can be solved by a proven numerical technique. This provides an approach for the solution of the Stackelberg game problem, which also applies in the case of state feedback, where no good solution technique is yet known. Decentralized control and game theory are topics that are apparently unrelated. Decentralized control is an important idea in large-scale systems theory, and game theory is used in various decision-making problems. Essentially, though, both topics deal with the efficient management of information.

This section is concerned with the problem of designing optimal output-feedback controllers for the cases of decentralized control (Sandell 1976), Nash games (Papavassilopoulos and Cruz, Jr. 1979), and Stackelberg games (Medanic 1978). In each case, a linear feedback is assumed, and the goal is the minimization of a quadratic cost index. To provide a unified treatment, we capitalize on the fact that all three cases arise from the same formulation but with different goals.

For decentralized control, the system under consideration is assumed to consist of subsystems that are to be controlled individually, and a single cost for the entire system is to be minimized. In the games problems, two players provide inputs to a system and the goal of each player is to minimize his own cost function. The Nash game deals with the case in which each player chooses his own strategy independently of the other player, and the Stackelberg game involves two players, one of whom knows the other player's strategy and one of whom does not know the other player's strategy.

In all three cases incomplete information is available, and so the necessary conditions are sets of coupled nonlinear matrix equations. Each section concludes with a form of the necessary conditions for optimal solution, which can be solved using the simple iterative technique presented in Table 8.1-2.

Problem Formulation

The decentralized control problem and the games problems may each be formulated in the following fashion, with different goals and interpretations of the matrices for each problem. Given the system

$$\dot{x} = Ax + B_1 u_1 + B_2 u_2 \tag{8.6-1}$$

with $x(0) = x_0$ given, and the outputs

$$y_i = C_i x_i, \quad i = 1, 2, \tag{8.6-2}$$

find linear output feedbacks of form

$$u_i = L_i y_i = -L_i C_i x, \quad i = 1, 2 \tag{8.6-3}$$

to minimize the following performance indices

$$J_i = \frac{1}{2} \int_0^\infty \left(x^T Q_i x + u_i^T R_{ii} u_i + u_j^T R_{ij} u_j \right) dt, \tag{8.6-4}$$

where $i = 1, 2$, $j \neq i$, and $R_{ij} \geq 0$, $R_{ii} > 0$, $Q_i \geq 0$. We assume C_1 and C_2 have full row rank.

Using (8.6-3), the closed-loop system has the form

$$\dot{x} = (A - B_1 L_1 C_1 - B_2 L_2 C_2)x \equiv A_c x, \tag{8.6-5}$$

and the performance criteria become

$$J_i = \frac{1}{2} \int_0^\infty x^{\mathrm{T}} (Q_i + C_i^{\mathrm{T}} L_i^{\mathrm{T}} R_{ii} L_i C_i + C_j^{\mathrm{T}} L_j^{\mathrm{T}} R_{ij} L_j C_j) x \, dt. \tag{8.6-6}$$

The performance indices are given by

$$J_i = \mathrm{tr}(M_i X), \tag{8.6-7}$$

with $X \equiv x_0 x_0^{\mathrm{T}}$, $\mathrm{tr}(\cdot)$ representing the trace of a matrix, and $M_i \geq 0$ the solution of

$$F_i \equiv A_c^{\mathrm{T}} M_i + M_i A_c + Q_i + C_i^{\mathrm{T}} L_i^{\mathrm{T}} R_{ii} L_i C_i + C_j^{\mathrm{T}} L_j^{\mathrm{T}} R_{ij} L_j C_j. \tag{8.6-8}$$

If it is desired to eliminate the dependence on the specific initial state, we may instead minimize the expected value $E(J_i)$ as we did before, in which case X becomes $E(x_0 x_0^{\mathrm{T}})$, the initial mean-square state, in (8.6-7) and in the remainder of the section.

Decentralized Linear Quadratic Regulator

The setting for the decentralized linear quadratic regulator is a system that is composed of two subsystems (Sandell 1976). The dynamics of the subsystems are coupled; however, it is desired to find a control u, for each subsystem that is based only on the output y of that subsystem and that minimizes a common quadratic cost function.

This problem may be easily extended to the case of output feedback. Since the feedback gains are being chosen to minimize a single performance index, we may take $Q_1 = Q_2$, $R_{11} = R_{21}$, $R_{12} = R_{22}$, so that $J_1 = J_2 = J$. Then $F_1 = F_2 = F$ and $M_1 = M_2 = M$. The goal is to pick the feedback gains L_1 and L_2 to minimize J, subject to the constraint (8.6-8).

Define the Hamiltonian function

$$H = \mathrm{tr}(MX) + \mathrm{tr}(FP), \tag{8.6-9}$$

where M, X, and F are defined by (8.6-7) and (8.6-8), and P is an undetermined matrix of multipliers. Then necessary conditions for the minimization of H (and hence (8.6-7) subject to (8.6-8)), are obtained by partial differentiation of H with respect to P, M, L_1, and L_2, respectively

$$0 = F^T = A_c^T M + M A_c + Q$$
$$+ C_1^T L_1^T R_{11} L_1 C_1 + C_2^T L_2^T R_{22} L_2 C_2 \qquad (8.6\text{-}10)$$

$$0 = A_c P + P A_c^T + X \qquad (8.6\text{-}11)$$

$$0 = -B_1^T M P C_1^T + R_{11} L_1 C_1 P C_1^T \qquad (8.6\text{-}12)$$

$$0 = -B_2^T M P C_2^T + R_{22} L_2 C_2 P C_2^T. \qquad (8.6\text{-}13)$$

If we partition the state according to the choice of subsystems and then let $C_1 = [I \ \ 0]$, $C_2 = [0 \ \ I]$, and specialize the input matrices to the forms $[B_1^T \ \ 0]^T$ and $[0 \ \ B_2^T]^T$, then equations (8.6-1), (8.6-2) reduce to

$$\dot{x} = Ax + \begin{bmatrix} B_1 \\ 0 \end{bmatrix} u_1 + \begin{bmatrix} 0 \\ B_2 \end{bmatrix} u_2 \qquad (8.6\text{-}14)$$

$$y_1 = [I \ \ 0]x, \quad y_2 = [0 \ \ I]x. \qquad (8.6\text{-}15)$$

That is the special case where each subsystem knows its own state component completely and has direct control only over its own state component (Sandell 1976).

A convergent algorithm is given in Table 8.1-2 for the solution of the coupled nonlinear equations arising in the linear quadratic regulator with output feedback (Levine and Athans 1970, Moerder and Calise 1985). The algorithm is fairly efficient numerically, is straightforward to program, and has been proven reliable. Equations (8.6-10)–(8.6-13) may be solved by this algorithm by combining (8.6-12) and (8.6-13) in the block form

$$0 = \begin{bmatrix} R_{11} & 0 \\ 0 & R_{22} \end{bmatrix} \begin{bmatrix} L_1 & 0 \\ 0 & L_2 \end{bmatrix} \begin{bmatrix} C_1 P C_1^T & 0 \\ 0 & C_2 P C_2^T \end{bmatrix} - \begin{bmatrix} B_1^T M & 0 \\ 0 & B_2^T M \end{bmatrix} \begin{bmatrix} P C_1^T & 0 \\ 0 & P C_2^T \end{bmatrix}. \qquad (8.6\text{-}16)$$

Now, the algorithm in Table 8.1-2 may be applied to equations (8.6-10), (8.6-11), and (8.6-14).

Although conditions for the existence and uniqueness of solutions, positive definite or otherwise, to (8.6-10)–(8.6-13) are not known, a few comments may be made. In the case of state feedback ($C_i = I$) there exists a unique positive definite solution M to (8.6-10) if $(A, \sqrt{Q})$ is observable and $(A, [B_1 \ B_2])$ is controllable. If $X = x_0 x_0$ with (A_c, x_0) controllable, then there exists a unique positive definite solution P to (8.6-11) if and only if A_c is stable. If $X = E(x_0 x_0^T)$ then what is required instead is (A_c, x_0) controllable.

The coupled equations (8.6-10)–(8.6-13) may have multiple positive definite solutions (Sandell 1976), each one associated with a local minimum of the performance index. Since the iterative algorithm in Table 8.1-2 finds only local minima, the solution may depend on the initial selection of a stabilizing output feedback gain to start the algorithm.

Nash Games

In a Nash game, two players independently try to minimize their own (possibly different) performance objectives. Neither has knowledge of the other's strategy. The problem may be formulated as follows.

Player i wishes to select his feedback gain (strategy) L_i, in order to minimize his cost function J_i. Defining the Hamiltonian equations

$$H_i = \text{tr}(M_i X) + \text{tr}(F_i P_i), \quad i = 1, 2, \tag{8.6-17}$$

where F_i and M_i are defined by (8.6-8) and P_i is a matrix of undetermined multipliers, necessary conditions for minimizing H_i are found by partial differentiation of H_i, with respect to P_i, M_i and L_i, respectively, to be

$$0 = F_i^T = A_c^T M_i + M_i A_c$$

$$+ Q_i + C_i^T L_i^T R_{ii} L_i C_i + C_j^T L_j R_{ij} L_j C_j \tag{8.6-18}$$

$$0 = A_c P_i + P_i A_c^T + X \tag{8.6-19}$$

$$0 = -B_1^T M_i P_i C_i^T + R_{ii} L_i C_i P_i C_i^T, \tag{8.6-20}$$

where $i = 1, 2$, $i \neq j$. Note that (8.6-19) implies $P_1 = P_2 \equiv P$. These results reduce to the conditions for state-variable feedback in (Papavassilopoulos and Cruz, Jr. 1979) if $C_1 = C_2 = I$.

To apply the solution method in Table 8.1-2, the necessary conditions stated above may be combined into two coupled Lyapunov equations and a constraint equation as follows:

$$0 = \begin{bmatrix} A_c & 0 \\ 0 & A_c \end{bmatrix}^T \begin{bmatrix} M_1 & 0 \\ 0 & M_2 \end{bmatrix} + \begin{bmatrix} M_1 & 0 \\ 0 & M_2 \end{bmatrix} \begin{bmatrix} A_c & 0 \\ 0 & A_c \end{bmatrix} + \begin{bmatrix} Q_1 & 0 \\ 0 & Q_2 \end{bmatrix}$$

$$+ \begin{bmatrix} L_1 C_1 & 0 \\ 0 & L_2 C_2 \end{bmatrix}^T \begin{bmatrix} R_{11} & 0 \\ 0 & R_{22} \end{bmatrix} \begin{bmatrix} L_1 C_1 & 0 \\ 0 & L_2 C_2 \end{bmatrix}$$

$$+ \begin{bmatrix} L_2 C_2 & 0 \\ 0 & L_1 C_1 \end{bmatrix}^T \begin{bmatrix} R_{12} & 0 \\ 0 & R_{21} \end{bmatrix} \begin{bmatrix} L_2 C_2 & 0 \\ 0 & L_1 C_1 \end{bmatrix} \tag{8.6-21}$$

$$0 = A_c P + P A_c^T + X \tag{8.6-22}$$

$$0 = \begin{bmatrix} R_{11} & 0 \\ 0 & R_{22} \end{bmatrix} \begin{bmatrix} L_1 & 0 \\ 0 & L_2 \end{bmatrix} \begin{bmatrix} C_1 PC_1^T & 0 \\ 0 & C_2 PC_2^T \end{bmatrix}$$
$$- \begin{bmatrix} B_1^T M_1 & 0 \\ 0 & B_2^T M_2 \end{bmatrix} \begin{bmatrix} PC_1^T & 0 \\ 0 & PC_2^T \end{bmatrix}. \tag{8.6-23}$$

Again, one must be aware that the algorithm in Table 8.1-2 may find a local minimum that may not be globally optimum.

Stackelberg Games

Stackelberg game strategies differ from Nash game strategies in that the players' objectives are no longer independent. In Stackelberg games, the players take on the roles of "leader" (u_2) and "follower" (u_1). The objective of the follower is to minimize his performance criterion, J_1, given only information about the strategy (linear output feedback) of the leader. The leader, on the other hand, attempts to minimize his performance criterion, J_2, given information about the strategy of the follower, as well as the overall objective of the follower (optimizing a quadratic performance index with output feedback). More exactly, in computing his optimal strategy, the leader must take into account the reaction of the follower to his actions. For our purposes, the game is represented by the system model (8.6-1), with the associated performance criteria, J_i, (8.6-4).

The follower's problem is presented first. The follower has no knowledge of the leader's strategy, so he assumes that L_2 is a fixed matrix. Under this assumption, L_1 is then chosen to minimize J_1 as follows.

Define the Hamiltonian

$$H_1 = \text{tr}(M_1 X) + \text{tr}(F_1 P_1), \tag{8.6-24}$$

where M_1 and F_1 are defined by (8.6-8) and P_1 is an undetermined matrix of multipliers. The necessary conditions for minimizing H_1 are then found by taking partial derivatives of H_1 with respect to P_1, M_1, and L_1, respectively, to give

$$0 = F_1^T = A_c^T M_1 + M_1 A_c + Q_1$$
$$+ C_1^T L_1^T R_{11} L_1 C_1 + C_2^T L_2^T R_{12} L_2 C_2 \tag{8.6-25}$$

$$0 = F_3 = A_c P_1 + P_1 A_c^T + X \tag{8.6-26}$$

$$0 = F_4 = -B_1^T M_1 P_1 C_1^T + R_{11} L_{11} C_1 P_1 C_1^T. \tag{8.6-27}$$

Note that these equations have the same form as (8.6-18)–(8.6-20) with $i = 1$, which is also the form of the necessary conditions for the linear quadratic regulator with output feedback.

The leader is now faced with the problem of minimizing his performance criterion, J_2, while simultaneously taking into account the reaction of the follower. Mathematically, this amounts to treating the equations that represent the follower's optimal solution (8.6-25)–(8.6-27) as constraints to be met while performing the minimization of J_2. The resulting Hamiltonian thus has the two terms that normally appear (e.g., (8.6-24)), as well as terms that represent these additional constraints. That is,

$$H_2 = \text{tr}(M_2 X) + \text{tr}(F_2 P_2) + \text{tr}(F_1 P_3)$$

$$+ \text{tr}(F_3 M_3) + \text{tr}(F_4 L_3^T) + \text{tr}(L_3 F_4^T), \qquad (8.6\text{-}28)$$

where M_2 and F_2 are defined in (8.6-8), F_1, F_3, and F_4 are defined in (8.6-25)–(8.6-27), and P_2, P_3, M_3, and L_3, are undetermined matrices of multipliers. (Note that F_4, and, hence, L_3, are not symmetric.) The choice of variable names with their subscripts results in greater clarity in understanding the necessary conditions, as we shall see.

The partial derivatives of H_2 may be broken into three classes. First, note that the partial derivatives of H_2, with respect to P_3, M_3, and L_3, simply reiterate conditions (8.6-25)–(8.6-27). Second, the conditions generated by taking partial derivatives with respect to the follower's variables P_1, M_1, and L_1, respectively, give

$$0 = A_c^T M_3 + M_3 A_c + C_1^T L_1^T R_{11} L_1 C_1 + C_1^T L_3^T R_{11} L_1 C_1$$

$$- M_1 B_1 L_3 C_1 - C_1^T L_3^T B_1^T M_1 \qquad (8.6\text{-}29)$$

$$0 = A_c P_3 + P_3 A_c^T - B_1 L_3 C_1 P_1 - P_1 C_1^T L_3^T B_1^T \qquad (8.6\text{-}30)$$

$$0 = -B_1^T M_1 P_3 C_1^T - B_1^T M_2 P_2 C_1^T - B_1^T M_3 P_1 C_1^T$$

$$+ R_{11} L_1 C_1 P_3 C_1^T + R_{21} L_1 C_1 P_2 C_1^T + R_{11} L_3 C_1 P_1 C_1^T. \qquad (8.6\text{-}31)$$

These equations may be interpreted as a "fictitious follower," with feedback gain of L_3, which is embedded in the leader's problem. Third are the conditions arising from the partial derivatives of H_2 with respect to the leader's variables P_2, M_2, and L_2, which are

$$0 = F_2^T = A_c^T M_2 + M_2 A_c + Q_2$$

$$+ C_2^T L_2^T R_{22} L_2 C_2 + C_1^T L_1^T R_{21} L_1 C_1 \qquad (8.6\text{-}32)$$

$$0 = A_c P_2 + P_2 A_c^T + X \qquad (8.6\text{-}33)$$

$$0 = -B_2^T M_2 P_2 C_2^T - B_2^T M_1 P_3 C_2^T - B_2^T M_3 P_1 C_2^T$$

$$R_{22} L_2 C_2 P_2 C_2^T + R_{12} L_2 C_2 P_3 C_2^T. \qquad (8.6\text{-}34)$$

Note that (8.6-26) and (8.6-33) imply $P_1 = P_2 \equiv P$. These conditions reduce to the conditions for state-variable feedback in Medanic (1978) by taking $C_1 = C_2 = I$.

There is no good approach to solving these equations even in the state-feedback case. We now outline a method that relies on the iterative approach in Table 8.1-2. To apply the solution method in Table 8.1-2, we must be able to solve at each iteration of the algorithm for the M_i, the P_i, and the gains L_i.

First, we discuss the solution for the M_i. Although (8.6-25) and (8.6-32) have positive definite solutions M_1 and M_2 when A_c is stable and all other terms are fixed, the same may not be said of (8.6-29) and M_3. This could lead to problems with the algorithm in Table 8.1-2. To correct this potential problem and allow the use of conventional Lyapunov equation solvers, we propose proceeding as follows. Introduce additional variables Q_3 and M_3 and the auxiliary equation

$$0 = A_c^T M_4 + M_4 A_c + Q_3 + 2C_1^T L_3^T R_{11} L_3 C_1, \tag{8.6-35}$$

which has a positive definite solution M_4 when all other variables are fixed, as long as $R_{11} > 0$, $Q_3 > 0$, and A_c is stable. An equation for

$$M = M_1 + M_3 + M_4 \tag{8.6-36}$$

may be found by adding (8.6-25), (8.6-29), and (8.6-35) to be

$$0 = A_c^T M + M A_c + Q_1 + C_1^T (L_1 + L_3)^T R_{11} (L_1 + L_3) C_1$$
$$+ C_2^T L_2^T R_{12} L_2 C_2$$
$$+ \begin{bmatrix} I \\ L_3 C_1 \end{bmatrix}^T \begin{bmatrix} Q_3 & -M_1 B_1 \\ -B_1^T M_1 & R_{11} \end{bmatrix} \begin{bmatrix} I \\ L_3 C_1 \end{bmatrix}, \tag{8.6-37}$$

which has a positive definite solution M when all other variables are fixed if A_c is stable and Q_3 in (8.6-35) is selected to make the block coefficient matrix in the middle of the last term positive definite. Let Q_3 be selected to ensure this. Then, when applying the algorithm in Table 8.1-2, to determine the M_1 at each iteration, solve in the order: (8.6-25) for M_1, (8.6-32) for M_2, (8.6-35) for M_4, (8.6-37) for M, and finally (8.6-36) for M_3.

Now, we consider the solution for the P_i. Equation (8.6-26) (or equivalently (8.6-33)) shows that $P \equiv P_1 = P_2$ is positive definite as long as A_c is stable and $(A, \sqrt{X})$ is controllable. Since a similar statement may not be made about (8.6-31) and P_3, we propose to proceed as follows. Introduce additional variables $Q_4 > 0$, $X_1 > 0$, and P_4 and the auxiliary equation

$$0 = A_c P_4 + P_4 A_c^T + X_1 + B_1 L_3 Q_4 L_3^T B_1^T. \tag{8.6-38}$$

Then an equation for

$$P_5 \equiv P + P_3 + P_4 \tag{8.6-39}$$

may be found by adding (8.6-26), (8.6-31), and (8.6-38) to be

$$0 = A_c P_5 + P_5 A_c^\mathsf{T} + X_1 + \begin{bmatrix} I \\ L_3^\mathsf{T} B_1^\mathsf{T} \end{bmatrix}^\mathsf{T} \begin{bmatrix} X & -PC_1^\mathsf{T} \\ -C_1 P & Q_4 \end{bmatrix} \begin{bmatrix} I \\ L_3^\mathsf{T} B_1^\mathsf{T} \end{bmatrix}, \qquad (8.6\text{-}40)$$

which has a positive definite solution P_5 when A_c is stable and Q_4 is selected to make the matrix in the middle of the last term positive definite. Let Q_4 be selected to ensure this. Then, when applying the algorithm in Table 8.1-2, to determine the P_i at each iteration solve in order: (8.6-26) for $P = P_1 = P_2$, (8.6-38) for P_4, (8.6-40) for P_5, and finally (8.6-39) for P_3.

To apply the algorithm in Table 8.1-2, at each iteration we must also solve for the gains L_i. Although (8.6-27) and (8.6-31) may be solved for L_1 and L_3, respectively, assuming all other variables are fixed, note that the form of (8.6-34) for L_2 is

$$R_{22} L_2 C_2 P_2 C_2^\mathsf{T} + R_{12} L_2 C_2 P_3 C_2^\mathsf{T} = \text{R.H.S.} \qquad (8.6\text{-}41)$$

so that a solution in terms of simple matrix inversions is not possible. This equation is a generalized Lyapunov equation of the sort studied in Golub, Nash, and Van Loan (1979), where conditions for its solution are given along with a solution technique based on the Bartles-Steward algorithm (Golub, Nash, and Van Loan 1979). Therefore, when applying the algorithm in Table 8.1-2, to determine the gain updates at each iteration we may solve in the order: (8.6-27) for L_1, (8.6-31) for L_3, and (8.6-41) for L_2 using the technique in Golub, Nash, and Van Loan (1979).

We may make no claim that the proposed algorithm will always converge; however, it seems to be superior from this standpoint to other proposed techniques. Even in the case of state variable feedback, $C_1 = I$ and $C_2 = I$, the proposed approach yields a methodical approach for solving the Stackelberg game problem, for which a good algorithm does not yet exist.

PROBLEMS

Section 8.1

8.1-1. Fill in the details in the derivation of the design equations in Table 8.1-1.

8.1-2. Output feedback design for scalar systems
a. Consider the case where $x(t)$, $u(t)$, $y(t)$ are all scalars. Show that the solution S to the second Lyapunov equation in Table 8.1-1 is not needed to determine the output-feedback gain K. Find an explicit solution for P and, hence, for the optimal gain K.
b. Repeat for the case where $x(t)$ and $y(t)$ are scalars, but $u(t)$ is an m-vector.

8.1-3. Use (583.28) to eliminate K in the Lyapunov equations of Table 8.1-1, hence deriving two coupled nonlinear equations that may be solved for the optimal auxiliary matrices S and P. Does this simplify the solution of the output-feedback design problem?

8.1-4. Software for output-feedback design. Write a program that finds the gain K minimizing the PI in Table 8.1-1 using the SIMPLEX algorithm in Press et al. (1986). Use it to verify the results of Example 8.1-1. Can you tune the elements of Q and R to obtain better closed-loop responses than the ones given?

8.1-5. For the system

$$\dot{x} = \begin{bmatrix} 0 & 1 \\ 0 & 0 \end{bmatrix} x + \begin{bmatrix} 0 \\ 1 \end{bmatrix} u, \quad y = [1 \quad 1]x \tag{1}$$

Find the output-feedback gain that minimizes the PI in Table 8.1-1 with $Q = I$. Try various values of R to obtain a good response. You will need the software from Problem 8.1-4. The closed-loop step response may be plotted using the *step.m* function from the Control System Toolbox. (Note that system (1) is nothing but Newton's law, since if $x = [p \quad v]^T$, then $\ddot{p} = u$, where $u(t)$ may be interpreted as an acceleration input F/m.)

8.1-6. Gradient-based Software for Output-Feedback Design. Write a MATLAB program that finds the gain K minimizing the PI in Table 8.1-1 using the Davidon-Fletcher-Powell algorithm in Press et al. (1986). Use it to verify the results of Example 8.1-1.

Section 8.2

8.2-1. Derive (8.2-31).

8.2-2. Derive the necessary conditions in Table 8.2-1.

8.2-3. In Example 8.2-1, use the observability matrix to verify that the original proposed value of $Q = H^T H$ has $(\sqrt{Q}, A0$ unobservable while the Q that contains a $(5, 5)$ element has $(\sqrt{Q}, A)$ observable.

8.2-4. Software for LQ output-feedback design. Write a MATLAB program to solve for the optimal gain K in Table 8.2-1 using the SIMPLEX algorithm in Press et al. (1986). Use it to verify Example 8.2-1.

8.2-5. In Example 8.2-1 we used an output with four components. There is an extra degree of freedom in the choice of control gains that may not be needed. Redo the example, using the software from Problem 8.2-4, with the output defined as $y = [\alpha_F \quad q \quad \varepsilon]^T$.

8.2-6. To see whether the angle-of-attack filter in Example 8.2-1 complicates the design, redo the example using $y = [\beta \ q \ e \ \delta]^T$.

8.2-7. Redo Example 8.2-1 using root-locus techniques like those in Chapter 3. Based on this, are the gains selected by the LQ algorithm sensible from the point of view of classical control theory?

8.2-8. Gradient-based software for LQ output-feedback design. Write a MATLAB program to solve for the optimal gain K in Table 8.2-1 using the

Davidon-Fletcher-Powell algorithm in Press et al. (1986). Use it to verify Example 8.2-1.

Section 8.3

8.3-1. Complete the design of Example 8.3-1. That is,
a. Select a value for α, and use Table 8.1-1 to find the regulator gain K. Tune the values of Q and R until the response to nonzero initial conditions is suitable.
b. Find the tracker control law. To verify the design, simulate the step response of the closed-loop system using *lsim.m* from MATLAB.

8.3-2. Regulator redesign servo for DC motor. Use the approach of this section to design a servo for the scalar DC motor model in Example 8.2-2. Simulate the step response of the closed-loop system.

8.3-3. Regulator redesign servo for DC motor. Use the approach of this section to design a servo for the armature-controlled DC motor model in Example 8.2-3. Simulate the step response of the closed-loop system.

8.3-4. Regulator redesign servo for inverted pendulum. Use the approach of this section to design a servo for the inverted pendulum. Simulate the step response of the closed-loop system.

Section 8.4

8.4-1. Find the reachability matrix of (8.4-16) to verify the tracking conditions for full state feedback relating to (8.4-22).

8.4-2. Derive the CGT for a system with an unknown disturbance $d(t)$ and verify that it is given by (8.4-21) with $\Delta(s)$ modified as in (8.4-30).

8.4-3. Complete the design in Example 8.4-1. That is, select $\omega_0 = 3000$ rpm and perform a regulator design on the augmented system using the equations in Table 8.1-1. Tune q to obtain suitable time responses of the augmented system to nonzero initial conditions. To verify the performance, simulate the CGT on the system using *lsim.m* in MATLAB.

8.4-4. Tracking with disturbance rejection. Redo Example 8.4-1 if there is a constant bias disturbance on the head position.

Section 8.5

8.5-1. Write the state variable description of Fig. 8.5-1, verifying that it is nothing but (8.5-28).

8.5-2. It is desired to make the scalar plant

$$\dot{x} = x + u, \quad y = x, \quad z = x$$

behave like the scalar model

$$\dot{x} = -2x + r, \quad y = x, \quad z = x$$

with reference input equal to the unit step. Use explicit model following to design a servosystem:

a. Draw the controller structure.

b. Select the control gains using LQR design on the augmented system. You will need to use the software written for the problems of Section 8.1.

8.5-3. A plant is described by Newton's law

$$\dot{x} = x_2, \quad \dot{x}_2 = u.$$

The velocity should follow the model output and measurements of position are taken so that

$$y = x_1, \quad z = x_2.$$

The prescribed model with desirable characteristics is given by

$$\dot{x} = -3x + r, \quad y = x, \quad z = x$$

with $r(t)$ the unit step. Use explicit model following to design a servosystem:

a. Draw the compensator structure.

b. Select the control gains using LQR design of the augmented system. You will need to use the software written for the problems of Section 8.1.

9

ROBUSTNESS AND MULTIVARIABLE FREQUENCY-DOMAIN TECHNIQUES

9.1 INTRODUCTION

Modeling Errors and Stability Robustness

In the design of control systems it is important to realize that the set of linear differential equations that are the basis of design are, most of the time, an approximation to the nonlinear system dynamics. Several systems have dynamics that are important at high frequencies that many times are neglected in the model design. These unmodeled high-frequency dynamics can act to destabilize a control system that may have quite suitable behavior in terms only of the system model.

Moreover, as the nonlinear system changes its equilibrium operation point, the linearized plant model describing its perturbed behavior changes. This parameter variation is a low-frequency effect that can also act to destabilize the system. To compensate for this variation, one may determine suitable controller gains for linearized models at several design equilibrium points over an operation envelope. Then these design gains may be scheduled in computer lookup tables for suitable controller performance over the whole envelope. For gain scheduling to work, it is essential for the controller gains at each design equilibrium point to guarantee stability for actual operation conditions near that equilibrium point. Thus, it is important to design controllers that have stability robustness, which is the ability to provide stability in spite of modeling errors due to high-frequency unmodeled dynamics and plant parameter variations.

Disturbances and Performance Robustness

It is often important to account for disturbances and also for sensor measurement noise. Disturbances can often cause unsatisfactory performancey in a system that has been designed without taking them into account. Thus, it is important to design controllers that have performance robustness, which is the ability to guarantee acceptable performance (in terms, for instance, of percent overshoot, settling time, and so on) even though the system may be subject to disturbances.

Classical Robust Design

In classical control, robustness may be designed into the system from the beginning by providing sufficient gain and phase margin to counteract the effects of inaccurate modeling or disturbances. In terms of the Bode magnitude plot, the loop gain should be high at low frequencies for performance robustness, but low at high frequencies, where unmodeled dynamics may be present, for stability robustness. The concept of bandwidth is important in this connection, as is the concept of the sensitivity function.

Classical controls design techniques are generally in the frequency domain, and so they afford a convenient approach to robust design for single-input/single-output (SISO) systems. However, the individual gain margins, phase margins, and sensitivities of all the SISO transfer functions in a multivariable or multiloop system have little to do with its overall robustness. Thus, there have been problems in extending classical robust design notions to multi-input/multi-output (MIMO) systems.

Modern Robust Design

Modern control techniques provide a direct way to design multiloop controllers for MIMO systems by closing all the loops simultaneously. Performance is guaranteed in terms of minimizing a quadratic performance index (PI) which, with a sensible problem formulation, generally implies closed-loop stability as well. All our work in previous chapters assumed that the model is exactly known and that there are no disturbances. In fact, this is rarely the case.

In this chapter we show that the classical frequency-domain robustness measures are easily extended to MIMO systems in a rigorous fashion by using the notion of the singular value. In Section 9.2 we develop the multivariable loop gain and sensitivity, and describe the multivariable Bode magnitude plot. In terms of this plot, we present bounds that guarantee both robust stability and robust performance for multivariable systems, deriving notions that are entirely analogous to those in classical control.

In Section 9.3 we give a design technique for robust multivariable controllers using modern output-feedback theory, showing how robustness may be guaranteed. The approach is a straightforward extension of classical techniques. We illustrate by designing a pitch rate control system that has good performance despite the presence of flexible modes and wind gusts.

A popular modern approach to the design of robust controllers is linear quadratic Gaussian/loop-transfer recovery (LQG/LTR). This approach has been used extensively by Honeywell in the design of advanced multivariable control systems. LQG/LTR relies on the separation principle, which involves designing a full state-variable feedback and then an observer to provide the state estimates for feedback purposes. The result is a dynamic compensator that is similar to those resulting from classical control approaches. The importance of the separation principle is that compensators can be designed for multivariable systems in a straightforward manner by solving matrix equations.

In Section 9.4 we discuss observers and the Kalman filter. In Section 9.5 we cover LQG/LTR design. In Section 9.6 we cover the H_∞ design approach in the state-space framework, introduced by Francis et al. (1984) and Doyle et al. (1989).

9.2   MULTIVARIABLE FREQUENCY-DOMAIN ANALYSIS

We shall deal with system uncertainties, as in classical control, using robust design techniques that are conveniently examined in the frequency domain. To this point, our work in modern control has been in the time domain, since the LQ performance index is a time-domain criterion.

One problem that arises immediately for MIMO systems is that of extending the SISO Bode magnitude plot. We are not interested in making several individual SISO frequency plots for various combinations of the inputs and outputs in the MIMO system and examining gain and phase margins. Such approaches have been tried and may not always yield much insight on the true behavior of the MIMO system. This is due to the coupling that generally exists between all inputs and all outputs of a MIMO system.

Thus, in this section we introduce the multivariable loop gain and sensitivity and the multivariable Bode magnitude plot, which will be nothing but the plot versus frequency of the singular values of the transfer-function matrix. This basic tool allows much of the rich experience of classical control theory to be applied to MIMO systems. Thus, we shall discover that for robust performance the minimum singular value of the loop gain should be large at low frequencies, where disturbances are present. On the other hand, for robust stability the maximum singular value of the loop gain should be small at high frequencies, where there are significant modeling inaccuracies. We shall also see that to guarantee stability in spite of parameter variations in the linearized model due to operating point changes, the maximum singular value should be below an upper limit.

Sensitivity and Cosensitivity

Figure 9.2-1 shows a standard feedback system of the sort that we have seen several times in our work to date. The plant is $G(s)$, and $K(s)$ is the feedback/feedforward compensator, which can be designed by any of the techniques we have covered. The plant output is $z(t) \in R^q$, the plant control input is $u(t) \in R^m$, and the reference input is $r(t) \in R^q$.

Perfect tracking may not be achieved unless the number m of control inputs $u(t)$ is greater than or equal to the number q of performance outputs $z(t)$ (Kwakernaak and Sivan 1972). Therefore, we shall assume that $m = q$ so that the plant $G(s)$ and compensator $K(s)$ are square. This is only a consequence of sensible design, and not a restriction on the sorts of plants that may be considered.

We have added a few items to the figure to characterize uncertainties. The signal $d(t)$ represents a disturbance acting on the system of the sort appearing in classical control. This could represent, for instance, wind gusts. The sensor measurement noise or errors are represented by $n(t)$. Both of these signals are generally vectors of dimension q. Typically, the disturbances occur at low frequencies, say, below some ω_d, while the measurement noise $n(t)$ has its predominant effect at high frequencies, say, above some value ω_n. Typical Bode plots for the magnitudes of these terms appear in Fig. 9.2-2 for the case that $d(t)$ and $n(t)$ are scalars. The reference input is generally also a low-frequency signal (e.g., the unit step).

The tracking error is

$$e(t) \equiv r(t) - z(t). \tag{9.2-1}$$

Due to the presence of $n(t)$, $e(t)$ may not be symbolized in Fig. 9.2-1. The signal $s(t)$ is in fact given by

$$s(t) = r(t) - z(t) - n(t) = e(t) - n(t). \tag{9.2-2}$$

Let us perform a frequency-domain analysis on the system to see the effects of the uncertainties on system performance. In terms of Laplace transforms we may write

$$Z(s) = G(s)K(s)S(s) + D(s), \tag{9.2-3}$$

$$S(s) = R(s) - Z(s) - N(s), \tag{9.2-4}$$

$$E(s) = R(s) - Z(s). \tag{9.2-5}$$

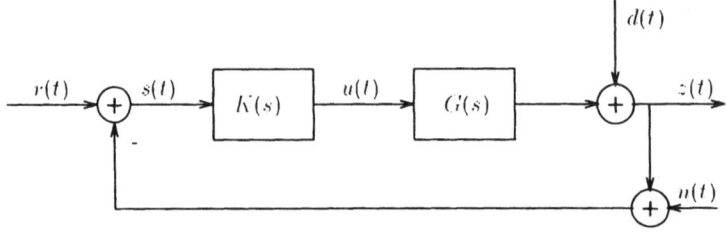

FIGURE 9.2-1   Standard feedback configuration.

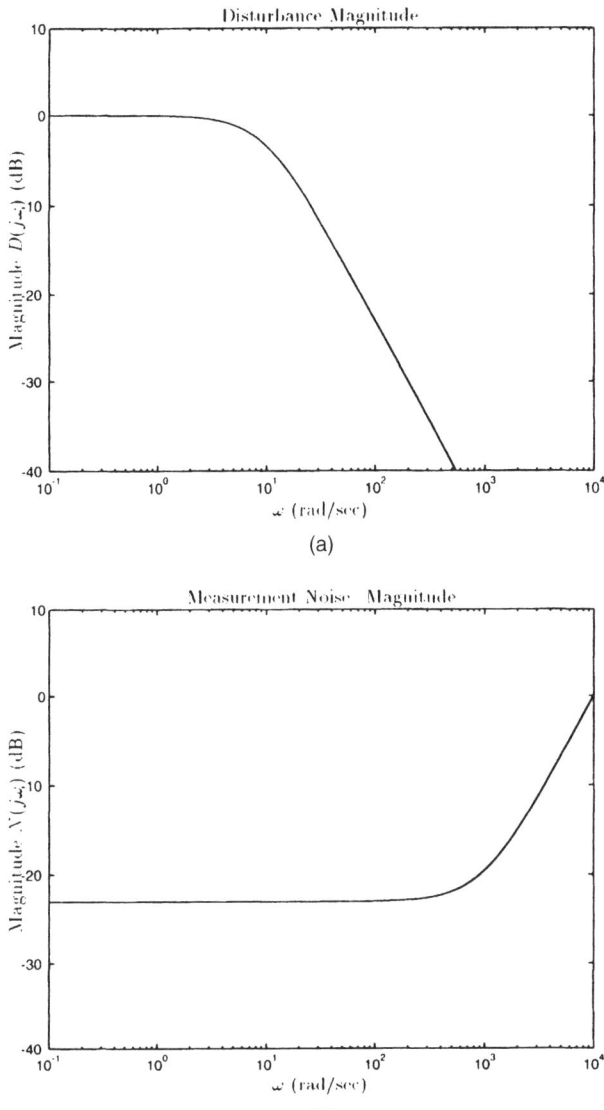

FIGURE 9.2-2 Typical Bode plots for the uncertain signals in the system. (a) Disturbance magnitude. (b) Measurement noise magnitude.

Now we may solve for $Z(s)$ and $E(s)$, obtaining the closed-loop transfer function relations

$$Z(s) = (I + GK)^{-1}GK(R - N) + (I + GK)^{-1}D, \qquad (9.2\text{-}6)$$

$$E(s) = [I - (I + GK)^{-1}GK]R + (I + GK)^{-1}GKN$$
$$- (I + GK)^{-1}D. \qquad (9.2\text{-}7)$$

It is important to note that, unlike the case for SISO systems, care must be taken to perform the matrix operations in the correct order. For instance, $GK \neq KG$. The multiplications by matrix inverses must also be performed in the correct order.

We can put these equations into a more convenient form. According to the matrix inversion lemma, (9.2-7) may be written as

$$E(s) = (I + GK)^{-1}(R - D) + (I + GK)^{-1}GKN. \qquad (9.2\text{-}8)$$

Moreover, since GK is square and invertible, we can write

$$(I + GK)^{-1}GK = [(GK)^{-1}(I + GK)]^{-1} = [(GK)^{-1} + I]^{-1}$$
$$= [(I + GK)(GK)^{-1}]^{-1} = GK(I + GK)^{-1}. \qquad (9.2\text{-}9)$$

Therefore, we may finally write $Z(s)$ and $E(s)$ as

$$Z(s) = GK(I + GK)^{-1}(R - N) + (I + GK)^{-1}D, \qquad (9.2\text{-}10)$$
$$E(s) = (I + GK)^{-1}(R - D) + GK(I + GK)^{-1}N. \qquad (9.2\text{-}11)$$

To simplify things a bit, define the system sensitivity

$$S(s) = (I + GK)^{-1} \qquad (9.2\text{-}12)$$

and

$$T(s) = GK(I + GK)^{-1} = (I + GK)^{-1}GK. \qquad (9.2\text{-}13)$$

Since

$$S(s) + T(s) = (I + GK)(I + GK)^{-1} = I, \qquad (9.2\text{-}14)$$

we call $T(s)$ the complementary sensitivity, or, in short, the consensitivity. Note that the return difference

$$L(s) = I + GK \qquad (9.2\text{-}15)$$

is the inverse of the sensitivity. The loop gain is given by $G(s)K(s)$.

These expressions extend the classical notions of loop gain, return difference, and sensitivity to multivariable systems. They are generally square transfer-function matrices of dimension $m \times m$. In terms of these new quantities, we have

$$Z(s) = T(s)(R(s) - N(s)) + S(s)D(s) \qquad (9.2\text{-}16)$$
$$E(s) = S(s)(R(s) - D(s)) + T(s)N(s). \qquad (9.2\text{-}17)$$

To ensure small tracking errors, we must have $S(j\omega)$ small at those frequencies ω where the reference input $r(t)$ and disturbance $d(t)$ are large. This will yield good disturbance rejection. On the other hand, for satisfactory sensor noise rejection, we should have $T(j\omega)$ small at those frequencies ω where $n(t)$ is large. Unfortunately, a glance at (9.2-14) reveals that $S(j\omega)$ and $T(j\omega)$ cannot simultaneously be small at any one frequency ω. According to Fig. 9.2-2, we should like to have $S(j\omega)$ small at low frequencies, where $r(t)$ and $d(t)$ dominate, and $T(j\omega)$ small at high frequencies, where $n(t)$ dominates. These are nothing but the multivariable generalizations of the well-known SISO classical notion that a large loop gain $GK(j\omega)$ is required at low frequencies for satisfactory performance and small errors, but a small loop gain is required at high frequencies where sensor noises are present.

Multivariable Bode Plot

These notions are not difficult to understand on a heuristic level. Unfortunately, it is not so straightforward to determine a clear measure for the "smallness" of $S(j\omega)$ and $T(j\omega)$. These are both square matrices of dimensions $m \times m$, with m the number of plant control inputs $u(t)$, which we assume is equal to the number of performance outputs $z(t)$ and reference inputs $r(t)$. They are complex functions of the frequency. Clearly, the classical notion of the Bode magnitude plot, which is defined only for scalar complex functions of ω, must be extended to the MIMO case.

Some work was done early on using the frequency-dependent eigenvalues of a square complex matrix as a measure of smallness (Rosenbrock 1974, MacFarlane 1970, and MacFarlane and Kouvaritakis 1977). However, note that the matrix

$$M = \begin{bmatrix} 0.1 & 100 \\ 0 & 0.1 \end{bmatrix} \tag{9.2-18}$$

has large and small components, but its eigenvalues are both at 0.1.

A better measure of the magnitude of square matrices is the singular value (SV) (Strang 1980). Given any matrix M we may write its singular value decomposition (SVD) as

$$M = U\Sigma V^*, \tag{9.2-19}$$

where U and V are square unitary matrices (i.e., $V^{-1} = V^*$, the complex conjugate transpose of V) and

$$\Sigma = \begin{bmatrix} \sigma_1 & & & & & & \\ & \sigma_2 & & & & & \\ & & \ddots & & & & \\ & & & \sigma_r & & & \\ & & & & 0 & & \\ & & & & & \ddots & \\ & & & & & & 0 \end{bmatrix} = \begin{bmatrix} \Sigma_r & 0 \\ 0 & 0 \end{bmatrix}, \tag{9.2-20}$$

with $r = \text{rank}(M)$. The singular values are the σ_i, which are ordered so that $\sigma_1 \geq \sigma_2 \geq \cdots \geq \sigma_{r-1} \geq \sigma_r$. The SVD may loosely be thought of as the extension to general matrices (which may be nonsquare or complex) of the Jordan form. If M is a function of $j\omega$, then so are U and V.

Since $MM^* = U\Sigma V^* V\Sigma U^* = U\Sigma^2 U^*$, it follows that the singular values of M are simply the (positive) square roots of the nonzero eigenvalues of MM^*. A similar proof shows that the nonzero eigenvalues of MM^* and those of M^*M are the same.

We note that the M given above has two singular values, namely $\sigma_1 = 100.0001$ and $\sigma_2 = 0.0001$. Thus, this measure indicates that M has a large and a small component. Indeed, note that

$$\begin{bmatrix} 0.1 & 100 \\ 0 & 0.1 \end{bmatrix} \begin{bmatrix} -1 \\ 0.001 \end{bmatrix} = \begin{bmatrix} 0 \\ 0.001 \end{bmatrix}, \tag{9.2-21}$$

while

$$\begin{bmatrix} 0.1 & 100 \\ 0 & 0.1 \end{bmatrix} \begin{bmatrix} 0.001 \\ 1 \end{bmatrix} = \begin{bmatrix} 100.0001 \\ 0.1 \end{bmatrix}. \tag{9.2-22}$$

Thus, the singular value σ_2 has the input direction

$$\begin{bmatrix} -1 \\ 0.001 \end{bmatrix}$$

associated with it for which the output contains the value σ_2. On the other hand, the singular value σ_1 has an associated input direction of

$$\begin{bmatrix} 0.001 \\ 1 \end{bmatrix},$$

for which the output contains the value σ_1.

There are many nice properties of the singular value that make it a suitable choice for defining the magnitude of matrix functions. Among these is the fact that the maximum singular value is an induced matrix norm, and norms have several useful attributes. The use of the SVs in the context of modern control was explored in Doyle and Stein (1981) and Safonov et al. (1981).

A major factor is that there are many good software packages that have good routines for computing the singular value (e.g., subroutine LSVDF in IMSL [1980] or MATLAB [1992]). Thus, plots like those we present may easily be obtained by writing only a computer program to drive the available subroutines. Indeed, since the SVD uses unitary matrices, its computation is numerically stable. An efficient technique for obtaining the SVs of a complex matrix as a function of frequency $j\omega$ is given in Laub (1981).

We note that a complete picture of the behavior of a complex matrix versus $j\omega$ must take into account the directions of the SVs as well as the multivariable phase, which may also be obtained from the SVD (Postlethwaite et al. 1981). Thus, complete MIMO generalizations of the Bode magnitude and phase plots are available. However, the theory relating to the phase portion of the plot is more difficult to use in a practical design technique, although a MIMO generalization of the Bode gain–phase relation is available (Doyle and Stein 1981). Therefore, we shall employ only plots of the SVs versus frequency, which correspond to the Bode magnitude plot for MIMO systems.

The magnitude of a square transfer-function matrix $H(j\omega)$ at any frequency $j\omega$ depends on the direction of the input excitation. Inputs in a certain direction in the input space will excite only the SV(s) associated with that direction. However, for any input, the magnitude of the transfer function $H(j\omega)$ at any given frequency $j\omega$ may be bounded above by its maximum singular value, denoted $\overline{\sigma}(H(j\omega))$, and below by its minimum singular value, denoted $\underline{\sigma}(H(j\omega))$. Therefore, all our results, as well as the plots we shall give, need take into account only these two constraining values of "magnitude."

Example 9.2-1. MIMO Bode Magnitude Plots

Consider the multivariable system

$$\dot{x} = \begin{bmatrix} -1 & -2 & 0 & 0 \\ 2 & -1 & 0 & 0 \\ 0 & 0 & -3 & -8 \\ 0 & 0 & 8 & -3 \end{bmatrix} x + \begin{bmatrix} 1 & 0 \\ 0 & 0 \\ 0 & 1 \\ 0 & 0 \end{bmatrix} u = Ax + bu \tag{1}$$

$$y = \begin{bmatrix} 1 & 0 & 0 & 0 \\ 0 & 0 & 1 & 0 \end{bmatrix} x = Cx, \tag{2}$$

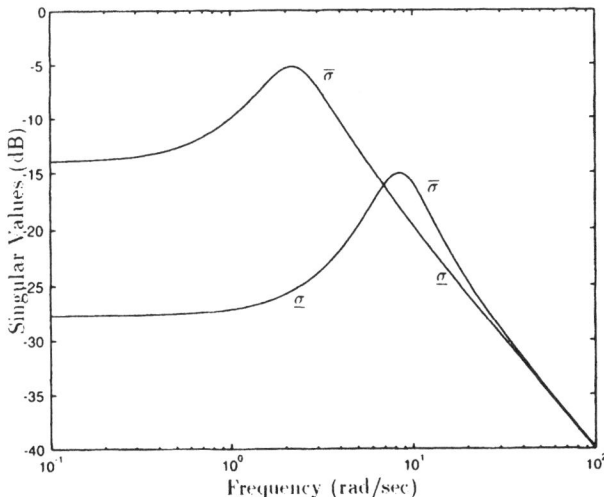

FIGURE 9.2-3 MIMO Bode magnitude plot of singular values versus frequency.

which has a 2×2 MIMO transfer function of

$$H(s) = C(sI - A)^{-1}B = N(s)D^{-1}(s) \tag{3}$$

with

$$D(s) = s^4 + 8s^3 + 90s^2 + 176s + 365, \tag{4}$$

$$N(s) = \begin{bmatrix} 1 & 0 \\ 0 & 1 \end{bmatrix} s^3 + \begin{bmatrix} 7 & 0 \\ 0 & 5 \end{bmatrix} s^2 + \begin{bmatrix} 79 & 0 \\ 0 & 11 \end{bmatrix} s + \begin{bmatrix} 73 & 0 \\ 0 & 15 \end{bmatrix}.$$

The SV plots versus frequency shown in Fig. 9.2-3 can be obtained by using the function *sigma.m* from the MATLAB Control System Toolbox. We call this the multivariable Bode magnitude plot for the MIMO transfer function $H(s)$.

Since $H(s)$ is 2×2, it has two singular values. Note that although each singular value is continuous, the maximum and minimum singular values are not. This is due to the fact that the singular values can cross over each other, as the figure illustrates. ∎

Example 9.2-2.   Singular Value Plots vs. Bode Plots

To illustrate the difference between the singular value plots and the individual SISO Bode plots of a multivariable system, let us consider the F-16 lateral dynamics given in Stevens and Lewis (1992). The transfer function of the system is a square 2×2 matrix. The individual SISO transfer functions in this 2-input/2-output open-loop system are

$$H_{11}(s) = \frac{14.8}{s(s + 0.0163)(s + 3.615)(s + 20.2)} \tag{1}$$

$$H_{12}(s) = \frac{-36.9s(s + 2.237)[(s + 0.55)^2 + 2.49^2]}{(s + 0.0163)(s + 1)(s + 3.615)(s + 20.2)[(s + 0.4225)^2 + 3.063^2]} \tag{2}$$

$$H_{21}(s) = \frac{-2.65(s + 2.573)(s - 2.283)}{s(s + 0.0163)(s + 3.615)(s + 20.2)[(s + 0.4225)^2 + 3.063^2]} \tag{3}$$

$$H_{22}(s) = \frac{-0.719.[(s + 0.139)^2 + 0.446^2]}{(s + 0.0163)(s + 1)(s + 20.2)[(s + 0.4225)^2 + 3.063^2]}. \tag{4}$$

The standard Bode magnitude plots for these SISO transfer functions are shown in Fig. 9.2-4.

On the other hand, shown in Fig. 9.2-5 are the singular values of this multivariable system. Note that it is not immediately evident how they relate to the SISO plots in Fig. 9.2-4. In the next section we shall see that bounds for guaranteed robustness are given for MIMO systems in terms of the minimum singular value being large at low frequencies (for performance robustness) and the maximum singular value being small at high frequencies (for stability robustness). The lack of any clear correspondence between Figs. 9.2-4 and 9.2-5 shows that these bounds cannot be expressed in terms of the individual SISO Bode plots.

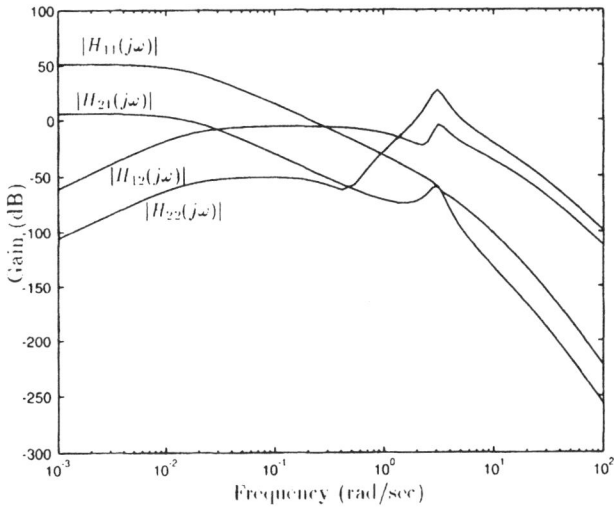

FIGURE 9.2-4   SISO Bode magnitude plots for F-16 lateral dynamics.

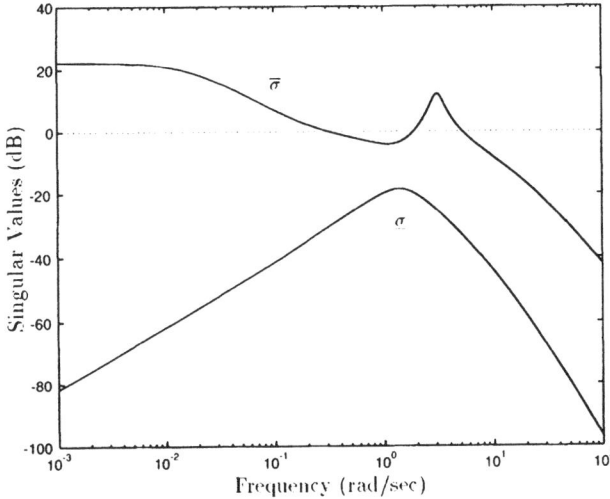

FIGURE 9.2-5   Singular values for F-16 lateral dynamics.   ■

Frequency-domain Performance Specifications

We have seen how to make a multivariable Bode magnitude plot of a square transfer-function matrix. It is now necessary to discuss performance specifications in the frequency domain in order to determine what a "desirable" Bode plot means in the MIMO case. The important point is that the low-frequency requirements are generally in terms of the minimum singular value being large, while the high-frequency requirements are in terms of the maximum singular value being small.

First, let us point out that the classical notion of bandwidth holds in the MIMO case. This is the frequency $j\omega$ for which the loop gain $GK(j\omega)$ passes through a value of 1, or 0 dB. If the bandwidth should be limited due to high-frequency noise considerations, the largest SV should satisfy $\bar{\sigma}(GK(j\omega)) = 1$, at the specified cutoff frequency $j\omega_c$.

$\mathscr{L}_2$ **Operator Gain**   To relate frequency-domain behavior to time-domain behavior, we may take into account the following considerations (Morari and Zafiriou 1989). Define the $\mathscr{L}_2$ norm of a vector time function $s(t)$ by

$$\|s\|_2 = \left[\int_0^\infty s^T(t)s(t)\,dt\right]^{1/2}. \tag{9.2-23}$$

This is related to the total energy in $s(t)$ and should be compared to the LQ performance index.

A linear time-invariant system has input $u(t)$ and output $z(t)$ related by the convolution integral

$$z(t) = \int_{-\infty}^\infty h(t-\tau)u(\tau)\,d\tau, \tag{9.2-24}$$

with $h(t)$ the impulse response. The $\mathscr{L}_2$ *operator gain*, denoted $\|H\|_2$, of such a system is defined as the smallest value of γ such that

$$\|z\|_2 \le \gamma \|u\|_2. \tag{9.2-25}$$

This is just the operator norm induced by the $\mathscr{L}_2$ vector norm. An important result is that the $\mathscr{L}_2$ operator gain is given by

$$\|H\|_2 = \max_\omega[\bar{\sigma}(H(j\omega))], \tag{9.2-26}$$

with $H(s)$ the system transfer function. That is, $\|H\|_2$ is nothing but the maximum value over ω of the maximum singular value of $H(j\omega)$. Thus, $\|H\|_2$ is an $\mathcal{H}_\infty$ norm in the frequency domain. This result gives increased importance to $\bar{\sigma}(H(j\omega))$, for if we are interested in keeping $z(t)$ small over a range of frequencies, then we should take care that $\bar{\sigma}(H(j\omega))$ is small over that range.

It is now necessary to see how this result may be used in deriving frequency-domain performance specifications. Some facts we use in this discussion are

$$\underline{\sigma}(GK) - 1 \le \underline{\sigma}(I + GK) \le \underline{\sigma}(GK) + 1, \tag{9.2-27}$$

$$\bar{\sigma}(M) = 1/\underline{\sigma}(M^{-1}), \tag{9.2-28}$$

$$\bar{\sigma}(AB) \le \bar{\sigma}(A)\bar{\sigma}(B) \tag{9.2-29}$$

for any matrices A, B, GK, M, with M nonsingular.

Before we begin a discussion of performance specifications, let us note the following. If $S(j\omega)$ is small, as desired at low frequencies, then

$$\overline{\sigma}(S = \overline{\sigma}[(I + GK)^{-1}] = \frac{1}{\overline{\sigma}(I + GK)} \approx \frac{1}{\underline{\sigma}(GK)}. \qquad (9.2\text{-}30)$$

That is, a large value of $\sigma(GK)$ guarantees a small value of $\overline{\sigma}(S)$. On the other hand, if $T(j\omega)$ is small, as is desired at high frequencies, then

$$\overline{\sigma}(T) = \overline{\sigma}[GK(I + GK)^{-1}] \approx \overline{\sigma}(GK). \qquad (9.2\text{-}31)$$

That is, a small value of $\overline{\sigma}(GK)$ guarantees a small value of $\overline{\sigma}(T)$.

This means that specifications that $S(j\omega)$ be small at low frequencies and $T(j\omega)$ be small at high frequencies may equally well be formulated in terms of $\sigma(GK)$ being large at low frequencies and $\overline{\sigma}(GK)$ being small at high frequencies. Thus, all of our performance specifications will be in terms of the minimum and maximum SVs of the loop gain $GK(j\omega)$. The practical significance of this is that we need compute only the SVs of $GK(j\omega)$, and not those of $S(j\omega)$ and $T(j\omega)$. These notions are symbolized in Fig. 9.2-6, where it should be recalled that $S + T = I$.

We first consider low-frequency specifications on the singular value plot, and then high-frequency specifications. According to our discussion relating to (9.2-17), the former will involve the reference input $r(t)$ and disturbances $d(t)$, while the latter will involve the sensor noise $n(t)$.

Low-frequency Specifications For low frequencies, let us suppose that the sensor noise $n(t)$ is zero so that (9.2.17) becomes

$$E(s) = S(s)(R(s) - D(s)). \qquad (9.2\text{-}32)$$

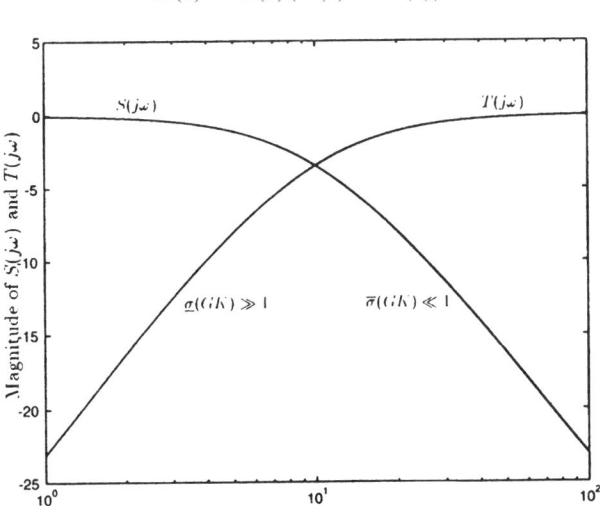

FIGURE 9.2-6 Magnitude specifications on $S(j\omega)$ and $T(j\omega)$, and $GK(j\omega)$.

Thus, to keep $||e(t)||_2$ small, it is only necessary to ensure that the $\mathcal{L}_2$ operator norm $||S||_2$ is small at all frequencies where $R(j\omega)$ and $D(j\omega)$ are appreciable. This may be achieved by ensuring that, at such frequencies, $\bar{\sigma}(S(j\omega))$ is small, As we have just seen, this may be guaranteed if we select

$$\underline{\sigma}(GK(j\omega)) \gg 1, \quad \text{for } \omega \le \omega_d, \tag{9.2-33}$$

where $D(s)$ and $R(s)$ are appreciable for $\omega \le \omega_d$.

Thus, exactly as in the classical case (Franklin et al. 1986), we are able to specify a low-frequency performance bound that guarantees performance robustness; that is, good performance in the face of low-frequency disturbances. For instance, to ensure that disturbances are attenuated by a factor of 0.01, (9.2-30) shows that we should ensure $\underline{\sigma}(GK(j\omega))$ is greater than 40 dB at low frequencies $\omega \le \omega_d$. At this point it is worth examining Fig. 9.2-6, which illustrates the frequency-domain performance specifications we are beginning to derive.

Another low-frequency performance bound may be derived from steady-state error considerations. Thus, suppose that $d(t) = 0$ and the reference input is a unit step of magnitude r so that $R(s) = r/s$. Then, according to (9.2-32) and the final value theorem (Franklin et al., 1986), the steady-state error e_∞ is given by

$$e_\infty = \lim_{s \to 0} sE(s) = rS(0). \tag{9.2-34}$$

To ensure that the largest component of e_∞ is less than a prescribed small acceptable value δ_∞, we should therefore select

$$\underline{\sigma}(GK(0)) > \frac{r}{\delta_\infty}. \tag{9.2-35}$$

The ultimate objective of all our concerns is to manufacture a compensator $K(s)$ in Fig. 9.2-1 that gives desirable performance. Let us now mention two low-frequency considerations that are important in the initial stages of the design of the compensator $K(s)$.

To make the steady-state error in response to a unit step at $r(t)$ exactly equal to zero, we may ensure that there is an integrator in each path of the system $G(s)$ so that it is of type 1 (Franklin et al. 1986). Thus, suppose that the system to be controlled is given by

$$\dot{x} = Ax + Bv \tag{9.2-36}$$

$$z = Hx.$$

To add an integrator to each control path, we may augment the dynamics so that

$$\begin{bmatrix} \dot{x} \\ \dot{\epsilon} \end{bmatrix} = \begin{bmatrix} A & B \\ 0 & 0 \end{bmatrix} \begin{bmatrix} x \\ \epsilon \end{bmatrix} + \begin{bmatrix} 0 \\ I \end{bmatrix} u \tag{9.2-37}$$

with the integrator outputs. See Fig. 9.2-7. The system $G(s)$ in Fig. 9.2-1 should now be taken as (9.2-37), which contains the integrators as a precompensator.

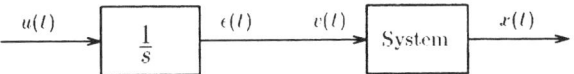

FIGURE 9.2-7   Plant augmented with integrators.

Although augmenting each control path with an integrator results in zero steady-state error, in some applications this may result in an unnecessarily complicated compensator. Note that the steady-state error may be made as small as desired without integrators by selecting $K(s)$ so that (9.2-35) holds.

A final concern on the low-frequency behavior of $G(s)$ needs to be addressed. It is desirable in many situations to have $\underline{\sigma}(GK)$ and $\overline{\sigma}(GK)$ close to the same value. Then, the speed of the responses will be nearly the same in all channels of the system. This is called the issue of balancing the singular values at low frequency. The SVs of $G(s)$ in Fig. 9.2-1 may be balanced at low frequencies as follows.

Suppose the plant has the state-variable description (9.2-36), and let us add a square constant precompensator gain matrix P, so that

$$v = Pu \tag{9.2-38}$$

is the relation between the control input $u(t)$ in Fig. 9.2-1 and the actual plant input $v(t)$. The transfer function of the plant plus precompensator is now

$$G(s) = H(sI - A)^{-1}BP. \tag{9.2-39}$$

As s goes to zero, this approaches

$$G(0) = H(-A)^{-1}BP,$$

as long as A has no poles at the origin. Therefore, we may ensure that $G(0)$ has all SVs equal to a prescribed value of γ by selecting

$$P = \gamma[H(-A)^{-1}B]^{-1}, \tag{9.2-40}$$

for then $G(0) = \gamma I$. The transfer function of (9.2-36) is

$$H(s) = H(sI - A)^{-1}B, \tag{9.2-41}$$

whence we see that the required value of the precompensator gain is

$$P = \gamma H^{-1}(0). \tag{9.2-42}$$

This is nothing but the (scaled) reciprocal DC gain.

Example 9.2-3.   Precompensator for Balancing and Zero Steady-state Error

Let us design a precompensator for the system in Example 9.2-1 using the notions just discussed. Substituting the values of A, B, and H in (9.2-40) with $\gamma = 1$ yields

$$B = [H(-A)^{-1}B]^{-1} = \begin{bmatrix} 5 & 0 \\ 0 & 24.3333 \end{bmatrix}. \tag{1}$$

To ensure zero steady-state error as well as equal singular values at low frequencies, we may incorporate integrators in each input channel along with the gain matrix P by writing the augmented system

$$\begin{bmatrix} \dot{x} \\ \dot{\varepsilon} \end{bmatrix} = \begin{bmatrix} A & B \\ 0 & 0 \end{bmatrix} \begin{bmatrix} x \\ \varepsilon \end{bmatrix} + \begin{bmatrix} 0 \\ P \end{bmatrix} u. \tag{2}$$

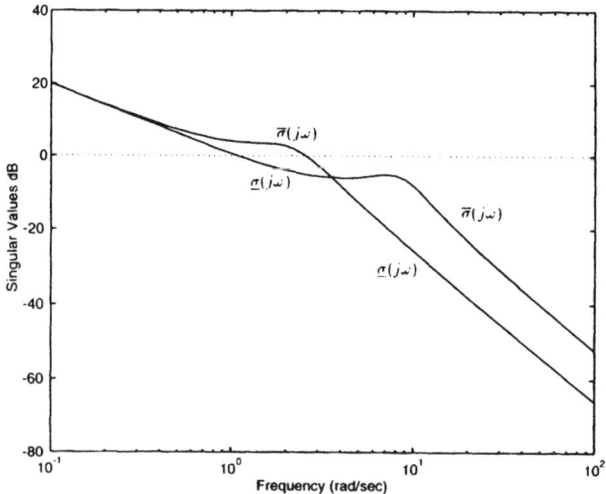

FIGURE 9.2-8   MIMO Bode magnitude plots for the augmented plant.

The singular-value plots for this plant plus precompensator appear in Fig. 9.2-8. At low frequencies there is now a slope of -20 dB/decade as well as equality of $\underline{\sigma}$ and $\overline{\sigma}$. Thus, the augmented system is both balanced and of type 1. Compare Fig. 9.2-8 with the singular value plot of the uncompensated system in Fig. 9.2-3. The remaining step is the selection of the feedback gain matrix for the augmented plant (2) so that the desired performance is achieved.   ∎

High-frequency Specifications   We now turn to a discussion of high-frequency performance specifications. The sensor noise is generally appreciable at frequencies above some known value ω_n (see Fig. 9.2-2). Thus, according to (9.2-17), to keep the tracking norm $||e(t)||_2$ small in the face of measurement noise we

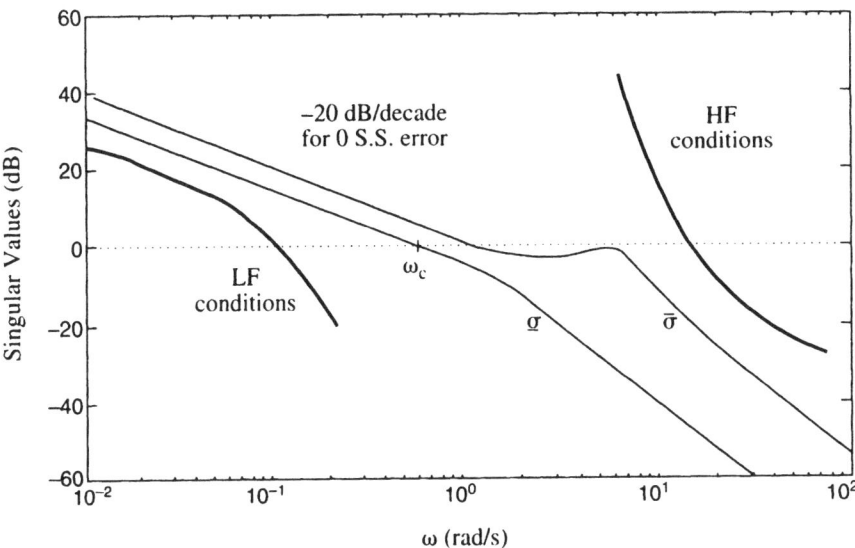

FIGURE 9.2-9 Frequency-domain performance specifications.

should ensure that the operator norm $||T||_2$ is small at high frequencies above this value. By (9.2-31) this may be guaranteed if

$$\overline{\sigma}(GK(j\omega)) \ll 1, \qquad \text{for } \omega \geq \omega_n. \qquad (9.2\text{-}43)$$

See Fig. 9.2-9. For instance, to ensure that sensor noise is attenuated by a factor of 0.1, we should guarantee that $\overline{\sigma}(GK(j\omega)) < -20$ dB for $\omega \geq \omega_n$.

One final high-frequency robustness consideration needs to be mentioned. It is unusual for the plant model to be exactly known. There are two basic sorts of modeling inaccuracies that concern us in controls. The first is plant parameter variation due to changes in the linearization equilibrium point of the nonlinear model. This is a low-frequency phenomenon and will be discussed in the next subsection. The second sort of inaccuracy is due to unmodeled high-frequency dynamics; this, we discuss here.

We are assuming a dynamical model for the purpose of controls design, and in so doing we, several times, neglect modes at high frequencies. Thus, although our design may guarantee closed-loop stability for the assumed mathematical model $G(s)$, stability is not assured for the actual plant $G'(s)$ with modes at high frequencies. To guarantee stability robustness in the face of plant parameter uncertainty, we may proceed as follows.

The model uncertainties may be of two types. The actual plant model G' and the assumed plant model G may differ by additive uncertainties so that

$$G'(j\omega) = G(j\omega) + \Delta G(j\omega), \qquad (9.2\text{-}44)$$

where the unknown discrepancy satisfies a known bound

$$\overline{\sigma}(\Delta G(j\omega)) < a(\omega), \tag{9.2-45}$$

with $a(\omega)$ known for all ω. On the other hand, the actual plant model $G'(s)$ and the assumed plant model $G(s)$ may differ by multiplicative uncertainties so that

$$G'(j\omega) = [I + M(j\omega)]G(j\omega), \tag{9.2-46}$$

where the unknown discrepancy satisfies a known bound

$$\overline{\sigma}(M(j\omega)) < m(\omega), \tag{9.2-47}$$

with $m(\omega)$ known for all ω. We shall show several ways of finding the bound $m(\omega)$. In Example 9.2-4 we show how to construct a reduced-order model for the system, which may then be used for controls design. There, $m(\omega)$ is determined from the neglected dynamics. In Example 9.3-1 we show how $m(\omega)$ may be determined in terms of the aircraft's neglected flexible modes. In the next subsection, we show how to determine $m(\omega)$ in terms of plant parameter variations in the linearized model due to operating point changes.

Since we may write (6.2.44) as

$$G'(j\omega) = [I + \Delta G(j\omega)G^{-1}(j\omega)]G(j\omega) = [I + M(j\omega)]G(j\omega), \tag{9.2-48}$$

we confine ourselves to a discussion of multiplicative uncertainties, following Doyle and Stein (1981). Suppose we have designed a compensator $K(s)$ so that the closed-loop system in Fig. 9.2-1 is stable. We should now like to derive a frequency-domain condition that guarantees the stability of the actual closed-loop system, which contains not $G(s)$, but $G'(s)$ satisfying (9.2-46), (9.2-47). For this, the multivariable Nyquist condition (Rosenbrock 1974) may be used.

Thus, it is required that the encirclement count of the map $|I + G'K|$ be equal to the negative number of unstable open-loop poles of $G'K$. By assumption, this number is the same as that of GK. Thus, the number of encirclements of $|I + G'K|$ must remain unchanged for all G' allowed by (9.2-47). This is assured if and only if $|I + G'K|$ remains nonzero as G is warped continuously toward G', or equivalently

$$0 < \underline{\sigma}[I + [I + \varepsilon M(s)]G(s)K(s)]$$

for all $0 \leq \varepsilon \leq$, all $M(s)$ satisfying (9.2-47), and all s on the standard Nyquist contour.

Since G' vanishes on the infinite radius segment of the Nyquist contour, and assuming for simplicity that no indentations are required along the $j\omega$-axis portion, this reduces to the following equivalent conditions:

$$0 < \underline{\sigma}[I + G(j\omega)K(j\omega) + \varepsilon M(j\omega)G(j\omega)K(j\omega)]$$

for all $0 \leq \varepsilon < 1$, $0 \leq \omega \leq \infty$, all M

$$\text{iff} \quad 0 \leq \underline{\sigma}[I + \varepsilon MGK(I + GK)^{-1}(I + GK)]$$

$$\text{iff} \quad 0 \leq \underline{\sigma}[I + MGK(I + GK)^{-1}]$$

all $0 \leq \omega < \infty$, and all M, if and only if

$$\underline{\sigma}[GK(I + GK)^{-1}] < \frac{1}{m(\omega)} \qquad (9.2\text{-}49)$$

for all $0 \leq \omega < \infty$. Thus, stability robustness translates into a requirement that the cosensitivity $T(j\omega)$ be bounded above by the reciprocal of the multiplicative modelling discrepancy bound $m(\omega)$.

In the case of high-frequency unmodeled dynamics, $1/m(\omega)$ is small at high ω, so that according to (9.2-31), we may simplify (9.2-49) by writing it in terms of the loop gain as

$$\bar{\sigma}(GK(j\omega)) < \frac{1}{m(\omega)}, \qquad (9.2\text{-}50)$$

for all ω such that $m(\omega) \gg 1$. This bound for stability robustness is illustrated in Fig. 9.2-9. An example will be useful at this point.

Example 9.2-4.   Model Reduction and Stability Robustness

In some situations we have a high-order aircraft model that is inconvenient to use for controller design. Examples occur in engine control and spacecraft control. In such situations, it is possible to compute a reduced-order model of the system, which may then be used for controller design. Here we show a convenient technique for model reduction as well as an illustration of the stability robustness bound $m(\omega)$. The technique described here is from Athans et al. (1986).

a. Model Reduction by Partial-fraction Expansion

Suppose the actual plant is described by

$$\dot{x} = Ax + Bu \qquad (1a)$$

$$z = Hx. \qquad (1b)$$

with $x \in R^n$. If A is simple with eigenvalues λ_i, right eigenvectors u_i, and left eigenvectors v_i so that

$$Au_i = \lambda_i u_i, \qquad v_i^T A = \lambda_i v_i^T, \qquad (2)$$

then the transfer function

$$G'(s) = H(sI - A)^{-1} B \qquad (3)$$

may be written as the partial-fraction expansion

$$G'(s) = \sum_{i=1}^{n} \frac{R_i}{s - \lambda_i}, \qquad (4)$$

with residue matrices given by

$$R_i = H u_i v_i^{\mathrm{T}} B. \qquad (5)$$

If the value of n is large, it may be desirable to find a reduced-order approximation to (1) for which a simplified compensator $K(s)$ in Fig. 9.2-1 may be designed. Then, if the approximation is a good one, the compensator $K(s)$ should work well when used on the actual plant $G'(s)$.

To find a reduced-order approximation $G(s)$ to the plant, we may proceed as follows. Decide which of the eigenvalues λ_i in (4) are to be retained in $G(s)$. This may be done using engineering judgment, by omitting high-frequency modes, by omitting terms in (4) that have small residues, and so on. Let the r eigenvalues to be retained in $G(s)$ be $\lambda_1, \lambda_2, \ldots, \lambda_1$.

Define the matrix

$$Q = \mathrm{diag}\{Q_i\}, \qquad (6)$$

where Q is an $r \times r$ matrix and the blocks Q_i are defined as

$$Q_i = \begin{cases} 1, & \text{for each real eigenvalue retained} \\[2mm] \begin{bmatrix} \frac{1}{2} & -\frac{j}{2} \\ \frac{1}{2} & \frac{j}{2} \end{bmatrix}, & \text{for each complex pair retained.} \end{cases} \qquad (7)$$

Compute the matrices

$$V \equiv Q^{-1} \begin{bmatrix} v_i^{\mathrm{T}} \\ \vdots \\ v_r^{\mathrm{T}} \end{bmatrix} \qquad (8)$$

$$U \equiv \begin{bmatrix} u_i^{\mathrm{T}} \cdots u_r^{\mathrm{T}} \end{bmatrix} Q. \qquad (9)$$

In terms of these constructions, the reduced-order system is nothing but a projection of (1) onto a space of dimension r with state defined by

$$w = Vx. \qquad (10)$$

The system matrices in the reduced-order approximate system

$$\dot{w} = Fw + Gu \qquad (11a)$$

$$z = Jw + Du \qquad (11b)$$

are given by

$$F = VAU$$

$$G = VB$$

$$J = HU, \tag{12}$$

with the direct-feed matrix given in terms of the residues of the neglected eigenvalues as

$$D = \sum_{i=r+1}^{n} -\frac{R_i}{\lambda_i} \tag{13}$$

The motivation for selecting such a D matrix is as follows.

The transfer function

$$G(s) = J(sI - F)^{-1}G + D \tag{14}$$

of the reduced system (11) is given as (verify!)

$$G(s) = \sum_{i-1}^{r} \frac{R_i}{s - \lambda_i} + \sum_{i=r+1}^{n} -\frac{R_i}{\lambda_i}. \tag{15}$$

Evaluating $G(j\omega)$ and $G'(j\omega)$ at $\omega = 0$, it is seen that they are equal at DC. Thus, the modeling errors induced by taking $G(s)$ instead of the actual $G'(s)$ occur at higher frequencies. Indeed, they depend on the frequencies of the neglected eigenvalues of (1).

To determine the $M(s)$ in (9.2-46) that is induced by the order reduction, note that

$$G' = (I + M)G \tag{16}$$

so that

$$M = (G' - G)G^{-1} \tag{17}$$

or

$$M(s) = \left[\sum_{i=r+1}^{n} -\frac{R_i}{\lambda_i} \frac{s}{s - \lambda_i} \right] G^{-1}(s). \tag{18}$$

Then, the high-frequency robustness bound is given in terms of

$$m(j\omega) = \bar{\sigma}(M(j\omega)). \tag{19}$$

Note that $M(j\omega)$ tends to zero as ω becomes small, reflecting our perfect certainty of the actual plant at DC.

b. An Example

Let us take an example to illustrate the model-reduction procedure and show also how to compute the upper bound $m(\omega)$ in (9.2-46), (9.2-47) on the high-frequency modeling errors thereby induced. To make it easy to see what is going on, we take a Jordan-form system.

Let there be prescribed the MIMO system

$$\dot{x} = \begin{bmatrix} -1 & 0 & 0 \\ 0 & -2 & 0 \\ 0 & 0 & -10 \end{bmatrix} x + \begin{bmatrix} 1 & 0 \\ 0 & 1 \\ 2 & 0 \end{bmatrix} u = Ax + bu \tag{19a}$$

$$z = \begin{bmatrix} 1 & 0 & 0 \\ 0 & 1 & 1 \end{bmatrix} x = Cx. \tag{19b}$$

The eigenvectors are given by $u_i = e_i$, $v_i = e_i$, $i = 1, 2, 3$, with e_i the ith column of the 3×3 identity matrix. Thus, the transfer function is given by the partial-fraction expansion

$$G'(s) = \frac{R_1}{s+1} + \frac{R_2}{s+2} + \frac{R_3}{s+10}, \tag{20}$$

with

$$R_1 = \begin{bmatrix} 1 & 0 \\ 0 & 0 \end{bmatrix}, \quad R_2 = \begin{bmatrix} 0 & 0 \\ 0 & 1 \end{bmatrix}, \quad R_3 = \begin{bmatrix} 0 & 0 \\ 2 & 0 \end{bmatrix}. \tag{21}$$

To find the reduced-order system that retains the poles at $\lambda = -1$ and $\lambda = -2$, define

$$Q = \begin{bmatrix} 1 & 0 \\ 0 & 1 \end{bmatrix}, \quad V = \begin{bmatrix} 1 & 0 & 0 \\ 0 & 1 & 0 \end{bmatrix}, \quad U = \begin{bmatrix} 1 & 0 \\ 0 & 1 \\ 0 & 0 \end{bmatrix} \tag{22}$$

and compute the approximate system

$$\dot{w} = \begin{bmatrix} -1 & 0 \\ 0 & -2 \end{bmatrix} w + \begin{bmatrix} 1 & 0 \\ 0 & 1 \end{bmatrix} u = Fw + Gu \tag{23a}$$

$$z = \begin{bmatrix} 1 & 0 \\ 0 & 1 \end{bmatrix} w + \begin{bmatrix} 0 & 0 \\ 0.2 & 0 \end{bmatrix} u = Jw + Du. \tag{23b}$$

This has a transfer function of

$$G(s) = \frac{R_1}{s+1} + \frac{R_2}{s+2} + D. \tag{24}$$

Singular value plots of the actual plant (19) and the reduced-order approximation (23) are shown in Fig. 9.2-10.

The multiplicative error is given by

$$M = (G' - G)G^{-1} = \begin{bmatrix} 0 & 0 \\ \dfrac{0.2s(s+1)}{s+10} & 0 \end{bmatrix}, \tag{25}$$

whence

$$m(\omega) = \bar{\sigma}(M(j\omega)) = \frac{0.2\omega\sqrt{\omega^2 + 1}}{\sqrt{\omega^2 + 100}}, \tag{26}$$

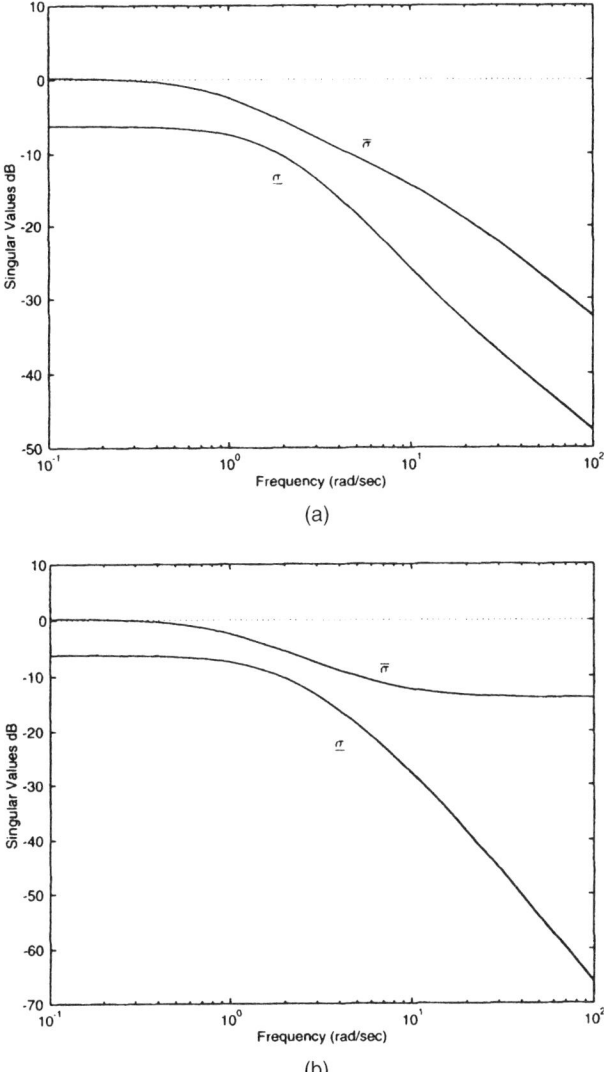

FIGURE 9.2-10   MIMO Bode magnitude plots of singular values. (a) Actual plant. (b) Reduced-order approximation.

and the high-frequency bound on the loop gain $GK(j\omega)$ is given by

$$\frac{1}{m(j\omega)} = \frac{5\sqrt{\omega^2 + 100}}{\omega\sqrt{\omega^2 + 1}}. \tag{27}$$

This bound is plotted in Fig. 9.2-11. Note that the modeling errors become appreciable (i.e., of magnitude one) at a frequency of 9.0 rads/sec. Above this frequency, we should

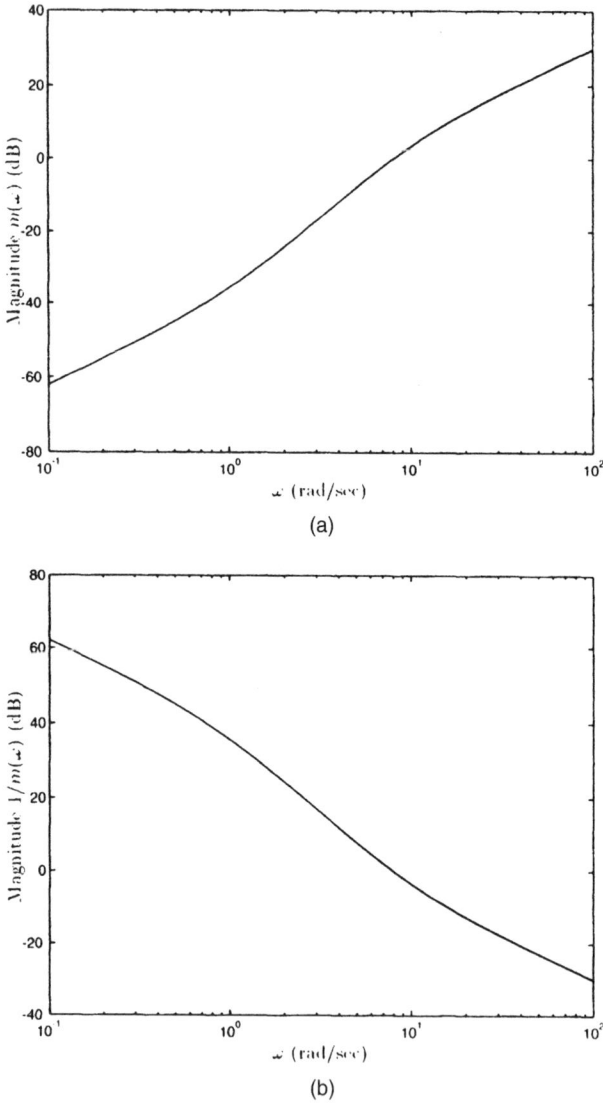

FIGURE 9.2-11   High-frequency stability robustness bound. (a) $m(\omega)$. (b) $1/m(\omega)$.

ensure that constraint (9.2-50) on the loop-gain magnitude holds to guarantee stability robustness. This will be a restriction on any compensator $K(s)$ designed using the reduced-order plant (23). ∎

Robustness Bounds for Plant Parameter Variations

Several dynamical systems are nonlinear; but for controller design we use linearized models obtained at some operating point. In practice, it is necessary to determine linear models at several design operating points over a specified operation envelope and determine optimal control gains for each one. Then, these design control gains are tabulated and scheduled using microprocessors, so that the gains most appropriate for the actual operating point of the plant are used in the controller. It is usual to determine which of the design operating points are closest to the actual operating point and use some sort of linear combination of the control gains corresponding to these design points.

It is important for the control gains to stabilize the system at all points near the design operating point for gain scheduling to be effective. In passing from operating point to operating point, the parameters of the state variable model vary. Using (9.2-49), we may design controllers that guarantee robust stability despite plant parameter variations.

Suppose the nominal perturbed model used for design is

$$\dot{x} = Ax + Bu \qquad (9.2\text{-}51)$$

$$y = Cx, \qquad (9.2\text{-}52)$$

which has the transfer function

$$G(s) = C(sI - A)^{-1}B. \qquad (9.2\text{-}53)$$

However, due to operating point changes the actual system is described by

$$\dot{x} = (A + \Delta A)x + (B + \Delta B)u \qquad (9.2\text{-}54)$$

$$y = (C + \Delta C)x, \qquad (9.2\text{-}55)$$

where the plant parameter variation matrices are ΔA, ΔB, ΔC. It is not difficult to show (see the problems at the end of the chapter) that this results in the transfer function

$$G'(s) = G(s) + \Delta G(s)$$

with

$$\Delta G(s) = C(sI - A)^{-1}\Delta B + \Delta C(sI - A^{-1})B$$
$$+ C(sI - A)^{-1} = \Delta A(sI - A)^{-1}B, \qquad (9.2\text{-}56)$$

where second-order effects have been neglected. Hence, (9.2-48) may be used to determine the multiplicative uncertainty bound $m(\omega)$. The consensitivity $T(j\omega)$ should then satisfy the upper bound (9.2-49) for guaranteed stability in the face of the parameter variations ΔA, ΔB, ΔC.

Since $(sI - A)^{-1}$ has a relative degree of at least one, the high-frequency roll-off of $\Delta G(j\omega)$ is at least -20 dB/decade. Thus, plant parameter variations yield an upper bound for the cosensitivity at low frequencies. Using (9.2-56) it is possible to design robust controllers over a range of operating points that do not require gain scheduling (Stevens and Lewis 1992).

9.3 ROBUST OUTPUT-FEEDBACK DESIGN

We should now like to incorporate the robustness concepts introduced in Section 9.2 into the LQ output-feedback design procedure for aircraft control systems. This may be accomplished using the following steps:

1. If necessary, augment the plant with added dynamics to achieve the required steady-state error behavior, or to achieve balanced singular values at DC. Use the techniques of Example 9.2-3.
2. Select a performance index, the PI weighting matrices Q and R.
3. Determine the optimal output-feedback gain K using, for instance, Table 8.1-1 or 8.1-2.
4. Simulate the time responses of the closed-loop system to verify that they are satisfactory. If not, select different Q, R and return to step 3.
5. Determine the low-frequency and high-frequency bounds required for performance robustness and stability robustness. Plot the loop gain singular values to verify that the bounds are satisfied. If they are not, select new Q, R and k and return to step 3.

An example will illustrate the robust output-feedback design procedure.

Example 9.3-1. Pitch Rate Control System Robust to Wind Gusts and Unmodeled Flexible Mode

Here we illustrate the design of a pitch rate control system that is robust in the presence of vertical wind gusts and the unmodeled dynamics associated with a flexible mode.

a. Control System Structure

The pitch rate CAS system is described as follows: the state and measured output are

$$
x = \begin{bmatrix} \alpha \\ q \\ \delta_e \\ \alpha_F \\ \varepsilon \end{bmatrix}, \quad y = \begin{bmatrix} \alpha_F \\ q \\ \varepsilon \end{bmatrix},
\tag{1}
$$

with α_F the filtered angle of attack and ε the output of the integrator added to ensure zero steady-state error. The performance output $z(t)$ that should track the reference input $r(t)$ is $q(t)$.

Linearizing the F-16 dynamics about the nominal flight condition yields in

$$
x = Ax + Bu + Gr
\tag{2}
$$

$$
y = Cx + Fr
\tag{3}
$$

$$
z = Hx,
\tag{4}
$$

with the system matrices

$$A = \begin{bmatrix} -1.01887 & 0.90506 & -0.00215 & 0 & 0 \\ 0.82225 & -1.07741 & -0.17555 & 0 & 0 \\ 0 & 0 & -20.2 & 0 & 0 \\ 10 & 0 & 0 & -10 & 0 \\ 0 & -57.2958 & 0 & 0 & 0 \end{bmatrix}$$

$$B = \begin{bmatrix} 0 \\ 0 \\ 20.2 \\ 0 \\ 0 \end{bmatrix}, \quad G = \begin{bmatrix} 0 \\ 0 \\ 0 \\ 0 \\ 1 \end{bmatrix}$$

$$C = \begin{bmatrix} 0 & 0 & 57.2958 & 0 & 0 \\ 0 & 57.2958 & 0 & 0 & 0 \\ 0 & 0 & 0 & 0 & 1 \end{bmatrix}, \quad F = \begin{bmatrix} 0 \\ 0 \\ 0 \\ 0 \end{bmatrix}$$

$$H = [0 \quad 57.2958 \quad 0 \quad 0 \quad 0].$$

The control input is

$$u = -Ky = -[k_\alpha \quad k_q \quad k_I]y = -k_\alpha \alpha_F - k_q q - k_I \varepsilon. \tag{5}$$

It is desired to select the control gains to guarantee a good response to a step command r in the presence of vertical wind gusts and the unmodeled dynamics of the first flexible mode.

b. Frequency-domain Robustness Bounds

The vertical wind gust noise is not white, but according to Stevens and Lewis (1992) has a spectral density given as

$$\Phi_w(s) = 2L\sigma^2 \frac{1 + 3L^2\omega^2}{(1 + L^2\omega^2)^2}, \tag{6}$$

with ω the frequency in rad/s, σ the turbulence intensity, and L the turbulence scale length divided by true airspeed. Using stochastic techniques like those in Example 9.4-2, the magnitude of the gust disturbance versus frequency can be found. It is shown in Fig. 9.3-1. We took $\sigma = 10$ ft/s and $L = 2.49$ s.

Let the transfer function of the rigid dynamics from $u(t)$ to $z(t)$ be denoted by $G(s)$. Then, the transfer function including the first flexible mode is given by Stevens and Lewis (1992)

$$G'(s) = G(s)F(s), \tag{7}$$

where

$$F(s) = \frac{\omega_n^2}{s^2 + 2\zeta\omega_n s + \omega_n^2} \tag{8}$$

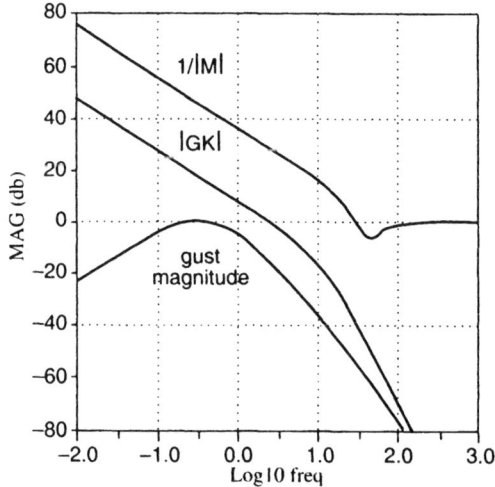

FIGURE 9.3-1   Frequency-domain magnitude plots and robustness bounds.

with $\omega_n = 40$ rad/sec and $\zeta = 0.3$. According to Section 9.2, therefore, the multiplicative uncertainty is given by

$$M(s) = F(s) - I = \frac{-s(s + 2\zeta\omega_n)}{s^2 + 2\zeta\omega_n s + \omega_n^2}. \qquad (9)$$

The magnitude of $1/M(j\omega)$ is shown in Fig. 9.3-1.

We should like to perform our controls design using only the rigid dynamics $G(s)$. Then, for performance robustness in the face of the gust disturbance and stability robustness in the face of the first flexible mode, the loop gain singular values should lie within the bounds implied by the gust disturbance magnitude and $1/|M(j\omega)|$.

c. Controls Design and Robustness Verification

In using the same design technique as in Example 8.2-1 we obtained the control gains

$$K = [-0.0807 \quad -0.4750 \quad 1.3610] \qquad (10)$$

The resulting step response is reproduced in Fig. 9.3-2, and the closed-loop poles were

$$s = -3.26 \pm j2.83$$

$$= -1.02 \qquad (11)$$

$$= -10.67, -14.09.$$

To verify that the robustness bounds hold for this design, it is necessary to find the loop gain $GK(s)$ of the closed-loop system. The magnitude of $GK(j\omega)$ is plotted in

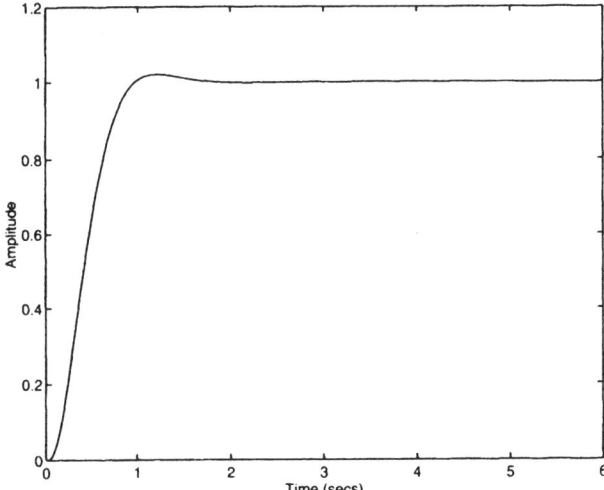

FIGURE 9.3-2 Optimal pitch-rate response.

Fig. 9.3-1. Note that the robustness bounds are satisfied. Therefore, this design is robust in the presence of vertical turbulence velocities up to 10 ft/s as well as the first flexible mode.   ∎

9.4   OBSERVERS AND THE KALMAN FILTER

The design equations for full state-variable feedback (see Chapter 3) are considerably simpler than those for output feedback. In fact, in state-variable design it is only necessary to solve the control matrix Riccati equation, for which there are many good techniques (ORACLS [Armstrong 1980], MATLAB [Control System Toolbox 1992]). By contrast, in output-feedback design it is necessary to solve three coupled nonlinear equations (see Table 8.1-2), which must generally be done using iterative techniques (Moerder and Calise 1985, Press et al. 1986).

Moreover, in the case of full state feedback, if the system (A, B) is reachable and $(\sqrt{Q}, A)$ is observable (with Q the state weighting in the PI), then the Kalman gain is guaranteed to stabilize the plant and yield a global minimum value for the PI. This is a fundamental result of modern control theory, and no such result yet exists for output feedback. The best that may be said is that if the plant is output stabilizable, then the algorithm of Table 8.3-2 yields a local minimum for the PI and a stable plant.

Another issue is that the LQ regulator with full state feedback enjoys some important robustness properties that are not guaranteed using output feedback. Specifically, as we shall see in Section 9.5, it has an infinite gain margin and $60°$ of phase margin.

Thus, state-feedback design offers some advantages over output feedback if the structure of the compensator is of no concern. Since all the states are seldom available, the first order of business is to estimate the full state $x(t)$, given only partial information in the form of the measured outputs $y(t)$. This is the observer design problem. Having estimated the state, we may then use the estimate of the state for feedback purposes, designing a feedback gain as if all the states were measurable. The combination of the observer and the state-feedback gain is then a dynamic regulator similar to those used in classical control, as we shall show in the last portion of this section. In the modern approach, however, it is straightforward to design multivariable regulators with desirable properties by solving matrix equations due to the fundamental separation principle, which states that the feedback gain and observer may be designed separately and then concatenated.

One of our prime objectives in this section is to discuss the linear quadratic Gaussian/loop-transfer recovery (LQG/LTR) technique for controls design. This is an important modern technique for the design of robust control systems. It relies on full state-feedback design, followed by the design of an observer that allows full recovery of the guaranteed robustness properties of the LQ regulator with state feedback.

Of course, observers and filters have important applications in system design in their own right. For instance, in aircraft control, the angle of attack is difficult to measure accurately; however, using an observer or Kalman filter it is not difficult to estimate the angle of attack very precisely by measuring pitch rate and normal acceleration (see Example 9.4-2).

Observer Design

In control design, all of the states are rarely available for feedback purposes. Instead, only the measured outputs are available. Using modern control theory, if the measured outputs capture enough information about the dynamics of the system, it is possible to use them to estimate or observe all the states. Then, these state estimates may be used for feedback purposes.

To see how a state observer can be constructed, consider the plant equations in state space form

$$\dot{x} = Ax + Bu \tag{9.4-1}$$

$$y = Cx, \tag{9.4-2}$$

with $x(t) \in R^n$ the state, $u(t) \in R^m$ the control input, and $y(t) \in R^p$ the available measured outputs. Let the estimate of $x(t)$ be $\hat{x}(t)$. We claim that the state observer is a dynamical system described by

$$\dot{\hat{x}} = A\hat{x} + Bu + L(y - C\hat{x}) \tag{9.4-3}$$

or

$$\dot{\hat{x}} = (A - LC)\hat{x} + Bu + Ly = A_o\hat{x} + Bu + Ly. \tag{9.4-4}$$

That is, the observer is a system with two inputs, namely $u(t)$ and $y(t)$, both of which are known.

Since $\hat{x}(t)$ is the state estimate, we could call

$$\hat{y} = C\hat{x} \qquad (9.4\text{-}5)$$

the estimated output. It is desired that $\hat{x}(t)$ be close to $x(t)$. Thus, if the observer is working properly, the quantity $y - \hat{y}$, which appears in (9.4-3), should be small. In fact,

$$\tilde{y} = y - \hat{y} \qquad (9.4\text{-}6)$$

is the output estimation error.

It is worth examining Fig. 9.4-1, which depicts the state observer. Note that the observer consists of two parts: a model of the system involving (A, B, C), and an error-correcting portion that involves the output error multiplied by L. We call matrix L the observer gain.

To demonstrate that the proposed dynamical system is indeed an observer, it is necessary to show that it manufactures an estimate $\hat{x}(t)$ that is close to the actual state $x(t)$. For this purpose, define the (state) estimation error as

$$\tilde{x} = x - \hat{x}. \qquad (9.4\text{-}7)$$

By differentiating (9.4-7) and using (9.4-1) and (9.4-4), it is seen that the estimation error has dynamics given by

$$\dot{\tilde{x}} = (A - LC)\tilde{x} = A_o\tilde{x}. \qquad (9.4\text{-}8)$$

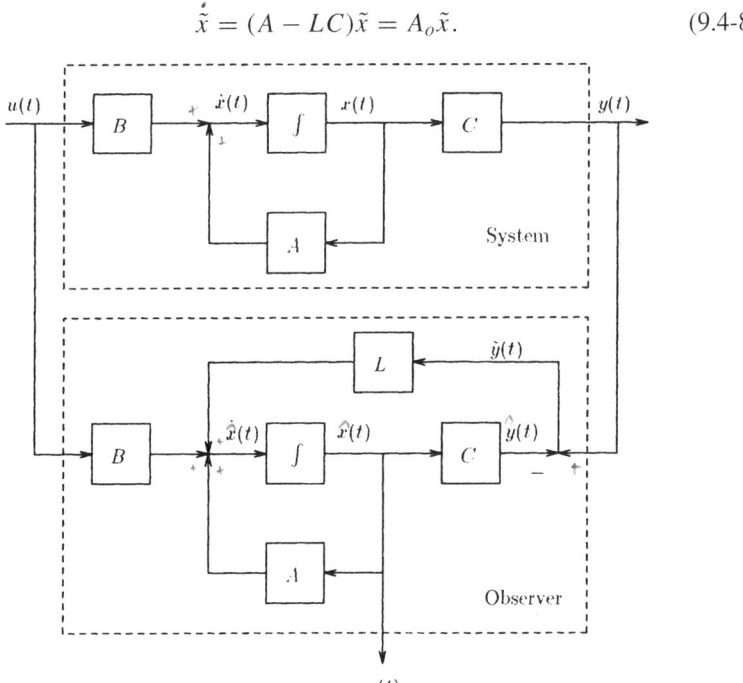

FIGURE 9.4-1   Standard feedback configuration.

The initial estimation error is $\tilde{x}(0) = x(0) - \hat{x}(0)$ the initial estimate, which is generally taken as zero.

It is required that the estimation error vanish with time for any $\tilde{x}(0)$, for then $\hat{x}(t)$ will approach $x(t)$. This will occur if $A_o = (A - LC)$ is asymptotically stable. Therefore, as long as we select the observer gain L so that $(A - LC)$ is stable, (9.4-3) is indeed an observer for the state in (9.4-1). The observer design problem is to select L so that the error vanishes suitably quickly. It is a well-known result of modern control theory that the poles of $(A - LC)$ may be arbitrarily assigned to desired locations if and only if (C, A) is observable.

Since, according to Fig. 9.4-1, we are injecting the output into the state derivative, L is called an output injection. Observers of the sort we are mentioning here are called output-injection observers, and their design could be called output-injection design. It is important to discuss the output-injection problem of selecting L so that $(A - LC)$ is stable, for it is a problem we have already solved under a different guise.

The state-feedback control law for system (9.4-1) is

$$u = -Kx, \tag{9.4-9}$$

which results in the closed-loop system

$$\dot{x} = (A - BK)x. \tag{9.4-10}$$

The state-feedback design problem is to select K for desired closed-loop properties. We have shown how this may be accomplished in Section 3.4. Thus, if we select the feedback gain as the Kalman gain

$$K = R^{-1}B^{T}P \tag{9.4-11}$$

with P the positive definite solution to the algebraic Riccati equation (ARE)

$$0 = A^{T}P + PA + Q - PBR^{-1}B^{T}P, \tag{9.4-12}$$

then, if (A, B) is reachable and $(\sqrt{Q}, A)$ is observable, the closed-loop system is guaranteed to be stable. The matrices Q and R are design parameters that will determine the closed-loop dynamics, as we have seen in the examples of Chapter 3.

Now, compare (9.4-8) and (9.4-10). They are very similar. In fact,

$$(A - LC)^{T} = A^{T} - C^{T}L^{T}, \tag{9.4-13}$$

which has the free matrix L^{T} to the right, exactly as in the state-feedback problem involving $(A - BK)$. This important fact is called duality; that is, state feedback and output injection are duals. (Note that $A - LC$ and $(A - LC)^{T}$ have the same poles.)

The important result of duality for us is that the same theory we have developed for selecting the state-feedback gain may be used to select the output-injection gain L. In fact, compare (9.4-13) with $(A - BK)$. Now, in the design equations (9.4-11) and (9.4-12) let us replace A, B, and K everywhere the occur by A^T, C^T, and L^T, respectively. The result is

$$L^T = R^{-1}CP$$

$$0 = AP + PA^T + Q - PC^T R^{-1}CP. \tag{9.4-14}$$

The first of these may be rewritten as

$$L = PC^T R^{-1}. \tag{9.4-15}$$

We call (9.4-14) the *observer* or *filter* ARE.

Let us note the following connection between reachability and observability. Taking the transpose of the reachability matrix yields

$$U^T = \begin{bmatrix} B & AB & A^2 B & \cdots & A^{n-1}B \end{bmatrix}^T$$

$$= \begin{bmatrix} B^T \\ B^T A^T \\ \vdots \\ B^T (A^T)^{n-1} \end{bmatrix}. \tag{9.4-16}$$

However, the observability matrix is

$$V = \begin{bmatrix} C \\ CA \\ \vdots \\ CA^{n-1} \end{bmatrix}. \tag{9.4-17}$$

Comparing U^T and V, it is apparent that they have the same form. In fact, since U and U^T have the same rank it is evident that (A, B) is reachable if and only if (B^T, A^T) is observable. This is another aspect of duality.

Taking into account these notions, an essential result on output injection is the following. It is the dual of the guaranteed stability using the Kalman gain discussed in Section 3.4. Due to its importance, we formulate it as a theorem.

Theorem 9.4-1. Let (C, A) be observable and $(A, \sqrt{Q})$ be reachable. Then the error system (9.4-9) using the gain L is given by (9.4-15), with P the unique positive definite solution to (9.4-14), is asymptotically stable. ∎

Stability of the error system guarantees that the state estimate $\hat{x}(t)$ will approach the actual state $x(t)$. By selecting L to place the poles of $(A - LC)$

far enough to the left in the s-plane, the estimation error $\tilde{x}(t)$ can be made to vanish as quickly as desired.

The significance of this theorem is that we may treat Q and R as design parameters that may be tuned until suitable observer behavior results for the gain computed from the observer ARE. As long as we select Q and R to satisfy the theorem, observer stability is assured. An additional factor, of course, is that software for solving the observer ARE is readily available (e.g., ORACLS [Armstrong 1980], MATLAB [Control System Toolbox 1992].

We have assumed that the system matrices (A, B, C) are exactly known. Unfortunately, in reality this is rarely the case. In several control problems for instance, (9.4-1), (9.4-2) represent a model of a nonlinear system at an equilibrium point. Variations in the operating point will result in variations in the elements of A, B, and C. However, if the poles of $(A - LC)$ are selected far enough to the left in the s-plane (i.e., fast enough), then the estimation error will be small in spite of uncertainties in the system matrices. That is, the observer has some robustness to modelling inaccuracies.

It is worth mentioning that there are many other techniques for the selection of the observer gain L. In the single-output case the observability matrix V is square. Then, Ackermann's formula (Franklin et al. 1986) may be used to compute L. If

$$\Delta_o(s) = |sI - (A - LC)| \qquad (9.4\text{-}18)$$

is the desired observer characteristic polynomial, then the required observer gain is given by

$$L = \Delta_o(A)V^{-1}e_n, \qquad (9.4\text{-}19)$$

with e_n the last column of the $n \times n$ identity matrix.

A general rule of thumb is that, for suitable accuracy in the state estimate $\hat{x}(t)$, the slowest observer pole should have a real part 5–10 times larger than the real part of the fastest system pole. That is, the observer time constants should be 5–10 times larger than the system time constants.

Example 9.4-1.   Observer Design for Double-integrator System

In Example 3.3-3 we discussed state-feedback design for systems obeying Newton's laws

$$\dot{x} = \begin{bmatrix} 0 & 1 \\ 0 & 0 \end{bmatrix} x + \begin{bmatrix} 0 \\ 1 \end{bmatrix} u = Ax + Bu, \qquad (1)$$

where the state is $x = [d \quad v]^{\mathrm{T}}$, with $d(t)$ the position and $v(t)$ the velocity, and the control $u(t)$ is an acceleration input. Let us take position measurements so that the measured output is

$$y = \begin{bmatrix} 1 \\ 0 \end{bmatrix} x = Cx. \qquad (2)$$

We should like to design an observer that will reconstruct the full state $x(t)$ given only position measurements. Let us note that simple differentiation of $y(t) = d(t)$ to obtain

$v(t)$ is unsatisfactory, since differentiation increases sensor noise. In fact, the observer is a low-pass filter that provides estimates while rejecting high-frequency noise. We shall discuss two techniques for observer design.

a. Riccati Equation Design

There is good software available in standard design packages for solving the filter ARE, e.g., MATLAB (Control System Toolbox 1992). However, in this example we want to analytically solve the ARE to show the relation between the design parameters Q and R and the observer poles.

Selecting $R = 1$ and $Q = \text{diag}\{q_d, q_v^2\}$, with q_d, q_v nonnegative, we may assume that

$$P = \begin{bmatrix} p_1 & p_2 \\ p_2 & p_3 \end{bmatrix} \tag{3}$$

for some scalars p_1, p_2, p_3 to be determined. The observer ARE (9.4-14) becomes

$$0 = \begin{bmatrix} 0 & 1 \\ 0 & 0 \end{bmatrix} \begin{bmatrix} p_1 & p_2 \\ p_2 & p_3 \end{bmatrix} + \begin{bmatrix} p_1 & p_2 \\ p_2 & p_3 \end{bmatrix} \begin{bmatrix} 0 & 0 \\ 1 & 0 \end{bmatrix} + \begin{bmatrix} q_d & 0 \\ 0 & q_v^2 \end{bmatrix}$$
$$- \begin{bmatrix} p_1 & p_2 \\ p_2 & p_3 \end{bmatrix} \begin{bmatrix} 0 & 1 \\ 0 & 0 \end{bmatrix} \begin{bmatrix} p_1 & p_2 \\ p_2 & p_3 \end{bmatrix}, \tag{4}$$

which may be multiplied out to obtain the three scalar equations

$$0 = 2p_2 - p_1^2 + q_d \tag{5a}$$

$$0 = p_3 - p_1 p_2 \tag{5b}$$

$$0 = -p_2^2 + q_v^2. \tag{5c}$$

Solving these equations gives

$$p_2 = q_v \tag{6a}$$

$$p_1 = \sqrt{2}\sqrt{q_v + \frac{q_d}{2}} \tag{6b}$$

$$p_3 = q_v \sqrt{2}\sqrt{q_v + \frac{q_d}{2}}, \tag{6c}$$

where we have selected the signs that make P positive definite.

According to (9.4-15), the observer gain is equal to

$$L = \begin{bmatrix} p_1 & p_2 \\ p_2 & p_3 \end{bmatrix} \begin{bmatrix} 1 \\ 0 \end{bmatrix}. \tag{7}$$

Therefore,

$$L = \begin{bmatrix} \sqrt{2}\sqrt{q_v + \frac{q_d}{2}} \\ q_v \end{bmatrix}. \tag{8}$$

Using (8), the error system matrix is found to be

$$A_o = A - LC = \begin{bmatrix} -\sqrt{2} \ \sqrt{q_v + \dfrac{q_d}{2}} & 1 \\ -q_v & 0 \end{bmatrix}. \tag{9}$$

Therefore, the observer characteristic polynomial is

$$\Delta_o(s) = |sI - A_o| = s^2 + 2\zeta\omega s + \omega^2, \tag{10}$$

with the observer natural frequency ω and damping ratio ζ given by

$$\omega = \sqrt{q_v}, \quad \zeta = \frac{1}{\sqrt{2}}\sqrt{1 + \frac{q_d}{2q_v}}. \tag{11}$$

It is now clear how selection of Q affects the observer behavior. Note that if $q_d = 0$, the damping ratio becomes the familiar $1/\sqrt{2}$. The reader should verify that the system is observable, and that $(A, \sqrt{Q})$ is reachable as long as $q_v \neq 0$. A comparison with Example 3.3-3, where a state feedback was designed for Newton's system, reveals some interesting aspects of duality.

b. Ackermann's Formula Design

Riccati-equation observer design is useful whether the plant has only one or multiple outputs. If there is only one output, we may use Ackermann's formula (9.4-19).
  Let the desired observer polynomial be

$$\Delta_o(s) = s^2 + 2\zeta\omega s + \omega^2 \tag{12}$$

for some specified damping ratio ζ and natural frequency ω. Then,

$$\Delta_o(A) = A^2 + 2\zeta\omega A + \omega^2 I = \begin{bmatrix} \omega^2 & 2\zeta\omega \\ 0 & \omega^2 \end{bmatrix} \tag{13}$$

$$V = \begin{bmatrix} C \\ CA \end{bmatrix}, \tag{14}$$

so that the observer gain is

$$L = \begin{bmatrix} 2\zeta\omega \\ \omega^2 \end{bmatrix}. \tag{15}$$

One may verify that the characteristic polynomial $A_o = A - LC$ is indeed (12).

c. Simulation

To design an observer with a complex pole pair having damping ratio of $\zeta = 1/\sqrt{2}$ and natural frequency of $\omega = 1$ rad/sec, the observer gain was selected as

$$L = \begin{bmatrix} \sqrt{2} \\ 1 \end{bmatrix}. \tag{16}$$

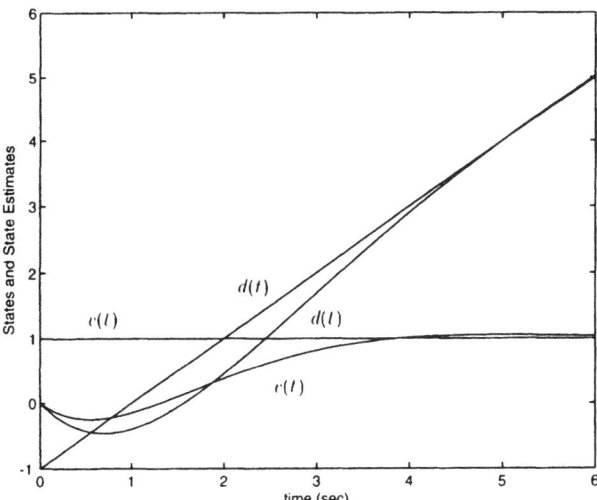

FIGURE 9.4-2   Actual and estimated states.

The resulting time histories of the actual states and their estimates are shown in Fig. 9.4-2. The initial conditions were $d(0) = -1$, $v(0) = 1$ and the input was $u(t) = 0$. The observer was started with initial states of $\hat{d}(0) = 0$, $\hat{v}(0) = 0$. The simulation was performed on SIMULINK.   ■

The Kalman Filter

Throughout this book we assume that the system is exactly known and that no modeling inaccuracies, disturbances, or noises are present. In fact, nature is seldom so cooperative. In Sections 9.2 and 9.3 we showed how to take account of uncertainties in the model and the environment using a robust frequency-domain approach. An alternative is to treat uncertainties using probability theory.

In this subsection we develop the Kalman filter, which is based on a probabilistic treatment of process and measurement noises. The Kalman filter is an observer that is used for navigation and other applications that require the reconstruction of the state from noisy measurements. Since it is fundamentally a low-pass filter, it has good noise-rejection capabilities. In Example 9.4-2 we show how to use the Kalman filter to estimate the angle of attack in the face of gust disturbances. In the next subsection we show how to use a Kalman filter along with a state-variable feedback to design a dynamic compensator for the plant. Then, in Section 9.5 we show how to use a state-variable feedback and a Kalman filter to design robust controllers by using the LQG/LTR technique.

We begin with a brief review of probability theory. It is not necessary to follow the derivation to use the Kalman filter: it is only necessary to solve the design equations in Table 9.4-1. Thus, one could skip the review that follows. However, an understanding of the theory will result in more sensible application of the filter. Supplemental references are Gelb (1974) and Lewis (1986).

Secor: R can be found by processing measurements when the output is held steady, remove the mean & only noise remain.

See papers/books by Dan Simon for help calculating Q (often done intuitively)

TABLE 9.4-1 The Kalman Filter

$$\dot{x} = Ax + Bu$$

$$y = Cx$$

$$x(0) \sim (x_0, P_0), \ w(t) \sim (0, Q), \ v(t) \sim (0, R)$$

Assumptions:

$w(t)$ and $v(t)$ are white noise processes orthogonal to each other and to $x(0)$. *i.e., they're independent processes*

Initialization:

$$\hat{x}(0) = \bar{x}_0$$

G just says which states are affected by noise are the input by the

Error covariance ARE:

$$AP + PA^T + GQG^T - PC^T R^{-1} CP = 0 \quad \text{solve for } P.$$

Kalman gain:

$$L = PC^T R^{-1}$$

see Fig. 9.4-1

Estimate dynamics (filter dynamics):

$$\dot{\hat{x}} = A\hat{x} + Bu + L(y - C\hat{x})$$

the whole point is to find L that gives the optimal balance between y (measured output) & Cx̂ (ŷ, estimated output based on plant model)

A Brief Review of Probability Theory   Suppose the plant is described by the stochastic dynamical equation

$$\dot{x} = Ax + Bu + Gw \tag{9.4-20}$$

$$y = Cx + v \tag{9.4-21}$$

with state $x(t) \in R^n$, control input $u(t) \in R^m$, and measured output $y(t) \in R^p$. Signal $w(t)$ is an unknown process noise that acts to disturb the plant. It could represent the effects of wind gusts, for instance, or unmodeled high-frequency plant dynamics. Signal $v(t)$ is an unknown measurement noise that acts to impair the measurements; it could represent sensor noise.

Since (9.4-20) is driven by process noise, the state $x(t)$ is now also a random process, as is $y(t)$. To investigate average properties of random processes we will require several concepts from probability theory (Papoulis 1984). The point is that although $w(t)$ and $v(t)$ represent unknown random processes, we do in fact know something about them that can help us in controls design. For instance, we may know their average values or total energy content. The concepts we now define allow us to incorporate this general sort of knowledge into our theory.

Given a random vector $z \in R^n$, we denote by $f_z(\zeta)$ the probability density function (PDF) of z. The PDF represents the probability that z takes on a value within the differential region $d\zeta$ centered at ζ. Although the value of z may be unknown, it is quite common in many situations to have a good feel for its PDF.

The expected value of a function $g(z)$ of a random vector z is defined as

$$E\{g(z)\} = \int_{-\infty}^{\infty} g(\zeta) f_z(\zeta) \, d\zeta. \tag{9.4-22}$$

The *mean* or *expected value* of z is defined by

$$E\{z\} = \int_{-\infty}^{\infty} \zeta f_z(\zeta)\,d\zeta, \tag{9.4-23}$$

which we symbolize by $\bar{z}$ to economize on notation. Note that $\bar{z} \in R^n$. The covariance of z is given by

$$P_z = E\{(z - \bar{z})(z - \bar{z})^T\}. \tag{9.4-24}$$

Note that P_z is an $n \times n$ constant matrix.

An important class of random vectors is characterized by the *Gaussian* or *normal* PDF

$$f_z(\zeta) = \frac{1}{\sqrt{(2\pi)^n |P_z|}} e^{-(\zeta-\bar{z})^T P_z^{-1}(\zeta-\bar{z})/2}. \tag{9.4-25}$$

In the scalar case $n = 1$ this reduces to the more familiar

$$f_z(\zeta) = \frac{1}{\sqrt{2\pi P_z}} e^{-(\zeta-\bar{z})^2/2P_z}, \tag{9.4-26}$$

which is illustrated in Fig. 9.4-3. Such random vectors take on values near the mean $\bar{z}$ with greatest probability, and have a decreasing probability of taking on values farther away from $\bar{z}$. Many naturally occurring random variables are Gaussian.

If the random vector is a time function, it is called a random process, symbolized as $z(t)$. Then, the PDF may also be time varying and we write $f_z(\zeta, t)$. One can imagine the PDF in Fig. 9.4-3 changing with time. In this situation, the

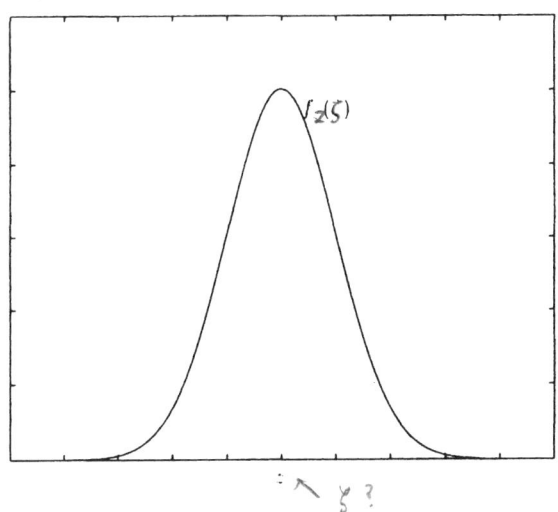

FIGURE 9.4-3   Gaussian PDF.

expected value and covariance matrix are also functions of time, so we write $\bar{z}(t)$ and $P_z(t)$.

Many random processes $z(t)$ of interest to us have a time-invariant PDF. These are stationary processes and, even though they are random time functions, they have a constant mean and covariance.

To characterize the relation between two random processes $z(t)$ and $x(t)$ we employ the joint PDF $f_{zx}(\zeta, \xi, t_1, t_2)$, which represents the probability that $(z(t_1), x(t_2))$ is within the differential area $d\zeta \times d\xi$ centered at (ζ, ξ). For our purposes, we assume that the processes $z(t), x(t)$ are jointly stationary, that is, the joint PDF is not a function of both times t_1 and t_2, but depends only on the difference $(t_1 - t_2)$.

In the stationary case, the expected value of the function of two variables $g(z, x)$ is defined as

$$E\{g(z(t_1), x(t_2))\} = \int_{-\infty}^{\infty} g(\zeta, \xi) f_{z,x}(\zeta, \xi, t_1 - t_2) \, d\zeta \, d\xi. \qquad (9.4\text{-}27)$$

In particular, the cross-correlation matrix is defined by

$$R_{zx}(\tau) = E\{z(t + \tau) x^T(t)\}. \qquad (9.4\text{-}28)$$

In the sequel, we shall briefly require the cross-correlation matrix of two nonstationary processes, which is defined as

$$R_{zx}(t, \tau) = E\{z(t) x^T(\tau)\}. \qquad (9.4\text{-}29)$$

Considering $z(t_1)$ and $z(t_2)$ as two jointly distributed random stationary processes, we may define the autocorrelation function of $z(t)$ as

$$R_z(\tau) = E\{z(t + \tau) z^T(t)\} \qquad (9.4\text{-}30)$$

The autocorrelation function gives us some important information about the random process $z(t)$. For instance,

$$\text{trace}[R_z(0)] = \text{trace}[E\{z(t) z^T(t)\}] = E\{\|z(t)\|\}$$

is equal to the total energy in the process $z(t)$. (In writing this equation recall that for any compatible matrices M and N, trace $(MN) = \text{trace}(NM)$.)

If

$$R_{zx}(\tau) = 0, \qquad (9.4\text{-}31)$$

we call $z(t)$ and $x(t)$ orthogonal. If

$$R_z(\tau) = P\delta(\tau), \qquad (9.4\text{-}32)$$

where P is a constant matrix and $\delta(t)$ is the Dirac delta, then $z(t)$ is orthogonal to $z(t + \tau)$ for any $\tau \neq 0$. What this means is that the value of the process $z(t)$ at one time t is unrelated to its value at another time $\tau \neq t$. Such a process is called white noise. An example is the thermal noise in an electric circuit, which is due to the thermal agitation of the electrons in the resistors.

Note that $P\delta(0)$ is the covariance of $z(t)$, which is unbounded. We call P a spectral density matrix. It is sometimes loosely referred to as a covariance matrix.

Derivation of the Kalman Filter   We may now return to system (9.4-20), (9.4-21). Neither the initial state $x(0)$, the process noise $w(t)$, nor the measurement noise $v(t)$ is exactly known. However, in practice we may have some feel for their general characteristics. Using the concepts we have just discussed, we may formalize this general knowledge so that it may be used in controls design.

The process noise is due to some sort of system disturbance such as wind gusts, the measurement noise is due to sensor inaccuracies, and the initial state is uncertain because of our ignorance. Since these are all unrelated, it is reasonable to assume that $x(0)$, $w(t)$, and $v(t)$ are mutually orthogonal. Some feeling for $x(0)$ may be present in that we may know its mean $\bar{x}_o$ and covariance P_0. We symbolize this as

$$\tilde{x}(0) \approx (\tilde{x}_0, P_0). \tag{9.4-33}$$

It is not unreasonable to assume that $w(t)$ and $v(t)$ have means of zero, since, for instance, there should be no bias on the measuring instruments. We shall also assume that the process noise and measurement noise are white noise processes so that

$$R_w(\tau) = E\{w(t + \tau)w^{\mathrm{T}}(t)\} = Q\delta(\tau) \tag{9.4-34}$$

$$R_v(\tau) = E\{v(t + \tau)v^{\mathrm{T}}(t)\} = R\delta(\tau). \tag{9.4-35}$$

Spectral density matrices Q and R will be assumed known. (Often, we have a good feel for the standard deviations of $w(t)$ and $v(t)$.) According to (9.4-30), Q and R are positive semi-definite. We shall assume in addition that R is nonsingular. In summary, we shall assume that

$$w(t) \quad (0, Q), \quad Q \geq 0 \tag{9.4-36}$$

$$v(t) \quad (0, R), \quad R > 0. \tag{9.4-37}$$

The assumption that $w(t)$ and $v(t)$ are white may in some applications be a bad one. For instance, wind gust noise is generally of low frequency. However, suppose that $w(t)$ is not white. Then, we can determine a system description

$$\dot{x}_w = A_w x_w + B_w n \tag{9.4-38}$$

$$w = C_w x_w + D_w n, \tag{9.4-39}$$

which has a white noise input $n(t)$ and output $w(t)$. This is called a noise-shaping filter. These dynamics may be combined with the plant (9.4-20), (9.4-21) to obtain the augmented dynamics

$$\begin{bmatrix} \dot{x} \\ \dot{x}_w \end{bmatrix} = \begin{bmatrix} A & GC_w \\ 0 & A_w \end{bmatrix} \begin{bmatrix} x \\ x_w \end{bmatrix} + \begin{bmatrix} B \\ 0 \end{bmatrix} u + \begin{bmatrix} CD_w \\ B_w \end{bmatrix} n \qquad (9.4\text{-}40)$$

$$y = [C0] \begin{bmatrix} x \\ x_w \end{bmatrix} + v. \qquad (9.4\text{-}41)$$

This augmented system does have a white process noise $n(t)$. A similar procedure may be followed if $v(t)$ is nonwhite. Thus, we can generally describe a plant with nonwhite noises in terms of an augmented system with white process and measurement noises.

The determination of a system (9.4-38), (9.4-39) that describes nonwhite noise $w(t)$ (or $v(t)$) is based on factoring the spectral density of the noise $w(t)$. For details see Lewis (1986). We illustrate the procedure in Example 9.4-2.

We should now like to design an estimator for the stochastic system (9.4-20), (9.4-21) under the assumptions just listed. We shall propose the output-injection observer, which has the form

$$\dot{\hat{x}} = A\hat{x} + Bu + L(y - \hat{y}) \qquad (9.4\text{-}42)$$

or

$$\dot{\hat{x}} = (A - LC)\hat{x} + Bu + Ly. \qquad (9.4\text{-}43)$$

The time function $\hat{x}(t)$ is the state estimate and

$$\hat{y} = E\{Cx + v\} = C\hat{x} \qquad (9.4\text{-}44)$$

is the estimate of the output $y(t)$. (This expected value is actually the conditional mean given the previous measurements. See Lewis [1986].)

The estimator gain L must be selected to provide an optimal estimate in the presence of the noises $w(t)$ and $v(t)$. To select L, we need to define the estimation error

$$\tilde{x}(t) = x(t) - \hat{x}(t). \qquad (9.4\text{-}45)$$

Using (9.4-20) and (9.4-42) we may derive the error dynamics to be

$$\dot{\tilde{x}} = (A - LC)\tilde{x} + Gw - Lv$$
$$\equiv A_o\tilde{x} + Gw - Lv. \qquad (9.4\text{-}46)$$

Note that the error system is driven by both the process and measurement noise. The output of the error system may be taken as $\tilde{y} = y - \hat{y}$ so that

$$\tilde{y} = C\tilde{x}. \qquad (9.4\text{-}47)$$

The error covariance is given by

$$P(t) = E\{\tilde{x}\tilde{x}^T\}, \tag{9.4-48}$$

which is time varying. Thus, $\tilde{x}(t)$ is a nonstationary random process. The error covariance is a measure of the uncertainty in the estimate, and smaller values for $P(t)$ mean that the estimate is better, since the error is more closely distributed about its mean value of zero if $P(t)$ is smaller.

If the observer is asymptotically stable and $w(t)$ and $v(t)$ are stationary processes, then the error $\tilde{x}(t)$ will eventually reach a steady state in which it is also stationary with constant mean and covariance. The gain L will be chosen to minimize the steady-state error covariance P. Thus, the optimal gain L will be a constant matrix of observer gains.

Before determining the optimal gain L, let us compute the mean and covariance of the estimation error $\tilde{x}(t)$. Using (9.4-46) and the linearity of the expectation operator,

$$E\{\dot{\tilde{x}}\} = A_o E\{\tilde{x}\} + GE\{w\}^{0} - LE\{v\}^{0}, \tag{9.4-49}$$

so that

$$\frac{d}{dt}E\{\tilde{x}\} = A_o E\{\tilde{x}\}. \tag{9.4-50}$$

Thus, $E\{\tilde{x}\}$ is a deterministic time-varying quantity that obeys a differential equation with system matrix A_o. If $A_o = A - LC$ is stable, then $E\{\tilde{x}\}$ eventually stabilizes at a steady-state value of zero, since the process and measurement noise are of zero mean. Since

$$E\{\tilde{x}\} = E\{x\} - E\{\hat{x}\} = E\{x\} - \hat{x}, \tag{9.4-51}$$

it follows that in this case the estimate $\hat{x}(t)$ approaches $E\{x(t)\}$. Then, the estimate is said to be unbiased. According also to (9.4-51), the mean of the initial error $\tilde{x}(0)$ is equal to zero if the observer (9.4-43) is initialized to $\hat{x}(0) = \bar{x}_0$, with $\bar{x}_0$ the mean of $x(0)$.

If the process noise $w(t)$ and/or measurement noise $v(t)$ have means that are not zero, then according to (9.4-49) the steady-state value of $E\{\tilde{x}\}$ is not equal to zero. In this case, $\hat{x}(t)$ does not tend asymptotically to the true state $x(t)$, but is offset from it by the constant value $-E\{\tilde{x}\}$. Then, the estimates are said to be biased. To determine the error covariance, note that the solution of (9.4-46) is given by

$$\tilde{x}(t) = e^{A_o t}\tilde{x}(0) - \int_0^t e^{A_o(t-\tau)}Lv(\tau)\,d\tau + \int_0^t e^{A_o(t-\tau)}Gw(\tau)\,d\tau. \tag{9.4-52}$$

We shall soon require the cross-correlation matrices $R_{v\tilde{x}}(t, t)$ and $R_{w\tilde{x}}(t, t)$. To find them, use (9.4-52) and the assumption that $x(0)$ (and hence $\tilde{x}(0)$), $w(t)$, and $v(t)$ are orthogonal. Thus,

$$R_{v\tilde{x}}(t, t) = E\{v(t)\tilde{x}^T(t)\}$$

$$= -\int_0^t E\{v(t)v^T(\tau)\}L^T e^{A_0^T(t-\tau)}d\tau. \qquad (9.4\text{-}53)$$

Note that

$$R_v(t, \tau) = R\delta(t - \tau), \qquad (9.4\text{-}54)$$

but the integral in (9.4-53) has an upper limit of t. Recall that the unit impulse can be expressed as

$$\delta(t) = \lim_{T \to 0} \frac{1}{T} \prod \left(\frac{t}{T}\right), \qquad (9.4\text{-}55)$$

where the rectangle function

$$\frac{1}{T} \prod \left(\frac{t}{T}\right) = \begin{cases} 1, & |t| < \frac{T}{2} \\ 0, & \text{otherwise} \end{cases} \qquad (9.4\text{-}56)$$

is centered at $t = 0$. Therefore, only half the area of $\delta(t - \tau)$ should be considered as being to the left of $\tau = t$. Hence, (9.4-53) is

$$R_{v\tilde{x}}(t, t) = -\tfrac{1}{2}RL^T. \qquad (9.4\text{-}57)$$

Similarly,

$$R_{w\tilde{x}}(t, t) = E\{w(t)\tilde{x}^T(t)\}$$

$$= \int_0^t E\{w(t)w^T(\tau)\}G^T e^{A_0^T(t-\tau)}d\tau, \qquad (9.4\text{-}58)$$

or

$$R_{w\tilde{x}}(t, t) = \tfrac{1}{2}QG^T. \qquad (9.4\text{-}59)$$

To find a differential equation for $P(t) = E\{\tilde{x}\tilde{x}^T\}$, write

$$\dot{P}(t) = E\left\{\frac{d\tilde{x}}{dt}\tilde{x}^T\right\} \overset{+\ ?}{=} E\left\{\tilde{x}\frac{d\tilde{x}^T}{dt}\right\}. \qquad (9.4\text{-}60)$$

According to the error dynamics (9.4-46) the first term is equal to

$$E\left\{\frac{d\tilde{x}}{dt}\tilde{x}^T\right\} = (A - LC)P + \tfrac{1}{2}LRL^T + \tfrac{1}{2}GQG^T, \qquad (9.4\text{-}61)$$

where we have used (9.4-57) and (9.4-59). To this equation add its transpose to obtain

$$\dot{P} = A_o P + P A_o^{\mathrm{T}} + LRL^{\mathrm{T}} + GQG^{\mathrm{T}}. \tag{9.4-62}$$

What we have derived in (9.4-62) is an expression for the error covariance when the observer (9.4-43) is used with a specific gain L. Given any L such that $(A - LC)$ is stable, we may solve (9.4-62) for $P(t)$, using as initial condition $P(0) = P_0$, with P_0 the covariance of the initial state, which represents the uncertainty in the initial estimate $\hat{x}(0) = \bar{x}_0$.

Clearly, gains that result in smaller error covariances $P(t)$ are better, for then the error $\tilde{x}(t)$ is generally closer to its mean of zero. That is, the error covariance is a measure of the performance of the observer, and smaller covariance matrices are indicative of better observers. We say that P is a measure of the uncertainty in the estimate. (Given symmetric positive semidefinite matrices P_1 and P_2, P_1 is less than P_2 if $(P_2 - P_1) \geq 0$.)

The error covariance $P(t)$ reaches a bounded steady-state value P as $t \to \infty$ as long as A_o is asymptotically stable. At steady state, $\dot{P} = 0$ so that (9.4-62) becomes the algebraic equation

$$0 = A_o P + P A_o^{\mathrm{T}} + LRL^{\mathrm{T}} + GQG^{\mathrm{T}}. \tag{9.4-63}$$

The steady-state error covariance is the positive semi-definite solution to (9.4-63). To obtain a constant observer gain, we may select L to minimize the steady-state error covariance P. Necessary conditions for L are now easily obtained after the same fashion that the output feedback gain K was obtained in Section 3.3.

Thus, define a performance index (PI)

$$J = \tfrac{1}{2} \mathrm{trace}(P). \tag{9.4-64}$$

(Note that trace(P) is the sum of the eigenvalues of P. Thus, a small J corresponds to a small P.) To select L so that J is minimized subject to the constraint (9.4-63), define the Hamiltonian

$$H = \tfrac{1}{2} \mathrm{trace}(P) + \tfrac{1}{2} \mathrm{trace}(gS), \tag{9.4-65}$$

where

$$g = A_o P + P A_o^{\mathrm{T}} + LRL^{\mathrm{T}} + GQG^{\mathrm{T}} \tag{9.4-66}$$

and S is an $n \times n$ undetermined (Lagrange) multiplier.

To minimize J subject to the constraint $g = 0$, we may equivalently minimize H with no constraints. Necessary conditions for a minimum are therefore given by

$$\frac{\delta H}{\delta S} = A_o P + P A_o^{\mathrm{T}} + LRL^{\mathrm{T}} + GQG^{\mathrm{T}} = 0 \tag{9.4-67}$$

$$\frac{\delta H}{\delta P} = A_o^T S + S A_o + I = 0 \tag{9.4-68}$$

$$\frac{1}{2}\frac{\delta H}{\delta L} = SLR - SPC^T = 0. \tag{9.4-69}$$

If A_o is stable, then the solution S to (9.4-68) is positive definite. Then, according to (9.4-69),

$$L = PC^T R^{-1}. \tag{9.4-70}$$

Substituting this value for L into (9.4-67) yields

$$(A - PC^T R^{-1} C)P + P(A - PC^T R^{-1} C)^T + PC^T R^{-1} CP + GQG^T = 0, \tag{9.4-71}$$

or

$$AP + PA^T + GQG^T - PC^T R^{-1} CP = 0. \tag{9.4-72}$$

To determine the optimal observer gain L, we may therefore proceed by solving (9.4-72) for the error covariance P and then using (9.4-70) to compute L. The matrix quadratic equation (9.4-72) is called the algebraic (filter) Riccati equation (ARE). There are several efficient techniques for solving the ARE for P (e.g., Armstrong [1980], IMSL [1980], MATLAB, Control System Toolbos [1992]).

The optimal gain L determined using (9.4-70) is called the (steady-state) Kalman gain, and the observer so constructed is called the (steady-state) Kalman filter. The term "steady state" refers to the fact that, although the optimal gain that minimizes $P(t)$ is generally time-varying, we have selected the optimal gain that minimizes the steady-state error covariance in order to obtain a constant observer gain. Since the gain must eventually be gain scheduled in actual flight controls applications, we require a constant gain to keep the number of parameters to be scheduled within reason.

The design equations for the Kalman filter are collected in Table 9.4-1. A block diagram appears in Fig. 9.4-1.

The steady-state Kalman filter is the best estimator with constant gains that has the dynamics of the form in the table. Such a filter is said to be linear. It can be shown (Lewis 1986) that if the process noise $w(t)$ and measurement noise $v(t)$ are Gaussian, then this is also the optimal steady-state estimator of any form.

The quantity

$$\tilde{y}(t) = y(t) - \hat{y}(t) = y(t) - C\hat{x}(t) \tag{9.4-73}$$

that drives the filter dynamics in the table is called the residual. For some interesting notions concerning the residual see Kailath (1980). For more information on the Kalman filter see Bryson and Ho (1975), Kwakernaak and Sivan (1972), and Lewis (1986).

The filter ARE should be compared to the ARE we discussed at the beginning of the section in connection with output injection design. There, no particular meaning was given to the auxiliary matrix P. In this stochastic setting, we have discovered that it is nothing but the error covariance. Small values of P generally indicate a filter with good estimation performance.

The theorem offered in connection with output-injection observer design also holds here. Thus, suppose (C, A) is observable and $(A, G\sqrt{Q})$ is reachable. Then the ARE has a unique positive definite solution P. Moreover, the error system (9.4-46) using the gain Kalman gain L given by (6.4-70), with P the unique positive definite solution to the ARE, is asymptotically stable.

One might be inclined to believe that the less noise in the system, the better. However, the actual situation is quite surprising. For existence of the Kalman filter it was necessary to assume that $R > 0$; that is, that the measurement noise corrupts all the measurements. If there are some noise-free measurements, a more complicated filter known as the Deyst filter must be used. Moreover, the assumption that $(A, G\sqrt{Q})$ is reachable means that the process noise should excite all the states.

Example 9.4-2.   *Kalman Filter Estimation of Angle of Attack in Gust Noise*

The short period approximation to the F-16 longitudinal dynamics is

$$\dot{x} = Ax + B\delta_e + Gw_g \tag{1}$$

with $x = [\alpha \quad q]^T$, α the angle of attack, q pitch rate, control input δ_e the elevator deflection, and w_g the vertical wind gust disturbance velocity. The plant matrices are found to be (see Stevens and Lewis [1992])

$$A = \begin{bmatrix} -1.01887 & 0.90506 \\ 0.82225 & -1.07741 \end{bmatrix}, \quad B = \begin{bmatrix} -0.00215 \\ -0.17555 \end{bmatrix}, \quad G = \begin{bmatrix} 0.00203 \\ -0.00164 \end{bmatrix}. \tag{2}$$

The vertical wind gust noise is not white, but according to Stevens and Lewis (1992) has a spectral density given as

$$\Phi_w(s) = 2L\sigma^2 \frac{1 + 3L^2\omega^2}{(1 + L^2\omega^2)^2}, \tag{3}$$

with ω the frequency in rad/s, σ the turbulence intensity, and L the turbulence scale length divided by true airspeed. Taking $\sigma = 10$ ft/sec and $L = (1750 \text{ ft})/(502 \text{ ft/sec}) = 3.49$ sec the gust spectral density is shown in Fig. 9.4-4.

a. Determination of Gust shaping Filter

Since w_g is not white, a noise-shaping filter of the form of (9.4-38), (9.4-39) must be determined by factoring $\Phi_w(s)$ (Lewis 1986). Note that

$$\Phi_w(s) = 2L\sigma^2 \frac{(1 + \sqrt{3}Lj\omega)(1 - \sqrt{3}Lj\omega)}{(1 + Lj\omega)^2(1 - Lj\omega)^2} \tag{4}$$

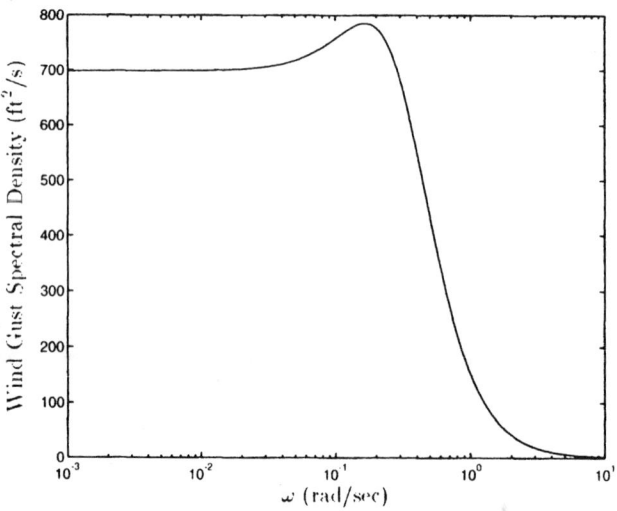

FIGURE 9.4-4   Vertical wind gust spectral density.

so that

$$\Phi_w(s) = H_w(s)H_w(-s) \tag{5}$$

with

$$H_w(s) = \sigma\sqrt{\frac{6}{L}}\,\frac{s + 1/L\sqrt{3}}{L(s + 1/L)^2} \tag{6}$$

$$= \sigma\sqrt{\frac{6}{L}}\,\frac{s + 1/L\sqrt{3}}{s^2 + 2s/L + 1/L^2}. \tag{7}$$

Now, a reachable canonical form realization of $H_w(s)$ (Kailath 1980) is given by

$$\dot{z} = \begin{bmatrix} 0 & 1 \\ -\dfrac{1}{L^2} & -\dfrac{2}{L} \end{bmatrix} z + \begin{bmatrix} 0 \\ 1 \end{bmatrix} w, \tag{8}$$

$$w_g = \gamma \begin{bmatrix} \dfrac{1}{L\sqrt{3}} & 1 \end{bmatrix} z, \tag{9}$$

where the gain is $\gamma = \sigma\sqrt{6/L}$. Using $\sigma = 10$, $L = 3.49$ yields

$$\dot{z} = \begin{bmatrix} 0 & 1 \\ -0.0823 & -0.5737 \end{bmatrix} z + \begin{bmatrix} 0 \\ 1 \end{bmatrix} w \equiv A_w z + B_w w \tag{10}$$

$$w_g = [2.1728 \quad 13.1192]z \equiv C_w z. \tag{11}$$

The shaping filter (10), (11) is a system driven by the white noise input $w(t) \approx (0, 1)$ that generates the gust noise $w_g(t)$ with spectral density given by (3).

b. Augmented Plant Dynamics

The overall system, driven by the white noise input $w(t) \approx (0, 1)$ and including an elevator actuator with transfer function $20.2/(s + 20.2)$ is given by (see (9.4-40))

$$\frac{d}{dt} \begin{bmatrix} \alpha \\ q \\ z_1 \\ z_2 \\ \delta_e \end{bmatrix}$$

$$= \begin{bmatrix} -1.01887 & 0.90506 & 0.00441 & 0.02663 & -0.00215 \\ 0.82225 & -1.07741 & -0.00356 & -0.02152 & -0.17555 \\ 0 & 0 & 0 & 1 & 0 \\ 0 & 0 & -0.08230 & -0.57370 & 0 \\ 0 & 0 & 0 & 0.57370 & -20.2 \end{bmatrix} \begin{bmatrix} \alpha \\ q \\ z_1 \\ z_2 \\ \delta_e \end{bmatrix}$$

$$+ \begin{bmatrix} 0 \\ 0 \\ 0 \\ 0 \\ 20.2 \end{bmatrix} u + \begin{bmatrix} 0 \\ 0 \\ 0 \\ 1 \\ 0 \end{bmatrix} w, \qquad (12)$$

with $u(t)$ the elevator actuator input. To economize on notation, let us symbolize this augmented system as

$$\dot{x} = Ax + Bu + Gw. \qquad (13)$$

c. Estimating Angle of Attack

Direct measurements of angle-of-attack α are noisy and biased. However, pitch rate q and normal acceleration n_z are convenient to measure. Using the results in Stevens and Lewis (1992),

$$n_z = 15.87875\alpha + 1.48113q. \qquad (14)$$

Therefore, let us select the measured output as

$$y = \begin{bmatrix} 15.87875 & 1.48113 & 0 & 0 & 0 \\ 0 & 1 & 0 & 0 & 0 \end{bmatrix} x = Cx + v, \qquad (15)$$

where $v(t)$ is measurement noise. A reasonable measurement noise covariance is

$$R = \begin{bmatrix} \dfrac{1}{20} & 0 \\ 0 & \dfrac{1}{60} \end{bmatrix}. \qquad (16)$$

The filter algebraic Riccati equation in Table 9.4-1 is solved using standard available software (Control System Toolbox, MATLAB, and the corresponding function *lqe.m*. The function *lqe2.m* is also recommended for more reliable results. For more details see the Control Systems Toolbox manual) to obtain the Kalman gain

$$
L = \begin{bmatrix}
0.0374 & -0.0041 \\
-0.0202 & 0.0029 \\
3.5985 & -0.2425 \\
1.9058 & -0.2873 \\
0.0000 & 0.0000
\end{bmatrix},
\tag{17}
$$

whence the Kalman filter is given by

$$
\dot{\hat{x}} = (A - LC)\hat{x} + Bu + Ly.
\tag{18}
$$

Note that the Kalman gain corresponding to the fifth state δ_e is zero. This is due to the fact that, according to (12), the gust noise $w(t)$ does not excite the actuator motor.

To implement the estimator we could use the state formulation (18) in a subroutine, or we could compute the transfer function to the angle-of-attack estimate given by

$$
H_\alpha(s) = [1 \quad 0 \cdots 0][sI - (A - LC)]^{-1}[B \quad L].
\tag{19}
$$

(Note that α is the first component of x.) Then, the angle-of-attack estimate is given by

$$
\hat{\alpha}(s) = H(s) \begin{bmatrix} U(s) \\ Y(s) \end{bmatrix},
\tag{20}
$$

so that $\alpha(t)$ may be estimated using $u(t)$ and $y(t)$, both of which are known. Similarly, the estimate of the wind gust velocity $w_g(t)$ may be recovered. Try to implement the Kalman filter and the systems using SIMULINK. ∎

Dynamic Regulator Design Using the Separation Principle

The fundamental approach to regulator and compensator design in this book involves selecting the compensator dynamics using the intuition of classical control and traditional design. Then, the adjustable compensator gains are computed using the output feedback design equations in Table 8.1-1, 8.1-2, or 8.2-1. The advantages of this approach include the following:

1. Good software for solving the design equations is available (e.g., the Davidon-Fletcher-Powell algorithm (Press et al. 1986).
2. General multi-input/multi-output controls design is straightforward.
3. If the design is sensible, the closed-loop system is generally stable for any choice of the weighting matrices Q and R.
4. All the intuition in classical controls design in the industry can be used to select the compensator structure.
5. Complicated compensator structures are avoided, which is important from the point of view of the pilot and also simplifies the gain-scheduling problem.

However, in complicated modern systems there may be no a priori guidelines for selecting the compensator structure. In this case, a combination of LQ state-feedback and observer/filter design proves very useful for controller design. This combination is known as linear quadratic Gaussian design and is explored next. In Section 9.5 we discuss the LQG/LTR technique for robust design, which has become popular in some aspects of control design.

Linear Quadratic Gaussian Design   The linear quadratic regulator (LQR) and the Kalman filter can be used together to design a dynamic regulator. This procedure is called linear quadratic Gaussian (LQG) design and will now be described. An important advantage of LQG design is that the compensator structure is given by the procedure, so that it need not be known beforehand. This makes LQG design useful in the control of complicated modern-day systems (e.g., space structures, aircraft engines), where an appropriate compensator structure may not be known.

Suppose the plant and measured output are given by

$$\dot{x} = Ax + Bu + Gw \qquad (9.4\text{-}74)$$

$$y = Cx + v \qquad (9.4\text{-}75)$$

with $x(t) \in R^n$, $u(t)$ the control input, $w(t)$ the process noise, and $v(t)$ the measurement noise. Suppose that the full state-feedback control

$$u = -Kx + r \qquad (9.4\text{-}76)$$

has been designed, with $r(t)$ the reference input command. That is, the state-feedback gain K has been selected by some technique, such as the LQR technique in Section 3.4. If the control (9.4-76) is substituted into (9.4-74), the closed-loop system is found to be

$$\dot{x} = (A - BK)x + Br + Gw. \qquad (9.4\text{-}77)$$

Full state feedback design is attractive because if the conditions in Section 3.4 hold, the closed-loop system is guaranteed stable. Such a strong result has not yet been shown for output feedback. Moreover, using full state feedback all the poles of $(A - BK)$ may be placed arbitrarily as desired. Finally, the state-feedback design equations are simpler than those for output feedback and may be solved using standard available routines. However, the control law (9.4-76) cannot be implemented since all the states are usually not available as measurements.

Now, suppose that an observer or Kalman filter

$$\dot{\hat{x}} = (A - LC)\hat{x} + Bu + Ly \qquad (9.4\text{-}78)$$

has been designed. That is, the filter gain L has been selected by any of the techniques discussed in this section to provide state estimates. Then, since all

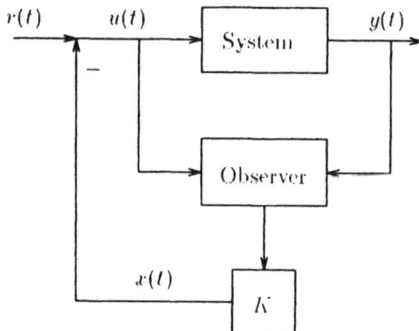

FIGURE 9.4-5   Regulator design using observer and state feedback.

the states are not measurable and the control (9.4-76) cannot be implemented in practice, we propose to feed back the estimate $\hat{x}(t)$ instead of the actual state $x(t)$. That is, let us examine the feedback law

$$u = -K\hat{x} + r. \tag{9.4-79}$$

The closed-loop structure using this controller is shown in Fig. 9.4-5. Due to the fact that the observer is a dynamical system, the proposed controller is nothing but a dynamical regulator of the sort seen in classical control theory. However, in contrast to classical design the theory makes it easy to design multivariable regulators with guaranteed stability even for complicated MIMO systems.

If K is selected using the LQR Riccati equation in Section 3.4 and L is selected using the Kalman filter Riccati equation in Table 9.4-1, this procedure is called LQG design. We propose to show that using this control:

1. The closed-loop poles are the same as if the full state feedback (9.4-76) had been used
2. The transfer function from $r(t)$ to $y(t)$ is the same as if (9.4-76) had been used.

The importance of these results is that the state feedback K and the observer gain L may be designed separately to yield desired closed-loop plant behavior and observer behavior. This is the separation principle, which is at the heart of modern controls design. Two important ramifications of the separation principle are that closed-loop stability is guaranteed, and good software is available to solve the matrix design equations that yield K and L.

The Separation Principle   To show the two important results just mentioned, define the estimation error (9.4-45) and examine the error dynamics (9.4-46). In terms of $\tilde{x}(t)$, we may write (9.4-79) as

$$u = -Kx + K\tilde{x} + r, \tag{9.4-80}$$

which when used in (9.4-74) yields

$$\dot{x} = (A - BK)x + BK\tilde{x} + Br + Gw. \qquad (9.4\text{-}81)$$

Now, write (9.4-81) and (9.4-46) as the augmented system

$$\begin{bmatrix} \dot{x} \\ \dot{\tilde{x}} \end{bmatrix} = \begin{bmatrix} A - BK & BK \\ 0 & A - LC \end{bmatrix} \begin{bmatrix} x \\ \tilde{x} \end{bmatrix} + \begin{bmatrix} B \\ 0 \end{bmatrix} r + \begin{bmatrix} G \\ G \end{bmatrix} w - \begin{bmatrix} 0 \\ L \end{bmatrix} v \qquad (9.4\text{-}82)$$

$$y = [C \quad 0] \begin{bmatrix} x \\ \tilde{x} \end{bmatrix} + v. \qquad (9.4\text{-}83)$$

Since the augmented system is block triangular, the closed-loop characteristic equation is

$$\Delta(s) = |sI - (A - BK)| \cdot |sI - (A - LC)| = 0. \qquad (9.4\text{-}84)$$

That is, the closed-loop poles are nothing but the plant poles that result by choosing K, and the desired observer poles that result by choosing L. Thus, the state-feedback gain K and observer gain L may be selected separately for desirable closed-loop behavior.

The closed-loop transfer function from $r(t)$ to $y(t)$ is given by

$$H_c(s) = [C \quad 0] \begin{bmatrix} sI - (A - BK) & -BK \\ 0 & sI - (A - LC) \end{bmatrix}^{-1} \begin{bmatrix} B \\ 0 \end{bmatrix}, \qquad (9.4\text{-}85)$$

and the triangular form of the system matrix makes it easy to see that

$$H_c(s) = C[sI - (A - BK)]^{-1}B. \qquad (9.4\text{-}86)$$

This, however, is exactly what results if the full state feedback (9.4-76) is used.

Of course, the initial conditions also affect the output $y(t)$. However, since the observer is stable, the effects of the initial error $\tilde{x}(0)$ will vanish with time. The observer poles (i.e., those of $(A - LC)$) should be chosen faster than the desired closed-loop plant poles (i.e., those of $(A - BK)$) for good closed-loop behavior.

Discussion   From our point of view, when possible it is usually better to design compensators using output feedback as we have demonstrated in the past chapters than to use separation principle design. To see why, let us examine the structure of the dynamic compensator in Fig. 9.4-5 in more detail.

The control input $u(t)$ may be expressed as

$$U(s) = H_y(s)Y(s) + H_u(s)U(s) \qquad (9.4\text{-}87)$$

where, according to (9.4-79) and (9.4-78), the transfer function from $y(t)$ to $u(t)$ is

$$H_y(s) = -K[sI - (A - LC)]^{-1}L \qquad (9.4-88)$$

and the transfer function from $u(t)$ to $u(t)$ is

$$H_u(s) = -K[sI - (A - LC)]^{-1}B. \qquad (9.4-89)$$

Now, note that the compensator designed by this technique has order equal to the order n of the plant. This means that it has too many parameters to be conveniently gain scheduled. Moreover, it has no special structure. This means that none of the classical controls intuition available in the industry has been used in its design.

It is possible to design reduced-order compensators using the separation principle. Three possible approaches are

1. First find a reduced-order model of the plant, then design a compensator for this reduced-order model.
2. First design a compensator for the full plant, then reduce the order of the compensator.
3. Design the reduced-order compensator directly from the full-order plant.

One technique for order-reduction is the partial-fraction-expansion technique in Example 9.2-3. Other techniques include principal component analysis (Moore 1982) and the frequency-weighted technique in Anderson and Liu (1989). It is important to realize that although the plant is minimal (i.e., reachable and observable), the LQ regulator may not be. That is, it may have unreachable or unobservable states.

In Section 9.5 we illustrate the design of a LQ regulator in robust design using the LQG/LTR approach.

9.5   LQG/LOOP-TRANSFER RECOVERY

We saw in Sections 9.2 and 9.3 how to use the multivariable Bode plot to design controllers guaranteeing performance robustness and stability robustness using outut feedback. In Section 9.4 we discussed the Kalman filter. In this section we propose to cover the linear quadratic Gaussian/loop-transfer recovery (LQG/LTR) design technique for robust controllers. This approach is quite popular in the current literature and has been used extensively to design multivariable control systems (Doyle and Stein 1981, Athans 1986). It is based on the fact that the linear quadratic regulator (LQR) using state variable feedback has certain guaranteed robustness properties.

Thus, suppose a state-feedback gain K has been computed using the ARE as in Section 3.4. This state feedback cannot be implemented since all of the states are not available as measurements; however, it can be used as the basis for the design of a dynamic LQ regulator by using a Kalman filter to provide state estimates for feedback purposes. We would like to discuss two issues. First, we show that state feedback, in contrast to output feedback, has certain guaranteed robustness properties in terms of gain and phase margins. Then, we see that the Kalman filter may be designed so that the dynamic regulator recovers the desirable robustness properties of full state feedback.

Guaranteed Robustness of the Linear Quadratic Regulator

We have discussed conditions for performance robustness and stability robustness for the general feedback configuration of the form shown in Fig. 9.2-1, where $G(s)$ is the plant and $K(s)$ is the compensator. The linear quadratic regulator using state feedback has many important properties, as we have seen Section 3.4. In this subsection we return to the LQR to show that it has certain guaranteed robustness properties that make it even more useful (Safonov and Athans 1977).

Thus, suppose that in Fig. 9.2-1 $K(s) = K$, the constant optimal LQ state-feedback gain determined using the algebraic Riccati equation (ARE) as in Section 3.4. Suppose moreover that

$$G(s) = (sI - A)^{-1}B \qquad (9.5\text{-}1)$$

is a plant in state-variable formulation.

For this subsection, it will be necessary to consider the loop gain referred to the control input $u(t)$ in Fig. 9.2-1. This is in contrast to the work in Section 9.2, where we referred the loop gain to the output $z(t)$, or equivalent to the signal $s(t)$ in the figure. Breaking the loop at $u(t)$ yields the loop gain

$$KG(s) = K(sI - A)^{-1}B. \qquad (9.5\text{-}2)$$

Our discussion will be based on the optimal return difference relation that holds for the LQR with state feedback (Lewis 1986, Grimble and Johnson 1988, Kwakernaak and Sivan 1972), namely,

$$[I + K(-sI - A)^{-1}B]^{\mathrm{T}}[I + K(sI - A)^{-1}B]$$
$$= I + \frac{1}{\rho}B^{\mathrm{T}}(-sI - A)^{-\mathrm{T}}Q(sI - A)^{-1}B. \qquad (9.5\text{-}3)$$

We have selected $R = I$.

Denoting the ith singular value of a matrix M as $\sigma_i(M)$, we note that by definition

$$\sigma_i(M) = \sqrt{\lambda_i(M^*M)}, \qquad (9.5\text{-}4)$$

with $\lambda_i(M^*M)$ the ith eigenvalue of matrix M^*M, and M^* the complex conjugate transpose of M. Therefore, according to (9.5-3) there results (Doyle and Stein 1981)

$$\sigma_i[I + KG(j\omega)] = \left[\lambda_i\left[I + \frac{1}{\rho}B^{\mathrm{T}}(-j\omega I - A)^{-\mathrm{T}}Q(j\omega I - A)^{-1}B\right]\right]^{1/2}$$

$$= \left[1 + \frac{1}{\rho}\lambda_i[B^{\mathrm{T}}(-j\omega I - A)^{-\mathrm{T}}Q(j\omega I - A)^{-1}B]\right]^{1/2}$$

or

$$\sigma_i[I + KG(j\omega)] = \left[1 + \frac{1}{\rho}\sigma_i^2[H(j\omega)]\right]^{1/2}, \qquad (9.5\text{-}5)$$

with

$$H(s) = H(sI - A)^{-1}B \qquad (9.5\text{-}6)$$

and $Q = H^{\mathrm{T}}H$. We could call (9.5-5) the optimal singular value relation of the LQR. It is important due to the fact that the right-hand side is known in terms of open-loop quantities before the optimal feedback gain is found by solution of the ARE, while the left-hand side is the closed-loop return difference. Thus, exactly as in classical control, we are able to derive properties of the closed-loop system in terms of properties of the open-loop system. According to this relation, for all ω the minimum singular value satisfies the LQ optimal singular value constraint

$$\underline{\sigma}[I + KG(j\omega)] \geq 1. \qquad (9.5\text{-}7)$$

Thus, the LQ regulator always results in a decreased sensitivity. Some important conclusions on the guaranteed robustness of the LQR may now be discovered using the multivariable Nyquist criterion (Postlethwaite et al. 1981), which we shall refer to the polar plot of the return difference $I + KG(s)$, where the origin is the critical point (Grimble and Johnson 1988). (Usual usage is to refer the criterion to the polar plot of the loop gain KG(s), where -1 is the critical point.) A typical polar plot of $\underline{\sigma}[I + KG(j\omega)]$ is shown in Figure 9.5-1, where the optimal singular value constraint appears as the condition that all the singular values remain outside the unit disc. To see how the end points of the plots were discovered, note that since $K(sI - A)^{-1}B$ has relative degree of at least one, its limiting value for $s = j\omega$ as $\omega \to \infty$ is zero. Thus, in this limit $I + KG(j\omega)$ tends to I. On the other hand, was $\omega \to 0$, the limiting value of $I + KG(j\omega)$ is determined by the DC loop gain, which should be large.

The multivariable Nyquist criterion says that the closed-loop system is stable if none of the singular value plots of $I + KG(j\omega)$ encircle the origin in the figure. Clearly, due to the optimal singular value constraint, no encirclements are possible. This constitutes a proof of the *guaranteed* stability of the LQR.

Multiplying the optimal feedback K by any positive scalar gain k results in a loop gain of $kKG(s)$, which has a minimum singular value plot identical to

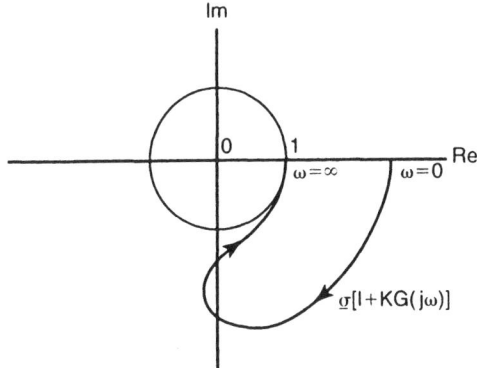

FIGURE 9.5-1 Typical polar plot for optimal LQ return difference (referred to the plant input).

the one in Fig. 9.5-1 except that it is scaled outward. That is, the $\omega \to 0$ limit (i.e., the DC gain) will be larger, but the $\omega \to \infty$ limit will still be 1. Thus, the closed-loop system will still be stable. In classical terms, the LQ regulator has an *infinite gain margin*.

The *phase margin* may be defined for multivariable systems as the angle marked "PM" in Fig. 9.5-2. As in the classical case, it is the angle through which the polar plot of $\sigma[I + KG(j\omega)]$ must be rotated (about the point 1) clockwise to make the plot go through the critical point.

Figure 9.5-3 combines Fig. 9.5-1 and Fig. 9.5-2. By using some simple geometry, we may find the value of the angle indicated as $60°$. Therefore, due to the LQ singular value constraint, the plot of $\sigma[I + KG(j\omega)]$ must be rotated through at least $60°$ to make it pass through the origin. The LQR thus has a guaranteed phase margin of at least $60°$. This means that a phase shift of up to $60°$ may be introduced in any of the m paths in Fig. 9.2-1, or in all paths simultaneously as long as the paths are not coupled to each other in the process.

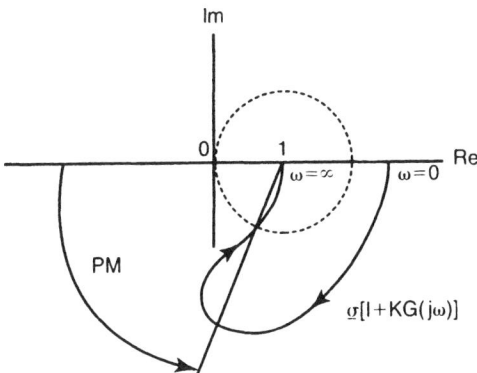

FIGURE 9.5-2 Definition of multivariable phase margin.

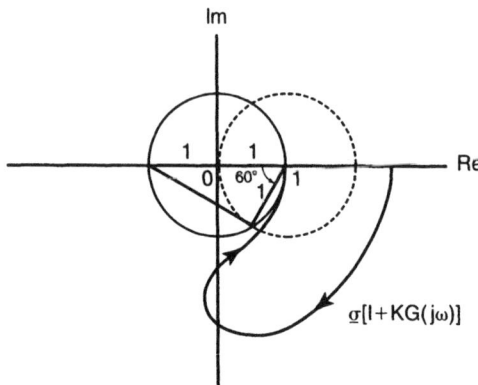

FIGURE 9.5-3   Guaranteed phase margin of the LQR.

This phase margin is excessive; it is higher than that normally required in classical control system design. this overdesign means that, in other performance aspects, the LQ regulator may have some deficiencies. One of these turns out to be that, at the crossover frequency (loop gain = 1), the slope of the multivariable Bode plot is −20 dB/decade, which is a relatively slow attenuation rate (Doyle and Stein 1981). By allowing a Q weighting matrix in the PI that is not positive semi-definite, it is possible to obtain better LQ designs that have higher roll-off rates at high frequencies (Shin and Chen 1974).

A stability robustness bound like (9.2-49) may be obtained for the loop gain referred to the input $u(t)$. It is

$$\overline{\sigma}[KG(I + KG)^{-1}] < \frac{1}{m(\omega)}. \tag{9.5-8}$$

The inverse of this is

$$m(\omega) < \frac{1}{\overline{\sigma}[KG(I + KG)^{-1}]} = \underline{\sigma}[KG(I + KG)^{-1}]. \tag{9.5-9}$$

It can be shown (see the problems at the end of the chapter) that (9.5-7) implies that

$$\underline{\sigma}[I + (KG(j\omega)) - 1] \geq \tfrac{1}{2}. \tag{9.5-10}$$

Therefore, the LQR remains stable for all multiplicative uncertainties in the plant transfer function that satisfy $m(\omega) < \tfrac{1}{2}$.

Loop-transfer Recovery

The controls design techniques we have discussed in Chapter 8 involve selecting a desirable compensator structure using classical controls intuition. Then, the compensator gains are adjusted using output-feedback design for suitable performance. Robustness may be guaranteed using the multivariable Bode plot as shown in Sections 9.2 and 9.3.

However, in some cases the plant may be so complex that there is little intuition available for selecting the compensator structure. In this event, the technique to be presented in this section may be useful for controller design, since it yields a suitable compensator structure automatically.

Let us examine here the plant

$$\dot{x} = Ax + Bu + Gw \tag{9.5-11}$$

$$y = Cx + v, \tag{9.5-12}$$

with process noise $w(t) \sim (0, M)$ and measurement $n(t) \sim (0, v^2 N)$ both white, $M > 0$, $N > 0$, and v a scalar parameter. We have seen that the full state feedback control

$$u = -Kx \tag{9.5-13}$$

has some extremely attractive features, including simplified design equations (Section 3.4) and some important guaranteed robustness properties. Unfortunately, these are not shared by an output-feedback control law, where the robustness must be checked independently. However, state feedback is usually impossible to use since all the states are seldom available for feedback in any practical application.

According to Fig. 9.5-4a, where the plant transfer function is

$$\Phi(s)B = (sI - A)^{-1}B, \tag{9.5-14}$$

the loop gain, breaking the loop at the input $u(t)$ is

$$L_s(s) = K\Phi B. \tag{9.5-15}$$

According to Section 9.4, if an observer or Kalman filter is used to produce a state estimate $\hat{x}(t)$, which is then used in the control law

$$u = -K\hat{x}, \tag{9.5-16}$$

the result is a regulator that, due to the separation principle, has the same transfer function as the state-feedback controller.

However, it is known that the guaranteed robustness properties of the full state-feedback controller are generally lost (Doyle 1978). In this section we assume that a state-feedback gain K has already been determined using, for instance, the algebraic Riccati-equation design technique in Section 3.4. This K yields suitable robustness properties of $K\Phi B$. We should like to present a technique for designing a Kalman filter that results in a regulator that recovers the guaranteed robustness properties of the full state-feedback control law as the design parameter v goes to zero. The technique is called LQG/loop-transfer recovery

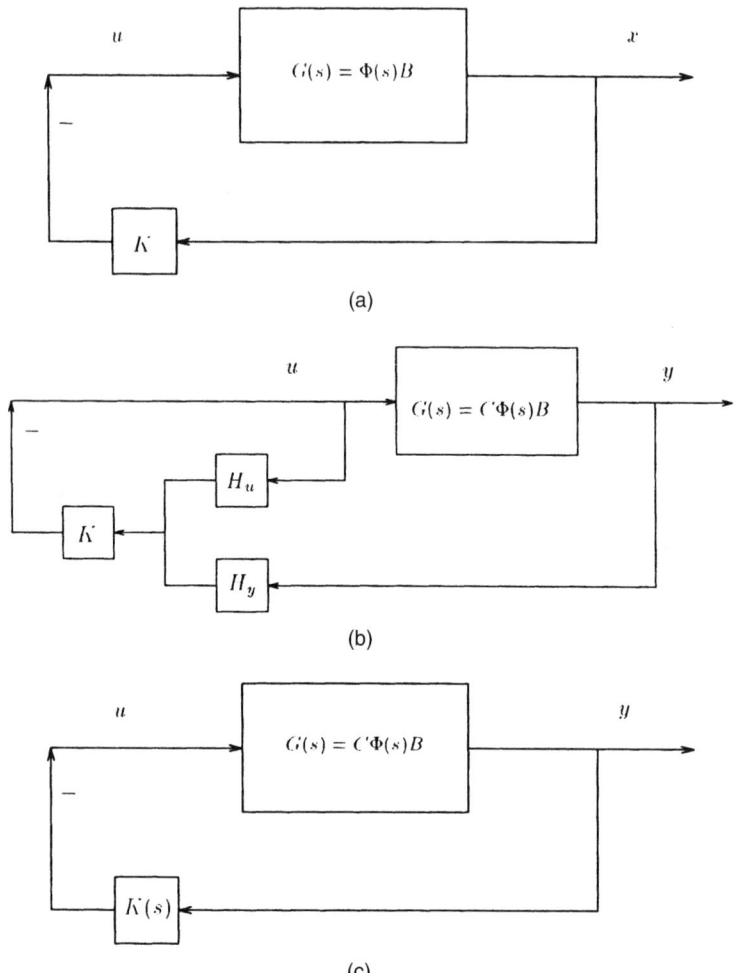

FIGURE 9.5-4   (a) Loop gain with full state feedback. (b) Regulator using observer and estimate feedback. (c) Regulator loop gain.

(LQG/LTR), since the loop gain (i.e., loop-transfer function) $K\Phi B$ of full state feedback is recovered in the regulator as $v \to 0$. As we shall see, the key to robustness using a stochastic regulator is in the selection of the noise spectral densities M and N.

Regulator Loop Gain   Using an observer or Kalman filter, the closed-loop system appears in Fig. 9.5-4b, where the regulator is given by (Section 9.4)

$$U(s) = -K(sI - A + LC)^{-1}BU(s) - K(sI - A + LC)^{-1}LY(s)$$
$$= -H_u(s)U(s) - H_y(s)Y(s) \qquad (9.5\text{-}17)$$

and L is the observer or Kalman gain. Denoting the observer resolvent matrix as

$$\Phi_o(s) = (sI - A + LC)^{-1}, \tag{9.5-18}$$

we write

$$H_u = K\Phi_o B, \quad H_y = K\Phi_o L. \tag{9.5-19}$$

To find an expression for $K(s)$ in Fig. 9.5-4c using the regulator, note that $(I + H_u)U = -H_y Y$, so that

$$U = -(I + H_u)^{-1} H_y Y = -K(s)Y. \tag{9.5-20}$$

However,

$$
\begin{aligned}
(I + H_u)^{-1} K &= [I + K(sI - (A - LC))^{-1} B]^{-1} K \\
&= [I - K(sI - (A - BK - LC))^{-1} B]K \\
&= K(sI - (A - BK - LC))^{-1}[(sI - (A - BK - LC)) - BK] \\
&= K(sI - (A - BK - LC))^{-1}\Phi_o^{-1},
\end{aligned}
$$

where the matrix inversion lemma was used in the second step. Therefore,

$$
\begin{aligned}
K(s) &= (I + Hu)^{-1} H_y \\
&= K[sI - (A - BK - LC)]^{-1}\Phi_o^{-1}\Phi_o L
\end{aligned}
$$

or

$$K(s) = K[sI - (A - BK - LC)]^{-1} L \equiv K\Phi_r L, \tag{9.5-21}$$

with $\Phi_r(s)$ the regulator resolvent matrix.

We now show how to make the loop gain (at the input) using the regulator

$$L_r(s) = K(s)G(s) = K\Phi_r LC\Phi B \tag{9.5-22}$$

approach the loop gain $L_s(s) = K\Phi B$ using full state feedback, which is guaranteed to be robust.

Recovery of State-feedback Loop Gain at the Input   To design the Kalman filter so that the regulator loop gain at the input $L_r(s)$ is the same as the state-feedback loop gain $L_s(s)$, we need to assume that the plant $C\Phi B$ is minimum phase (i.e., with stable zeros), with B and C of full rank and $\dim(u) = \dim(y)$. The references for this subsection are Doyle and Stein (1979, 1981), Athans (1986), and Stein and Athans (1987).

Let us propose $G = I$ and the process noise spectral density matrix

$$M = v^2 M_o + BB^{\mathrm{T}}, \tag{9.5-23}$$

with $M_o > 0$. Then, according to Table 9.4-1,

$$L = PC^{\mathrm{T}}(\nu^2 N)^{-1} \tag{9.5-24}$$

and the Kalman filter ARE becomes

$$0 = AP + PA^{\mathrm{T}} + (\nu^2 M_o + BB^{\mathrm{T}}) - PC^{\mathrm{T}}(\nu^2 N)^{-1} CP. \tag{9.5-25}$$

According to Kwakernaak and Sivan (1972), if the aforementioned assumptions hold, then $P \to 0$ as $\nu \to 0$, so that

$$L(\nu^2 N)L^{\mathrm{T}} = PC^{\mathrm{T}}(\nu^2 N)^{-1} CP \to BB^{\mathrm{T}}.$$

The general solution of this equation is

$$L \to \frac{1}{\nu} BUN^{-1/2}, \tag{9.5-26}$$

with U any unitary matrix. We claim that in this situation $L_r(s) \to L_s(s)$ as $\nu \to 0$. Indeed, defining the full state-feedback resolvent as

$$\Phi_c(s) = (sI - (A - BK))^{-1} \tag{9.5-27}$$

we may write

$$
\begin{aligned}
L_r(s) = K(s)G(s) &= K[sI - (A - BK - LC)]^{-1}LC\Phi B \\
&= K[\Phi_c^{-1} + LC]^{-1}LC\Phi B \\
&= K[\Phi_c - \Phi_c L(I + C\Phi_c L)^{-1}C\Phi_c]LC\Phi B \\
&= K\Phi_c L[I - (I + C\Phi_c L)^{-1}C\Phi_c L]C\Phi B \\
&= K\Phi_c L[(I + C\Phi_c L) - C\Phi_c L](I + C\Phi_c L)^{-1}C\Phi B \\
&= K\Phi_c L(I + C\Phi_c L)^{-1}C\Phi B \\
&\to K\Phi_c B(C\Phi_c B)^{-1}C\Phi B \\
&= K\Phi B(I + K\Phi B)^{-1}[C\Phi B(I + K\Phi B)^{-1}]^{-1}C\Phi B \\
&= [K\Phi B(C\Phi B)^{-1}]C\Phi B = K\Phi B. \tag{9.5-28}
\end{aligned}
$$

The matrix inversion lemma was used in going from line 2 to line 3, and from line 7 to line 8. The limiting value (9.5-26) for L was used at the arrow. What we have shown is that, using $G = I$ and the process noise given by (9.5-23), as $\nu \to 0$ the regulator loop gain using a Kalman filter approaches the loop gain using full state feedback. This means that, as $\nu \to 0$, all the robustness properties of the full state feedback control law are recovered in the stochastic regulator. The LQG/LTR design procedure is thus as follows:

1. Use the control ARE in Section 3.4 to design a state-feedback gain K with desirable properties. This may involve iterative design varying the PI weighting matrices Q and R.
2. Select $G = I$, process noise spectral density $M = v^2 M_0 + BB^T$, and noise spectral density $v^2 N$ for some $M_0 > 0$ and $N > 0$. Fix the design parameter v and use the Kalman filter ARE to solve for the Kalman gain L.
3. Plot the maximum and minimum singular values of the regulator loop gain $L_r(s)$ and verify that the robustness bounds are satisfied. If they are not, decrease v and return to 2.

A *reduced-order* regulator with suitable robustness properties may be designed by the LQG/LTR approach using the notions at the end of Section 9.4. That is, either a regulator may be designed for a reduced-order model of the plant, or the regulator designed for the full-order plant may then have its order reduced. In using the first approach, a high-frequency bound characterizing the unmodeled dynamics should be used to guarantee stability robustness.

An interesting aspect of the LQR/LTR approach is that the recovery process may be viewed as a *frequency-domain linear quadratic* technique that trades off the smallness of the sensitivity $S(j\omega)$ and the consensitivity $T(j\omega)$ at various frequencies. These notions are explored in Stein and Athans (1987) and Safonov et al. (1981).

Nonminimum-phase Plants and Parameter Variations   The limiting value of $K(s)$ is given by the bracketed term in (9.5-28). Clearly, as $v \to 0$ the regulator inverts the plant transfer function $C\Phi B$. If the plant is of minimum phase, with very stable zeros, the LQG/LTR approach generally gives good results. On the other hand, if the plant is nonminimum phase or has stable zeros with large time constants, the approach can be unsuitable.

In some applications, however, even if the plant is nonminimum phase, the LQG/LTR technique can produce satisfactory results (Athans 1986). In this situation, better performance may result if the design parameter v is not nearly zero. If the right-half-plane zeros occur at high frequencies where the loop gain is small, the LQG/LTR approach works quite well. An additional defect of the LQG/LTR approach appears when there are plant parameter variations. As seen in Section 9.2, stability in the presence of parameter variations requires that the loop gain singular values be below some upper bound at low frequencies. However, this bound is not taken into account in the LQG/LTR derivation. Thus, LQG/LTR can yield problems for controls design, where gain scheduling is required. The H_∞ design approach (Francis et al. 1984, Francis 1986, Doyle et al. 1989) has been used with success to overcome this problem.

Recovery of Robust Loop Gain at the Output   We have shown that by designing the state feedback first and then computing the Kalman filter gain using a specific choice of noise spectral densities, the stochastic regulator recovers the robustness of the loop gain $K(s)G(s)$ referred to the input $u(t)$ in Fig. 9.5-4. However, in Section 9.2 we saw that for a small tracking error the robustness should be studied in terms of the loop gain $G(s)K(s)$ referred to the error, or equivalently to the system output. Here, we show how to design a stochastic regulator that recovers a robust loop gain $G(s)K(s)$. Thus, suppose we first design a Kalman filter with gain L using Table 9.4-1. By duality theory, one may see that the Kalman filter loop gain

$$L_k(s) = C\Phi L \tag{9.5-29}$$

enjoys exactly the same guaranteed robustness properties as the state feedback loop gain $K\Phi B$ that were described earlier in this section. The regulator loop gain referred to the output is

$$L_r^o(s) = G(s)K(s) = C\Phi BK\Phi_r L. \tag{9.5-30}$$

Thus, we should like to determine how to design a state-feedback gain K so that $L_r^o(s)$ approaches $C\Phi L$. The key to this is in the selection of the PI weighting matrices Q and R in Section 3.4. To determine K let us propose the PI

$$J = \frac{1}{2}\int_0^\infty (x^TQx + \rho^2 u^TRu)du, \tag{9.5-31}$$

with

$$Q = \rho^2 Q_0 + C^TC, \tag{9.5-32}$$

with $Q_0 > 0$. By using techniques dual to those above, we may demonstrate that as $\rho \to 0$, the state-feedback gain determined using Section 3.4 approaches

$$K \to \frac{1}{\rho}R^{-1/2}WC, \tag{9.5-33}$$

with W a unitary matrix. Using this fact, it may be shown that

$$L_r^o(s) = G(s)K(s) \to C\Phi L. \tag{9.5-34}$$

The LQG/LTR design technique for loop gain recovery at the output is, therefore, exactly dual to that for recovery at the input. Specifically, the Kalman gain L is first determined using Table 9.4-1 for desired robustness properties. Then, Q and R are selected, with Q of the special form (9.5-32). For a small value of ρ, the state-feedback gain K is determined using the results of Section 3.4. If the singular value Bode plots of $L_r^o(s)$ do not show acceptable robustness, then ρ is decreased and a new K is determined.

If the plant $C\Phi B$ is minimum phase, all is well as ρ is decreased. However, if there are zeros in the right-half plane there could be problems as ρ becomes too small, although with care the LQG/LTR technique often still produces good results for suitable ρ.

Example 9.5-1. LQG/LTR Design of Aircraft Lateral Control System

We shall illustrate the loop-transfer recovery technique on a lateral aircraft control design. All computations, including solving for the state-feedback gains and Kalman filter gains, were carried out very easily using MATLAB, Control System Toolbox.

a. Control Objective

The tracking control system shown in Fig. 9.5-5 is meant to provide coordinated turns by causing the bank angle $\phi(t)$ to follow a desired command while maintaining the sideslip angle $\beta(t)$ at zero. It is a two-channel system with control input $u = [u_\phi \quad u_\beta]^T$. The reference command is $r = [r_\phi \quad r_\beta]^T$. The control system should hold ϕ at the commanded value of r_ϕ and $\beta(t)$ at the commanded value of r_β, which is equal to zero. The tracking error is $e = [e_\phi \quad e_\beta]^T$, with

$$e_\omega = r_\phi - \Phi$$

$$e_\beta = r_\beta - \beta. \tag{1}$$

The negatives of the errors appear in the figure since a minus sign appears in $u = -K\hat{x}$, as is standard for LQG design.

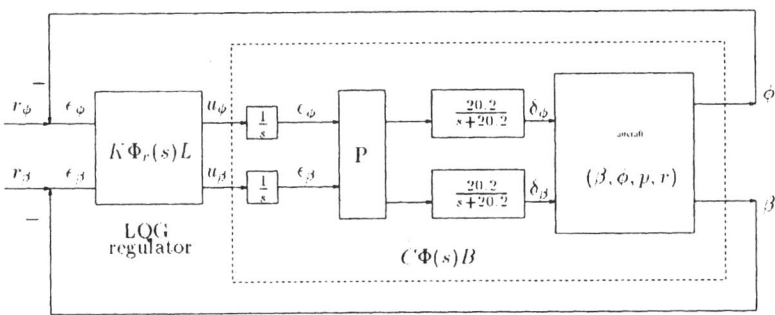

FIGURE 9.5-5 Aircraft turn coordinator control system.

b. State Equations of Aircraft and Basic Compensator Dynamics

The nonlinear F-16 model was linearized at the nominal flight condition as in Stevens and Lewis (1992), retaining as the states sideslip β, bank angle ϕ, roll rate p, and yaw rate r. Additional states δ_a and δ_r are introduced by the aileron and rudder actuators, both of which are modeled as having approximate transfer functions of $20.2/(s + 20.2)$. The aileron deflection is δ_a and the rudder deflection is δ_r.

The singular values versus frequency of the basic aircraft with actuators are shown in Fig. 9.5-6. Clearly, the steady-state error will be large in closed-loop since the loop gain has neither integrator behavior nor large singular values at DC. Moreover, the singular values are widely separated at DC, so that they are not balanced.

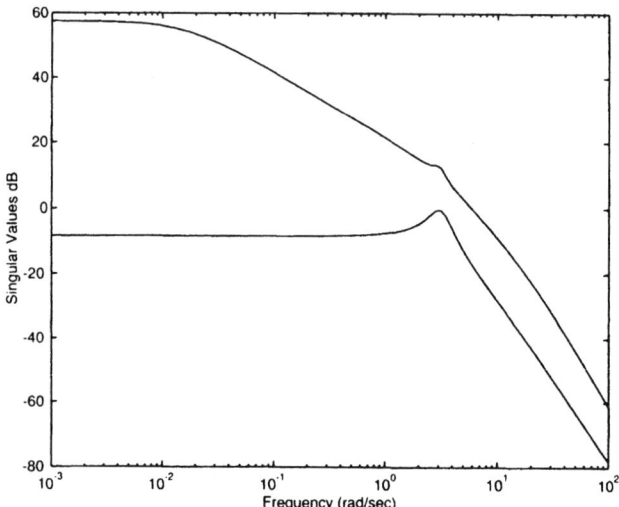

FIGURE 9.5-6   Singular values of the basic aircraft dynamics.

To correct these deficiencies we may use the techniques of Example 9.2-3. The DC gain of the system is given by

$$H(0) = \begin{bmatrix} -727.37 & -76.94 \\ -2.36 & 0.14 \end{bmatrix}. \tag{2}$$

First, the dynamics are augmented by integrators in each control channel. We denote the integrator outputs by ε_ϕ, ε_β.

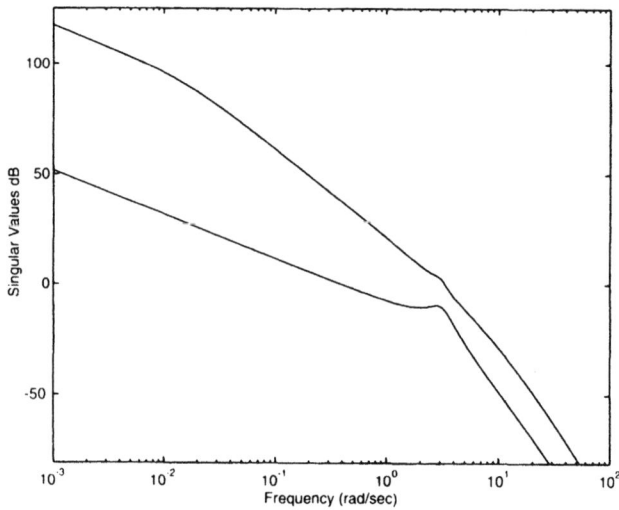

FIGURE 9.5-7   Singular values of aircraft augmented by integrators.

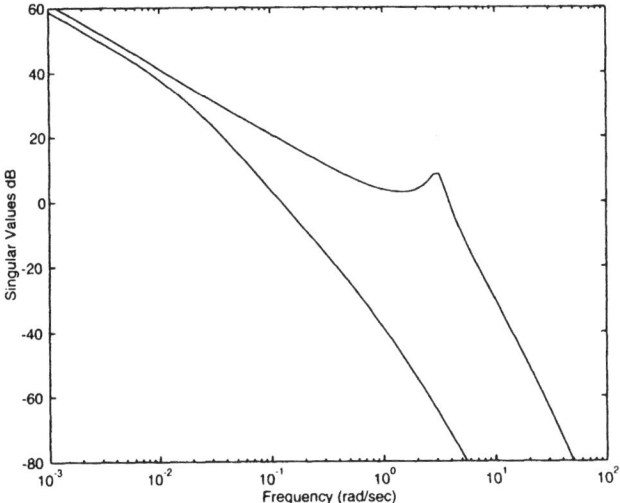

FIGURE 9.5-8   Singular values of aircraft augmented by integrators and inverse dc gain matrix P.

The singular value plots including the integrators are shown in Fig. 9.5-7. The DC slope is now -20 dB/decade, so that the closed-loop steady-state error will be zero. Next, the system was augmented by $P = H^{-1}(0)$ to balance the singular values at DC. The net result is shown in Fig. 9.5-8, which is very suitable.

The entire state vector, including aircraft states and integrator states, is

$$x = [\beta \quad \phi \quad p \quad r \quad \delta_a \quad \delta_r \quad \varepsilon_\phi \quad \varepsilon_\beta]^{\mathrm{T}}. \tag{3}$$

The full state-variable model of the aircraft plus actuators and integrators is of the form

$$\dot{x} = Ax + Bu, \tag{4}$$

with

$A =$

$$\begin{bmatrix} -0.3220 & 0.0640 & 0.0364 & -0.9917 & 90.0003 & 0.0008 & 0 & 0 \\ 0 & 0 & 1 & 0.0037 & 0 & 0 & 0 & 0 \\ -30.6492 & 0 & -3.6784 & 0.6646 & -0.7333 & 0.1315 & 0 & 0 \\ 8.5395 & 0 & -0.0254 & -0.4764 & -0.0319 & -0.0620 & 0 & 0 \\ 0 & 0 & 0 & 0 & -20.2 & 0 & -0.01 & -5.47 \\ 0 & 0 & 0 & 0 & 0 & -0.2 & -0.168 & 51.71 \\ 0 & 0 & 0 & 0 & 0 & 0 & 0 & 0 \\ 0 & 0 & 0 & 0 & 0 & 0 & 0 & 0 \end{bmatrix}, \tag{5}$$

$$
B = \begin{bmatrix} 0 & 0 \\ 0 & 0 \\ 0 & 0 \\ 0 & 0 \\ 0 & 0 \\ 0 & 0 \\ 1 & 0 \\ 0 & 1 \end{bmatrix}. \tag{6}
$$

The output is given by $y = [\phi\beta]^T$, or

$$
y = \begin{bmatrix} 0 & 57.2958 & 0 & 0 & 0 & 0 & 0 & 0 \\ 57.2958 & 0 & 0 & 0 & 0 & 0 & 0 & 0 \end{bmatrix} x = Cx, \tag{7}
$$

where the factor of 57.2958 converts radians to degrees. Then,

$$
e = r - y. \tag{8}
$$

c. Frequency-domain Robustness Bounds

We now derive the bounds on the loop-gain MIMO Bode magnitude plot that guarantee robustness of the closed-loop system. Consider first the high-frequency bound. Let us assume that the aircraft model is accurate to within 10% up to a frequency of 2 rad/sec, after which the uncertainty grows without bound at the rate of 20 dB/decade. The uncertainty could be due to actuator modeling inaccuracies, aircraft flexible modes, and so on. This behavior is modeled by

$$
m(\omega) = \frac{s+2}{20}. \tag{9}
$$

We asume $m(\omega)$ to be a bound on the multiplicative uncertainty in the aircraft transfer function (Section 9.2). For stability robustness in spite of the modeling errors, we saw in Section 9.2 that the loop-gain referred to the output should satisfy

$$
\underline{\sigma}(GK(j\omega)) < 1/m(\omega) = \left| \frac{20}{s+2} \right| \tag{10}
$$

when $1/m(\omega) \ll 1$. The function $1/m(\omega)$ is plotted in Fig. 9.5-9.

Turning to the low-frequency bound on the closed-loop loop gain, the closed-loop system should be robust to wind gust disturbances. Using techniques like those in Example 9.3-1, the gust magnitude plot shown in Fig. 9.3-1 may be obtained. According to Section 9.2, for robust performance in spite of wind gusts, the minimum loop-gain singular value $\underline{\sigma}(GK(j\omega))$ should be above this bound.

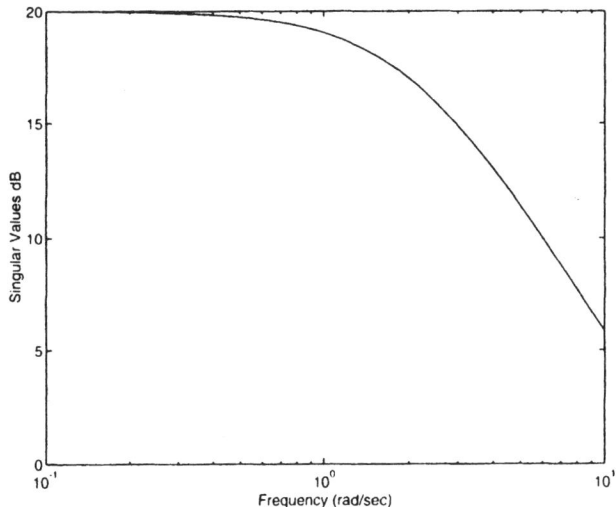

FIGURE 9.5-9 Multiplicative uncertainty bound $1/m(\omega)$ for the aircraft dynamical model.

d. Target Feedback Loop Design

The robustness bounds just derived are expressed in terms of the singular value plots referred to $e(t)$. To recover the loop gain $GK(j\omega)$ at $e(t)$, or equivalently at the output, the Kalman filter should be designed first, we should employ LQG/LTR algorithm number two.

In standard applications of the LQG/LTR technique, the regulator is designed for robustness, but the time responses are not even examined until the design has been completed. It is difficult to obtain decent time responses using this approach. In this example we emphasize the fact that it is not difficult to obtain good time responses as well as robustness using LQG/LTR. It is only necessary to select the Kalman gain L in Table 9.4-1 for good robustness properties as well as suitable step responses of the target feedback loop $C\Phi(s)L$, where $\Phi(s) = (sI - A)^{-1}$.

Using MATLAB, the Kalman filter design equations in Table 9.4-1 were solved using

$$Q = \mathrm{diag}\{0.01, 0.01, 0.01, 0.010, 0, 1, 1\}, \tag{11}$$

$R = r_f I$, and various values of r_f. The maximum and minimum singular values of the filter open-loop gain $C\Phi(s)L$ for $r_f = 1$ are shown in Fig. 9.5-10a, which also depicts the robustness bounds. The singular values for several values of r_f are shown in Fig. 9.5-10b.

Note how the singular value magnitudes increase as r_f decreases, reflecting improved rejection of low-frequency disturbances. The figures show that the robustness bounds are satisfied for $r_f = 1$ and $r_f = 10$, but that the high-frequency bound is violated for $r_f = 0.1$.

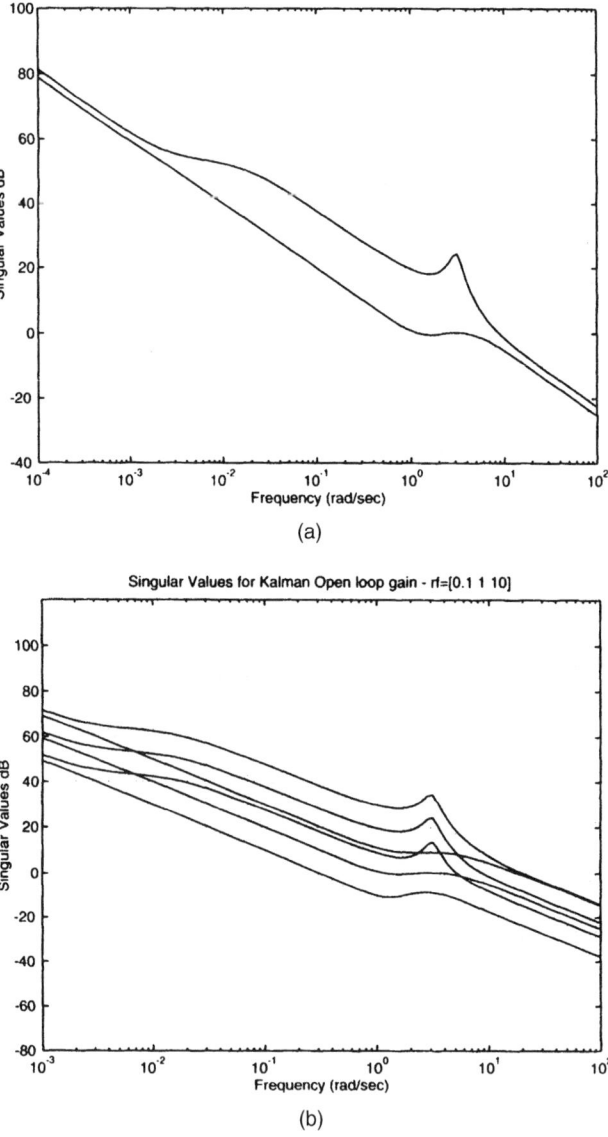

FIGURE 9.5-10   Singular values of Kalman filter open-loop gain $C\Phi L$: (a) for $r_f = 1$; (b) for various values of r_f.

The associated step responses of $C\Phi(s)L$ with reference commands of $r_\phi = 1$, $r_\beta = 0$ are shown in Fig. 9.5-11. The response rate $r_f = 10$ is unsuitable, while the response for $r_f = 0.1$ is too fast and would not be appreciated by the pilot. On the other hand, the response for $r_f = 1$ shows suitable time-of-response and overshoot characteristics, as well as good decoupling between the bank angle $\phi(t)$ and the sideslip $\beta(t)$. Therefore, the

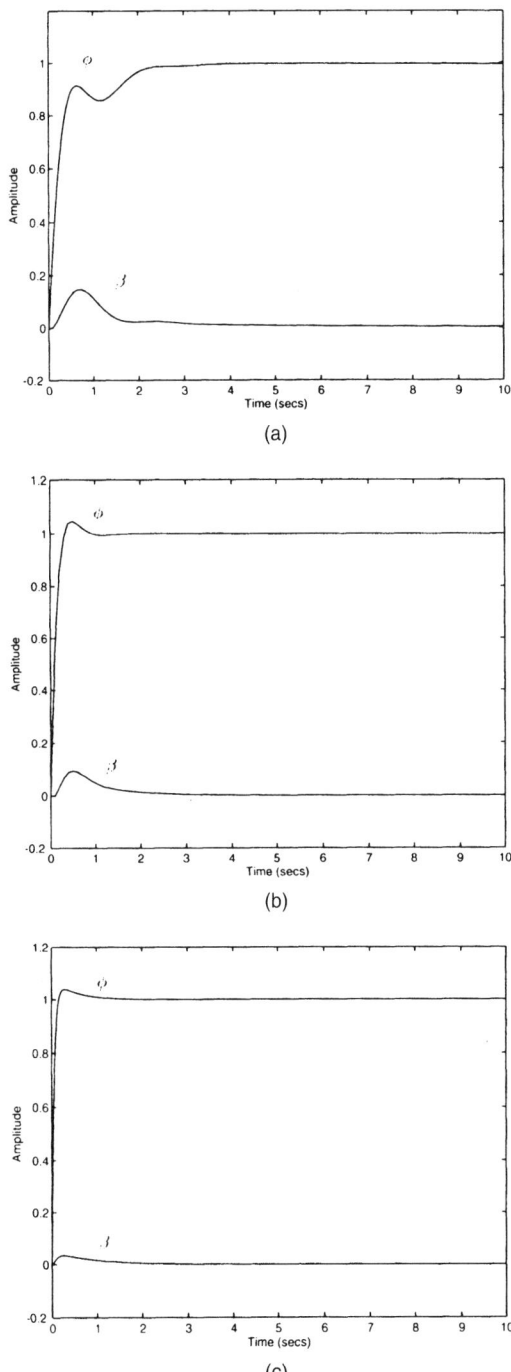

FIGURE 9.5-11   Step responses of target feedback loop $C\Phi L$: (a) $r_f = 10$; (b) $r_f = 1$; (c) $r_f = 0.1$.

target feedback loop was selected as $C\Phi(s)L$ with $r_f = 1$, since this results in a design that has suitable robustness properties and step responses. The corresponding Kalman gain is given by

$$L = \begin{bmatrix} -0.007 & 0.097 \\ 0.130 & -0.007 \\ 0.199 & -0.198 \\ -0.093 & -0.020 \\ -0.197 & -0.185 \\ 1.858 & 1.757 \\ 0.685 & -0.729 \\ 0.729 & 0.684 \end{bmatrix}. \tag{12}$$

The Kalman filter poles (e.g., those of $A - LC$) are given by

$$s = -0.002, -0.879, -1.470,$$
$$- 3.952 \pm j3.589,$$
$$- 7.205, -20.2, -20.2. \tag{13}$$

Although there is a slow pole, the step response is good, so this pole evidently has a small residue.

It is of interest to discuss how the frequency and time responses were plotted. For the frequency response, we used the open-loop system

$$\dot{\hat{x}} = A\hat{x} + Le$$
$$\hat{y} = C\hat{x}, \tag{14}$$

which has a transfer function of $C\Phi(s)L = C(sI - A)^{-1}L$. The program *sigma.m* was used which plots the singular values versus frequency for a system given in state-space form. This yielded Fig. 9.5-10.

For the step response, it is necessary to examine the closed-loop system. In this case, the loop is closed by using $e = r - \hat{y}$ in (14), obtaining

$$\dot{\hat{x}} = (A - LC)\hat{x} + Lr$$
$$y = C\hat{x}. \tag{15}$$

Using these dynamics in program *step.m* (MATLAB, Control System Toolbox) with $r = [1\,0]^T$ produces the step response lot.

A word on the choice for Q is in order. The design parameters Q and R should be selected so that the target feedback loop $C\Phi(s)L$ has good robustness and time-response properties. It is traditional to select $Q = BB^T$, which accounts for the last two diagonal entries of (11). However, in this example it was impossible to obtain good step responses using this selection for Q. Motivated by the fact that the process noise in the aircraft excites the first four states as well, we experimented with different values for Q, plotting in each case the singular values and step responses. After a few iterations, the final choice (11) was made.

e. Loop Transfer Recovery at the Output

The target feedback loop $C\Phi(s)L$ using $r_f = 1$ has good properties in both the frequency and time domains. Unfortunately, the closed-loop system with LQG regulator has a loop gain referred to the output of $C\Phi(s)BK\Phi_r(s)L$, with the regulator resolvent given by

$$\Phi_r(s) = [sI - (A - LC - BK)]^{-1}. \tag{16}$$

On the other hand, LQG/LTR algorithm shows how to select a state feedback gain K so that the LQG regulator loop gain approaches the ideal loop gain $C\Phi(s)L$. Let us now select such a feedback gain matrix.

Using MATLAB, the LQR design problem in Section 3.4 was solved with $Q = C^T C$, $R = \rho^2 I$, and various values for $r_c \equiv \rho^2$ to obtain different feedback gains K. Some representative singular values of the LQG loop gain $C\Phi(s)BK\Phi_r(s)L$ are plotted in Fig. 9.5-12, where L is the target-loop Kalman gain (12). Note how the actual singular values approach the target singular values in Fig. 9.5-10a as r_c decreases. A good match is obtained for $r_c = 10^{-11}$.

Figure 9.5-12c also depicts the robustness bounds, which are satisfied for this choice of $r_c = 10^{-11}$. The corresponding step responses are given in Fig. 9.5-13. A suitable step response that matches well the target response of Fig. 9.5-11b results when $r_c = 10^{-11}$.

It is of interest to discuss how these plots were obtained. For the LQG singular value plots, the complete dynamics are given by

$$\dot{x} = Ax + Bu$$
$$\dot{\hat{x}} = (A - LC)\hat{x} + Bu + Lw$$
$$u = -K\hat{x}, \tag{17}$$

where $w(t) = -e(t)$. These may be combined into the augmented system

$$\begin{bmatrix} \dot{x} \\ \dot{\hat{x}} \end{bmatrix} = \begin{bmatrix} A & -BK \\ 0 & A - LC - BK \end{bmatrix} \begin{bmatrix} x \\ \hat{x} \end{bmatrix} + \begin{bmatrix} 0 \\ L \end{bmatrix} w \tag{18}$$

$$y = \begin{bmatrix} C & 0 \end{bmatrix} \begin{bmatrix} x \\ \hat{x} \end{bmatrix}, \tag{19}$$

which has transfer function $C\Phi(s)BK\Phi_r(s)L$. The singular values are now easily plotted.

For the step responses, the closed-loop system must be studied. To close the loop, set $w = y - r$ in (18) to obtain the closed-loop dynamics

$$\begin{bmatrix} \dot{x} \\ \dot{\hat{x}} \end{bmatrix} = \begin{bmatrix} A & -BK \\ LC & A - LC - BK \end{bmatrix} \begin{bmatrix} x \\ \hat{x} \end{bmatrix} + \begin{bmatrix} 0 \\ -L \end{bmatrix} r \tag{20}$$

$$y = \begin{bmatrix} C & 0 \end{bmatrix} \begin{bmatrix} x \\ \hat{x} \end{bmatrix}. \tag{21}$$

These are used with *lsim.m* to obtain Fig. 9.5-13.

The final LQG regulator is given by the Kalman gain L in (12) and the feedback gain K corresponding to $r_c = 10^{-11}$.

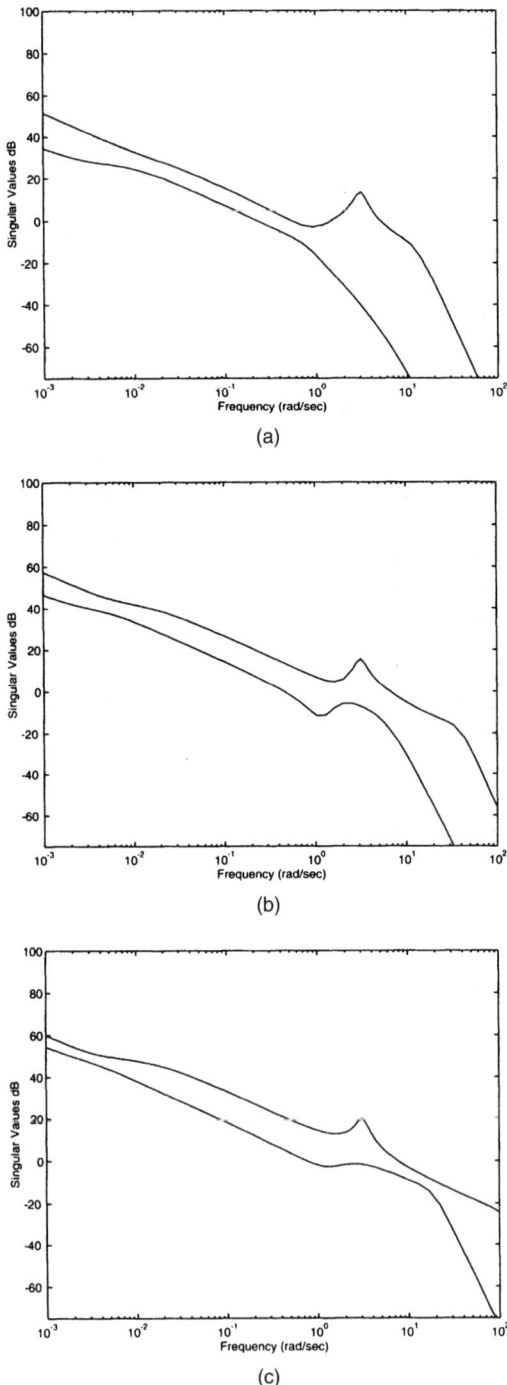

FIGURE 9.5-12   Singular values plots for the LQG regulator. (a) LQG with $r_c = 10^{-3}$. (b) LQG with $r_c = 10^{-7}$. (c) LQG with $r_c = 10^{-11}$.

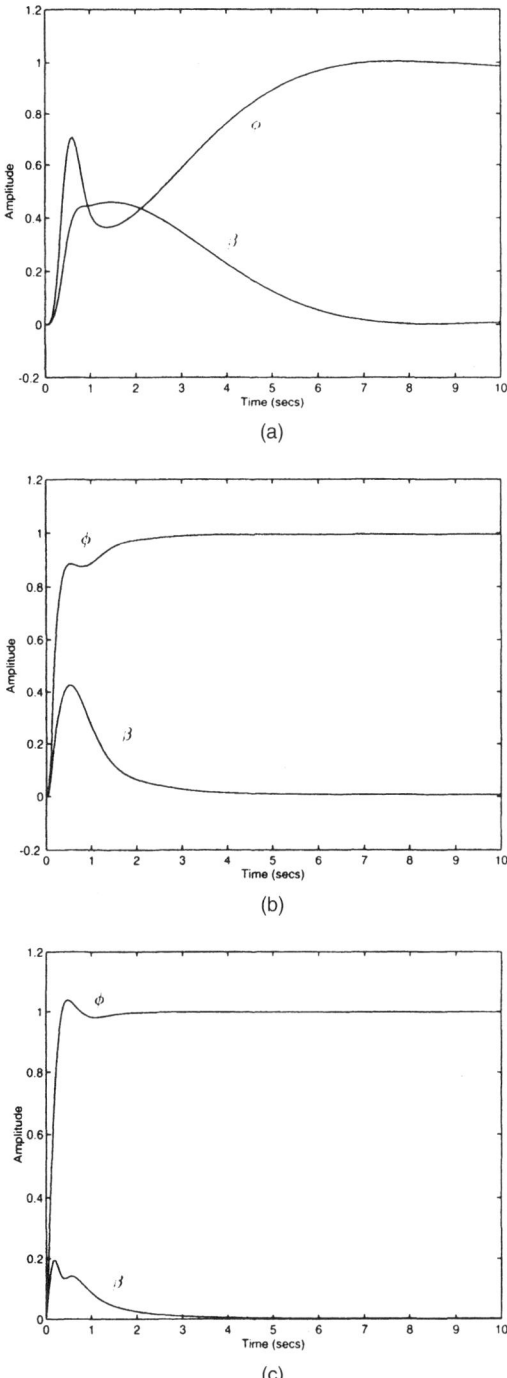

FIGURE 9.5-13 Closed loop step responses of the LQG regulator. (a) LQG with $r_c = 10^{-3}$. (b) LQG with $r_c = 10^{-7}$. (c) LQG with $_c = 10^{-11}$.

f. Reduced-order Regulator

The LQG regulator just designed has order $n = 8$, the same as the plant. This is excessive for an aircraft lateral control system. A reduced-order regulator that produces very good results may easily be determined using the partial-fraction-expansion approach in Example 9.2-4, principal component analysis (Moore 1982), or other techniques. This is easily accomplished using MATLAB. The singular value plots and step response using the reduced-order regulator should be examined to verify robustness and suitable performance.   ∎

9.6 H_∞ DESIGN

Polynomial Techniques

In the previous sections we have seen the design of MIMO systems for stability robustness in the presence of uncertainties in the model. Another approach to the problem of robust design is the worst-case design. When the system has a disturbance input of unknown or uncertain (statistical) nature, the effect on the output is required to be minimized. The worst-case design is the preferred method of optimization as the controller is designed to account for the worst-case disturbance input. In this section we approach the problem of the synthesis of the controller using the polynomial approach to the design of the H_∞ controller. In Section 9.2 we have seen that the design specifications are related to the sensitivity function of the system. Specifically, the sensitivity function is required to be small at low frequencies for robust performance and the co-sensitivity function to be small at high frequencies for robust stabilization. Therefore, we can incorporate these performance criteria into the H_∞ design problem by introducing a suitable weighting function. The procedure for the choice of the weighting function for the multivariable systems has been elaborated in Doyle and Stein (1982) and McFarlane and Glover (1992). A typical plot for the weighting function is shown in Figure 9.6-1. Note that for multivariable systems the choice of the weighting function is done using the singular value plot. The figure shows the frequency plot of the maximum singular value of the weighting function. The standard H_∞ problem is to synthesize a controller that minimizes the H_∞ norm of the closed-loop system. When the weighting functions are introduced, they can be absorbed into the plant, and the optimization is done for the weighted plant. The standard plant and feedback configuration is shown Figure 9.6-2.

The closed-loop system is given by

$$T_{zw} = G_{11} + G_{12}K(I - G_{22}K)^{-1}G_{21}. \qquad (9.6\text{-}1)$$

The objective is to find all the controllers that stabilize the plant and minimize the H_∞ norm of the transfer function T_{zw} from the disturbance inputs $w(t)$ to the regulated outputs $z(t)$ using the control input $u(t)$ and the measurement outputs $y(t)$. We note that the optimal controllers may be nonunique. The optimal

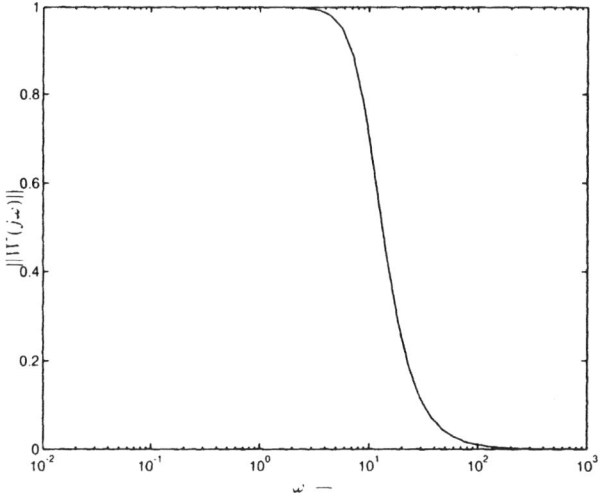

FIGURE 9.6-1   Typical weighting function for MIMO system.

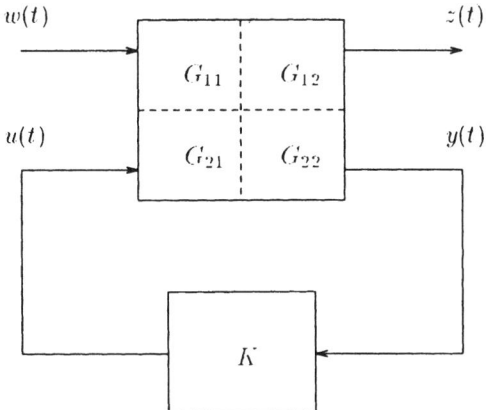

FIGURE 9.6-2   Standard feedback configuration.

controllers are difficult to synthesize; therefore, we will focus on the study of the suboptimal problem where the H_∞ norm of the system satisfies

$$\|T_{zw}\|_\infty < \gamma, \quad \gamma > 0. \tag{9.6-2}$$

Clearly, $\gamma > \gamma_0$ where γ_0 is the minimum value of the cost function.

The frequency-domain solution to the problem can obtained through a related problem of model matching. In fact, many problems, such as the regulation problem and the robust stabilization problem, can be reformulated in terms of the model-matching problem. The details of the reformulation can be found in Francis (1986).

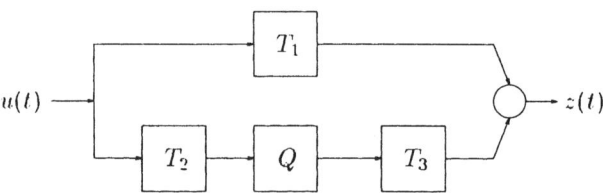

FIGURE 9.6-3   Standard feedback configuration.

Model-matching Problem   Consider the configuration shown in Figure 9.6-3. The plants T_1, T_2, T_3 are stable and the model-matching problem is to find a stable filter so that the H_∞ norm of the error function is minimized. That is, find stable Q so that

$$\|T_1 - T_2 Q T_3\|_\infty \qquad (9.6\text{-}3)$$

is minimized.

Once again we deal with the suboptimal problem of achieving an upper bound on the norm of the error function. Furthermore, we may assume without loss of generality that the upper bound is 1 (since this may be achieved by scaling T_1, T_2). We will confine our attention to the SISO case to see how the H_∞ design problem can be related to the familiar LQG design techniques. For the SISO problem T_1, T_2, T_3 are scalars, and, hence, without loss of generality we may assume that $T_3 = 1$. Now, if T_2 is minimum phase then the solution is trivial and is given by

$$Q = T_2^{-1} T_1. \qquad (9.6\text{-}4)$$

In this case, the Q is optimal and the error is zero. On the other hand, if T_2 has zeros in the right half-plane, then the problem becomes considerably harder. To solve this problem we introduce the notion of inner–outer factorizations. An inner function $T(s)$ satisfies

$$T(-s)T(s) = 1. \qquad (9.6\text{-}5)$$

That is, its magnitude on the $j\omega$ axis is 1. An outer function has no zeros in the right half-plane. Now, any stable function can be factored into inner and outer functions. Therefore,

$$T_2(s) = T_{2i}(s)T_{2o}(s). \qquad (9.6\text{-}6)$$

Now,

$$\|T_1 - T_2 Q\|_\infty = \left\| T_{2i}(T_{2i}^{-1} T_1 - T_{2o} Q \right\|_\infty, \qquad (9.6\text{-}7)$$

$$= \left\| T_{2i}^{-1} T_1 - T_{2o} Q \right\|_\infty, \qquad (9.6\text{-}8)$$

$$= \|R - X\|_\infty, \qquad (9.6\text{-}9)$$

since $|T_{2i}(j\omega)| = 1$. Here $R = T_{2i}^{-1}T_1$ and has no poles on the $j\omega$ axis and $X = T_{2o}Q$ and is stable since T_{2o}^{-1} is stable. Furthermore, there is a one-to-one correspondence between X and Q. Now, the problem reduces to finding the distance between the function R and the space of stable functions. This problem has been solved as the Nehari problem, and a procedure for the solution comprises solving the Lyapunov equations corresponding to the stable part of R. The optimal Q can be constructed from the procedure. The details of the procedure entail the definitions of Hankel operators and are beyond the scope of this chapter. We refer to Francis (1986) for details of the solution and the solution to the multivariable case. Another approach to the frequency-domain solution to the model-matching problem is through the *J–spectral* factorization approach given by Kimura et al. (1991).

State-space Techniques

The polynomial techniques described in the earlier section provide the basis for the problem of the H_∞ design. Some connections between the H_∞ problem and the LQR problem can be developed in the state-space domain, which are intuitively appealing. In this section we develop a method for the solution to the H_∞ problem. It will be seen that the solution can be approached through the techniques developed for the LQR design procedure.

In Section 9.2 we have seen that the H_∞ norm of a transfer function is equal to the induced norm of the system. Specifically, for a stable system $z = Tw$,

$$\|T\|_\infty = \max_{z \in L_2} \frac{\|z\|_2}{\|w\|_2}. \qquad (9.6\text{-}10)$$

Therefore, the H_∞ criterion of bounding the H_∞ norm by γ can be translated into the time domain as follows:

$$\max_{z \in L_2} \frac{\|z\|_2^2}{\|w\|_2^2} < \gamma^2 \Leftrightarrow \|z\|_2^2 - \gamma^2 \|w\|_2^2 \leq -\varepsilon \|w\|_2^2, \qquad \varepsilon > 0. \qquad (9.6\text{-}11)$$

We first consider the full-information problem where the state and the disturbance inputs are available to the controller. The state-space representation of the plant is given by

$$\dot{x} = Ax + B_1 w + B_2 u, \qquad (9.6\text{-}12)$$

$$z = C_1 x + D_{12} u, \qquad (9.6\text{-}13)$$

$$y = \begin{bmatrix} x \\ w \end{bmatrix}. \qquad (9.6\text{-}14)$$

Note that the term D_{11} is zero. This assumption on the system simplifies the computations and can be assumed without loss of generality. The general case can be solved by converting the system to this form by using the so called "loop-shifting" techniques elaborated in Green and Limebeer (1993). Furthermore, we assume

$$C_1^T D_{12} = 0, \qquad (9.6\text{-}15)$$

$$D_{12}^T D_{12} = 1. \qquad (9.6\text{-}16)$$

These assumptions ensure that there is no cross-weighting in the state and the input and the weighting of the input energy is normalized. Furthermore, we assume that (A, B_2) is controllable and (C_1, A) is observable. These conditions ensure that the system is stabilizable. Now, the objective is to design a control law so that the condition (9.6-11) is satisfied for all disturbance inputs $w(t)$. Therefore, we seek a linear controller of the form

$$u = K \begin{bmatrix} x \\ w \end{bmatrix}. \qquad (9.6\text{-}17)$$

The H_∞ criterion can be rewritten as the inequality (9.6-11). Therefore, we define the cost function $J(u, w)$ as

$$J(u, w) = \|z\|_2^2 - \gamma^2 \|w\|_2^2, \qquad u = Ky. \qquad (9.6\text{-}18)$$

Let w^* be the worst-case input to the system in the sense that it maximizes the induced norm. Then, for any disturbance input $w(t)$ we have the relationship,

$$J(u, w) = J(u, w^*) + \|z\|_2^2 - \|z^*\|_2^2, \qquad (9.6\text{-}19)$$

$$\leq J(u, w^*). \qquad (9.6\text{-}20)$$

Therefore, we build a controller that achieves the H_∞ criterion for the worst-case input. But we choose the control input so that the cost function is minimized for the worst-case input. Therefore, using the standard arguments for first-order variation of the cost function, we obtain

$$\dot{x} = Ax + B_1 w^* + B_2 u, \qquad x(0) = 0, \qquad (9.6\text{-}21)$$

$$\dot{\lambda} = -A^T \lambda - C^T C x, \qquad \lambda(\infty) = 0, \qquad (9.6\text{-}22)$$

$$u - -B_2^T \lambda, \qquad (9.6\text{-}23)$$

$$w^* = \gamma^{-2} B_1^T \lambda. \qquad (9.6\text{-}24)$$

Clearly, this corresponds to the Riccati equation

$$A^T X_\infty + X_\infty A - X_\infty (B_2 B_2^T - \gamma^{-2} B_1 B_1^T) X_\infty + C_1^T C_1 = 0. \qquad (9.6\text{-}25)$$

However, note that the quadratic term of the Riccati equation is indefinite unlike in the case of the LQR problem, and, hence, the solution to the Riccati equation is not guaranteed to exist always. Therefore, the solution to the original

H_∞ problem exists if the solution to the Riccati equation is positive semi-definite and stabilizing. Therefore, the H_∞ control law (when it exists) is given by the state feedback

$$u^*(t) = -B_2 B_2^T X_\infty x(t). \tag{9.6-26}$$

Furthermore, we note that the worst-case input that maximizes the induced norm is also linear combination of the state given by

$$w^* = \gamma^{-2} B_1 B_1^T X_\infty x(t). \tag{9.6-27}$$

Also, note that the solution to the Riccati equation approaches the solution to the LQR problem as γ tends to infinity. In other words, as the constraint on the norm is relaxed, the effect of the worst-case disturbance is reduced and the cost function is optimized only over the control input leading to the LQR solution. Therefore, we see that the solution to the H_∞ problem can be obtained by using the techniques developed earlier. The H_∞ controllers are nonunique, and a parameterization of all the controllers can be obtained. We will not go into the details of the parametrization, but refer to Green and Limebeer (1993) for a more detailed exposition where the output feedback problem has been treated.

PROBLEMS

Section 9.2

9.2-1. Derive in detail the multivariable expressions (9.2-16) and (9.2-17) for the performance output and the tracking error.

9.2-2. Prove (9.2-56). You will need to neglect any terms that contain second-order terms in the parameter variation matrices and use the fact that, for small X, $(I - X)^{-1} \approx (I + X)$

9.2-3. Multivariable closed-loop transfer relations. In Fig. 9.2-1, the plant $G(s)$ is described by

$$\dot{x} = \begin{bmatrix} 0 & 1 & 0 \\ 0 & -3 & 0 \\ 0 & 0 & 0 \end{bmatrix} x + \begin{bmatrix} 0 & 0 \\ 1 & 0 \\ 0 & 1 \end{bmatrix} u, \qquad z = \begin{bmatrix} 1 & 0 & 0 \\ 0 & 0 & 1 \end{bmatrix} x$$

and the compensator is $K(s) = 2I_2$.
a. Find the multivariable loop gain and return difference.
b. Find the sensitivity and cosensitivity.
c. Find the closed-loop transfer function from $r(t)$ to $z(t)$, and, hence, the closed-loop poles.

9.2-4. For the continuous-time system in Example 9.2-1, plot the individual SISO Bode magnitude plots from input one to outputs one and two, and from input two to outputs one and two. Compare them with the MIMO Bode plot to see that there is no obvious relation. Thus, the robustness bounds cannot be given in terms of the individual SISO Bode plots.

9.2-5. Multivariable bode plot. For the system in Problem 9.2-3, plot the multivariable Bode magnitude plots for:
a. The loop gain GK.
b. The sensitivity S and cosensitivity T. For which frequency ranges do the plots for $GK(j\omega)$ match those for $S(j\omega)$? For $T(j\omega)$?

9.2-6. Bode plots for F-16 lateral regulator. Plot the loop gain multivariable Bode magnitude plot for the F-16 lateral regulator designed in Example 8.1-1.

9.2-7. Balancing and zero steady-state error. Find a precompensator for balancing the singular values at low frequency and ensuring zero steady-state error for the system

$$\dot{x} = \begin{bmatrix} 0 & 1 & 0 \\ -2 & -3 & 0 \\ 0 & 0 & -3 \end{bmatrix} x + \begin{bmatrix} 0 & 0 \\ 1 & 0 \\ 0 & 1 \end{bmatrix} u, \qquad z = \begin{bmatrix} 1 & 0 & 0 \\ 0 & 0 & 1 \end{bmatrix} x$$

Plot the singular values of the original and precompensated system.

Section 9.4

9.4-1. Nonzero-mean noise. Use (9.4-49) to write down the best estimate for $x(t)$ in terms of the filter state $\hat{x}(t)$ if the process noise $w(t)$ and measurement noise $v(t)$ have nonzero means of $\overline{w}$ and $\overline{v}$, respectively.

9.4-2. Observer for angle of Attack. In Example 8.2-1 a low-pass filter of $10/(s + 10)$ was used to smooth out the angle-of-attack measurements to design a pitch rate CAS. An alternative is to use an observer to reconstruct α. This completely avoids measurements of the angle of attack.
a. Design an observer that uses measurements of $q(t)$ to provide estimates of $\alpha(t)$. The observer should have $\zeta = 1/\sqrt{2}$ and $\omega_n = 10$ rad/sec. Use Ackermann's formula to find the output-injection matrix L.
b. Delete the α filter in Example 8.2-1, replacing it by the dynamics of the second-order observer just designed. With the new augmented dynamics, perform the LQ design of Example 8.2-1.
c. Compare the performance of this pitch rate CAS to the one using the α filter.

9.4-3. Kalman filter. Software for solving the Kalman filter ARE is available in the MATLAB Control Systems Toolbox (*lqe.m*). Alternatively, the Kalman filter gain L can be found using the software for Table 8.1-1 on the dual plant (A^T, C^T, B^T) with $B = 1$. Repeat Example 9.4-2 if the wind gusts have a turbulence intensity of 20 ft/sec.

Section 9.5

9.5-1. Show that (9.5-7) implies (9.5-10).

9.5-2. LQG/LTR design. Note that the state-feedback gain K can be found using the software for Table 8.1-1 with $C = I$. Likewise, the Kalman filter gain L can be found using the software for Table 8.1-1 on the dual plant (A^T, C^T, B^T), with $B = I$.

a. In Problem 9.4-3, plot the loop gain singular values assuming full state feedback.

b. Now, angle-of-attack measurements are not allowed. Design a Kalman filter for various values of the design parameter. In each case, plot the closed-loop step response, as well as the loop gain singular values. Compare the step response and the singular values with the case for full state feedback as ρ becomes small.

10

DIFFERENTIAL GAMES

This chapter presents some basic ideas of differential games. Modern day society relies on the operation of complex systems, including aircraft, automobiles, electric power systems, economic entities, business organizations, banking and finance systems, computer networks, manufacturing systems, and industrial processes. Decision and control are responsible for ensuring that these systems perform properly and meet prescribed performance objectives. Networked dynamical agents have cooperative team-based goals as well as individual selfish goals, and their interplay can be complex and yield unexpected results in terms of emergent teams. Cooperation and conflict of multiple decision-makers for such systems can be studied within the field of cooperative and noncooperative game theory.

We discussed linear quadratic Nash games and Stackelberg games in Section 8.6. There, a unified treatment was given for linear games and decentralized control using structured design methods that are familiar from output feedback design. Controller designs were given based on solving coupled matrix equations. Here, we discuss nonlinear and linear differential games using a Hamiltonian function formulation.

In the first part of the book we used the calculus of variations to develop optimal control results such as those in Tables 3.2-1 and 3.3-1. In this chapter we use an approach based on the Hamiltonian function, Pontryagin's minimum principle (PMP) (Pontryagin et al. 1962), and the Bellman equation to develop optimal game theoretic solutions. PMP was discussed in Section 5.2 in connection with constrained input problems. First we show how to use PMP to derive the solutions to the nonlinear and linear optimal control problems for a single agent given in Tables 3.2-1 and 3.3-1. The Bellman equation is defined in terms of the Hamiltonian function and appears naturally in that development. Then we use PMP and the Bellman equation to solve multi agent decision problems. First we study 2-player noncooperative Nash zero-sum games, which have applications to H-infinity robust control. Finally, we use PMP and the Bellman equation to solve multiplayer nonzero sum games, which can have a cooperative team component as well as noncooperative components. In this chapter we include some proofs that show the relationship between performance criteria, value functions, and Lyapunov functions. The proofs also show the way of thinking in this field of research.

10.1 OPTIMAL CONTROL DERIVED USING PONTRYAGIN'S MINIMUM PRINCIPLE AND THE BELLMAN EQUATION

In this section we rederive the Hamilton-Jacobi-Bellman (HJB) equation from Section 6.3 and the Riccati equation solution of Table 3.3-1 using PMP and the Bellman equation.

Optimal Control for Nonlinear Systems

Consider the continuous-time nonlinear affine system

$$\dot{x} = F(x, u) = f(x) + g(x)u, \qquad (10.1\text{-}1)$$

with state $x(t) \in R^n$ and control $u(t) \in R^m$. Let the drift dynamics $f(x)$ be locally Lipschitz and $f(0) = 0$ so that $x = 0$ is an equilibrium point. With this system, associate the performance index or value integral

$$V(x(t)) = \int_t^\infty r(x, u)d\tau = \int_t^\infty (Q(x) + u^{\mathrm{T}}Ru)d\tau \qquad (10.1\text{-}2)$$

with $Q(x) > 0$, $R > 0$. The quadratic form in the control makes the development here simpler, but similar results hold for more general positive definite control weightings. A control $u(t)$ is said to be *admissible* if it is continuous, stabilizes the system (10.1-1), and makes the value (10.1-2) finite.

The optimal control problem is to select the control input $u(t)$ in (10.1-1) to minimize the value (10.1-2). Differentiating $V(x(t))$ using Leibniz's formula, one sees that a differential equivalent to the value integral is given in terms of the Hamiltonian function $H(\cdot)$ by the *Bellman equation*

$$0 = r(x, u) + \dot{V} = r(x, u) + \left(\frac{\partial V}{\partial x}\right)^{\mathrm{T}} \dot{x}$$

or

$$0 = Q(x) + u^{\mathrm{T}}Ru + \left(\frac{\partial V}{\partial x}\right)^{\mathrm{T}} (f(x) + g(x)u) \equiv H\left(x, \frac{\partial V}{\partial x}, u\right). \qquad (10.1\text{-}3)$$

That is, given any stabilizing feedback control policy $u(x)$ yielding finite value, the positive definite solution to the Bellman equation (10.1-3) is the value given by (10.1-2). The Bellman equation is a partial differential equation for the value. The initial condition is $V(0) = 0$.

It is desired to select the control input to minimize the value. A necessary condition for this is Pontryagin's minimum principle

$$H\left(x^*, \frac{\partial V^*}{\partial x}, u^*\right) \leq H\left(x^*, \frac{\partial V^*}{\partial x}, u\right) \qquad (10.1\text{-}4)$$

for all admissible controls $u(t)$, where * denotes the optimal (minimizing) state, value, and control. Since the control $u(t)$ is unconstrained this is equivalent to the stationarity condition

$$\frac{\partial H}{\partial u} = 0. \qquad (10.1\text{-}5)$$

Applying PMP to (10.1-3) yields

$$u(x) = u(V(x)) \equiv -\frac{1}{2}R^{-1}g^{\mathrm{T}}(x)\frac{\partial V}{\partial x} = -\frac{1}{2}R^{-1}g^{\mathrm{T}}(x)\nabla V(x), \qquad (10.1\text{-}6)$$

where $\nabla V(x) = \partial V/\partial x \in R^n$ is the gradient, taken as a vector. Substitute this into the Bellman equation (10.1-3) to obtain

$$0 = Q(x) + \nabla V^{\mathrm{T}}(x)f(x) - \tfrac{1}{4}\nabla V^{\mathrm{T}}(x)g(x)R^{-1}g^{\mathrm{T}}(x)\nabla V(x). \qquad V(0) = 0 \qquad (10.1\text{-}7)$$

This is the Hamilton-Jacobi-Bellman (HJB) equation from Section 6.3. It can also be written as

$$0 = \min_u[H(x, \nabla V^*, u)]. \qquad (10.1\text{-}8)$$

This can be written in terms of the optimal Hamiltonian as

$$0 = H(x^*, \nabla V^*, u^*), \qquad (10.1\text{-}9)$$

where $u^* = u(V^*(x)) = -\tfrac{1}{2}R^{-1}g^{\mathrm{T}}(x)\nabla V^*(x)$.

To solve the optimal control problem, one solves the HJB equation (10.1-7) for the optimal value $V^* > 0$, then the optimal control is given as a state variable feedback $u(V^*(x))$ in terms of the HJB solution by (10.1-6).

Note that in the calculus of variations derivation of Table 3.2-1, the Hamiltonian function $H(x, \nabla V, u)$ plays a central role, and its various partial derivatives define the flows of the state equation and the costate equation, as well as the optimal control through the stationarity condition. Yet, there is no mention there that the Hamiltonian should be equal to zero. The Hamiltonian is set to zero in the Bellman equation (10.1-3), which arises from differentiation of the value function (10.1-2), to get a differential equivalent to (10.1-2). Note further that the costate of Table 3.2-1 is the gradient vector $\nabla V(x)$.

Now we give a formal proof that the solution to the HJB equation provides the optimal control solution. The following key fact is instrumental. It shows that the Hamiltonian function is quadratic in the control deviations from a certain key control value.

Lemma 10.1-1. For any admissible control policy $u(x)$, let $V(x) \geq 0$ be the corresponding solution to the Bellman equation (10.1-3). Define $u^* = u(V(x))$ by (10.1-6) in terms of $V(x)$. Then

$$H(x, \nabla V, u) = H(x, \nabla V, u^*) + (u - u^*)^{\mathrm{T}} R(u - u^*). \tag{10.1-10}$$

Proof: The Hamiltonian function is

$$H(x, \nabla V, u) = Q(x) + u^{\mathrm{T}} Ru + \nabla V^{\mathrm{T}}(f + gu).$$

Complete the squares to write

$$H(x, \nabla V, u) = \nabla V^{\mathrm{T}} f + Q(x) + (\tfrac{1}{2}\nabla V^{\mathrm{T}} g R^{-1} + u^{\mathrm{T}}) R(\tfrac{1}{2} R^{-1} g^{\mathrm{T}} \nabla V + u)$$
$$- \tfrac{1}{4}\nabla V^{\mathrm{T}} g R^{-1} g^{\mathrm{T}} \nabla V. \qquad\blacksquare$$

The next result shows that under certain conditions the HJB solution solves the optimal control problem. Set $Q(x) = h^{\mathrm{T}}(x)h(x)$. System

$$\dot{x} = f(x) + g(x)u(x), \quad y = h(x) \tag{10.1-11}$$

is said to be zero-state observable if $u(t) \equiv 0$, $y(t) \equiv 0 \Rightarrow x(t) = 0$.

Theorem 10.1-2. Solution to Optimal Control Problem.   Consider the optimal control problem for (10.1-1), (10.1-2) with $Q(x) = h^{\mathrm{T}}(x)h(x)$. Suppose $V^*(x) \in C^1 : R^n \to R$ is a smooth positive definite solution to the HJB equation (10.1-7). Define control $u^* = u(V^*(x))$ as given by (10.1-6). Assume (10.1-11) is zero-state observable. Then the closed-loop system

$$\dot{x} = f(x) + g(x)u^* = f(x) - \tfrac{1}{2}gR^{-1}g^{\mathrm{T}}\nabla V^* \tag{10.1-12}$$

is locally asymptotically stable. Moreover, $u^* = u(V^*(x))$ minimizes the performance index (10.1-2) over all admissible controls, and the optimal value on $[0, \infty)$ is given by $V^*(x(0))$.

Proof:

a. Stability.

Note that for any C^1 function $V(x) : R^n \to R$ one has, along the system trajectories,

$$\frac{dV}{dt} = \frac{\partial V}{\partial t} + \frac{\partial V}{\partial x}^{\mathrm{T}} \dot{x} = \frac{\partial V}{\partial x}^{\mathrm{T}}(f + gu),$$

so that

$$\frac{dV}{dt} + h^{\mathrm{T}}h + u^{\mathrm{T}}Ru = H(x, \nabla V, u).$$

Suppose now that $V(x)$ satisfies the HJB equation $H(x^*, \nabla V^*, u^*) = 0$. Then according to (10.1-10) one has

$$H(x, \nabla V^*, u) = H(x^*, \nabla V^*, u^*) + (u - u^*)^{\mathrm{T}}R(u - u^*)$$
$$= (u - u^*)^{\mathrm{T}}R(u - u^*).$$

Therefore,

$$\frac{dV}{dt} + h^{\mathrm{T}}h + u^{\mathrm{T}}Ru = (u - u^*)^{\mathrm{T}}R(u - u^*).$$

Selecting $u = u^*$ yields

$$\frac{dV}{dt} + h^{\mathrm{T}}h + u^{\mathrm{T}}Ru = 0$$

$$\frac{dV}{dt} \leq -(h^{\mathrm{T}}h + u^{\mathrm{T}}Ru).$$

Now LaSalle's extension shows that the state goes to a region of R^n wherein $\dot{V} = 0$. However, zero-state observable means $u(t) \equiv 0$, $y(t) \equiv 0 \Rightarrow x(t) = 0$. Therefore, the system is locally asymptotically stable with Lyapunov function $V(x) > 0$.

b. Optimality.

For any C^1 smooth function $V(x) : R^n \to R$ and $T > 0$ one can write the performance index (10.1-2) as

$$V(x(0), u) = \int_0^T \left(h^{\mathrm{T}}h + u^{\mathrm{T}}Ru \right) dt + \int_0^T \dot{V} dt - V(x(T)) + V(x(0)).$$

$$= \int_0^T \left(h^{\mathrm{T}}h + u^{\mathrm{T}}Ru \right) dt + \int_0^T \nabla V^{\mathrm{T}}(f + gu) dt - V(x(T)) + V(x(0))$$

$$= \int_0^T H(x, \nabla V, u) dt - V(x(T)) + V(x(0)).$$

Now, suppose $V(x)$ satisfies the HJB equation (10.1-7). Then $0 = H(x^*, \nabla V^*, u^*)$ and (10.1-10) yields

$$V(x(0), u) = \int_0^T \left((u - u^*)^{\mathrm{T}}R(u - u^*) \right) dt - V^*(x(T)) + V^*(x(0)).$$

Assuming $u(t)$ is admissible one has $V^*(x(\infty)) = 0$ and can write

$$V(x(0), u) = \int_0^\infty \left((u - u^*)^T R(u - u^*) \right) dt + V^*(x(0)),$$

which shows that u^* is an optimal control and the optimal value is $V^*(x(0))$. ∎

Solution of HJB Equations

The HJB equation may not have smooth solutions, but may have the so-called viscosity solutions (Bardi and Capuzzo-Dolcetla 1997). Under certain local reachability and observability assumptions, it has a local smooth solution (van der Schaft 1992). Various other assumptions guarantee existence of smooth solutions, such as that the dynamics not be bilinear and the value not contain cross-terms in the state and control input.

The HJB equation is difficult to solve for nonlinear systems. An algorithm for finding an approximate smooth solution to the HJB is given by Abu-Khalaf and Lewis (2005). The solution to optimal control problems is generally obtained by solving the HJB equation offline. This does not allow the objectives to change in real time. The work of Vamvoudakis and Lewis (2010a) shows how to solve the HJB equation online in real time using data measured along the system trajectories. This allows solution of the optimal control problem if $Q(x)$, R in (10.1-2) change slowly with time. Vrabie and Lewis (2009) has provided algorithms for solving the HJB equations online in real time without knowing the system drift dynamics $f(x)$.

In Chapter 11 we show how to solve these equations online in real time using adaptive control techniques (Vrabie and Lewis 2009, Vamvoudakis and Lewis 2010a). We developed there a class of adaptive controllers that converge online to optimal control solutions by measuring data along the system trajectories. The approach is based on techniques of policy iteration from reinforcement learning and uses two approximator structures: one critic neural network to solve the Bellman equation (10.1-3), and one actor neural network to compute the control using (10.1-6). Note that if these two equations are solved simultaneously, then one has effectively solved the HJB equation (10.1-7).

Linear Quadratic Regulator

For the linear quadratic regulator (LQR) one has the linear dynamics

$$\dot{x} = F(x, u) = Ax + Bu, \tag{10.1-13}$$

with state $x(t) \in R^n$ and control $u(t) \in R^m$, and quadratic value integral

$$V(x(t)) = \int_t^\infty r(x, u)d\tau = \frac{1}{2}\int_t^\infty (x^T Qx + u^T Ru)d\tau, \tag{10.1-14}$$

with $Q \geq 0$, $R > 0$. Then, the value is quadratic in the state so that

$$V(x) = \tfrac{1}{2}x^T S x \tag{10.1-15}$$

for some matrix $S > 0$. Substituting this into (10.1-3) gives the Bellman equation

$$0 = \tfrac{1}{2}(x^T Q x + u^T R u) + x^T S(Ax + Bu) \equiv H(x, \nabla V, u). \tag{10.1-16}$$

Using now a linear state variable feedback (SVFB)

$$u = -Kx \tag{10.1-17}$$

gives

$$0 = (A - BK)^T S + S(A - BK) + Q + K^T RK. \tag{10.1-18}$$

It has been assumed that the Bellman equation holds for all initial conditions, and the state $x(t)$ has been cancelled in writing (10.1-18). It is seen that, for the LQR, the Bellman equation is equivalent to a Lyapunov equation for S in terms of the prescribed SVFB K. If $(A - BK)$ is stable and $(A, \sqrt{Q})$ observable, there is a positive definite solution $S > 0$, and then (10.1-15) is the value (10.1-14) for that selected feedback K.

The LQR optimal control that minimizes (10.1-14) is (10.1-6), or

$$u(x) = -R^{-1}B^T S x = -Kx. \tag{10.1-19}$$

Substituting this into (10.1-18) yields the algebraic Riccati equation (ARE)

$$0 = A^T S + SA + Q - SBR^{-1}B^T S \tag{10.1-20}$$

exactly as given in Table 3.3-1.

To solve the optimal control problem, one solves the ARE equation (10.1-20) for the optimal value kernel $S > 0$, then the optimal control is given as a state variable feedback in terms of the ARE solution by (10.1-19). There exists a solution $S > 0$ if (A, B) is stabilizable and $(A, \sqrt{Q})$ is observable.

The work of Vamvoudakis and Lewis (2010a) shows how to solve the ARE equation online in real time using data measured along the system trajectories. This allows solution of the optimal control problem if the performance index weighting matrices Q, R change slowly with time. Vrabie and Lewis (2009) has provided algorithms for solving the ARE equation online in real time without knowing the system matrix A. In Chapter 11 we show how to solve the ARE equation online in real time using adaptive control techniques based on this work.

10.2 TWO-PLAYER ZERO-SUM GAMES

In this section we use the Bellman equation approach to solve the 2-player zero-sum (ZS) games (Basar and Olsder 1999). ZS games refer to the fact that whatever one player gains, the other loses. The Nash solution of the 2-player ZS game is important in feedback control because it provides a solution to the bounded L_2-gain problem, and, hence, allows solution of the H-infinity disturbance rejection problem (Zames 1981, van der Schaft 1992, Knobloch et al. 1993).

Zero-sum Game

Consider the nonlinear time-invariant dynamical system given by

$$\dot{x} = f(x) + g(x)u + k(x)d, \qquad (10.2\text{-}1)$$

where state $x(t) \in R^n$. This system has two inputs or players, known as the control input $u(t) \in R^m$ and the disturbance input $d(t) \in R^q$. Let $f(x)$ be locally Lipschitz and $f(0) = 0$. Define the performance index

$$J(x(0), u, d) = \int_0^{\infty} \left(Q(x) + u^{\mathrm{T}} R u - \gamma^2 \|d\|^2 \right) d\tau \equiv \int_t^{\infty} r(x, u, d) d\tau, \quad (10.2\text{-}2)$$

where $Q(x) = h^{\mathrm{T}}(x)h(x) \geq 0$ for some function $h(x)$, $R = R^{\mathrm{T}} > 0$, and $\gamma > 0$. This is a function of the initial state and the functions $u(\tau), d(\tau) : 0 \leq \tau$.

Define the 2-player zero-sum differential game

$$V^*(x(0)) = \min_u \max_d J(x, u, d) = \min_u \max_d \int_0^{\infty} \left(h^{\mathrm{T}}h + u^{\mathrm{T}} R u - \gamma^2 \|d\|^2 \right) dt,$$

$$(10.2\text{-}3)$$

where the control player seeks to minimize the value and the disturbance to maximize it. This game puts the control and disturbance at odds with one another so that anything one gains in its objective is lost by the other. Therefore, it is termed a zero-sum game. This is equivalent to defining two performance measures $J_1(x(0), u, d) = J(x(0), u, d) = -J_2(x(0), u, d)$ and solving the game where the control player seeks to minimize J_1 and the disturbance player seeks to minimize J_2.

This game has a unique solution if a game theoretic saddle point (u^*, d^*) exists; that is, if

$$V^*(x_0) = \min_u \max_d J(x(0), u, d) = \max_d \min_u J(x(0), u, d). \qquad (10.2\text{-}4)$$

A saddle point is shown in Figure 10.2-1. The associated value V^* is called the value of the game. This is equivalent to the Nash equilibrium condition

$$J(x(0), u^*, d) \leq J(x(0), u^*, d^*) \leq J(x(0), u, d^*) \qquad (10.2\text{-}5)$$

holding for all policies u, d. These conditions depend on the performance integral (10.2-2) and the dynamics (10.2-1). According to the Nash condition, if both

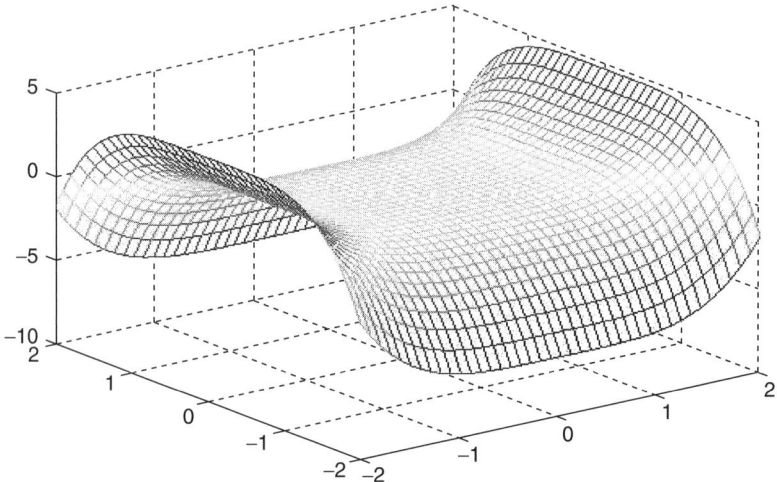

FIGURE 10.2-1   Saddle-point equilibrium.

players are at equilibrium, then neither has an incentive to change his policy unilaterally, since unilateral changes make one's performance worse.

For fixed control and disturbance feedback policies $u(x), d(x)$ define the value function

$$V(x(t), u, d) = \int_t^\infty \left(h^T h + u^T R u - \gamma^2 \|d\|^2\right) d\tau \equiv \int_t^\infty r(x, u, d) d\tau, \quad (10.2-6)$$

which is only a function of the initial state $x(t)$. When the value is finite, a differential equivalent to this is found by differentiating using Leibniz's formula. The result is the nonlinear *ZS game Bellman equation* given in terms of the Hamiltonian function as

$$0 = r(x, u, d) + \dot{V} = r(x, u, d) + \nabla V^T \dot{x}$$

or

$$0 = h^T h + u^T R u - \gamma^2 \|d\|^2 + \nabla V^T (f + gu + kd) \equiv H(x, \nabla V, u, d). \quad (10.2-7)$$

The boundary condition for this partial differential equation is $V(0) = 0$. A minimal solution such that $V(x) \geq 0$ to (10.2-7) is the value (10.2-6) for the given feedback policy $u(x)$ and disturbance policy $d(x)$.

A necessary condition for Nash condition (10.2-4) is Isaacs' condition

$$\min_u \max_d H(x, \nabla V, u, d) = \max_d \min_u H(x, \nabla V, u, d) \quad (10.2-8)$$

or, equivalently,

$$H(x, \nabla V, u^*, d) \leq H(x, \nabla V, u^*, d^*) \leq H(x, \nabla V, u, d^*) \quad (10.2-9)$$

for all policies u, d. These conditions hold pointwise in time. Under certain regularity conditions they guarantee (10.2-4). The Isaacs condition generalizes the PMP (10.1-4) to ZS games.

At equilibrium, one has the two stationarity conditions

$$\frac{\partial H}{\partial u} = 0, \quad \frac{\partial H}{\partial d} = 0. \tag{10.2-10}$$

Note from (10.2-7) that $\partial^2 H/\partial^2 u = 2R > 0$, $\partial^2 H/\partial^2 d = -2\gamma^2 < 0$, so that at the stationary point the Hamiltonian attains a minimum in u and a maximum in d, that is, a saddle point.

Applying (10.2-10) to the Hamiltonian (10.2-7) yields the control and disturbance policies

$$u = u(V(x)) \equiv -\tfrac{1}{2} R^{-1} g^T(x) \nabla V \tag{10.2-11}$$

$$d = d(V(x)) \equiv \frac{1}{2\gamma^2} k^T(x) \nabla V. \tag{10.2-12}$$

Substituting these into the Bellman equation (10.2-7) yields the Hamilton-Jacobi-Isaacs (HJI) equation

$$0 = h^T h + \nabla V^T(x) f(x) - \frac{1}{4} \nabla V^T(x) g(x) R^{-1} g^T(x) \nabla V(x)$$

$$+ \frac{1}{4\gamma^2} \nabla V^T(x) k k^T \nabla V(x), \quad V(0) = 0, \tag{10.2-13}$$

which can be written

$$0 = H(x, \nabla V^*, u^*, d^*), \qquad V(0) = 0. \tag{10.2-14}$$

The minimum positive semi-definite (PSD) solution V^* of this equation gives the Nash value, and the Nash equilibrium solution $(u^*, d^*) = (u(V^*(x)), d(V^*(x)))$ is given by (10.2-11), (10.2-12) in terms of ∇V^*.

For the HJI equation to have a PSD solution, the gain γ should be chosen large enough (Başar and Olsder 1999, van der Schaft 1992). It is shown that there exists a critical gain γ^* such that the HJI has a PSD solution for any $\gamma > \gamma^*$. The critical value γ^* is known as the H-infinity gain for system (10.2-1). The H-infinity gain can be explicitly computed for linear systems (Chen et al. 2004).

Instrumental to the analysis of ZS games is the following key fact, which shows that the Hamiltonian is quadratic in deviations in the control and disturbance about some critical policies.

Lemma 10.2-1. For any policies $u(x), d(x)$ yielding a finite value, let $V(x) \geq 0$ be the corresponding solution to the Bellman equation 10.2-7. Define $(u^*, d^*) = (u(V(x)), d(V(x)))$ by (10.2-11), (10.2-12) in terms of $V(x)$. Then

$$H(x, \nabla V, u, d) = H(x, \nabla V, u^*, d^*) + (u - u^*)^T R(u - u^*) - \gamma^2 \left\| d - d^* \right\|^2. \tag{10.2-15}$$

∎

Exercise 10.2-1.

Prove (10.2-15) by completing the squares in (10.2-7). This result immediately shows that the Isaacs condition (10.2-9) holds for solutions V^* to the HJI equation (10.2-13), for according to (10.2-15), when (10.2-14) holds one has

$$H(x, \nabla V^*, u, d) = (u - u^*)^T R(u - u^*) - \gamma^2 \|d - d^*\|^2, \tag{10.2-16}$$

which satisfies (10.2-9).

The next result shows that under certain conditions the HJI solution satisfies the Nash condition (10.2-4) and so solves the ZS game (Başar and Olsder 1999, van der Schaft 1992). System

$$\dot{x} = f(x) + g(x)u(x), \quad y = h(x) \tag{10.2-17}$$

is said to be zero-state observable if $u(t) \equiv 0, y(t) \equiv 0 \Rightarrow x(t) = 0$. ∎

Theorem 10.2-2. Solution to 2-player ZS Game. Assume the game (10.2-3) has a finite value. Select $\gamma > \gamma^* > 0$. Suppose $V^*(x) \in C^1 : R^n \to R$ is a smooth positive semi-definite solution to the HJI equation (10.2-13) such that closed-loop system

$$\dot{x} = f(x) + g(x)u^* + k(x)d^* = f(x) - \frac{1}{2}gR^{-1}g^T\nabla V^* + \frac{1}{2\gamma^2}k(x)k^T(x)\nabla V^* \tag{10.2-18}$$

is locally asymptotically stable. Assume (10.2-17) is zero-state observable. The Nash condition (10.2-5) is satisfied for control $u^* = u(V^*(x))$ given by (10.2-11) and $d^* = d(V^*(x))$ given by (10.2-12) in terms of $V^*(x)$. Then the system is in Nash equilibrium, the game has a value, and (u^*, d^*) is a saddle-point equilibrium solution among policies in $L_2[0, \infty)$. Moreover, the value of the game is given by the HJI solution $V^*(x(0))$.

Proof: One has for any C^1 smooth function $V(x) : R^n \to R$ and $T > 0$

$$J_T(x(0), u, d) \equiv \int_0^T \left(h^Th + u^TRu - \gamma^2 \|d\|^2\right) dt$$

$$= \int_0^T \left(h^Th + u^TRu - \gamma^2 \|d\|^2\right) dt \quad + \int_0^T \dot{V} dt \quad - V(x(T)) + V(x(0))$$

$$= \int_0^T \left(h^Th + u^TRu - \gamma^2 \|d\|^2\right) dt$$

$$+ \int_0^T \nabla V^T(f + gu + kd) dt \quad - V(x(T)) + V(x(0))$$

$$= \int_0^T H(x, \nabla V, u, d) dt \quad - V(x(T)) + V(x(0)).$$

Now, suppose $V^*(x)$ satisfies the HJI equation (10.2-13). Then $0 = H(x, \nabla V^*, u^*, d^*)$ and (10.2-15) yields

$$J_T(x(0), u, d) = \int_0^T \left((u - u^*)^\mathrm{T} R(u - u^*) - \gamma^2 \left\| d - d^* \right\|^2 \right) dt$$
$$- V^*(x(T)) + V^*(x(0)).$$

Since $u(t)$, $d(t) \in L_2[0, \infty)$, and since the game has a finite value as $T \to \infty$, this implies that $x(t) \in L_2[0, \infty)$; therefore, zero state observability implies $x(t) \to 0$, $V^*(x(\infty)) = 0$ and

$$J(x(0), u, d) = \int_0^\infty \left((u - u^*)^\mathrm{T} R(u - u^*) - \gamma^2 \left\| d - d^* \right\|^2 \right) dt + V^*(x(0)),$$

which implies saddle-point equilibrium, e.g., (10.2-5). One obtains

$$J(x(0), u^*, d^*) = \min_u \max_d J(x(0), u, d) = V^*(x(0)),$$

which shows the Nash solution is $u(t) = u^*(t), d(t) = d^*(t)$ and the Nash value is

$$V^*(x(0)). \qquad \blacksquare$$

Solution of HJI Equations

A minimum positive semi-definite solution $V^*(x) \geq 0$ to the HJI is one for which there is no other solution $V(x) \geq 0$ such that $V^*(x) \geq V(x) \geq 0$. The minimum positive semi-definite HJI solution is the unique HJI solution such that the closed-loop system (10.2-18) is locally asymptotically stable (Başar and Olsder 1999, van der Schaft 1992). More information is provided below on the discussion on linear quadratic zero-sum games.

The HJI equation may not have smooth solutions, but may have the so-called viscosity solutions (Bardi and Capuzzo-Dolcetla 1997). Under certain local reachability and observability assumptions, it has a local smooth solution (van der Schaft 1992).

The HJI equation is difficult to solve for nonlinear systems. An algorithm for finding an approximate minimum PSD solution to the HJI is given in Abu-Khalaf et al. (2006, 2008). The solution to ZS games is generally obtained by solving the HJI equation offline. This does not allow the objectives to change in real time as the players learn from each other. The work of Vamvoudakis and Lewis (2010b) shows how to solve the HJI equation online in real time using data measured along the system trajectories. This allows the weights R, γ in (10.2-2) to change slowly as the game develops. Vrabie and Lewis (2010a,b) has provided algorithms for solving the HJI equations online in real time without knowing the

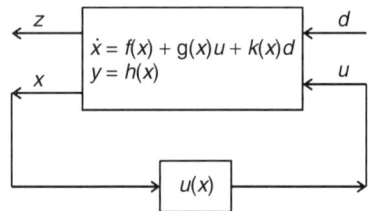

FIGURE 10.3-1   Bounded L_2-gain control.

system drift dynamics $f(x)$. These references extend the reinforcement learning techniques presented in Chapter 11 to online solution of zero-sum games.

10.3   APPLICATION OF ZERO-SUM GAMES TO H_∞ CONTROL

Consider the system (10.2.1) with output $y = h(x)$ and a performance output $z(t) = [u^T(t) \quad y^T(t)]^T$. This setup is shown in Figure 10.3-1. In the bounded L_2-gain problem (Zames 1981, van der Schaft 1992, Başar and Olsder 1999), one desires to find a feedback control policy $u(x)$ such that, when $x(0) = 0$ and for all disturbances $d(t) \in L_2[0, \infty)$ one has

$$\frac{\int_0^T \|z(t)\|^2 \, dt}{\int_0^T \|d(t)\|^2 \, dt} = \frac{\int_0^T (h^T h + u^T R u) \, dt}{\int_0^T \|d(t)\|^2 \, dt} \le \gamma^2 \qquad (10.3\text{-}1)$$

for a prescribed $\gamma > 0$ and $R = R^T > 0$ and for all $T > 0$. That is, the L_2-gain from the disturbance to the performance output is less than or equal to γ.

The H-infinity control problem is to find, if it exists, the smallest value $\gamma^* > 0$ such that for any $\gamma > \gamma^*$ the bounded L_2-gain problem has a solution. In the linear case an explicit expression can be provided for the H-infinity gain γ^* (Chen et al. 2004).

To solve the bounded L_2-gain problem, one may use the machinery of 2-player ZS games just developed. In the 2-player ZS game, both inputs can be controlled, with the control input seeking to minimize a performance index and the disturbance input seeking to maximize it. By contrast, here $d(t)$ is a disturbance that cannot be controlled, and $u(t)$ is the control input used to offset the deleterious effects of the disturbance.

The next result provides a solution to the bounded L_2-gain problem in terms of a solution to the HJI equation.

Theorem 10.3-1. Solution to Bounded L_2-gain Problem. Select $\gamma > \gamma^* > 0$. Suppose $V^*(x) > 0 \in C^1 : R^n \to R$ is a smooth positive definite solution to the HJI equation 10.2-13. Assume (10.2-17) is zero-state observable. Then the closed-loop system

$$\dot{x} = f(x) + g(x)u^* = f(x) - \tfrac{1}{2}gR^{-1}g^{\mathrm{T}}\nabla V^* \qquad (10.3\text{-}2)$$

is locally asymptotically stable with control input $u^* = u(V^*(x))$ given by (10.2-11) in terms of $V^*(x)$. Moreover, for this choice of control input, (10.3-1) holds for all disturbances $d(t) \in L_2[0, \infty)$.

Proof: Note that for any C^1 function $V(x) : R^n \to R$ one has, along the system trajectories,

$$\frac{dV}{dt} = \frac{\partial V}{\partial t} + \frac{\partial V}{\partial x}^{\mathrm{T}} \dot{x} = \frac{\partial V}{\partial x}^{\mathrm{T}} (f + gu + kd),$$

so that

$$\frac{dV}{dt} + h^{\mathrm{T}}h + u^{\mathrm{T}}Ru - \gamma^2 d^{\mathrm{T}}d = H(x, \nabla V, u, d).$$

Suppose now that $V(x)$ satisfies the Hamilton-Jacobi-Isaacs (HJI) equation (10.2-13), (10.2-14). Then according to (10.2-15) one has

$$H(x, \nabla V, u, d) = -\gamma^2 \|d - d*\|^2 + (u - u^*)^{\mathrm{T}} R(u - u^*).$$

Therefore,

$$\frac{dV}{dt} + h^{\mathrm{T}}h + u^{\mathrm{T}}Ru - \gamma^2 d^{\mathrm{T}}d = -\gamma^2 \|d - d*\|^2 + (u - u^*)^{\mathrm{T}} R(u - u^*).$$

Selecting $u = u^*$ yields

$$\frac{dV}{dt} + h^{\mathrm{T}}h + u^{\mathrm{T}}Ru - \gamma^2 d^{\mathrm{T}}d \le 0 \qquad (10.3\text{-}3)$$

for all $d(t)$. To show asymptotic stability of the closed-loop system (10.3-2) set $d = 0$ and note that

$$\frac{dV}{dt} \le -(h^{\mathrm{T}}h + u^{\mathrm{T}}Ru) = - \|z\|^2 .$$

LaSalle's extension now shows that the state goes to a region of R^n wherein $\dot{V} = 0$. However, zero-state observable means $u(t) \equiv 0$, $y(t) \equiv 0 \Rightarrow x(t) = 0$. Therefore, the system is locally asymptotically stable with Lyapunov function $V(x) > 0$.

  Now, integrating (10.3-3) yields

$$V(x(T)) - V(x(0)) + \int_0^T \left(h^{\mathrm{T}}h + u^{\mathrm{T}}Ru - \gamma^2 d^{\mathrm{T}}d \right) dt \le 0.$$

Select $x(0) = 0$. Noting that $V(0) = 0$ and $V(x(T)) > 0$ one has

$$\int_0^T \left(h^\mathsf{T} h + u^\mathsf{T} R u \right) dt \leq \gamma^2 \int_0^T \left(d^\mathsf{T} d \right) dt,$$

so the L_2 gain is less than γ. ∎

The disturbance $d^* = d(V^*) = (1/2\gamma^2)k^\mathsf{T}(x)\nabla V^*$ provided by the ZS game solution (10.2-12) is known as the worst-case disturbance.

Linear Quadratic Zero-sum Game

In the LQ ZS game one has linear dynamics

$$\dot{x} = Ax + Bu + Dd, \tag{10.3-4}$$

with $x(t) \in R^n$. The value is the integral quadratic form

$$V(x(t), u, d) = \frac{1}{2} \int_t^\infty \left(x^\mathsf{T} H^\mathsf{T} H x + u^\mathsf{T} R u - \gamma^2 \|d\|^2 \right) d\tau \equiv \int_t^\infty r(x, u, d) d\tau, \tag{10.3-5}$$

where $R = R^\mathsf{T} > 0$, and $\gamma > 0$. Assume that the value is quadratic in the state so that

$$V(x) = \tfrac{1}{2} x^\mathsf{T} S x \tag{10.3-6}$$

for some matrix $S > 0$. Select state feedbacks for the control and disturbance so that

$$u = -Kx \tag{10.3-7}$$
$$d = Lx. \tag{10.3-8}$$

Substituting this into Bellman equation (10.2-7) gives

$$S(A - BK + DL) + (A - BK + DL)^\mathsf{T} S + H^\mathsf{T} H + K^\mathsf{T} R K - \gamma^2 L^\mathsf{T} L. \tag{10.3-9}$$

It has been assumed that the Bellman equation holds for all initial conditions, and the state $x(t)$ has been canceled. This is a Lyapunov equation for S in terms of the prescribed SVFB policies K and L. If $(A - BK + DL)$ is stable, (A, H) is observable, and $\gamma > \gamma^* > 0$, then there is a positive definite solution $S \geq 0$, and then (10.3-6) is the value (10.3-5) for the selected feedback policies K and L.

The stationary point control (10.2-11) and disturbance (10.2-12) are given by

$$u(x) = -R^{-1} B^\mathsf{T} S x = -Kx \tag{10.3-10}$$
$$d(x) = \frac{1}{\gamma^2} D^\mathsf{T} S x = Lx. \tag{10.3-11}$$

Substituting these into (10.3-9) yields the game algebraic Riccati equation (GARE)

$$0 = A^\mathrm{T}S + SA + H^\mathrm{T}H - SBR^{-1}B^\mathrm{T}S + \frac{1}{\gamma^2}SDD^\mathrm{T}S. \qquad (10.3\text{-}12)$$

To solve the ZS game problem, one solves the GARE equation for the optimal value kernel $S \geq 0$, then the optimal control is given as a state-variable feedback in terms of the ARE solution by (10.3-10) and the worst-case disturbance by (10.3-11). There exists a solution $S > 0$ if $(A, \ B)$ is stabilizable, $(A, \ \sqrt{Q})$ is observable, and $\gamma > \gamma^*$, the H-infinity gain. Since we have developed a solution to the problem, it is verified that assumption (10.3-6) holds.

There may more than one PSD solution to the GARE. To solve the ZS game problem, one requires gains such that the poles of $(A - BK + DL)$ are in the open left-half plane. The minimum PSD solution of the GARE is the unique stabilizing solution. It is shown in van der Schaft (1992), and Başar and Olsder (1999) that the stabilizing solution of the GARE corresponds to the stable eigenspace of a certain Hamiltonian matrix, similarly to the discussion presented in Section 3.4 for the stabilizing ARE solution. A method for finding the minimum PSD solution is given in (Abu Khalaf et al. 2006).

The work of Vamvoudakis and Lewis (2010b) shows how to solve the GARE online in real time using data measured along the system trajectories. This allows the performance index weights R, γ to change slowly as the game develops. Vrabie and Lewis (2010a) has provided algorithms for solving the HJB equations online in real time without knowing the system matrix A. These references extend the reinforcement learning techniques presented in Chapter 11 to online solution of LQ zero-sum games.

10.4   MULTIPLAYER NON-ZERO-SUM GAMES

In zero-sum games, whatever one player gains, the rest lose. Cooperation and competition in nature and in human enterprises are fascinating topics. Most often, there is a balance between cooperation toward team goals and competition among players to achieve individual goals. This interplay is very nicely captured by the machinery of non-zero-sum (NZS) multiplayer games (Başar and Olsder 1999).

Nonlinear NZS Games

Consider the nonlinear time-invariant dynamical system given by

$$\dot{x} = f(x) + \sum_{j=1}^{N} g_j(x)u_j, \qquad (10.4\text{-}1)$$

with state $x(t) \in R^n$ and controls $u_j(t) \in R^{m_j}$. This system has N inputs or players, all of whom influence each other through their joint effects on the overall system state dynamics. Let $f(x)$ be locally Lipschitz, $f(0) = 0$, and $g_j(x)$ be continuous. Define the performance index of player i as

$$J_i(x(0), u_1, u_2, \ldots, u_N) = \int_0^\infty \left(Q_i(x) + \sum_{j=1}^N u_j^T R_{ij} u_j \right) dt$$

$$\equiv \int_0^\infty r_i(x(t), u_1, u_2, \ldots, u_N) \, dt, \, i \in N, \qquad (10.4\text{-}2)$$

where function $Q_i(x) \geq 0$ is generally nonlinear, and $R_{ii} > 0$, $R_{ij} \geq 0$ are symmetric matrices. The notation $i \in N$ means $i = 1, \ldots, N$.

Define the multiplayer differential game

$$V_i^*(x(t), u_1, u_2, \ldots, u_N) = \min_{u_i} \int_t^\infty \left(Q_i(x) + \sum_{j=1}^N u_j^T R_{ij} u_j \right) d\tau, \quad \forall i \in N.$$

$$(10.4\text{-}3)$$

This game implies that all the players have the same competitive hierarchical level and seek to attain a Nash equilibrium as given by the following definition.

Definition 10.4-1. Nash equilibrium.

Policies $\{u_1^*(x), u_2^*(x), \ldots, u_N^*(x)\}$ are said to constitute a Nash equilibrium solution for the N-*player* game if

$$J_i^* \triangleq J_i(u_1^*, u_2^*, u_i^*, \ldots, u_N^*) \leq J_1(u_1^*, u_2^*, u_i, \ldots, u_N^*), \quad \forall u_i, \forall i \in N. \quad (10.4\text{-}4)$$

The N-*tuple* $\{J_1^*, J_2^*, \ldots, J_N^*\}$ is known as a Nash equilibrium value set or outcome of the N-player game.

The implication of this definition is that if any player unilaterally changes his control policy while the policies of all other players remain the same, then that player will obtain worse performance. For fixed stabilizing feedback control policies $u_j(x)$ define the value function

$$V_i(x(t)) = \int_t^\infty \left(Q_i(x) + \sum_{j=1}^N u_j^T R_{ij} u_j \right) d\tau = \int_t^\infty r(x, u_1, \ldots, u_N) d\tau, i \in N,$$

$$(10.4\text{-}5)$$

which is only a function of the initial state $x(t)$. When the value is finite, a differential equivalent to this is found by differentiating using Leibniz's formula.

This yields the nonlinear NZS *game Bellman equations* given in terms of the Hamiltonian functions as

$$0 = Q_i(x) + \sum_{j=1}^{N} u_j^{\mathrm{T}} R_{ij} u_j + (\nabla V_i)^{\mathrm{T}} \left(f(x) + \sum_{j=1}^{N} g_j(x) u_j \right)$$

$$\equiv H_i(x, \nabla V_i, u_1, \ldots, u_N), \ i \in N. \tag{10.4-6}$$

The initial conditions for these partial differential equations are $V_i(0) = 0$.

At equilibrium, one has the stationarity conditions

$$\frac{\partial H_i}{\partial u_i} = 0, \quad i \in N, \tag{10.4-7}$$

which yield the policies

$$u_i(x) = u_i(V_i(x)) = -\tfrac{1}{2} R_{ii}^{-1} g_i^{\mathrm{T}}(x) \nabla V_i, \quad i \in N. \tag{10.4-8}$$

Note that $\partial^2 H_i / \partial^2 u_i = 2R_{ii} > 0$, so that at the stationary point the Hamiltonian H_i attains a minimum in control policy $u_i(x)$. Using these control policies the closed-loop system is

$$\dot{x} = f(x) - \frac{1}{2} \sum_{j=1}^{N} g_j(x) R_{jj}^{-1} g_j^{\mathrm{T}}(x) \nabla V_j. \tag{10.4-9}$$

Substituting (10.4-8) into (10.4-6) one obtains the N-coupled Hamilton-Jacobi (HJ) equations

$$0 = (\nabla V_i)^{\mathrm{T}} \left(f(x) - \frac{1}{2} \sum_{j=1}^{N} g_j(x) R_{jj}^{-1} g_j^{\mathrm{T}}(x) \nabla V_j \right)$$

$$+ Q_i(x) + \frac{1}{4} \sum_{j=1}^{N} \nabla V_j^{\mathrm{T}} g_j(x) R_{jj}^{-\mathrm{T}} R_{ij} R_{jj}^{-1} g_j^{\mathrm{T}}(x) \nabla V_j, \tag{10.4-10}$$

with $V_i(0) = 0$. These coupled HJ equations are in closed-loop form. The equivalent open-loop form is

$$0 = \nabla V_i^{\mathrm{T}} f(x) + Q_i(x) - \frac{1}{2} \nabla V_i^{\mathrm{T}} \sum_{j=1}^{N} g_j(x) R_{jj}^{-1} g_j^{\mathrm{T}}(x) \nabla V_j$$

$$+ \frac{1}{4} \sum_{j=1}^{N} \nabla V_j^{\mathrm{T}} g_j(x) R_{jj}^{-\mathrm{T}} R_{ij} R_{jj}^{-1} g_j^{\mathrm{T}}(x) \nabla V_j. \tag{10.4-11}$$

These equations can be written as

$$H_i(x, \nabla V_i^*, u_1^*, \ldots, u_N^*) = 0, \qquad (10.4\text{-}12)$$

where $u_i^*(x) = u_i(V_i^*(x))$ as given by (10.4-8).

The next key result shows that the Hamiltonian function has a specific dependence on control deviations from certain key values.

Lemma 10.4-2. For any control policies $u_i(x), i = 1, N$ that yield finite values (10.4-5), let $V_i(x) \geq 0$ be the corresponding solutions to the Bellman equations (10.4-6). Define $u_i^*(x) = u_i(V_i(x))$ according to (10.4-8) in terms of $V_i(x)$. Then

$$H_i(x, \nabla V_i, u_1, \ldots, u_N x)$$

$$= H_i(x, \nabla V_i, u_1^*, \ldots, u_N^*) + \sum_j (u_j - u_j^*)^{\mathrm{T}} R_{ij}(u_j - u_j^*)$$

$$+ \nabla V_i^{\mathrm{T}} \sum_j g_j(u_j - u_j^*) + 2 \sum_j (u_j^*)^{\mathrm{T}} R_{ij}(u_j - u_j^*). \qquad (10.4\text{-}13)$$

∎

Exercise 10.4-1.

Prove Lemma 10.4-1 by completing the squares on (10.4-6). This is simplified if one writes

$$H_i(x, \nabla V_i, u_1^*, \ldots, u_N^*) = \nabla V_i^{\mathrm{T}} f(x) + Q_i(x) + \nabla V_i^{\mathrm{T}} \sum_{j=1}^{N} g_j(x) u_j^* + \sum_{j=1}^{N} (u_j^*)^{\mathrm{T}} R_{ij} u_j^*.$$

$$(10.4\text{-}14)$$

The next result shows when the coupled HJ solutions solve the multiplayer game (Basar and Olsder 1999). ∎

Theorem 10.4-2. Stability and Solution of NZS Game Nash Equilibrium.
Let $Q_i(x) > 0$ in the performance index (10.4-2). Let $V_i(x) > 0 \in C^1, i = 1, N$ be smooth solutions to HJ equations (10.4-11), and control policies $u_i^*, i \in N$ be given by (10.4-8) in terms of these solutions V_i. Then

a. System (10.4-9) is asymptotically stable and $V_i(x)$ serve as Lyapunov functions.

b. $\{u_i^*, i = 1, N\}$ are in Nash equilibrium and the corresponding game values are

$$J_i^*(x(0)) = V_i, i \in N.$$

$$(10.4\text{-}15)$$

Proof:

a. If $V_i > 0$ satisfies (10.4-11) then it also satisfies (10.4-6). Take the time derivative to obtain

$$\dot{V}_i = \nabla V_i^T \dot{x} = \nabla V_i^T \left(f(x) + \sum_j g_j(x) u_j \right)$$

$$= -\frac{1}{2} \left(Q_i(x) + \sum_j u_j^T R_{ij} u_j \right),$$

(10.4-16)

which is negative definite since $Q_i > 0$. Therefore, $V_i(x)$ is a Lyapunov function for x and systems (10.4-9) are asymptotically stable.

b. According to part a, $x(t) \to 0$ for the selected control policies. Define u_{-i} as the set of control policies of all nodes other than node i. For any smooth functions $V_i(x), i \in N$, such that $V_i(0) = 0$, by setting $V_i(x(\infty)) = 0$ one can write (10.4-2) as

$$J_i(x(0), u_i, u_{-i}) = \frac{1}{2} \int_0^\infty (Q_i(x) + \sum_j u_j^T R_{ij} u_j) \, dt + \int_0^\infty \dot{V}_i dt + V_i(x(0))$$

or

$$J_i(x(0), u_i, u_{-i}) = \frac{1}{2} \int_0^\infty (Q_i(x) + \sum_j u_j^T R_{ij} u_j) \, dt$$

$$+ \int_0^\infty \nabla V_i^T (f + \sum_j g_j u_j) \, dt + V_i(x(0))$$

$$J_i(x(0), u_i, u_{-i}) = \frac{1}{2} \int_0^\infty H_i(x, \nabla V_i, u_1, \ldots, u_N) + V_i(x(0)).$$

Now let V_i satisfy (10.4-11) and u_i^*, u_{-i}^* be the optimal controls given by (10.4-8). By Lemma 10.4-1 one has

$$J_i(x(0), u_i, u_{-i}) = V_i(x(0)) + \int_0^\infty \left(\sum_j (u_j - u_j^*)^T R_{ij} (u_j - u_j^*)^T \right.$$

$$\left. - \nabla V_i^T \sum_j B_j (u_j - u_j^*) + \sum_j u_j^{*T} R_{ij} (u_j - u_j^*) \right) dt$$

(10.4-17)

$$J_i(x(0), u_i, u_{-i}) = V_i(x(0)) + \int_0^\infty (u_i - u_i^*)^\mathrm{T} R_{ii} (u_i - u_i^*)^\mathrm{T} dt$$

$$+ \int_0^\infty \left(\sum_{j \neq i} (u_j - u_j^*)^\mathrm{T} R_{ij} (u_j - u_j^*)^\mathrm{T} dt \right.$$

$$\left. - \nabla V_i^\mathrm{T} \sum_j B_j (u_j - u_j^*) + \sum_j u_j^{*\mathrm{T}} R_{ij} (u_j - u_j^*) \right) dt.$$

At the equilibrium point $u_i = u_i^*$ and $u_j = u_j^*, \forall j$ so

$$J_i^*(x(0), u_i^*, u_{-i}^*) = V_i(x(0)).$$

Define

$$J_i(u_i, u_{-i}^*) = V_i(x(0)) + \frac{1}{2} \int_0^\infty (u_i - u_i^*)^\mathrm{T} R_{ii} (u_i - u_i^*) \, dt$$

$$\tag{10.4-18}$$

and $J_i^* = V_i(x(0))$. Then clearly J_i^* and $J_i(u_i, u_{-i}^*)$ satisfy (10.4-4). ∎

Note that Lemma 10.4-1 shows that the values are not quadratic in all the control policies. In the proof this carries over to the fact that the performance indices (10.4-17) are not quadratic in all the control policies. However, Definition 10.4-1 of Nash equilibrium requires that the optimal value J_i^* be quadratic in u_i when all other policies are held at Nash. This definition means that the nonquadratic terms in (10.4-13) are equal to zero at the equilibrium and do not cause trouble. Then, (10.4-18) is quadratic in u_i, as required.

Solution of the Coupled HJ Equations

The coupled HJ equations are difficult to solve for nonlinear systems and the solution to multiplayer games is generally obtained by solving these equations offline. This does not allow the objectives to change in real time as the players learn from each other. The work of Vamvoudakis (2011) shows how to solve the coupled HJ equations online in real time using data measured along the system trajectories. This allows the weight functions $Q_i(x)$, R_{ij} in (10.4-2) to change slowly as the game develops. Vrabie (2010) has provided algorithms for solving the coupled HJ equations online in real time without knowing the system drift dynamics $f(x)$. These references extend the reinforcement learning techniques presented in Chapter 11 to online solution of non-zero-sum games.

Cooperative and Competitive Aspects of NZS Games   The multiplayer game formulation allows for considerable freedom of each agent within the framework of overall team Nash equilibrium. Each agent has a performance objective (10.4-2) that can embody team objectives as well as individual objectives. In fact, the performance objective of each agent can be written as

$$J_i = \frac{1}{N} \sum_{j=1}^{N} J_j + \frac{1}{N} \sum_{j=1}^{N} (J_i - J_j) \equiv J_{\text{team}} + J_{\text{conflict}}^i, \tag{10.4-19}$$

where J_{team} is the overall cooperative (center of gravity) performance objective of the networked team, and J_{conflict}^i is the conflict of interest or competitive objective. J_{team} measures how much the players are vested in common goals, and J_{conflict}^i expresses to what extent their objectives differ. The objective functions can be chosen by the individual players, or they may be assigned to yield some desired team behavior.

In the case of N-player zero-sum games, one has $J_{\text{team}} = 0$ and there are no common objectives. One case is the 2-player zero-sum game, where $J_2 = -J_1$, and one has $J_{\text{team}} = 0$ and $J_{\text{conflict}}^i = J_i$.

Linear Quadratic Multiplayer Games   In linear systems of the form

$$\dot{x} = Ax + \sum_{j=1}^{N} B_j u_j, \tag{10.4-20}$$

with quadratic performance indices

$$J_i(x(0), u_1, u_2, \ldots, u_N) = \frac{1}{2} \int_0^\infty (x^T Q_i x + \sum_{j=1}^{N} u_j^T R_{ij} u_j) \, dt, \tag{10.4-21}$$

the HJ equations (10.4-11) become the N-coupled generalized algebraic Riccati equations

$$0 = P_i A_c + A_c^T P_i + Q_i + \sum_{j=1}^{N} P_j B_j R_{jj}^{-T} R_{ij} R_{jj}^{-1} B_j^T P_j, \quad i \in N, \tag{10.4-22}$$

where $A_c = A - \sum_{i=1}^{N} B_i R_{ii}^{-1} B_i^T P_i$. It is shown in Başar and Olsder (1999) that if there exist solutions to (10.4.22) further satisfying the conditions that for each $i \in N$ the pairs

$$\left(A - \sum_{j \neq i} B_j R_{jj}^{-1} B_j^T P_j, \quad B_i \right),$$

$$\left(A - \sum_{j \neq i} B_j R_{jj}^{-1} B_j^T P_j, \quad \sqrt{Q_i + \sum_{j \neq i} P_j B_j R_{jj}^{-T} R_{ij} R_{jj}^{-1} B_j^T P_j} \right)$$

are respectively stabilizable and detectable, then the N-tuple of stationary feedback policies $u_i^*(x) = -K_i x = -\frac{1}{2} R_{ii}^{-1} B_i^T P_i x$, $i \in N$ provides a Nash equilibrium solution for the linear quadratic N-player differential game. Furthermore, the resulting system dynamics, described by $\dot{x} = A_c x$, are asymptotically stable.

Vamvoudakis and Lewis (2011) show how to solve the coupled HJ equations online in real time using data measured along the system trajectories. Vrabie and Lewis (2010a,b) have provided algorithms for solving these equations online in real time without knowing the system A matrix. These references use reinforcement learning techniques based on the development in Chapter 11.

11

REINFORCEMENT LEARNING
AND OPTIMAL ADAPTIVE CONTROL

In this book we have presented a variety of methods for the analysis and design of optimal control systems. Design has generally been based on solving matrix design equations assuming full knowledge of the system dynamics. Optimal control is fundamentally a backward-in-time problem, as we have seen especially clearly in Chapter 6 Dynamic Programming. Optimal controllers are normally designed offline by solving Hamilton-Jacobi-Bellman (HJB) equations, for example, the Riccati equation, using complete knowledge of the system dynamics. The controller is stored and then implemented online in real time. In the linear quadratic case, this means the feedback gains are stored. Determining optimal control policies for nonlinear systems requires the offline solution of nonlinear HJB equations, which are often difficult or impossible to solve analytically.

In this book, we have developed many matrix design equations whose solutions yield various sorts of optimal controllers. This includes the Riccati equation, the HJB equation, and the design equations in Chapter 10 for differential games. These equations are normally solved offline. In this chapter we give practical methods for solving these equations online in real time using data measured along the system trajectories. Some of the methods do not require knowledge of the system dynamics.

In practical applications, it is often important to be able to design controllers online in real time without having complete knowledge of the plant dynamics. Modeling uncertainties may exist, including inaccurate parameters, unmodeled high-frequency dynamics, and disturbances. Moreover, both the system dynamics and the performance objectives may change with time. A class of controllers known as adaptive controllers learn online to control unknown systems using data measured in real time along the system trajectories. While learning the control solutions, adaptive controllers are able to guarantee stability and system performance. Adaptive control and optimal control represent different philosophies for designing feedback controllers. Adaptive controllers are not usually designed to be optimal in the sense of minimizing user-prescribed performance functions. Indirect adaptive controllers use system identification techniques to identify the system parameters, then use the obtained model to solve optimal design equations (Ioannou and Fidan 2006). Adaptive controllers may satisfy certain inverse optimality conditions, as shown in Li and Krstic (1997).

In this chapter we show how to design optimal controllers online in real time using data measured along the system trajectories. We present several adaptive control algorithms that converge to optimal control solutions. Several of these algorithms do not require full knowledge of the plant dynamics. In the LQR case, for instance, this amounts to using adaptive control techniques to learn the solution of the algebraic Riccati equation online without knowing the plant matrix A. In the nonlinear case, these algorithms allow the approximate solution of complicated HJ equations that cannot be exactly solved using analytic means. Design of optimal controllers online allows the performance objectives, such as the LQR weighting matrices, to change slowly in real time as control objectives change.

The framework we use in this chapter is the theory of Markov decision processes (MDP). It is shown here that MDP provide a natural framework that connects reinforcement learning, optimal control, adaptive control, and cooperative control (Tsitsiklis 1984, Jadbabaie et al. 2003, Olfati-Saber and Murray 2004). Some examples are given of cooperative decision and control of dynamical systems on communication graphs.

11.1 REINFORCEMENT LEARNING

Dynamic programming (Chapter 6) is a method for determining optimal control solutions using Bellman's principle (Bellman 1957) by working backward in time from some desired goal states. Designs based on dynamic programming yield offline solution algorithms, which are then stored and implemented online forward in time. In this chapter we show that techniques based on reinforcement learning allow the design of optimal decision systems that learn optimal solutions online and forward in time. This allows both the system dynamics and the performance objectives to vary slowly with time. The methods studied here depend on solving a certain equation, known as Bellman's equation (Sutton and Barto 1998), whose solution both evaluates the performance of current control policies and provides methods for improving those policies.

Reinforcement learning (RL) refers to a class of learning methods that allow the design of adaptive controllers that learn online, in real time, the solutions to user-prescribed optimal control problems. In machine learning, reinforcement learning (Mendel and MacLaren 1970, Werbos 1991, Werbos 1992, Bertsekas and Tsitsiklis 1996, Sutton and Barto 1998, Powell 2007, Cao 2007, Busoniu et al. 2009) is a method for solving optimization problems that involves an actor or agent that interacts with its environment and modifies its actions, or control policies, based on stimuli received in response to its actions. Reinforcement

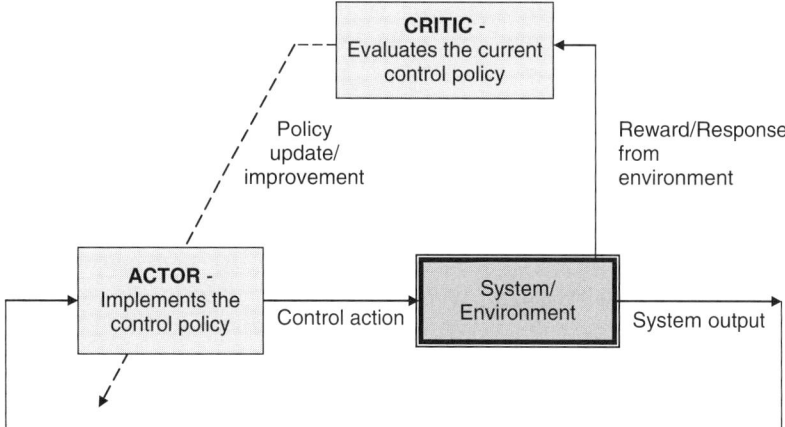

FIGURE 11.1-1 Reinforcement learning with an actor–critic structure. This structure provides methods for learning optimal control solutions online based on data measured along the system trajectories.

learning is inspired by natural learning mechanisms, where animals adjust their actions based on reward and punishment stimuli received from the environment. Other reinforcement learning mechanisms operate in the human brain, where the dopamine neurotransmitter in the basal ganglia acts as a reinforcement informational signal that favors learning at the level of the neuron (Doya et al. 2001, Schultz 2004).

The actor–critic structures shown in Figure 11.1-1 (Barto et al. 1983) are one type of reinforcement learning algorithm. These structures give forward-in-time algorithms for computing optimal decisions that are implemented in real time where an actor component applies an action, or control policy, to the environment, and a critic component assesses the value of that action. The learning mechanism supported by the actor–critic structure has two steps, namely, policy evaluation, executed by the critic, followed by policy improvement, performed by the actor. The policy evaluation step is performed by observing from the environment the results of applying current control actions. These results are evaluated using a performance index that quantifies how close to optimal the current action is. Performance can be defined in terms of optimality objectives, such as minimum fuel, minimum energy, minimum risk, or maximum reward. Based on the assessment of the performance, one of several schemes can then be used to modify or improve the control policy in the sense that the new policy yields a performance value that is improved relative to the previous value. In this scheme, reinforcement learning is a means of learning optimal behaviors by observing the real-time responses from the environment to nonoptimal control policies.

Direct adaptive controllers tune the controller parameters to directly identify the controller. Indirect adaptive controllers identify the system, and the identified model is then used in design equations to compute a controller. Actor–critic schemes are a logical extension of this sequence in that they identify the performance value of the current control policy, and then use that information to update the controller.

This chapter presents the main ideas and algorithms of reinforcement learning and applies them to design adaptive feedback controllers that converge online to optimal control solutions relative to prescribed cost metrics. Using these techniques, we can solve online in real time the Riccati equation, the HJB equation, and the design equations in Chapter 10 for differential games. Some of the methods given here do not require knowledge of the system dynamics.

We start from a discussion of MDP and specifically focus on a family of techniques known as approximate or adaptive dynamic programming (ADP) (Werbos 1989, 1991, 1992) or neurodynamic programming (Bertsekas and Tsitsiklis 1996). We show that the use of reinforcement learning techniques provides optimal control solutions for linear or nonlinear systems using adaptive control techniques.

This chapter shows that reinforcement learning methods allow the solution of HJB design equations online, forward in time, and without knowing the full system dynamics. Specifically, the drift dynamics is not needed, but the input coupling function is needed. In the linear quadratic case, these methods determine the solution to the algebraic Riccati equation online, without solving the equation and without knowing the system A matrix. This chapter presents an expository development of ideas from reinforcement learning and ADP and their applications in feedback control systems. Surveys of ADP are given in Si et al. (2004), Lewis, Lendaris, and Liu (2008), Balakrishnan et al. (2008), Wang et al. (2009), and Lewis and Vrabie (2009).

11.2 MARKOV DECISION PROCESSES

A natural framework for studying RL is provided by Markov decision processes (MDP). Many dynamical decision problems can be cast into the framework of MDP. Included are feedback control systems for human engineered systems, feedback regulation mechanisms for population balance and survival of species (Darwin 1859, Luenberger 1979), decision-making in multiplayer games, and economic mechanisms for regulation of global financial markets. Therefore, we provide a development of MDP here. References for this material include (Bertsekas and Tsitsiklis 1996, Sutton and Barto 1998, Busoniu et al. 2009).

Consider the Markov decision process (MDP) (X, U, P, R), where X is a set of states and U is a set of actions or controls. The transition probabilities $P: X \times U \times X \to [0, 1]$ give for each state $x \in X$ and action $u \in U$ the conditional probability $P_{x,x'}^{u} = \Pr\{x'|x, u\}$ of transitioning to state $x' \in X$ given the MDP is in state x and takes action u. The cost function $R: X \times U \times X \to \mathbb{R}$ gives the expected immediate cost $R_{xx'}^{u}$ paid after transition to state $x' \in X$ given the MDP starts in state $x \in X$ and takes action $u \in U$. The Markov property refers to the fact that transition probabilities $P_{x,x'}^{u}$ depend only on the current state x and not on the history of how the MDP attained that state. An MDP is shown in Figure 11.2-1.

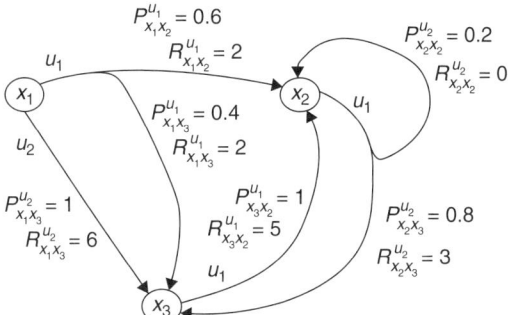

FIGURE 11.2-1 MDP shown as a finite state machine with controlled state transitions and costs associated with each transition.

The basic problem for MDP is to find a mapping $\pi\colon X \times U \to [0, 1]$ that gives for each state x and action u the conditional probability $\pi(x, u) = \Pr\{u | x\}$ of taking action u given the MDP is in state x. Such a mapping is termed a (closed-loop) control or action strategy or policy.

The strategy or policy $\pi(x, u) = \Pr\{u | x\}$ is called *stochastic* or mixed if there is a nonzero probability of selecting more than one control when in state x. We can view mixed strategies as probability distribution vectors having as component i the probability of selecting the ith control action while in state $x \in X$. If the mapping $\pi\colon X \times U \to [0, 1]$ admits only one control (with probability 1) when in every state x, it is called a *deterministic* policy. Then, $\pi(x, u) = \Pr\{u | x\}$ corresponds to a function mapping states into controls $\mu(x)\colon X \to U$.

Most work on reinforcement learning has been done for MDP that have finite state and action spaces. These are termed finite MDP.

Optimal Sequential Decision Problems

Dynamical systems evolve causally through time. Therefore, we consider sequential decision problems and impose a discrete stage index k such that the MDP takes an action and changes states at nonnegative integer stage values k. The stages may correspond to time or more generally to sequences of events. We refer to the stage value as the time. Denote state values and actions at time k by x_k, u_k. MDP traditionally evolve in discrete time.

Naturally occurring systems, including biological organisms and living species, have available a limited set of resources for survival and increase. Natural systems are therefore optimal in some sense. Likewise, human engineered systems should be optimal in terms of conserving resources such as cost, time, fuel, energy. Thus, it is important to capture the notion of optimality in selecting control policies for MDP.

Define, therefore, a *stage cost* at time k by $r_k = r_k(x_k, u_k, x_{k+1})$. Then $R_{xx'}^u = E\{r_k | x_k = x, u_k = u, x_{k+1} = x'\}$, with $E\{\cdot\}$ the expected value operator. Define a performance index as the sum of future costs over time interval $[k, k+T]$

$$J_{k,T} = \sum_{i=0}^{T} \gamma^i r_{k+i} = \sum_{i=k}^{k+T} \gamma^{i-k} r_i, \qquad (11.2\text{-}1)$$

where $0 \le \gamma < 1$ is a discount factor that reduces the weight of costs incurred further in the future. T is a planning horizon over which decisions are to be made.

Traditional usage of MDP in the fields of computational intelligence and economics consider r_k as a reward incurred at time k, also known as *utility*, and $J_{k,T}$ as a discounted return, also known as strategic reward. We refer instead to stage costs and discounted future costs to be consistent with objectives in the control of dynamical systems. We may sometimes loosely call r_k the utility.

Consider that an agent selects a control policy $\pi_k(x_k, u_k)$ and uses it at each stage k of the MDP. We are primarily interested in stationary policies, where the conditional probabilities $\pi_k(x_k, u_k)$ are independent of k. Then $\pi_k(x, u) = \pi(x, u) = \Pr\{u|x\}$, for all k. Nonstationary deterministic policies have the form $\pi = \{\mu_0, \mu_1, \ldots\}$, where each entry is a function $\mu_k(x): X \to U; k = 0, 1, \ldots$. Stationary deterministic policies are independent of time so that $\pi = \{\mu, \mu, \ldots\}$.

Select a fixed stationary policy $\pi(x, u) = \Pr\{u|x\}$. Then the ("closed-loop") MDP reduces to a Markov chain with state space X. That is, the transition probabilities between states are fixed with no further freedom of choice of actions. The transition probabilities of this Markov chain are given by

$$p_{x,x'} \equiv P_{x,x'}^{\pi} = \sum_{u} \Pr\{x'|x, u\} \Pr\{u|x\} = \sum_{u} \pi(x, u) P_{x,x'}^{u}, \qquad (11.2\text{-}2)$$

where we have used the Chapman-Kolmogorov identity.

Under the assumption that the Markov chain corresponding to each policy (with transition probabilities given as in (11.2-2)) is ergodic, it can be shown that every MDP has a stationary deterministic optimal policy (Wheeler and Narendra 1986, Bertsekas and Tsitsiklis 1996). A Markov chain is *ergodic* if all states are positive recurrent and aperiodic (Luenberger 1979). Then, for a given policy there exists a stationary distribution $p_\pi(x)$ over X that gives the steady-state probability the Markov chain is in state x. We shall soon discuss more about the closed-loop Markov chain.

The *value* of a policy is defined as the conditional expected value of future cost when starting in state x at time k and following policy $\pi(x, u)$ thereafter,

$$V_k^{\pi}(x) = E_{\pi}\{J_{k,T}|x_k = x\} = E_{\pi}\left\{\sum_{i=k}^{k+T} \gamma^{i-k} r_i \big| x_k = x\right\}. \qquad (11.2\text{-}3)$$

Here, $E_{\pi}\{\cdot\}$ is the expected value given that the agent follows policy $\pi(x, u)$. $V^{\pi}(x)$ is known as the *value function* for policy $\pi(x, u)$. It tells the value of being in state x given that the policy is $\pi(x, u)$.

An important objective of MDP is to determine a policy $\pi(x, u)$ to minimize the expected future cost

$$\pi^*(x, u) = \arg\min_{\pi} V_k^{\pi}(s) = \arg\min_{\pi} E_{\pi} \left\{ \sum_{i=k}^{k+T} \gamma^{i-k} r_i | x_k = x \right\}. \quad (11.2\text{-}4)$$

This is termed the *optimal policy*, and the corresponding *optimal value* is given as

$$V_k^*(x) = \min_{\pi} V_k^{\pi}(x) = \min_{\pi} E_{\pi} \left\{ \sum_{i=k}^{k+T} \gamma^{i-k} r_i | x_k = x \right\}. \quad (11.2\text{-}5)$$

In computational intelligence and economics, when we talk about utilities and rewards, we are interested in maximizing the expected performance index.

A Backward Recursion for the Value

By using the Chapman-Kolmogorov identity and the Markov property we may write the value of policy $\pi(x, u)$ as

$$V_k^{\pi}(x) = E_{\pi}\{J_k | x_k = x\} = E_{\pi} \left\{ \sum_{i=k}^{k+T} \gamma^{i-k} r_i | x_k = x \right\} \quad (11.2\text{-}6)$$

$$V_k^{\pi}(x) = E_{\pi} \left\{ r_k + \gamma \sum_{i=k+1}^{k+T} \gamma^{i-(k+1)} r_i | x_k = x \right\} \quad (11.2\text{-}7)$$

$$V_k^{\pi}(x) = \sum_{u} \pi(x, u) \sum_{x'} P_{xx'}^u \left[R_{xx'}^u + \gamma E_{\pi} \left\{ \sum_{i=k+1}^{k+T} \gamma^{i-(k+1)} r_i | x_{k+1} = x' \right\} \right].$$
$$(11.2\text{-}8)$$

Therefore, the value function for policy $\pi(x, u)$ satisfies

$$V_k^{\pi}(x) = \sum_{u} \pi(x, u) \sum_{x'} P_{xx'}^u \left[R_{xx'}^u + \gamma V_{k+1}^{\pi}(x') \right]. \quad (11.2\text{-}9)$$

This provides a backward recursion for the value at time k in terms of the value at time $k + 1$.

Dynamic Programming

The optimal cost can be written as

$$V_k^*(x) = \min_{\pi} V_k^{\pi}(x) = \min_{\pi} \sum_{u} \pi(x, u) \sum_{x'} P_{xx'}^u \left[R_{xx'}^u + \gamma V_{k+1}^{\pi}(x') \right]. \quad (11.2\text{-}10)$$

Bellman's optimality principle (Bellman 1957) states that "an optimal policy has the property that no matter what the previous control actions have been, the remaining controls constitute an optimal policy with regard to the state resulting from those previous controls." Therefore, we may write

$$V_k^*(x) = \min_{\pi} \sum_u \pi(x, u) \sum_{x'} P_{xx'}^u \left[R_{xx'}^u + \gamma V_{k+1}^*(x') \right]. \qquad (11.2\text{-}11)$$

Suppose we now apply an arbitrary control u at time k and the optimal policy from time $k + 1$ on. Then Bellman's optimality principle says that the optimal control at time k is given by

$$u_k^* = \arg\min_{\pi} \sum_u \pi(x, u) \sum_{x'} P_{xx'}^u \left[R_{xx'}^u + \gamma V_{k+1}^*(x') \right]. \qquad (11.2\text{-}12)$$

Under the assumption that the Markov chain corresponding to each policy (with transition probabilities given as in (11.2-2)) is ergodic, every MDP has a stationary deterministic optimal policy. Then we can equivalently minimize the conditional expectation over all actions u in state x. Therefore,

$$V_k^*(x) = \min_{u} \sum_{x'} P_{xx'}^u \left[R_{xx'}^u + \gamma V_{k+1}^*(x') \right], \qquad (11.2\text{-}13)$$

$$u_k^* = \arg\min_{u} \sum_{x'} P_{xx'}^u \left[R_{xx'}^u + \gamma V_{k+1}^*(x') \right]. \qquad (11.2\text{-}14)$$

The backward recursion (11.2-11), (11.2-13) forms the basis for dynamic programming (DP), which gives offline methods for working backward in time to determine optimal policies. DP was discussed in Chapter 6. It is an offline procedure for finding the optimal value and optimal policies that requires knowledge of the complete system dynamics in the form of transition probabilities $P_{x,x'}^u = \Pr\{x'|x, u\}$ and expected costs $R_{xx'}^u = E\{r_k | x_k = x, u_k = u, x_{k+1} = x'\}$. Once the optimal control has been found offline using DP, it is stored and implemented on the system online forward in time.

Bellman Equation and Bellman Optimality Equation (HJB)

Dynamic programming is a backward-in-time method for finding the optimal value and policy. By contrast, reinforcement learning is concerned with finding optimal policies based on causal experience by executing sequential decisions that improve control actions based on the observed results of using a current policy. This requires the derivation of methods for finding optimal values and optimal policies that can be executed forward in time. Here we develop the Bellman equation, which is the basis for such methods.

To derive forward-in-time methods for finding optimal values and optimal policies, set now the time horizon T to infinity and define the infinite-horizon cost

$$J_k = \sum_{i=0}^{\infty} \gamma^i r_{k+i} = \sum_{i=k}^{\infty} \gamma^{i-k} r_i. \qquad (11.2\text{-}15)$$

The associated (infinite-horizon) value function for policy $\pi(x, u)$ is

$$V^{\pi}(x) = E_{\pi}\{J_k | x_k = x\} = E_{\pi}\left\{\sum_{i=k}^{\infty} \gamma^{i-k} r_i | x_k = x\right\}. \quad (11.2\text{-}16)$$

By using (11.2-8) with $T = \infty$ we see that the value function for policy $\pi(x, u)$ satisfies the *Bellman equation*

$$V^{\pi}(x) = \sum_{u} \pi(x, u) \sum_{x'} P_{xx'}^{u} \left[R_{xx'}^{u} + \gamma V^{\pi}(x')\right]. \quad (11.2\text{-}17)$$

This equation is of extreme importance in reinforcement learning. It is important that the same value function appears on both sides. This is due to the fact that the infinite-horizon cost was used. Therefore, (11.2-17) can be interpreted as a consistency equation that must be satisfied by the value function at each time stage. The Bellman equation expresses a relation between the current value of being in state x and the value(s) of being in the next state x' given that policy $\pi(x, u)$ is used.

The Bellman equation forms the basis for a family of reinforcement learning algorithms for finding optimal policies by using causal experiences received stagewise forward in time. In this context, the meaning of the Bellman equation is shown in Figure 11.2-2, where $V^{\pi}(x)$ may be considered as a predicted performance, $\sum_{u} \pi(x, u) \sum_{x'} P_{xx'}^{u} R_{xx'}^{u}$ the observed one-step reward, and $V^{\pi}(x')$

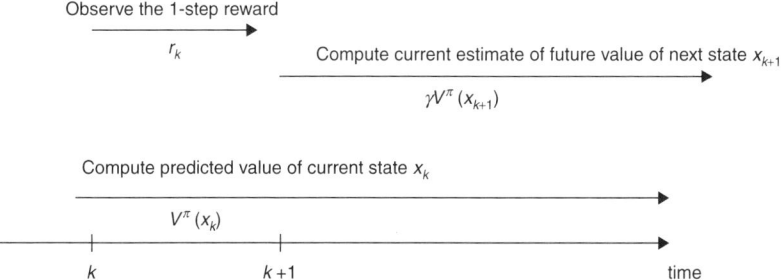

1. Apply control action

Observe the 1-step reward

r_k

Compute current estimate of future value of next state x_{k+1}

$\gamma V^{\pi}(x_{k+1})$

Compute predicted value of current state x_k

$V^{\pi}(x_k)$

k $k+1$ time

2. Update predicted value to satisfy the Bellman equation

$V^{\pi}(x_k) = r_k + \gamma V^{\pi}(x_{k+1})$

3. Improve control action

FIGURE 11.2-2 Temporal difference interpretation of Bellman equation, showing how the Bellman equation captures the action, observation, evaluation, and improvement mechanisms of reinforcement learning.

a current estimate of future behavior. These notions are capitalized on in the subsequent discussion of temporal difference learning, which uses them to develop adaptive control algorithms that can learn optimal behavior online in real-time applications.

If the MDP is finite and has N states, then the Bellman equation (11.2-17) is a system of N simultaneous linear equations for the value $V^\pi(x)$ of being in each state x given the current policy $\pi(x, u)$. The optimal value satisfies

$$V^*(x) = \min_\pi V^\pi(x) = \min_\pi \sum_u \pi(x, u) \sum_{x'} P_{xx'}^u \left[R_{xx'}^u + \gamma V^\pi(x') \right]. \quad (11.2\text{-}18)$$

Bellman's optimality principle then yields the *Bellman optimality equation*

$$V^*(x) = \min_\pi V^\pi(x) = \min_\pi \sum_u \pi(x, u) \sum_{x'} P_{xx'}^u \left[R_{xx'}^u + \gamma V^*(x') \right]. \quad (11.2\text{-}19)$$

Equivalently, under the ergodicity assumption on the Markov chains corresponding to each policy, we have

$$V^*(x) = \min_u \sum_{x'} P_{xx'}^u \left[R_{xx'}^u + \gamma V^*(x') \right]. \quad (11.2\text{-}20)$$

If the MDP is finite and has N states, then the Bellman optimality equation is a system of N nonlinear equations for the optimal value $V^*(x)$ of being in each state. The optimal control is given by

$$u^* = \arg\min_u \sum_{x'} P_{xx'}^u \left[R_{xx'}^u + \gamma V^*(x') \right]. \quad (11.2\text{-}21)$$

Though the ideas just introduced may not seem familiar to the control engineer, it is shown in the next examples that they correspond to some familiar notions in feedback control system theory.

Example 11.2-1. Bellman Equation for Discrete-time Linear Quadratic Regulator (DT LQR)

This example studies the Bellman equation for the discrete-time LQR and shows that it is closely related to ideas developed in Chapter 2.

a. MDP Dynamics for Deterministic DT Systems

Consider the discrete-time (DT) linear quadratic regulator (LQR) problem where the MDP is deterministic and satisfies the state transition equation

$$x_{k+1} = Ax_k + Bu_k, \quad (11.2\text{-}22)$$

with k the discrete time index. The associated infinite horizon performance index has deterministic stage costs and is

$$J_k = \frac{1}{2} \sum_{i=k}^\infty r_i = \frac{1}{2} \sum_{i=k}^\infty \left(x_i^T Q x_i + u_i^T R u_i \right). \quad (11.2\text{-}23)$$

In this example, the state space $X = R^n$ and action space $U = R^m$ are infinite and continuous.

b. Bellman Equation for DT LQR: The Lyapunov Equation

The performance index J_k depends on the current state x_k and all future control inputs $u_k, u_{k+1}, \ldots$. Select a fixed stabilizing policy $u_k = \mu(x_k)$ and write the associated value function as

$$V(x_k) = \frac{1}{2} \sum_{i=k}^{\infty} r_i = \frac{1}{2} \sum_{i=k}^{\infty} \left(x_i^T Q x_i + u_i^T R u_i \right). \tag{11.2-24}$$

The value function for a fixed policy depends only on the initial state x_k. A difference equation equivalent to this infinite sum is given by

$$V(x_k) = \frac{1}{2} \left(x_k^T Q x_k + u_k^T R u_k \right) + \frac{1}{2} \sum_{i=k+1}^{\infty} \left(x_i^T Q x_i + u_i^T R u_i \right)$$

$$= \frac{1}{2} \left(x_k^T Q x_k + u_k^T R u_k \right) + V(x_{k+1}). \tag{11.2-25}$$

That is, the positive definite solution $V(x_k)$ to this equation that satisfies $V(0) = 0$ is the value given by (11.2-24). Equation (11.2-25) is exactly the Bellman equation (11.2-17) for the LQR.

Assuming the value is quadratic in the state so that

$$V_k(x_k) = \tfrac{1}{2} x_k^T P x_k, \tag{11.2-26}$$

for some kernel matrix P yields the Bellman equation form

$$2V(x_k) = x_k^T P x_k = x_k^T Q x_k + u_k^T R u_k + x_{k+1}^T P x_{k+1}, \tag{11.2-27}$$

which, using the state equation, can be written as

$$2V(x_k) = x_k^T Q x_k + u_k^T R u_k + (A x_k + B u_k)^T P (A x_k + B u_k). \tag{11.2-28}$$

Assuming a constant (that is, stationary) state feedback policy $u_k = \mu(x_k) = -K x_k$ for some stabilizing gain K, we write

$$2V(x_k) = x_k^T P x_k = x_k^T Q x_k + x_k^T K^T R K x_k + x_k^T (A - BK)^T P (A - BK) x_k. \tag{11.2-29}$$

Since this holds for all state trajectories, we have

$$(A - BK)^T P (A - BK) - P + Q + K^T R K = 0. \tag{11.2-30}$$

This is a Lyapunov equation. That is, the Bellman equation (11.2-17) for the DT LQR is equivalent to a Lyapunov equation. Since the performance index is undiscounted ($\gamma = 1$), we must select a stabilizing gain K, that is, a stabilizing policy.

The formulations (11.2-25), (11.2-27), (11.2-29), (11.2-30) for the Bellman equation are all equivalent. Note that forms (11.2-25) and (11.2-27) do not involve the system dynamics (A, B). On the other hand, the Lyapunov equation (11.2-30) can be used only if we know the state dynamics (A, B). Optimal controls design using matrix equations is the standard procedure in control systems theory. Unfortunately, by assuming that (11.2-29) holds for all trajectories and going to (11.2-30), we lose all possibility of applying any sort of reinforcement learning algorithms to solve for the optimal control and value online by observing data along the system trajectories. By contrast, it will be shown that by employing the form (11.2-25) or (11.2-27) for the Bellman equation, we can devise RL algorithms for learning optimal solutions online by using temporal difference methods. That is, RL allows us to solve the Lyapunov equation online without knowing A or B.

c. Bellman Optimality Equation for DT LQR: The Algebraic Riccati Equation

The DT LQR Hamiltonian function is

$$H(x_k, u_k) = x_k^\mathrm{T} Q x_k + u_k^\mathrm{T} R u_k + (Ax_k + Bu_k)^\mathrm{T} P(Ax_k + Bu_k) - x_k^\mathrm{T} P x_k. \qquad (11.2\text{-}31)$$

This is known as the temporal difference error in MDP. A necessary condition for optimality is the stationarity condition $\partial H(x_k, u_k)/\partial u_k = 0$, which is equivalent to (11.2-21). Solving this yields the optimal control

$$u_k = -K x_k = -(B^\mathrm{T} PB + R)^{-1} B^\mathrm{T} PA x_k.$$

Putting this into (11.2-31) yields the DT algebraic Riccati equation (ARE)

$$A^\mathrm{T} PA - P + Q - A^\mathrm{T} PB(B^\mathrm{T} PB + R)^{-1} B^\mathrm{T} PA = 0. \qquad (11.2\text{-}32)$$

This is exactly the Bellman optimality equation (11.2-19) for the DT LQR. ∎

The Closed-loop Markov Chain

References for this section include (Wheeler and Narendra 1986, Bertsekas and Tsitsiklis 1996, Luenberger 1979). Consider a finite MDP with N states. Select a fixed stationary policy $\pi(x, u) = \Pr\{u|x\}$ Then the "closed-loop" MDP reduces to a Markov chain with state space X. The transition probabilities of this Markov chain are given by (11.2-2). Enumerate the states using index $i = 1, \ldots, N$ and denote the transition probabilities (11.2-2) by $p_{ij} = p_{x=i,x'=j}$. Define the transition matrix $P = [p_{ij}] \in R^{N \times N}$. Denote the expected costs $r_{ij} = R_{ij}^\pi$ and define the cost matrix $R = [r_{ij}] \in R^{N \times N}$. Array the scalar values of the states $V(i)$ into a vector $V = [V(i)] \in R^N$.

With this notation, we may write the Bellman equation (11.2-17) as

$$V = \gamma PV + (P \odot R)\underline{1}, \qquad (11.2\text{-}33)$$

where $\odot$ is the Hadamard (element-by-element) matrix product, $P \odot R = [p_{ij}r_{ij}] \in R^{N \times N}$ is the cost-transition matrix, and $\underline{1}$ is the N-vector of 1's. This is a system of N linear equations for the value vector V of using the selected policy.

Transition matrix P is stochastic, that is, all row sums are equal to one. If the Markov chain is irreducible, that is, all states communicate with each other with nonzero probability, then P has a single eigenvalue at 1 and all other eigenvalues inside the unit circle (Gershgorin theorem). Then, if the discount factor is less than one, $I - \gamma P$ is nonsingular and there is a unique solution to (11.2-33) for the value, namely $V = (I - \gamma P)^{-1}(P \odot R)\underline{1}$.

Define vector $p \in R^N$ as a probability distribution vector (that is, its elements sum to 1) with element $p(i)$ being the probability that the Markov chain is in state i. Then, the evolution of p is given by

$$p_{k+1}^{\mathrm{T}} = p_k^{\mathrm{T}} P, \tag{11.2-34}$$

with k the time index and starting at some initial distribution p_0. The solution to this equation is

$$p_k^{\mathrm{T}} = p_0^{\mathrm{T}} P^k. \tag{11.2-35}$$

The limiting value of this recursion is the invariant or steady-state distribution, given by solving

$$p_\infty^{\mathrm{T}} = p_\infty^{\mathrm{T}} P, \quad p_\infty^{\mathrm{T}}(I - P) = 0. \tag{11.2-36}$$

Thus, p_∞ is the left eigenvector of the eigenvalue $\lambda = 1$ of P. Since P has all row sums equal one, the right eigenvector of $\lambda = 1$ is given by $\underline{1}$. If the Markov chain is irreducible, the eigenvalue $\lambda = 1$ is simple, and vector p_∞ has all elements positive. That is, all states have a nonzero probability of being occupied in steady state.

Define the average cost per stage and its value under the selected policy as

$$\overline{V}(i) = \lim_{T \to \infty} \frac{1}{T} E \left(\sum_{k=0}^{T-1} r_k | x_0 = i \right). \tag{11.2-37}$$

Define the vector $\overline{V} = [\overline{V}(i)] \in R^N$. Then we have

$$\overline{V} = (P \odot R)\underline{1}. \tag{11.2-38}$$

Define the expected value over all the states as $\overline{\overline{V}} = E_i\{\overline{V}(i)\}$. Then

$$\overline{\overline{V}} = p_\infty^{\mathrm{T}}(P \odot R)\underline{1}. \tag{11.2-39}$$

The left eigenvector $p_\infty = [p(i)]$ for $\lambda = 1$ has received a great deal of attention in the literature about cooperative control (Jadbabaie et al. 2003, Olfati-Saber and Murray 2004). The meaning of its elements $p(i)$ is interesting. Element $p(i)$ is the probability that the Markov chain is in state i at steady state, and also the percentage of time the Markov chain is in state i at steady state. Starting in state i, the time at which the Markov chain first returns to state i is a random variable known as the *hitting time* or first return time. Its expected value T_i is known as the *expected return time* to state i. If the Markov chain is irreducible, $T_i > 0, \forall i$; then $p(i) = 1/T_i$.

11.3 POLICY EVALUATION AND POLICY IMPROVEMENT

Given a current policy $\pi(x, u)$, we can determine its value (11.2-16) by solving the Bellman equation (11.2-17). As will be discussed, there are several methods of doing this, several of which can be implemented online in real time. This procedure is known as *policy evaluation*.

Moreover, given the value function for any policy $\pi(x, u)$, we can always use it to find another policy that is better, or at least no worse. This step is known as *policy improvement*. Specifically, suppose $V^{\pi}(x)$ satisfies (11.2-17). Then define a new policy $\pi'(x, u)$ by

$$\pi'(x, u) = \arg\min_{u} \sum_{x'} P_{xx'}^{u} \left[R_{xx'}^{u} + \gamma V^{\pi}(x') \right]. \qquad (11.3\text{-}1)$$

Then it is easy to show that $V^{\pi'}(x) \leq V^{\pi}(x)$ (Bertsekas and Tsitsiklis 1996, Sutton and Barto 1998). The policy determined as in (11.3-1) is said to be *greedy* with respect to value function $V^{\pi}(x)$.

In the special case that $V^{\pi'}(x) = V^{\pi}(x)$ in (11.3-1), then $V^{\pi'}(x)$, $\pi'(x, u)$ satisfy (11.2-20) and (11.2-21) so that $\pi'(x, u) = \pi(x, u)$ is the optimal policy and $V^{\pi'}(x) = V^{*}(x)$ the optimal value. That is, (only) an optimal policy is greedy with respect to its own value. In computational intelligence, *greedy* refers to quantities determined by optimizing over short or 1-step horizons, without regard to potential impacts far into the future.

Now let us consider algorithms that repeatedly interleave the two procedures.

Policy evaluation by Bellman equation:

$$V^{\pi}(x) = \sum_{u} \pi(x, u) \sum_{x'} P_{xx'}^{u} \left[R_{xx'}^{u} + \gamma V^{\pi}(x') \right], \quad \text{for all } x \in S \subseteq X. \quad (11.3\text{-}2)$$

Policy improvement:

$$\pi'(x, u) = \arg\min_{u} \sum_{x'} P_{xx'}^{u} \left[R_{xx'}^{u} + \gamma V^{\pi}(x') \right], \quad \text{for all } x \in S \subseteq X. \quad (11.3\text{-}3)$$

S is a suitably selected subspace of the state space, to be discussed later. We call an application of (11.3-2) followed by an application of (11.3-3) one *step*. This is in contrast to the decision time stage k defined above.

At each step of such algorithms, one obtains a policy that is no worse than the previous policy. Therefore, it is not difficult to prove convergence under fairly mild conditions to the optimal value and optimal policy. Most such proofs are based on the Banach fixed-point theorem. Note that (11.2-20) is a fixed-point equation for $V^{*}(\cdot)$. Then the two equations (11.3-2), (11.3-3) define an associated map that can be shown under mild conditions to be a contraction map (Bertsekas and Tsitsiklis 1996 and Powell 2007). Then, it converges to the solution of (11.2-20) for any initial policy.

There is a large family of algorithms that implement the policy evaluation and policy improvement procedures in various ways, or interleave them differently, or select subspace $S \subseteq X$ in different ways, to determine the optimal value and optimal policy. We shall soon outline some of them.

The importance for feedback control systems of this discussion is that these two procedures can be implemented for dynamical systems online in real time by observing data measured along the system trajectories. This yields a family of adaptive control algorithms that converge to optimal control solutions. These algorithms are of the *actor–critic* class of reinforcement learning systems, shown in Figure 11.1-1. There, a critic agent evaluates the current control policy using methods based on (11.3-2). After this has been completed, the action is updated by an actor agent based on (11.3-3).

Policy Iteration

One method of reinforcement learning for using (11-3.2), (11-3.3) to find the optimal value and optimal policy is *policy iteration (PI)*.

POLICY ITERATION (PI) ALGORITHM

Initialize.

Select an initial policy $\pi_0(x, u)$. Do for $j = 0$ until convergence:

Policy evaluation (value update):

$$V_j(x) = \sum_u \pi_j(x, u) \sum_{x'} P_{xx'}^u \left[R_{xx'}^u + \gamma V_j(x') \right], \quad \text{for all } x \in X. \quad (11.3\text{-}4)$$

Policy improvement (policy update):

$$\pi_{j+1}(x, u) = \arg\min_u \sum_{x'} P_{xx'}^u \left[R_{xx'}^u + \gamma V_j(x') \right], \quad \text{for all } x \in X. \quad (11.3\text{-}5)$$

At each step j the PI algorithm determines the solution of the Bellman equation (11.3-4) to compute the value $V_j(x)$ of using the current policy $\pi_j(x, u)$. This value corresponds to the infinite sum (11.2-16) for the current policy. Then the policy is improved using (11.3-5). The steps are continued until there is no change in the value or the policy.

Note that j is not the time or stage index k, but a PI step iteration index. It will be seen how to implement PI for dynamical systems online in real time by observing data measured along the system trajectories. Generally, data for multiple times k is needed to solve the Bellman equation (11.3-4) at each step j.

If the MDP is finite and has N states, then the policy evaluation equation (11.3-4) is a system of N simultaneous linear equations, one for each state. The PI algorithm must be suitably initialized to converge. The initial policy $\pi_0(x, u)$ and value V_0 must be selected so that $V_1 \leq V_0$. Then, for finite Markov chains with N states, PI converges in a finite number of steps (less than or equal to N) because there are only a finite number of policies.

The Bellman equation (11.3-4) is a system of simultaneous equations. Instead of directly solving the Bellman equation, we can solve it by an iterative policy evaluation procedure. Note that (11.3-4) is a fixed-point equation for $V_j(\cdot)$. It defines the iterative policy evaluation map

$$V_j^{i+1}(x) = \sum_u \pi_j(x, u) \sum_{x'} P_{xx'}^u \left[R_{xx'}^u + \gamma V_j^i(x') \right], \quad i = 1, 2, \dots \quad (11.3\text{-}6)$$

which can be shown to be a contraction map under rather mild conditions. By the Banach fixed-point theorem the iteration can be initialized at any non-negative value of $V_j^1(\cdot)$ and it will converge to the solution of (11.3-4). Under certain conditions, this solution is unique. A good initial value choice is the value function $V_{j-1}(\cdot)$ from the previous step $j - 1$. On (close enough) convergence, we set $V_j(\cdot) = V_j^i(\cdot)$ and proceed to apply (11.3-5).

Index j in (11.3-6) refers to the step number of the PI algorithm. By contrast i is an iteration index. It is interesting to compare iterative policy evaluation (11.3-6) to the backward-in-time recursion (11.2-9) for the finite-horizon value. In (11.2-9), k is the time index. By contrast, in (11.3-6), i is an iteration index. Dynamic programming is based on (11.2-9) and proceeds backward in time. The methods for online optimal adaptive control described in this chapter proceed forward in time and are based on PI and similar algorithms.

The usefulness of these concepts is shown in the next example, where we use the theory of MDP and iterative policy evaluation to derive the relaxation algorithm for solution of Poisson's equation.

Example 11.3-1.   Solution of Partial Differential Equations: Relaxation Algorithms

Consider Poisson's equation in two dimensions

$$\Delta V(x, y) = \nabla^2 V(x, y) = \left(\frac{\partial^2}{\partial x^2} + \frac{\partial^2}{\partial y^2} \right) V(x, y) = f(x, y), \quad (11.3\text{-}7)$$

where $\Delta = \nabla^2$ is the Laplacian operator and ∇ the gradient. Function $f(x, y)$ is a forcing function, often specified on the boundary of a region. Discretizing the equation on a uniform mesh as shown in Figure 11.3-1 with grid size h in the (x, y) plane we write in terms of the forward difference

$$\frac{\partial V(x, y)}{\partial x} = \frac{1}{h} (V(x + h, y) - V(x, y)) + O(h),$$

$$\frac{\partial^2 V(x, y)}{\partial x^2} = \frac{1}{h^2} (V(x + h, y) + V(x - h, y) - 2V(x, y)) + O(h^2),$$

$$\left(\frac{\partial^2}{\partial x^2} + \frac{\partial^2}{\partial x^2} \right) V(x, y) = \frac{1}{h^2} (V(x + h, y) + V(x - h, y) + V(x, y + h)$$

$$+ V(x, y - h) - 4V(x, y)) \approx f(x, y).$$

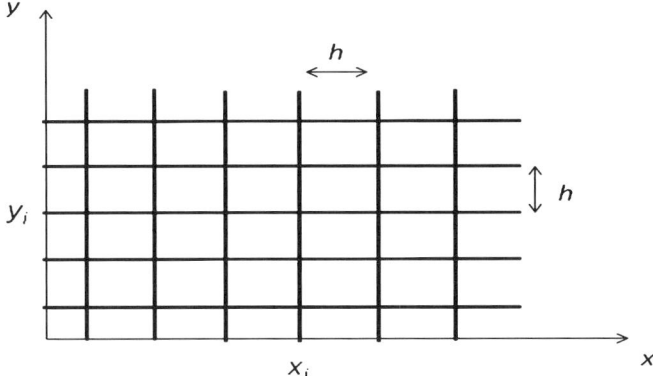

FIGURE 11.3-1 Sampling in the (x, y) plane with a uniform mesh of size h.

Indexing the (x, y) positions with an index i and denoting $X_i = (x_i, y_i)$ we have approximately to order h^2

$$\frac{1}{h^2}\left(V(X_{i,R}) + V(X_{i,L}) + V(X_{i,u}) + V(X_{i,D}) - 4V(X_i)\right) = f(X_i), \qquad (11.3\text{-}8)$$

for states not on the boundary and a similar equation for boundary states. $X_{i,R}$ denotes the (x, y) location of the state to the right of X_i, and similarly for states to the left, up, and down relative to X_i. This is a set of N simultaneous equations.

We can interpret (11.3-8) as a Bellman equation (11.2-17) for a properly defined underlying MDP. Specifically, define an MDP with stage costs equal to zero and $\gamma = 1$. Then, we may interpret the state transitions as deterministic and the control as the equiprobable control with probabilities of $1/4$ (for nonboundary nodes) for moving up, right, down, and left. Alternatively, we may interpret the control as deterministic and the state transition probabilities as equi-probable with probabilities of $1/4$ for moving up, right, down, and left. In either case, the Bellman equation for the MDP is (11.3-8). Functions $f(X_i)$ may be interpreted as stage costs.

Now the iterative policy evaluation method of solution (11.3-6) performs the iterations

$$V^{m+1}(X_i) = \tfrac{1}{4}V^m(X_{i,U}) + \tfrac{1}{4}V^m(X_{i,R}) + \tfrac{1}{4}V^m(X_{i,D}) + \tfrac{1}{4}V^m(X_{i,L}) - \tfrac{h^2}{4}f(X_i).$$
$$(11.3\text{-}9)$$

The theory of MDP guarantees that this algorithm will converge to the solution of (11.3-8). These updates may be done for all nodes or states simultaneously. Then, this is nothing but the relaxation method for numerical solution of Poisson's equation. The relaxation algorithm converges, but may do so slowly. Variants have been developed to speed it up. ∎

Value Iteration

A second method for using (11.3-2), (11.3-3) in reinforcement learning is *value iteration (VI)*, which is easier to implement than policy iteration.

VALUE ITERATION (VI) ALGORITHM

Initialize.

Select an initial policy $\pi_0(x, u)$. Do for $j = 0$ until convergence—

Value update:

$$V_{j+1}(x) = \sum_u \pi_j(x, u) \sum_{x'} P_{xx'}^u \left[R_{xx'}^u + \gamma V_j(x') \right], \quad \text{for all } x \in S_j \subseteq X.$$

$$(11.3\text{-}10)$$

Policy improvement:

$$\pi_{j+1}(x, u) = \arg \min_u \sum_{x'} P_{xx'}^u \left[R_{xx'}^u + \gamma V_{j+1}(x') \right], \quad \text{for all } x \in S_j \subseteq X.$$

$$(11.3\text{-}11)$$

We may combine the value update and policy improvement into one equation to obtain the equivalent form for VI

$$V_{j+1}(x) = \min_\pi \sum_u \pi(x, u) \sum_{x'} P_{xx'}^u \left[R_{xx'}^u + \gamma V_j(x') \right], \quad \text{for all } x \in S_j \subseteq X.$$

$$(11.3\text{-}12)$$

or, equivalently under the ergodicity assumption, in terms of deterministic policies

$$V_{j+1}(x) = \min_u \sum_{x'} P_{xx'}^u \left[R_{xx'}^u + \gamma V_j(x') \right], \quad \text{for all } x \in S_j \subseteq X. \quad (11.3\text{-}13)$$

Note that, now, (11.3-10) is a simple one-step recursion, not a system of linear equations as is (11.3-4) in the PI algorithm. In fact, VI uses simply one iteration of (11.3-6) in its value update step. It does not find the value corresponding to the current policy, but takes only one iteration toward that value. Again, j is not the time index, but the VI step index.

It will be seen later how to implement VI for dynamical systems online in real time by observing data measured along the system trajectories. Generally, data for multiple times k is needed to solve the update (11.3-10) for each step j.

Asynchronous Value Iteration.   Standard VI takes the update set as $S_j = X$, for all j. That is, the value and policy are updated for all states simultaneously. *Asynchronous* VI methods perform the updates on only a subset of the states at each step. In the extreme case, one may perform the updates on only one state at each step.

It is shown in Bertsekas and Tsitsiklis (1996) that standard VI ($S_j = X$, for all j) converges for finite MDP for any initial conditions when the discount factor satisfies $0 < \gamma < 1$. When $S_j = X$, for all j and $\gamma = 1$ an absorbing state is added and a "properness" assumption is needed to guarantee convergence to the optimal value. When a single state is selected for value and policy updates at each step, the algorithm converges, for any choice of initial value, to the optimal cost and policy if each state is selected for update infinitely often. More general algorithms result if value update (11.3-10) is performed multiple times for various choices of S_j prior to a policy improvement. Then updates (11.3-10) and (11.3-11) must be performed infinitely often for each state, and a monotonicity assumption must be satisfied by the initial starting value.

Considering (11.2-20) as a fixed-point equation, VI is based on the associated iterative map (11.3-10), (11.3-11), which can be shown under certain conditions to be a contraction map. In contrast to PI, which converges under certain conditions in a finite number of steps, VI generally takes an infinite number of steps to converge (Bertsekas and Tsitsiklis 1996). Consider finite MDP and consider the transition probability graph having probabilities (11.2-2) for the Markov chain corresponding to an optimal policy $\pi^*(x, u)$. If this graph is acyclic for some $\pi^*(x, u)$, then VI converges in at most N steps when initialized with a large value.

Having in mind the dynamic programming equation (11.2-9) and examining the VI value update (11.3-10), we can interpret $V_j(x')$ as an approximation or estimate for the future stage cost-to-go from the future state x'. See Figure 11.2-2. Those algorithms wherein the future cost estimate are themselves costs or values for some policy are called *rollout algorithms* in Bertsekas and Tsitsiklis (1996). These policies are forward looking and self-correcting. They are closely related to receding horizon control, as shown in Zhang et al. (2009).

MDP, policy iteration, and value iteration provide connections between optimal control, decisions on finite graphs, and cooperative control of networked systems, as amplified in the next examples. The first example shows how to use VI to derive the Bellman Ford algorithm for finding the shortest path in a graph to a destination node. The second example uses MDP and iterative policy evaluation to develop control protocols familiar in cooperative control of distributed dynamical systems. The third example shows that for the discrete-time LQR, policy iteration and value iteration can be used to derive algorithms for solution of the optimal control problem that are quite familiar in the feedback control systems community, including Hewer's algorithm.

Example 11.3-2. **Deterministic Shortest Path Problems: The Bellman Ford Algorithm**

A special case of finite MDP is the *shortest path problems* (Bertsekas and Tsitsiklis 1996), which are undiscounted $\gamma = 1$, and have a cost-free termination or absorbing state. These effectively have the time horizon T finite since the number of states is finite. The setup is shown in Figure 11.3-2.

Consider a directed graph $G = (V, E)$ with a nonempty finite set of N nodes $V = \{v_1, \ldots, v_N\}$ and a set of edges or arcs $E \subseteq V \times V$. We assume the graph is simple, that is, it has no repeated edges and $(v_i, v_i) \notin E, \forall i$ no self-loops. Let the edges have weights $e_{ij} \geq 0$, with $e_{ij} > 0$ if $(v_i, v_j) \in E$ and $e_{ij} = 0$ otherwise. Note $e_{ii} = 0$. The set of neighbors of a node v_i is $N_i = \{v_j : (v_i, v_j) \in E\}$, that is, the set of nodes with arcs coming out from v_i.

Consider an agent moving through the graph along edges between nodes. Interpret the control at each node as the decision on which edge to follow leading out of that

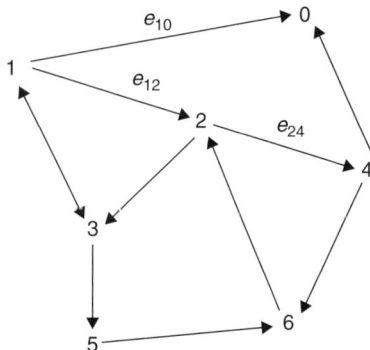

FIGURE 11.3-2 Sample graph for shortest path routing problem. The absorbing state 0 has only incoming edges. The objective is to find the shortest path from all nodes to node 0 using only local neighborhood computations.

node. Assume deterministic controls and state transitions. Interpret the edge weights e_{ij} as deterministic costs incurred by moving along that link. Then, the value iteration algorithm (11.3-13), (11.3-11) in this general digraph is

$$V_i(k+1) = \min_{j \in N_i}(e_{ij} + V_j(k)), \tag{11.3-14}$$

$$u_i(k+1) = \arg\min_{j \in N_i}(e_{ij} + V_j(k)), \tag{11.3-15}$$

with k an iteration index. We have changed the notation to conform to existing practice in cooperative control theory. Endow the graph with one node, v_0, which uses as its update law $V_0(k+1) = V_0(k) = 0$. This corresponds to an absorbing state in MDP parlance. It is interpreted as a node having only incoming edges, none outgoing.

The VI algorithm in this scenario is nothing but the Bellman-Ford algorithm, which finds the shortest path from any node in the graph to the absorbing node v_0. In the terminology of cooperative systems, each node v_i is endowed with two state variables. The node state variable $V_i(k)$ keeps track of the value, that is, the shortest path length to node v_0, while node state variable $u_i(k)$ keeps track of which direction to follow while leaving the ith node in order to follow the shortest path.

The VI update iterations may be performed on all nodes simultaneously. Alternatively, updates may be performed on one state at a time, as in asynchronous VI. With simultaneous updates at all nodes, according to results about VI (Bertsekas and Tsitsiklis 1996) it is known that this algorithm converges to the solution to the shortest path problem in a finite number of steps (less than or equal to N) if there is a path from every node to node v_0 and the graph is acyclic. It is known further that the algorithm converges, possibly in an infinite number of iterations, if there are no cycles with net negative gain (that is, the product of gains around the cycle is not negative). With asynchronous updates, the algorithm converges under these connectivity conditions if each node is selected for update infinitely often.

What Is a State?

Note that in the terminology of MDP, the nodes are termed states, so that the state space is finite. On the other hand, in the terminology of feedback control systems theory, the state (variable) of each node is $V_i(m)$, which is a real number so that the state space is continuous.   ∎

Example 11.3-3.   Cooperative Control Systems

Consider the directed graph $G = (V, E)$ with N nodes $V = \{v_1, \ldots, v_N\}$ and edge weights $e_{ij} \geq 0$, with $e_{ij} > 0$ if $(v_i, v_j) \in E$ and $e_{ij} = 0$ otherwise. The set of neighbors of a node v_i is $N_i = \{v_j \colon (v_i, v_j) \in E\}$, that is, the set of nodes with arcs coming out from v_i. Define the out-degree of node i in graph G as $d_i = \sum_{j \in N_i} e_{ij}$. Figure 11.3-3 shows a representative graph topology.

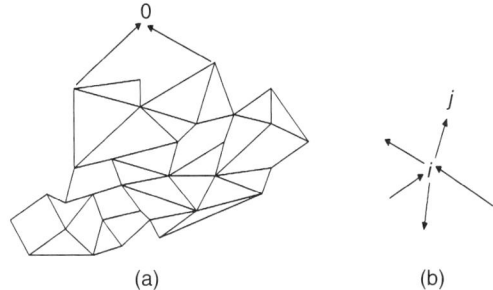

(a)   (b)

FIGURE 11.3-3   Cooperative control of multiple systems linked by a communication graph structure. The objective is for all nodes to reach consensus to the value of control node 0 by using only local neighborhood communication. (a) Representative graph structure. (b) Local neighborhood of node i. Each node has edges incoming and outgoing. The outgoing edges show the states reached from node i in one step.

On the graph G, define a MDP that has a stochastic policy $\pi(v_i, u_{ij}) = \Pr\{u_{ij} | v_i\}$, where u_{ij} means the control action that takes the MDP from node v_i to node v_j. Let the transition probabilities be deterministic. Endow the graph with one absorbing node v_0 that is connected to a few of the existing nodes. If there is an edge from node v_i to node v_0, define $b_i \equiv e_{i0}$ as its edge weight. Let actions u_{ij} have probabilities $u_{ii} = 1/(1 + d_i + b_i)$, $u_{ij} = e_{ij}/(1 + d_i + b_i)$, that is, the MDP may return to the same state in one step. Let the stage costs all be zero. Then, the Bellman equation (11.2-17) is

$$V_i = \frac{1}{1 + d_i + b_i} \left[V_i + \sum_{j \in N_i} e_{ij} V_j + b_i V_0 \right], \qquad (11.3\text{-}16)$$

where V_i is the value of node v_i. This is a set of N simultaneous equations in the values $V_i, i = 1, N$ of the nodes.

Iterative policy evaluation (11.3-6) can be used to solve this set of equations. The iterative policy evaluation algorithm is written here as

$$V_i(k+1) = \frac{1}{1 + d_i + b_i} \left[V_i(k) + \sum_{j \in N_i} e_{ij} V_j(k) + b_i V_0 \right], \qquad (11.3\text{-}17)$$

with k the iteration index, which can be thought of here as a time index. Node v_0 keeps its value constant at V_0.

The updates in (11.3-17) may be done simultaneously at all nodes. Then the theory of MDP shows that it is guaranteed to converge to the solution of the set of equations (11.3-16). Alternatively, one may use the theory of asynchronous VI or generalized PI to motivate other update schemes. For instance, if only one node is updated at each value of k, the theory of MDP shows that the algorithm still converges as long as each node is selected for update infinitely often.

Algorithm (11.3-17) can be written as the local control protocol

$$V_i(k+1) = V_i(k) + \frac{1}{1 + d_i + b_i} \left[\sum_{j \in N_i} e_{ij} \left(V_j(k) - V_i(k) \right) + b_i \left(V_0 - V_i(k) \right) \right].$$

$$(11.3\text{-}18)$$

On convergence $V_i(k+1) = V_i(k)$ so that the term in square brackets converges to zero. Assuming all nodes have a path to the absorbing node v_0, it is easy to show that this guarantees that $\| V_j(k) - V_i(k) \| \to 0$, $\| V_0 - V_i(k) \| \to 0$, for all i, j; that is, all nodes reach the same consensus value, namely the value of the absorbing node (Jadbabaie et al. 2003, Olfati-Saber and Murray 2004).

The term in square brackets in (11.3-18) is known as the *temporal difference error* for this MDP.

Routing Graph vs. Control Graph

The graphs used in routing problems have edges from node i to node j if the transition probabilities (11.2-2) in the MDP using a fixed policy $\pi(x_i, u)$, namely $p_{ij} \equiv P^{\pi}_{x_i, x_j} = \sum_u \pi(x_i, u) P^u_{x_i, x_j}$, are nonzero. The edge weights are taken as $e_{ij} = p_{ij}$. Then, e_{ij} is nonzero if there is an edge coming *out* of node i to node j. By contrast, the edge weights a_{ij} for graphs in cooperative control problems are generally taken as nonzero if there is an edge coming *in* from node j to node i, that is, $a_{ij} > 0$ iff $(v_j, v_i) \in E$ This is interpreted to mean that information from node j is available to node i for its decision process in computing its control input.

It is convenient to think of the former as motion or routing graphs, and the latter as information flow or decision and control graphs. In fact, the routing graph is the reverse of the decision graph. The reverse graph of a given graph $G = (V, E)$ is the graph with the same node set, but all edges reversed. Note that the Bellman-Ford protocols (11.3-14), (11.3-15) assume that node i gets information from its neighbor node j, but that the shortest path problem is, in fact, solved for motion in the reverse graph having edges e_{ij}.

Define the matrix of transition probabilities $P = [p_{ij}]$ and the adjacency matrix $A = [a_{ij}]$. Then $A = P^T$.   ∎

Example 11.3-4.   Policy Iteration and Value Iteration for the DT LQR

The Bellman equation (11.2-17) is equivalent for the DT LQR to all the formulations (11.2-25), (11.2-27), (11.2-29), (11.2-30) in Example 11.2-1. We may use any of these to implement policy iteration and value iteration.

a. Policy Iteration: Hewer's Algorithm

With step index j, and using superscripts to denote algorithm steps and subscripts to denote the time k, the PI policy evaluation step (11.3-4) applied on (11.2-25) in

Example 11.2-1 yields

$$V^{j+1}(x_k) = \tfrac{1}{2}\left(x_k^{\mathrm{T}} Q x_k + u_k^{\mathrm{T}} R u_k\right) + V^{j+1}(x_{k+1}). \tag{11.3-19}$$

PI applied on (11.2-27) yields

$$x_k^{\mathrm{T}} P^{j+1} x_k = x_k^{\mathrm{T}} Q x_k + u_k^{\mathrm{T}} R u_k + x_{k+1}^{\mathrm{T}} P^{j+1} x_{k+1}, \tag{11.3-20}$$

and PI on (11.2-30) yields the Lyapunov equation

$$0 = (A - BK^j)^{\mathrm{T}} P^{j+1}(A - BK^j) - P^{j+1} + Q + (K^j)^{\mathrm{T}} R K^j. \tag{11.3-21}$$

In all cases the PI policy improvement step is

$$\mu^{j+1}(x_k) = K^{j+1} x_k = \arg\min(x_k^{\mathrm{T}} Q x_k + u_k^{\mathrm{T}} R u_k + x_{k+1}^{\mathrm{T}} P^{j+1} x_{k+1}), \tag{11.3-22}$$

which can be written explicitly as

$$K^{j+1} = -(B^{\mathrm{T}} P^{j+1} B + R)^{-1} B^{\mathrm{T}} P^{j+1} A. \tag{11.3-23}$$

PI algorithm format (11.3-21), (11.3-23) relies on repeated solutions of Lyapunov equations at each step, and is nothing but Hewer's algorithm, well known in control systems theory. It was proven in Hewer (1971) to converge to the solution of the Riccati equation (11.2-32) in Example 11.2-1 if (A, B) is reachable and $(A, \sqrt{Q})$ is observable. It is an offline algorithm that requires complete knowledge of the system dynamics (A, B) to find the optimal value and control. It requires that the initial gain K^0 be stabilizing.

b. Value Iteration: Lyapunov Recursions

Applying VI (11.3-10) to Bellman equation format (11.2-27) in Example 11.2-1 yields

$$x_k^{\mathrm{T}} P^{j+1} x_k = x_k^{\mathrm{T}} Q x_k + u_k^{\mathrm{T}} R u_k + x_{k+1}^{\mathrm{T}} P^j x_{k+1}, \tag{11.3-24}$$

and on format (11.2-30) in Example 11.2-1 yields the Lyapunov recursion

$$P^{j+1} = (A - BK^j)^{\mathrm{T}} P^j (A - BK^j) + Q + (K^j)^{\mathrm{T}} R K^j. \tag{11.3-25}$$

In both cases the policy improvement step is still given by (11.3-22), (11.3-23).

VI algorithm format (11.3-25), (11.3-23) is simply a Lyapunov recursion, which is easy to implement and does not, in contrast to PI, require Lyapunov equation solutions. This algorithm was shown in Lancaster and Rodman (1995) to converge to the solution of the Riccati equation (11.2-32) in Example 11.2-1. It is an offline algorithm that requires complete knowledge of the system dynamics (A, B) to find the optimal value and control. It does not require that the initial gain K^0 be stabilizing, and can be initialized with any feedback gain.

c. Online Solution of the Riccati Equation without Knowing Plant Matrix A

Hewer's algorithm and the Lyapunov recursion algorithm are both offline methods for solving the algebraic Riccati equation (11.2-32) in Example 11.2-1. Full knowledge of the plant dynamics (A, B) is needed to implement these algorithms. By contrast, it will be seen that both PI Algorithm format (11.3-20), (11.3-22) and VI Algorithm format (11.3-24), (11.3-22) can be implemented online to determine the optimal value and control *in real time* using data measured along the system trajectories, and without knowing the system A matrix. This is accomplished through the temporal difference methods to be presented. That is, RL allows the solution of the algebraic Riccati equation online without knowing the system A matrix.

d. Iterative Policy Evaluation

Given a fixed policy K, the iterative policy evaluation procedure (11.3-6) becomes

$$P^{j+1} = (A - BK)^{\mathrm{T}} P^j (A - BK) + Q + K^{\mathrm{T}} RK. \qquad (11.3\text{-}26)$$

This recursion converges to the solution to the Lyapunov equation $P = (A - BK)^{\mathrm{T}} P(A - BK) + Q + K^{\mathrm{T}} RK$ if $(A - BK)$ is stable, for any choice of initial value P^0. ∎

Generalized Policy Iteration

In PI one fully solves the system of linear equations (11.3-4) at each step to compute the value (11.2-16) of using the current policy $\pi_j(x, u)$. This can be accomplished by running iterations (11.3-6) until convergence at each step j. By contrast, in VI one takes only one iteration of (11.3-6) in the value update step (11.3-10). Generalized policy iteration (GPI) algorithms make several iterations (11.3-6) in their value update step.

Generally, PI converges to the optimal value in fewer steps j, since it does more work in solving equations at each step. On the other hand, VI is the easiest to implement as it takes only one iteration of a recursion as per (11.3-10). GPI provides a suitable compromise between computational complexity and convergence speed. GPI is a special case of the VI algorithm given above, where we select $S_j = X$, for all j and perform value update (11.3-10) multiple times before each policy update (11.3-11).

Q Function

The conditional expected value in (11.2-13)

$$Q_k^*(x, u) = \sum_{x'} P_{xx'}^u \left[R_{xx'}^u + \gamma V_{k+1}^*(x') \right] = E_\pi \{ r_k + \gamma V_{k+1}^*(x') | x_k = x, u_k = u \}.$$

$$(11.3\text{-}27)$$

is known as the *optimal Q (quality) function* (Watkins 1989, Watkins and Dayan 1992). This has also been called the action-value function (Sutton and Barto 1998). It is equal to the expected return for taking an arbitrary action u at time k in state x and thereafter following an optimal policy. It is a function of the current state x and the action u.

In terms of the Q function, the Bellman optimality equation has the particularly simple form

$$V_k^*(x) = \min_u Q_k^*(x, u),$$ (11.3-28)

$$u_k^* = \arg\min_u Q_k^*(x, u).$$ (11.3-29)

Given any fixed policy $\pi(x, u)$, define the Q function for that policy as

$$Q_k^\pi(x, u) = E_\pi\{r_k + \gamma V_{k+1}^\pi(x') | x_k = x, u_k = u\} = \sum_{x'} P_{xx'}^u \left[R_{xx'}^u + \gamma V_{k+1}^\pi(x') \right],$$ (11.3-30)

where we have used (11.2-9). This is equal to the expected return for taking an arbitrary action u at time k in state x and thereafter following the existing policy $\pi(x, u)$. The meaning of the Q function is elucidated by the next example.

Example 11.3-5. *Q Function for the DT LQR*

The Q function following a given policy $u_k = \mu(x_k)$ is defined in (11.3-30). For the DT LQR in Example 11.2-1 the Q function is

$$Q(x_k, u_k) = \tfrac{1}{2} \left(x_k^T Q x_k + u_k^T R u_k \right) + V(x_{k+1}),$$ (11.3-31)

where the control u_k is arbitrary and the policy $u_k = \mu(x_k)$ is followed for $k + 1$ and subsequent times. Writing

$$Q(x_k, u_k) = x_k^T Q x_k + u_k^T R u_k + (A x_k + B u_k)^T P (A x_k + B u_k),$$ (11.3-32)

with P the Riccati solution yields the Q function for the DT LQR:

$$Q(x_k, u_k) = \frac{1}{2} \begin{bmatrix} x_k \\ u_k \end{bmatrix}^T \begin{bmatrix} A^T P A + Q & B^T P A \\ A^T P B & B^T P B + R \end{bmatrix} \begin{bmatrix} x_k \\ u_k \end{bmatrix}.$$ (11.3-33)

Define

$$Q(x_k, u_k) \equiv \frac{1}{2} \begin{bmatrix} x_k \\ u_k \end{bmatrix}^T S \begin{bmatrix} x_k \\ u_k \end{bmatrix} = \frac{1}{2} \begin{bmatrix} x_k \\ u_k \end{bmatrix}^T \begin{bmatrix} S_{xx} & S_{xu} \\ S_{ux} & S_{uu} \end{bmatrix} \begin{bmatrix} x_k \\ u_k \end{bmatrix},$$ (11.3-34)

for kernel matrix S.

Applying $\partial Q(x_k, u_k)/\partial u_k = 0$ to (11.3-34) yields

$$u_k = -S_{uu}^{-1} S_{ux} x_k,$$ (11.3-35)

and to (11.3-33) yields

$$u_k = -(B^T P B + R)^{-1} B^T P A x_k.$$ (11.3-36)

The latter equation requires knowledge of the system dynamics (A, B) to perform the policy improvement step of either PI or VI. On the other hand, the former equation requires knowledge only of the Q-function matrix kernel S. We shall subsequently show how to use RL temporal difference methods to determine the kernel matrix S online in real time without knowing the system dynamics (A, B) using data measured along the system trajectories. This provides a family of Q learning algorithms that can solve the algebraic Riccati equation online without knowing the system dynamics (A, B).   ■

Note that $V_k^\pi(x) = Q_k^\pi(x, \pi(x, u))$ so that (11.3-30) may be written as the backward recursion in the Q function:

$$Q_k^\pi(x, u) = \sum_{x'} P_{xx'}^u \left[R_{xx'}^u + \gamma Q_{k+1}^\pi(x', \pi(x', u')) \right]. \tag{11.3-37}$$

The Q function is a 2-dimensional function of both the current state x and the action u. By contrast, the value function is a 1-dimensional function of the state. For finite MDP, the Q function can be stored as a 2-D lookup table at each state/action pair. Note that direct minimization in (11.2-11), (11.2-12) requires knowledge of the state transition probabilities $P_{xx'}^u$ (system dynamics) and costs $R_{xx'}^u$. By contrast, the minimization in (11.3-28), (11.3-29) requires knowledge only of the Q function and not the system dynamics.

The importance of the Q function is twofold. First, it contains information about control actions in every state. As such, the best control in each state can be selected using (11.3-29) by knowing only the Q function. Second, the Q function can be estimated online in real time directly from data observed along the system trajectories, without knowing the system dynamics information (that is, the transition probabilities). We shall see how this is accomplished later.

The infinite horizon Q function for a prescribed fixed policy is given by

$$Q^\pi(x, u) = \sum_{x'} P_{xx'}^u \left[R_{xx'}^u + \gamma V^\pi(x') \right]. \tag{11.3-38}$$

The Q function also satisfies a Bellman equation. Note that, given a fixed policy $\pi(x, u)$,

$$V^\pi(x) = Q^\pi(x, \pi(x, u)), \tag{11.3-39}$$

whence according to (11.3-38) the Q function satisfies the Bellman equation

$$Q^\pi(x, u) = \sum_{x'} P_{xx'}^u \left[R_{xx'}^u + \gamma Q^\pi(x', \pi(x', u')) \right]. \tag{11.3-40}$$

It is important that the same Q function Q^π appears on both sides of this equation. The Bellman optimality equation for the Q function is

$$Q^*(x, u) = \sum_{x'} P_{xx'}^u \left[R_{xx'}^u + \gamma Q^*(x', \pi^*(x', u')) \right], \tag{11.3-41}$$

$$Q^*(x, u) = \sum_{x'} P^u_{xx'} \left[R^u_{xx'} + \gamma \min_{u'} Q^*(x', u') \right]. \tag{11.3-42}$$

Compare (11.2-20) and (11.3-42), where the minimum operator and the expected value operator occurrences are reversed.

Based on Bellman equation (11.3-40), PI and VI are especially easy to implement in terms of the Q function, as follows.

PI USING Q FUNCTION

Policy evaluation (value update):

$$Q_j(x, u) = \sum_{x'} P^u_{xx'} \left[R^u_{xx'} + \gamma Q_j(x', \pi_j(x', u')) \right], \quad \text{for all } x \in X. \tag{11.3-43}$$

Policy improvement:

$$\pi_{j+1}(x, u) = \arg \min_u Q_j(x, u), \quad \text{for all } x \in X. \tag{11.3-44}$$

VI USING Q FUNCTION

Value update:

$$Q_{j+1}(x, u) = \sum_{x'} P^u_{xx'} \left[R^u_{xx'} + \gamma Q_j(x', \pi_j(x', u')) \right], \quad \text{for all } x \in S_j \subseteq X. \tag{11.3-45}$$

Policy improvement:

$$\pi_{j+1}(x, u) = \arg \min_u Q_{j+1}(x, u), \quad \text{for all } x \in S_j \subseteq X. \tag{11.3-46}$$

Combining both steps of VI yields the form

$$Q_{j+1}(x, u) = \sum_{x'} P^u_{xx'} \left[R^u_{xx'} + \gamma \min_{u'} Q_j(x', u') \right], \quad \text{for all } x \in S_j \subseteq X, \tag{11.3-47}$$

which should be compared to (11.3-13)

As shall be seen, the importance of the Q function is that these algorithms can be implemented online in real time, without knowing the system dynamics, by measuring data along the system trajectories. They yield optimal adaptive control algorithms—that is, adaptive control algorithms that converge online to optimal control solutions.

Methods for Implementing PI and VI

There are different methods for performing the value and policy updates for PI and VI (Bertsekas and Tsitsiklis 1996, Sutton and Barto 1998, Powell 2007). The main three are exact computation, Monte Carlo methods, and temporal difference (TD) learning. The last two methods can be implemented without knowledge of the system dynamics. TD learning is the means by which optimal adaptive control algorithms may be derived for dynamical systems. Therefore, TD is covered in the next section.

Exact computation. PI requires solution at each step of Bellman equation (11.3-4) for the value update. For a finite MDP with N states, this is a set of linear equations in N unknowns (the values of each state). VI requires performing the one-step recursive update (11.3-10) at each step for the value update. Both of these can be accomplished exactly if we know the transition probabilities $P_{x,x'}^u = \Pr\{x'|x, u\}$ and costs $R_{xx'}^u$ of the MDP. This corresponds to knowing full system dynamics information. Likewise, the policy improvements (11.3-5), (11.3-11) can be explicitly computed if the dynamics are known. It was shown in Example 11.2-1 that, for the DT LQR, the exact computation method for computing the optimal control yields the Riccati equation solution approach. As shown in Example 11.3-4, PI and VI boil down to repetitive solutions of Lyapunov equations or Lyapunov recursions. In fact, PI becomes nothing but Hewer's method (Hewer 1971), and VI becomes a well-known Lyapunov recursion scheme that was shown to converge in Lancaster and Rodman (1995). These are offline methods relying on matrix equation solutions and requiring complete knowledge of the system dynamics.

Monte Carlo learning is based on the definition (11.2-16) for the value function, and uses repeated measurements of data to approximate the expected value. The expected values are approximated by averaging repeated results along sample paths. An assumption on the ergodicity of the Markov chain with transition probabilities (11.2-2) for the given policy being evaluated is implicit. This is suitable for episodic tasks, with experience divided into episodes (Sutton and Barto 1998)—namely, processes that start in an initial state and run until termination, and are then restarted at a new initial state. For finite MDP, Monte Carlo methods converge to the true value function if all states are visited infinitely often. Therefore, to ensure good approximations of value functions, the episode sample paths must go through all the states $x \in X$ many times. This is called the problem of maintaining exploration. There are several ways of ensuring this, one of which is to use *exploring starts*, in which every state has nonzero probability of being selected as the initial state of an episode.

Monte Carlo techniques are useful for dynamic systems control because the episode sample paths can be interpreted as system trajectories beginning in a prescribed initial state. However, no updates to the value function estimate or the control policy are made until after an episode terminates. In fact, Monte Carlo learning methods are closely related to repetitive or *iterative learning control (ILC)* (Moore 1993). They do not learn in real time along a trajectory, but learn as trajectories are repeated.

11.4 TEMPORAL DIFFERENCE LEARNING AND OPTIMAL ADAPTIVE CONTROL

The temporal difference (TD) method (Sutton and Barto 1998) for solving Bellman equations leads to a family of optimal adaptive controllers—that is, adaptive controllers that learn online the solutions to optimal control problems without knowing the full system dynamics. TD learning is true online reinforcement learning, wherein control actions are improved in real time based on estimating their value functions by observing data measured along the system trajectories. In the context of TD learning, the interpretation of the Bellman equation is shown in Figure 11.2-2, where $V^\pi(x)$ may be considered a predicted performance, $\sum_u \pi(x,u) \sum_{x'} P^u_{xx'} R^u_{xx'}$, the observed one-step reward, and $V^\pi(x')$ a current estimate of future behavior.

Temporal Difference Learning along State Trajectories

PI requires solution at each step of N linear equations (11.3-4). VI requires performing the recursion (11.3-10) at each step. Temporal difference RL methods are based on the Bellman equation, and solve equations such as (11.3-4) and (11.3-10) without using systems dynamics knowledge, but using and data observed along a single trajectory of the system. This makes them extremely applicable for feedback control applications. TD updates the value at each time step as observations of data are made along a trajectory. Periodically, the new value is used to update the policy. TD methods are related to adaptive control in that they adjust values and actions online in real time along system trajectories.

TD methods can be considered to be stochastic approximation techniques, whereby the Bellman equation (11.2-17), or its variants (11.3-4), (11.3-10), is replaced by its evaluation along a single sample path of the MDP. This turns the Bellman equation, into a deterministic equation, which allows the definition of a so-called *temporal difference error*.

Equation (11.2-9) was used to write the Bellman equation (11.2-17) for the infinite-horizon value (11.2-16). According to (11.2-7)–(11.2-9), an alternative form for the Bellman equation is

$$V^\pi(x_k) = E_\pi\{r_k|x_k\} + \gamma E_\pi\{V^\pi(x_{k+1})|x_k\}. \qquad (11.4\text{-}1)$$

This equation forms the basis for TD learning.

Temporal difference RL uses one sample path, namely the current system trajectory, to update the value. That is (11.4-1) is replaced by the deterministic Bellman equation

$$V^\pi(x_k) = r_k + \gamma V^\pi(x_{k+1}), \qquad (11.4\text{-}2)$$

which holds for each observed data experience set (x_k, x_{k+1}, r_k) at each time stage k. This set consists of the current state x_k, the observed cost incurred r_k, and the next state x_{k+1}. The temporal difference error is defined as

$$e_k = -V^\pi(x_k) + r_k + \gamma V^\pi(x_{k+1}), \tag{11.4-3}$$

and the value estimate is updated to make the temporal difference error small.

The Bellman equation can be interpreted as a consistency equation, which holds if the current estimate for the value $V^\pi(x_k)$ is correct. Therefore, TD methods update the value estimate $\hat{V}^\pi(x_k)$ to make the TD error small. The idea is that if the deterministic version of Bellman's equation is used repeatedly, then on average one will converge toward the solution of the stochastic Bellman equation.

11.5 OPTIMAL ADAPTIVE CONTROL FOR DISCRETE-TIME SYSTEMS

A family of optimal adaptive control algorithms can now be developed for dynamical systems. Physical analysis of dynamical systems using Lagrangian mechanics, Hamiltonian mechanics, etc. produces system descriptions in terms of nonlinear ordinary differential equations. Discretization yields nonlinear difference equations. The bulk of research in RL has been conducted for systems that operate in discrete time (DT). Therefore, we cover DT dynamical systems first, then continuous-time systems.

TD learning is a stochastic approximation technique based on the deterministic Bellman's equation (11.4-2). Therefore, we lose little by considering deterministic systems here. Consider a class of discrete-time systems described by deterministic nonlinear dynamics in the affine state space difference equation form

$$x_{k+1} = f(x_k) + g(x_k)u_k, \tag{11.5-1}$$

with state $x_k \in R^n$ and control input $u_k \in R^m$. We use this form because its analysis is convenient. The following development can be generalized to the general sampled-data form $x_{k+1} = F(x_k, u_k)$.

A control policy is defined as a function from state space to control space $h(\cdot): R^n \to R^m$. That is, for every state x_k, the policy defines a control action

$$u_k = h(x_k). \tag{11.5-2}$$

That is, a policy is simply a feedback controller.

Define a cost function that yields the value function

$$V^h(x_k) = \sum_{i=k}^{\infty} \gamma^{i-k} r(x_i, u_i) = \sum_{i=k}^{\infty} \gamma^{i-k} \left(Q(x_i) + u_i^T R u_i \right), \tag{11.5-3}$$

with $0 < \gamma \le 1$ a discount factor, $Q(x_k) > 0$, $R > 0$, and $u_k = h(x_k)$ a prescribed feedback control policy. The stage cost

$$r(x_k, u_k) = Q(x_k) + u_k^T R u_k. \tag{11.5-4}$$

is taken to be quadratic in u_k to simplify developments, but can be any positive definite function of the control. We assume the system is stabilizable on some set $\Omega \in R^n$; that is, there exists a control policy $u_k = h(x_k)$ such that the closed-loop system $x_{k+1} = f(x_k) + g(x_k)h(x_k)$ is asymptotically stable on Ω. A control policy $u_k = h(x_k)$ is said to be *admissible* if it is stabilizing and yields a finite cost $V^h(x_k)$ for trajectories in Ω.

For the deterministic DT system, the optimal value is given by Bellman's optimality equation

$$V^*(x_k) = \min_{h(\cdot)} \left(r(x_k, h(x_k)) + \gamma V^*(x_{k+1}) \right). \tag{11.5-5}$$

This is just the discrete-time Hamilton-Jacobi-Bellman (HJB) equation. One then has the optimal policy as

$$h^*(x_k) = \arg\min_{h(\cdot)} \left(r(x_k, h(x_k)) + \gamma V^*(x_{k+1}) \right). \tag{11.5-6}$$

In this setup, the deterministic Bellman's equation (11.4-2) is

$$V^h(x_k) = r(x_k, u_k) + \gamma V^h(x_{k+1}) = Q(x_k) + u_k^T R u_k + \gamma V^h(x_{k+1}), \ V^h(0) = 0. \tag{11.5-7}$$

This is nothing but a difference equation equivalent of the value (11.5-3). That is, instead of evaluating the infinite sum (11.5-3), one can solve the difference equation (11.5-7), with boundary condition $V(0) = 0$, to obtain the value of using a current policy $u_k = h(x_k)$.

The DT Hamiltonian function is

$$H(x_k, h(x_K), \Delta V_k) = r(x_k, h(x_k)) + \gamma V^h(x_{k+1}) - V^h(x_k), \tag{11.5-8}$$

where $\Delta V_k = \gamma V_h(x_{k+1}) - V_h(x_k)$ is the forward difference operator. The Hamiltonian function captures the energy content along the trajectories of a system. In fact, the Hamiltonian is the temporal difference error (11.4-3). The Bellman equation requires that the Hamiltonian be equal to zero for the value associated with a prescribed policy.

For the DT linear quadratic regulator (DT LQR) case,

$$x_{k+1} = Ax_k + Bu_k, \tag{11.5-9}$$

$$V^h(x_k) = \frac{1}{2} \sum_{i=k}^{\infty} \gamma^{i-k} \left(x_i^T Q x_i + u_i^T R u_i \right), \tag{11.5-10}$$

and the Bellman equation is written in several ways, as seen in Example 11.2-1.

Policy Iteration and Value Iteration Using Temporal Difference Learning

It has been seen that two forms of reinforcement learning can be based on policy iteration and value iteration. For TD learning, PI is written as follows in terms of the deterministic Bellman equation.

POLICY ITERATION USING TD LEARNING

Initialize.

Select any admissible control policy $h_0(x_k)$. Do for $j = 0$ until convergence—

Policy evaluation:

$$V_{j+1}(x_k) = r(x_k, h_j(x_k)) + \gamma V_{j+1}(x_{k+1}). \qquad (11.5\text{-}11)$$

Policy improvement:

$$h_{j+1}(x_k) = \arg\min_{h(\cdot)} \left(r(x_k, h(x_k)) + \gamma V_{j+1}(x_{k+1}) \right), \qquad (11.5\text{-}12)$$

or equivalently

$$h_{j+1}(x_k) = -\frac{\gamma}{2} R^{-1} g^T(x_k) \nabla V_{j+1}(x_{k+1}), \qquad (11.5\text{-}13)$$

where $\nabla V(x) = \partial V(x)/\partial x$ is the gradient of the value function, interpreted here as a column vector.

VI is similar, but the policy evaluation procedure is performed as follows.

VALUE ITERATION USING TD LEARNING

Value update step:

Update the value using

$$V_{j+1}(x_k) = r(x_k, h_j(x_k)) + \gamma V_j(x_{k+1}). \qquad (11.5\text{-}14)$$

Also, in VI we may select any initial control policy $h_0(x_k)$, not necessarily admissible or stabilizing.

It has been shown in Example 11.2-1 that for DT LQR the Bellman equation (11.5-7) is nothing but a linear Lyapunov equation and that (11.5-5) is the DT algebraic Riccati equation (ARE). In Example 11.3-4 it was seen that for the DT LQR the policy evaluation step (11.5-11) in PI is a Lyapunov equation and PI exactly corresponds to Hewer's algorithm (Hewer 1971) for solving the DT ARE. Hewer proved that it converges under stabilizability and detectability assumptions. For DT LQR, VI is a Lyapunov recursion, which has been shown to converge to the solution to the DT ARE under the stated assumptions by (Lancaster and Rodman 1995).

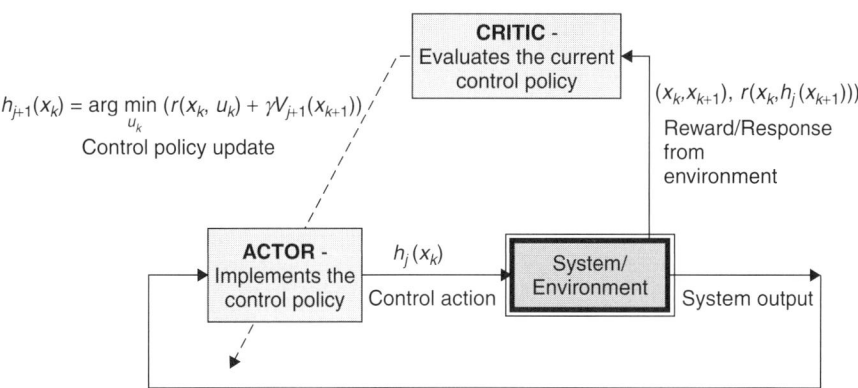

Value update using Bellman equation

$$V_{j+1}(x_k) = r(x_k, h_j(x_k)) + \gamma V_{j+1}(x_{k+1})$$

Use RLS until convergence

CRITIC - Evaluates the current control policy

$(x_k, x_{k+1}), r(x_k, h_j(x_{k+1})))$

Reward/Response from environment

$$h_{j+1}(x_k) = \arg\min_{u_k} (r(x_k, u_k) + \gamma V_{j+1}(x_{k+1}))$$

Control policy update

ACTOR - Implements the control policy

$h_j(x_k)$

Control action

System/ Environment

System output

FIGURE 11.5-1 Temporal difference learning using policy iteration. At each time one observes the current state, the next state, and the cost incurred. This is used to update the value estimate. Based on the new value, the action is updated.

The online implementation of PI using temporal difference learning is shown in Figure 11.5-1. PI and VI using TD learning have an actor–critic structure, as shown in Figure 11.1-1. The critic evaluates the current policy by solving the Bellman equation (11.5-11). Then, the actor updates the policy using (11.5-12).

Value Function Approximation

PI and VI can be implemented for finite MDP by storing and updating look-up tables. The key to practical schemes for implementing PI and VI online for dynamical systems with infinite state and action spaces is to approximate the value function by a suitable approximator structure in terms of unknown parameters. Then, the unknown parameters are tuned online exactly as in system identification. This idea of *value function approximation (VFA)* was used by Werbos (1989, 1991, 1992) and called approximate dynamic programming (ADP) or adaptive dynamic programming. It was used by Bertsekas and Tsitsiklis (1996) and called neuro-dynamic programming (NDP). See Powell (2007) and Busoniu et al. (2009).

In the LQR case it is known that the value is quadratic in the state so that

$$V(x_k) = \tfrac{1}{2} x_k^T P x_k = \tfrac{1}{2} (\text{vec}(P))^T (x_k \otimes x_k) \equiv \overline{p}^T \overline{x}_k \equiv \overline{p}^T \phi(x_k), \qquad (11.5\text{-}15)$$

for some kernel matrix P. The Kronecker product $\otimes$ (Brewer 1978) allows us to write this quadratic form as linear in the parameter vector $\overline{p} = \text{vec}(P)$, which is formed by stacking the columns of the P matrix. The vector $\phi(x_k) = \overline{x}_k = x_k \otimes$

x_k is the quadratic polynomial vector containing all possible pairwise products of the n components of x_k. Noting that P is symmetric and has only $n(n+1)/2$ independent elements, we remove the redundant terms in $x_k \otimes x_k$ to define a quadratic basis set $\phi(x_k)$ with $n(n+1)/2$ independent elements.

For general nonlinear systems (11.5-1) the value function contains higher-order nonlinearities. Then, we assume the Bellman equation (11.5-7) has a local smooth solution (van der Schaft 1992). According to the Weierstrass higher-order approximation theorem, there exists a dense basis set $\{\varphi_i(x)\}$ such that

$$V(x) = \sum_{i=1}^{\infty} w_i \varphi_i(x) = \sum_{i=1}^{L} w_i \varphi_i(x) + \sum_{i=L+1}^{\infty} w_i \varphi_i(x) \equiv W^T \phi(x) + \varepsilon_L(x),$$

(11.5-16)

where basis vector $\phi(x) = \begin{bmatrix} \varphi_1(x) & \varphi_2(x) & \cdots & \varphi_L(x) \end{bmatrix}: R^n \to R^L$, and $\varepsilon_L(x)$ converges uniformly to zero as the number of terms retained $L \to \infty$. In the Weierstrass Theorem, standard usage takes a polynomial basis set. In the neural network (NN) community, approximation results have been shown for various other basis sets including sigmoid, hyperbolic tangent, Gaussian radial basis functions, etc. There, standard results show that the NN approximation error $\varepsilon_L(x)$ is bounded by a constant on a compact set. L is referred to as the number of hidden-layer neurons, $\varphi_i(x)$ as the NN activation functions, and w_i as the NN weights.

Optimal Adaptive Control Algorithms for DT Systems

We are now in a position to present several adaptive control algorithms based on TD RL that converge online to the optimal control solution.

The parameters in $\overline{p}$ or W are unknown. Substituting the value function approximation into the value update (11.5-11) in PI we obtain the following algorithm.

OPTIMAL ADAPTIVE CONTROL USING A POLICY ITERATION ALGORITHM

Initialize.

Select any admissible control policy $h_0(x_k)$. Do for $j = 0$ until convergence—

Policy evaluation step:

Determine the least-squares solution W_{j+1} to

$$W_{j+1}^T (\phi(x_k) - \gamma\phi(x_{k+1})) = r(x_k, h_j(x_k)) = Q(x_k) + h_j^T(x_k)Rh_j(x_k).$$

(11.5-17)

Policy improvement step:

Determine an improved policy using

$$h_{j+1}(x_k) = -\frac{\gamma}{2} R^{-1} g^T(x_k) \nabla\phi^T(x_{k+1}) W_{j+1}.$$

(11.5-18)

This algorithm is easily implemented online by standard system identification techniques (Ljung 1999). In fact, note that (11.5-17) is a scalar equation, whereas the unknown parameter vector $W_{j+1} \in R^L$ has L elements. Therefore, data from multiple time steps is needed for its solution. At time $k + 1$ one measures the previous state x_k, the control $u_k = h_j(x_k)$, the next state x_{k+1}, and computes the resulting utility $r(x_k, h_j(x_k))$. This gives one scalar equation. This is repeated for subsequent times using the same policy $h_j(\cdot)$ until we have at least L equations, at which point we may determine the LS solution W_{j+1}. We may use batch LS for this.

Alternatively, note that equations of the form (11.5-17) are exactly those solved by recursive least-squares (RLS) techniques (Ljung 1999) Therefore, we may run RLS online until convergence. Write (11.5-17) as

$$W_{j+1}^T \Phi(k) \equiv W_{j+1}^T \left(\phi(x_k) - \gamma \phi(x_{k+1}) \right) = r(x_k, h_j(x_k)), \qquad (11.5\text{-}19)$$

with $\Phi(k) \equiv (\phi(x_k) - \gamma \phi(x_{k+1}))$ a regression vector. At step j of the PI algorithm, one fixes the control policy at $u = h_j(x)$. Then, at each time k one measures the data set $(x_k, x_{k+1}, r(x_k, h_j(x_k)))$. One step of RLS is then performed. This is repeated for subsequent times until convergence to the parameters corresponding to the value $V_{j+1}(x) = W_{j+1}^T \phi(x)$.

Note that for RLS to converge, the regression vector $\Phi(k) \equiv (\phi(x_k) - \gamma \phi(x_{k+1}))$ must be persistently exciting.

As an alternative to RLS, we could use a gradient descent tuning method such as

$$W_{j+1}^{i+1} = W_{j+1}^i - \alpha \Phi(k) \left(\left(W_{j+1}^i \right)^T \Phi(k) - r(x_k, h_j(x_k)) \right), \qquad (11.5\text{-}20)$$

with $\alpha > 0$ a tuning parameter. The step index j is held fixed, and index i is incremented at each increment of the time index k. Note that the quantity inside the large brackets is just the temporal difference error.

Once the value parameters have converged, the control policy is updated according to (11.5-18). Then, the procedure is repeated for step $j + 1$. This entire procedure is repeated until convergence to the optimal control solution.

This provides an online reinforcement learning algorithm for solving the optimal control problem using policy iteration by measuring data along the system trajectories. Likewise, an online reinforcement learning algorithm can be given based on value iteration. Substituting the value function approximation into the value update (11.5-14) in VI we obtain the following algorithm.

OPTIMAL ADAPTIVE CONTROL USING A VALUE ITERATION ALGORITHM

Initialize

Select any control policy $h_0(x_k)$, not necessarily admissible or stabilizing. Do for $j = 0$ until convergence—

Value update step:

Determine the least-squares solution W_{j+1} to

$$W_{j+1}^{T}\phi(x_k) = r(x_k, h_j(x_k)) + W_j^{T}\gamma\phi(x_{k+1}). \tag{11.5-21}$$

Policy improvement step:

Determine an improved policy using (11.5-18).

To solve (11.5-21) in real-time we can use batch LS, RLS, or gradient-based methods based on data $(x_k, x_{k+1}, r(x_k, h_j(x_k)))$ measured at each time along the system trajectories. Then the policy is improved using (11.5-18).

Note that the old weight parameters are on the right-hand side of (11.5-21). Thus, the regression vector is now $\phi(x_k)$, which must be persistently exciting for convergence of RLS.

Online Solution of Lyapunov and Riccati Equations

It is important to note that the PI and VI adaptive optimal control algorithms just given actually solve the Bellman equation (11.5-7) and the HJB equation (11.5-5) online in real time by using data measured along the system trajectories. The system drift function $f(x_k)$ (or the A matrix in the LQR case) is not needed in these algorithms. That is, these algorithms solve the Riccati equation online in real time without knowledge of the A matrix. The online implementation of PI is shown in Figure 11.5-1.

According to Example 11.3-4, for DT LQR policy iteration, this means that the Lyapunov equation

$$0 = (A - BK^j)^{T} P^{j+1}(A - BK^j) - P^{j+1} + Q + (K^j)^{T} RK^j, \tag{11.5-22}$$

has been replaced by (11.5-17) or

$$\bar{p}_{j+1}^{T}(\bar{x}_{k+1} - \bar{x}_k) = r(x_k, h_j(k)) = x_k^{T}(Q + K_j^{T} RK_j)x_k, \tag{11.5-23}$$

which is solved for the parameters $\bar{p}_{j+1} = \text{vec}(P^{j+1})$ using, for instance, RLS by measuring the data set $(x_k, x_{k+1}, r(x_k, h_j(x_k)))$ at each time. For this step the dynamics (A, B) can be unknown, as they are not needed. For DT LQR value iteration, the Lyapunov recursion

$$P^{j+1} = (A - BK^j)^{T} P^j (A - BK^j) + Q + (K^j)^{T} RK^j \tag{11.5-24}$$

has been replaced by (11.5-21) or

$$\bar{p}_{j+1}^{T}\bar{x}_{k+1} = r(x_k, h_j(k)) + \bar{p}_j^{T}\bar{x}_k = x_k^{T}(Q + K_j^{T} RK_j)x_k + \bar{p}_j^{T}\bar{x}_k, \tag{11.5-25}$$

which may be solved for the parameters $\bar{p}_{j+1} = \text{vec}(P_{j+1})$ using RLS without knowing A, B.

Introduction of a Second "Actor" Neural Network

Using value function approximation allows standard system identification techniques to be used to find the value function parameters that approximately solve the Bellman equation. The approximator structure just described that is used for approximation of the value function is known as the critic NN (neural network), as it determines the value of using the current policy. Using VFA, the PI and VI reinforcement learning algorithm solve the Bellman equation during the value update portion of each iteration step j by observing only the data set $\left(x_k, x_{k+1}, r(x_k, h_j(x_k)) \right)$ at each time along the system trajectory and solving (11.5-17) or (11.5-21).

However, according to Example 11.3-4, in the LQR case, the policy update (11.5-18) is given by

$$K^{j+1} = -(B^{\mathrm{T}} P^{j+1} B + R)^{-1} B^{\mathrm{T}} P^{j+1} A, \qquad (11.5\text{-}26)$$

which requires full knowledge of the dynamics (A, B). Note further that the embodiment (11.5-18) cannot easily be implemented in the nonlinear case because it is implicit in the control, since x_{k+1} depends on u_k and is the argument of a nonlinear activation function.

These problems are both solved by introducing a *second neural network* for the control policy, known as the actor NN (Werbos, 1989, 1991, 1992). Introduce a parametric approximator structure for the control action

$$u_k = h(x_k) = U^{\mathrm{T}} \sigma(x_k), \qquad (11.5\text{-}27)$$

with $\sigma(x)$: $R^n \to R^M$ a vector of M activation or basis functions and $U \in R^{M \times m}$ a matrix of weights or unknown parameters. After convergence of the critic NN parameters to W_{j+1} in PI or VI, it is required to perform the policy update (11.5-18). To achieve this we may use a gradient descent method for tuning the actor weights U such as

$$U_{j+1}^{i+1} = U_{j+1}^{i} - \beta \sigma(x_k) \left(2R \left(U_{j+1}^{i} \right)^{\mathrm{T}} \sigma(x_k) + \gamma g(x_k)^{\mathrm{T}} \nabla \phi^{\mathrm{T}}(x_{k+1}) W_{j+1} \right)^{\mathrm{T}},$$
$$(11.5\text{-}28)$$

with $\beta > 0$ a tuning parameter. The tuning index i may be incremented with the time index k. On convergence, set the updated policy to $h_{j+1}(x_k) = (U_{j+1}^{i})^{\mathrm{T}} \sigma(x_k)$.

Several items are worthy of note at this point. First, the tuning of the actor NN requires observations at each time k of the data set (x_k, x_{k+1}), that is, the current state and the next state. However, as per the formulation (11.5-27), the actor NN yields the control u_k at time k in terms of the state x_k at time k. The next state x_{k+1} is not needed in (11.5-27). Thus, after (11.5-28) has converged, (11.5-27) is a legitimate feedback controller. Second, in the LQR case, the actor NN (11.5-27) embodies the feedback gain computation (11.5-26). This is highly intriguing, for the latter contains the state internal dynamics A, but the former does not. This means that the A matrix is not needed to compute the feedback control. The reason is that the actor NN has learned information about A in its weights, since (x_k, x_{k+1}) are used in its tuning.

Finally, note that only the input function $g(\cdot)$ (in the LQR case, the B matrix) is needed in (11.5-28) to tune the actor NN. Thus, introducing a second actor NN has completely avoided the need for knowledge of the state drift dynamics $f(\cdot)$ (or A in the LQR case).

Example 11.5-1. Discrete-time Optimal Adaptive Control of Power System Using Value Iteration

In this simulation we show how to use DT value iteration to solve the DT ARE online without knowing the system matrix A. We simulated the online VI algorithm (11.5-21), (11.5-28) for load frequency control of an electric power system. Power systems are complex nonlinear systems. However, during normal operation the system load, which gives the nonlinearity, has only small variations. As such, a linear model can be used to represent the system dynamics around an operating point specified by a constant load value. A problem rises from the fact that in an actual plant the parameter values are not precisely known, as reflected in an unknown system A matrix, yet an optimal control solution is sought.

The model of the system that is considered here is $\dot{x} = Ax + Bu$, where

$$
A = \begin{bmatrix} -1/T_p & K_p/T_p & 0 & 0 \\ 0 & -1/T_T & 1/T_T & 0 \\ -1/RT_G & 0 & -1/T_G & -1/T_G \\ K_E & 0 & 0 & 0 \end{bmatrix}, \quad B = \begin{bmatrix} 0 \\ 0 \\ 1/T_G \\ 0 \end{bmatrix}.
$$

The system state is $x(t) = \begin{bmatrix} \Delta f(t) & \Delta P_g(t) & \Delta X_g(t) & \Delta E(t) \end{bmatrix}^{\mathrm{T}}$, where $\Delta f(t)$ is the incremental frequency deviation (Hz), $\Delta P_g(t)$ is the incremental change in generator output (p.u. MW), $\Delta X_g(t)$ is the incremental change in governor position (p.u. MW), and $\Delta E(t)$ is the incremental change in integral control. The system parameters are T_G, the governor time constant; T_T, turbine time constant; T_P, plant model time constant; K_P, planet model gain; R, speed regulation due to governor action; and K_E, integral control gain.

The values of the CT system parameters were randomly picked within specified operating ranges so that

$$
A = \begin{bmatrix} -0.0665 & 8 & 0 & 0 \\ 0 & -3.663 & 3.663 & 0 \\ -6.86 & 0 & -13.736 & -13.736 \\ 0.6 & 0 & 0 & 0 \end{bmatrix}, \quad B = \begin{bmatrix} 0 & 0 & 13.7355 & 0 \end{bmatrix}.
$$

The discrete-time dynamics was obtained using the zero-order hold method with a sampling period of $T = 0.01$ sec. The solution to the DT ARE with cost function weights $Q = I$, $R = I$, and $\gamma = 1$ is

$$
P_{\mathrm{DARE}} = \begin{bmatrix} 0.4750 & 0.4766 & 0.0601 & 0.4751 \\ 0.4766 & 0.7831 & 0.1237 & 0.3829 \\ 0.0601 & 0.1237 & 0.0513 & 0.0298 \\ 0.4751 & 0.3829 & 0.0298 & 2.3370 \end{bmatrix}.
$$

In this simulation, only the time constant T_G of the governor, which appears in the B matrix, is considered to be known, while the values for all the other parameters appearing in the system A matrix are not known. That is, the A matrix is needed only to simulate the system and obtain the data and is not needed by the control algorithm.

For the DT LQR, the value is quadratic in the states $V(x) = \frac{1}{2}x^{\mathrm{T}}Px$, as in (11.5-15). Therefore, the basis functions for the critic NN in (11.5-16) are selected as the quadratic polynomial vector in the state components. Since there are $n = 4$ states, this vector has $n(n+1)/2 = 10$ components. The control is linear in the states $u = -Kx$, so the basis functions for the actor NN (11.5-27) are taken as the state components.

The online implementation of VI may be done by setting up a batch least-squares problem to solve for the 10 critic NN parameters, that is the Riccati solution entries $\overline{p}_{j+1} \equiv W_{j+1}$ in (11.5-21), for each step j. In this simulation the matrix P^{j+1} is determined after collecting 15 points of data $(x_k, x_{k+1}, r(x_k, u_k))$ for each least-squares problem. Therefore, a least-squares problem for the critic weights is solved each 0.15 sec. Then the actor NN parameters (that is, the feedback gain matrix entries) are updated using (11.5-28). The simulations were performed over a time interval of 60 sec.

The system states trajectories are given in Figure 11.5-2, which shows that they are regulated to zero as desired. The convergence of the Riccati matrix parameters is shown in Figure 11.5-3. The final values of the critic NN parameter estimates are

$$
P_{\text{critic NN}} =
\begin{bmatrix}
0.4802 & 0.4768 & 0.0603 & 0.4754 \\
0.4768 & 0.7887 & 0.1239 & 0.3834 \\
0.0603 & 0.1239 & 0.0567 & 0.0300 \\
0.4754 & 0.3843 & 0.0300 & 2.3433
\end{bmatrix}.
$$

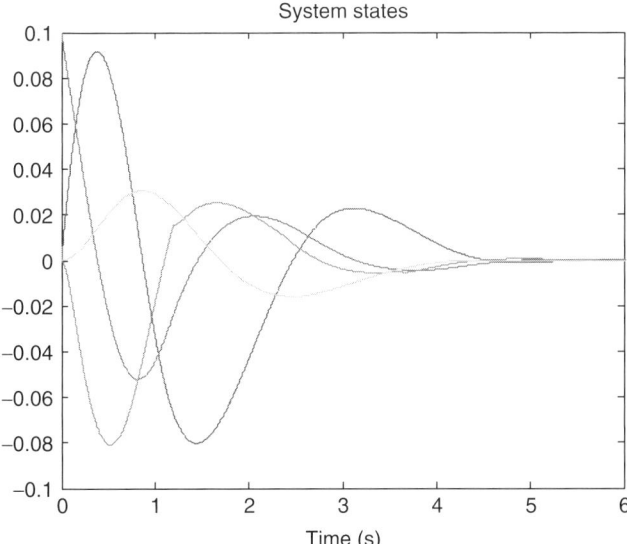

FIGURE 11.5-2 System states during the first 6 sec. This figure shows that even though the A matrix of the power system is unknown, the adaptive controller based on value iteration keeps the states stable and regulates them to zero.

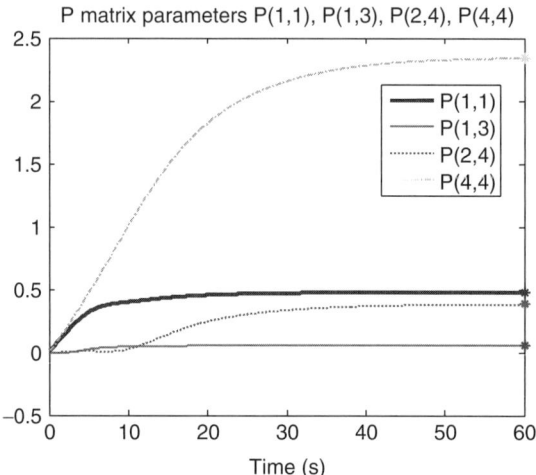

FIGURE 11.5-3 Convergence of selected ARE solution parameters. This figure shows that the adaptive controller based on VI converges to the Riccati-equation solution in real time without knowing the system A matrix.

The optimal adaptive control VI algorithm has converged to the optimal control solution, as given by the ARE solution. This has been accomplished in real time without knowing the system A matrix. ∎

Actor–Critic Implementation of DT Optimal Adaptive Control

Two algorithms for optimal adaptive control of DT systems based on RL have been given. A PI algorithm is implemented by solving (11.5-17) using RLS or batch LS to determine the value of the current policy, then the policy is updated by running (11.5-28). A VI algorithm is implemented by solving (11.5-21) using RLS or batch LS to determine the value of the current policy, then the policy is updated by running (11.5-28).

   The implementation of reinforcement learning using two NNs, one as a critic and one as an actor, yields the actor–critic RL structure shown in Figure 11.1-1. In this control system, the critic and the actor are tuned online using the observed data $(x_k, x_{k+1}, r(x_k, h_j(x_k)))$ along the system trajectory. The critic and actor are tuned sequentially in both the PI and the VI algorithms. That is, the weights of one NN are held constant while the weights of the other are tuned until convergence. This procedure is repeated until both NN have converged. Then, the controller has learned the optimal controller online. Thus, this is an online adaptive optimal control system wherein the value function parameters are tuned online and the convergence is to the optimal value and control. The convergence of value iteration using two NN for the DT nonlinear system (11.5-1) was proven in Al-Tamimi et al. (2008).

Q Learning for Optimal Adaptive Control

The Q learning RL method gives an adaptive control algorithm that converges online to the optimal control solution for completely unknown systems. That is, it solves the Bellman equation (11.5-7) and the HJB equation (11.5-5) online in real time by using data measured along the system trajectories, without any knowledge of the dynamics $f(x_k), g(x_k)$.

Q learning is a simple method for reinforcement learning that works for unknown MDP, that is, for systems with completely unknown dynamics. It was developed by Watkins (1989) and Watkins and Dayan (1992) and Werbos (1989, 1991, 1992), who called it action-dependent heuristic dynamic programming (ADHDP), since the Q function depends on the control input. Q learning learns the Q function (11.3-38) using TD methods by performing an action u_k and measuring at each time stage the resulting data experience set (x_k, x_{k+1}, r_k) consisting of the current state, the next state, and the resulting stage cost. Writing the Q function Bellman equation (11.3-40) along a sample path gives

$$Q^\pi(x_k, u_k) = r(x_k, u_k) + \gamma Q^\pi(x_{k+1}, h(x_{k+1})), \qquad (11.5\text{-}29)$$

which defines a TD error

$$e_k = -Q^\pi(x_k, u_k) + r(x_k, u_k) + \gamma Q^\pi(x_{k+1}, h(x_{k+1})). \qquad (11.5\text{-}30)$$

The VI algorithm for Q function is given as (11.3-47). Based on this, the Q function is updated using the algorithm

$$Q_k(x_k, u_k) = Q_{k-1}(x_k, u_k) + \alpha_k\left[r(x_k, u_k) + \gamma \min_u Q_{k-1}(x_{k+1}, u) - Q_{k-1}(x_k, u_k)\right].$$
$$(11.5\text{-}31)$$

This Q learning algorithm is similar to stochastic approximation methods of adaptive control or parameter estimation used in control systems. It was developed for finite MDP and the convergence proven by Watkins (1989) using stochastic approximation methods. It was shown that the algorithm converges for finite MDP provided that all state–action pairs are visited infinitely often and

$$\sum_{k=1}^{\infty} \alpha_k = \infty, \quad \sum_{k=1}^{\infty} \alpha_k^2 < \infty, \qquad (11.5\text{-}32)$$

that is, standard stochastic approximation conditions. On convergence, the TD error is (approximately) equal to zero. For finite MDP, Q learning requires storing a 2-D lookup table in terms of all the states x and actions u.

The requirement that all state–action pairs are visited infinitely often translates to the problem of maintaining sufficient exploration during learning. Let us now derive methods for Q learning for dynamical systems that yield adaptive control algorithms that converge to optimal control solutions. PI and VI algorithms have been given using the Q function in (11.3-43)–(11.3-47). A Q learning algorithm is easily developed for DT dynamical systems using Q function approximation (Werbos 1989, 1991, 1992; Bradtke et al. 1994). It was shown in Example 11.3-5 that for DT LQR the Q function is a quadratic form in terms of $z_k \equiv [x_k^T \ u_k^T]^T \in R^{n+m}$. Assume, therefore, that for nonlinear systems the Q function is parameterized as

$$Q(x, u) = W^T \phi(z),$$

for some unknown parameter vector W and basis set vector $\phi(z)$. For the DT LQR, $\phi(z)$ is the quadratic basis set formed from the state and input components. Substituting the Q function approximation into the TD error (11.5-30) yields

$$e_k = -W^T \phi(z_k) + r(x_k, u_k) + \gamma W^T \phi(z_{k+1}), \qquad (11.5\text{-}33)$$

upon which either PI or VI algorithms can be based. Considering the PI algorithm (11.3-43), (11.3-44) yields the Q function evaluation step

$$W_{j+1}^T (\phi(z_k) - \gamma \phi(z_{k+1})) = r(x_k, h_j(x_k)) \qquad (11.5\text{-}34)$$

and the policy improvement step

$$h_{j+1}(x_k) = \arg \min_u \left(W_{j+1}^T \phi(x_k, u) \right), \quad \text{for all } x \in X. \qquad (11.5\text{-}35)$$

Q learning using VI (11.3-45) is given by

$$W_{j+1}^T \phi(z_k) = r(x_k, h_j(x_k)) + \gamma W_j^T \phi(z_{k+1}), \qquad (11.5\text{-}36)$$

and (11.5-35). These equations do not require knowledge of the dynamics $f(\cdot), g(\cdot)$.

For online implementation, batch LS or RLS can be used to solve (11.5-34) for the parameter vector W_{j+1} given the regression vector $(\phi(z_k) - \gamma \phi(z_{k+1}))$, or (11.5-36) using regression vector $\phi(z_k)$. The observed data at each time instant is $(z_k, z_{k+1}, r(x_k, u_k))$ with $z_k \equiv [x_k^T \ u_k^T]^T$. We take $z_{k+1} \equiv [x_{k+1}^T \ u_{k+1}^T]^T$, with $u_{k+1} = h_j(x_{k+1})$ and $h_j(\cdot)$ the current policy. Probing noise must be added to the control input to obtain persistence of excitation.

After convergence of the Q function parameters, the action update (11.5-35) is performed. This is easily accomplished without knowing the system dynamics due to the fact that the Q function contains u_k as an argument so that $\partial \left(W_{j+1}^T \phi(x_k, u) \right) / \partial u$ can be explicitly computed. In fact,

$$\frac{\partial Q(x, u)}{\partial u} = \left(\frac{\partial z}{\partial u} \right)^T \left(\frac{\partial \phi(z)}{\partial z} \right)^T W = \begin{bmatrix} 0_{m,n} & I_m \end{bmatrix} \nabla \phi^T W, \qquad (11.5\text{-}37)$$

where $0_{m,n} \in R^{m \times n}$ is a matrix of zeros. For the specific case of the DT LQR, for instance, from Example 11.3-5 we have

$$Q(x, u) \equiv \frac{1}{2} \begin{bmatrix} x \\ u \end{bmatrix}^{\mathrm{T}} S \begin{bmatrix} x \\ u \end{bmatrix} = \frac{1}{2} z^{\mathrm{T}} S z = \frac{1}{2} \mathrm{vec}^{\mathrm{T}}(S)(z \otimes z) \equiv W^{\mathrm{T}} \phi(z), \quad (11.5\text{-}38)$$

with $\otimes$ the Kronecker product (Brewer 1978) and $\mathrm{vec}(S) \in R^{(n+m)^2}$ the vector formed by stacking the columns of the S matrix. The basis vector $\phi(z) = z \otimes z \in R^{(n+m)^2}$ is the quadratic polynomial vector containing all possible pairwise products of the $n + m$ components of z. Define $N = n + m$. Then,

$$\nabla \phi^{\mathrm{T}} = \frac{\partial \phi^{\mathrm{T}}}{\partial z} = (I_N \otimes z + z \otimes I_N)^{\mathrm{T}} \in R^{N \times N^2} \qquad (11.5\text{-}39)$$

Using these equations we obtain a form equivalent to the control update (11.5-35) given in Example 11.3-5, which can be done knowing the Q function parameters S without knowing system matrices A,B.

If the control can be computed explicitly from the action update (11.5-35), as in the DT LQR case, an actor NN is not needed for Q learning and it can be implemented using only one critic NN for Q function approximation.

11.6 INTEGRAL REINFORCEMENT LEARNING FOR OPTIMAL ADAPTIVE CONTROL OF CONTINUOUS-TIME SYSTEMS

Reinforcement learning is considerably more difficult for continuous-time (CT) systems than for discrete-time systems, and its development has lagged. See (Abu-Khalaf et al. 2006) for the development of a PI method for CT systems. Using a method known as *integral reinforcement learning* (IRL) (Vrabie et al. Vrabie and Lewis 2009) allows the application of RL to formulate online optimal adaptive control methods for CT systems.

Consider the continuous-time nonlinear dynamical system

$$\dot{x} = f(x) + g(x)u, \qquad (11.6\text{-}1)$$

with state $x(t) \in R^n$, control input $u(t) \in R^m$, and the usual assumptions required for existence of unique solutions and an equilibrium point at $x = 0$, e.g., $f(0) = 0$ and $f(x) + g(x)u$ Lipschitz on a set $\Omega \subseteq R^n$ that contains the origin. We assume the system is stabilizable on Ω; that is, there exists a continuous control function $u(t)$ such that the closed-loop system is asymptotically stable on Ω.

Define a performance measure or cost function that has the value associated with the feedback control policy $u = \mu(x)$ given by

$$V^{\mu}(x(t)) = \int_{t}^{\infty} r(x(\tau), u(\tau)) \, d\tau, \qquad (11.6\text{-}2)$$

with utility $r(x, u) = Q(x) + u^T R u$, $Q(x)$ positive definite, that is, $Q(x) > 0$ for all x and $x = 0 \Rightarrow Q(x) = 0$, and $R > 0$ a positive definite matrix. For the CT linear quadratic regulator (LQR) we have

$$\dot{x} = Ax + Bu, \tag{11.6-3}$$

$$V^\mu(x(t)) = \frac{1}{2} \int_t^\infty (x^T Q x + u^T R u) \, d\tau. \tag{11.6-4}$$

A policy is called admissible if it is continuous, stabilizes the system, and has a finite associated cost. If the cost is smooth, then an infinitesimal equivalent to (11.6-2) can be found by differentiation to be the nonlinear equation

$$0 = r(x, \mu(x)) + (\nabla V^\mu)^T (f(x) + g(x)\mu(x)), \quad V^\mu(0) = 0, \tag{11.6-5}$$

where ∇V^μ (a column vector) denotes the gradient of the cost function V^μ with respect to x. This is the CT Bellman equation. It is defined based on the CT Hamiltonian function

$$H(x, \mu(x), \nabla V^\mu) = r(x, \mu(x)) + (\nabla V^\mu)^T (f(x) + g(x)\mu(x)). \tag{11.6-6}$$

The optimal value satisfies the CT Hamilton-Jacobi-Bellman (HJB) equation (Chapter 6),

$$0 = \min_\mu H(x, \mu(x), \nabla V^*), \tag{11.6-7}$$

and the optimal control satisfies

$$\mu^* = \arg \min_\mu H(x, \mu(x), \nabla V^*). \tag{11.6-8}$$

We now see the problem with CT systems immediately. Compare the CT Bellman Hamiltonian (11.6-6) to the DT Hamiltonian (11.5-8). The former contains the full system dynamics $f(x) + g(x)u$, while the DT Hamiltonian does not. This means the CT Bellman equation (11.6-5) cannot be used as a basis for reinforcement learning unless the full dynamics are known.

Reinforcement learning methods based on (11.6-5) can be developed (Baird 1994, Doya 2000, Murray et al. 2001, Mehta and Meyn 2009, Hanselmann et al. 2007). These have limited use for adaptive control purposes because the system dynamics must be known. In another approach, one can use Euler's method to discretize the CT Bellman equation (Baird 1994). Noting that

$$0 = r(x, \mu(x)) + (\nabla V^\mu)^T (f(x) + g(x)\mu(x)) = r(x, \mu(x)) + \dot{V}^\mu, \tag{11.6-9}$$

We use Euler's method to discretize this to obtain

$$0 = r(x_k, u_k) + \frac{V^\mu(x_{k+1}) - V^\mu(x_k)}{T} \equiv \frac{r_S(x_k, u_k)}{T} + \frac{V^\mu(x_{k+1}) - V^\mu(x_k)}{T}, \tag{11.6-10}$$

with sample period T so that $t = kT$. The discrete sampled utility is $r_S(x_k, u_k) = r(x_k, u_k)T$, where it is important to multiply the CT utility by the sample period.

Now note that the discretized CT Bellman equation (11.6-10) has the same form as the DT Bellman equation (11.5-7). Therefore, all the reinforcement learning methods just described for DT systems can be applied.

However, this is an approximation only. An alternative exact method for CT reinforcement learning was given by Vrabie et al. (2009) and Vrabie and Lewis (2009). This is termed integral reinforcement learning (IRL). Note that we may write the cost (11.6-2) in the integral reinforcement form

$$V^\mu(x(t)) = \int_t^{t+T} r(x(\tau), u(\tau))\, d\tau + V^\mu(x(t+T)),\qquad (11.6\text{-}11)$$

for any $T > 0$. This is exactly in the form of the DT Bellman equation (11.5-7). According to Bellman's principle, the optimal value is given in terms of this construction as (Chapter 6)

$$V^*(x(t)) = \min_{\bar{u}(t:t+T)} \left(\int_t^{t+T} r(x(\tau), u(\tau))\, d\tau + V^*(x(t+T)) \right),$$

where $\bar{u}(t: t+T) = \{u(\tau): t \leq \tau < t+T\}$. The optimal control is

$$\mu^*(x(t)) = \arg\min_{\bar{u}(t:t+T)} \left(\int_t^{t+T} r(x(\tau), u(\tau))\, d\tau + V^*(x(t+T)) \right).$$

It is shown in (Vrabie et al. 2009) that the nonlinear equation (11.6-5) is exactly equivalent to the integral reinforcement form (11.6-11). That is, the positive definite solution of both that satisfies $V(0) = 0$ is the value (11.6-2) of the policy $u = \mu(x)$. Therefore, integral reinforcement form (11.6-11) also serves as a Bellman equation for CT systems and serves is a fixed-point equation. Thus, we can define the temporal difference error for CT systems as

$$e(t: t+T) = \int_t^{t+T} r(x(\tau), u(\tau))\, d\tau + V^\mu(x(t+T)) - V^\mu(x(t)).\quad (11.6\text{-}12)$$

This does not involve the system dynamics.

Now, policy iteration and value iteration can be directly formula for CT systems. The following algorithms are termed integral reinforcement learning for CT systems (Vrabie et al. 2009). They both give optimal adaptive controllers for CT systems, that is, adaptive control algorithms that converge to optimal control solutions.

IRL OPTIMAL ADAPTIVE CONTROL USING POLICY ITERATION (PI)

Initialize

Select any admissible control policy $\mu_0(x)$. Do for $j = 0$ until convergence—

Policy evaluation step:

Solve for $V_{j+1}(x(t))$ using

$$V_{j+1}(x(t)) = \int_t^{t+T} r(x(s), \mu_j(x(s)))\, ds + V_{j+1}(x(t+T)), \quad \text{with } V_{j+1}(0) = 0.$$

$$(11.6\text{-}13)$$

Policy improvement step:

Determine an improved policy using

$$\mu_{j+1} = \arg\min_u [H(x, u, \nabla V_{j+1})], \qquad (11.6\text{-}14)$$

which explicitly is

$$\mu_{j+1}(x) = -\tfrac{1}{2} R^{-1} g^{\mathrm{T}}(x) \nabla V_{j+1}. \qquad (11.6\text{-}15)$$

IRL OPTIMAL ADAPTIVE CONTROL USING VALUE ITERATION (VI)

Initialize

Select any control policy $\mu_0(x)$, not necessarily stabilizing. Do for $j = 0$ until convergence—

Policy evaluation step:

Solve for $V_{j+1}(x(t))$ using

$$V_{j+1}(x(t)) = \int_t^{t+T} r(x(s), \mu_j(x(s)))\, ds + V_j(x(t+T)). \qquad (11.6\text{-}16)$$

Policy improvement step:

Determine an improved policy using (11.6-15).

Note that neither algorithm requires knowledge about the system drift dynamics function $f(x)$. That is, they work for partially unknown systems. Convergence of PI is proved in Vrabie et al. (2009).

Online Implementation of IRL—A Hybrid Optimal Adaptive Controller

Both of these IRL algorithms may be implemented online by reinforcement learning techniques using value function approximation $V(x) = W^T\phi(x)$ in a critic approximator network. Using VFA in the PI algorithm (11.6-13) yields

$$W_{j+1}^T [\phi(x(t)) - \phi(x(t + T))] = \int_{t}^{t+T} r(x(s), \mu_j(x(s)))\, ds. \qquad (11.6\text{-}17)$$

Using VFA in the VI algorithm (11.6-16) yields

$$W_{j+1}^T \phi(x(t)) = \int_{t}^{t+T} r(x(s), \mu_j(x(s)))\, ds + W_j^T \phi(x(t + T)). \qquad (11.6\text{-}18)$$

RLS or batch LS can be used to update the value function parameters in these equations. On convergence of the value parameters, the action is updated using (11.6-15). The implementation is shown in Figure 11.6-1. This is an optimal adaptive controller, that is, an adaptive controller that measures data along the system trajectories and converges to optimal control solutions. Note that only the system input coupling dynamics $g(x)$ are needed to implement these algorithms, since it appears in action update (11.6-15). The drift dynamics $f(x)$ are not needed.

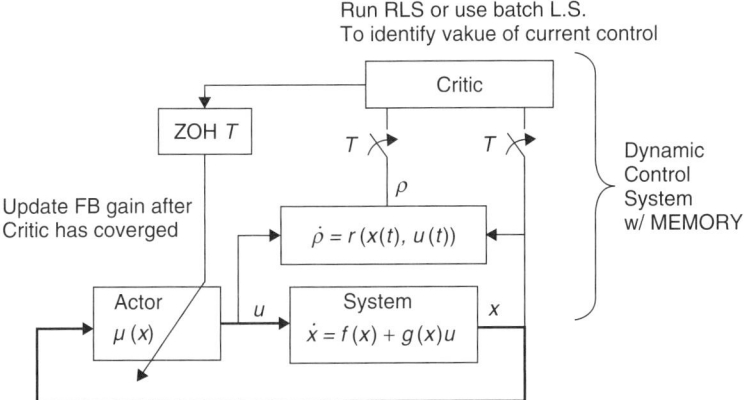

FIGURE 11.6-1 Hybrid optimal adaptive controller based on integral reinforcement learning (IRL), showing the two-time scale hybrid nature of the IRL controller. The integral reinforcement signal is added as an extra state and functions as the memory of the controller. The critic runs on a slow time scale and learns the value of using the current control policy. When the critic has converged, the actor control policy is updated to obtain an improved value.

The time is incremented at each iteration by the reinforcement learning time interval T. This time interval need not be the same at each iteration. T can be changed depending on how long it takes to get meaningful information from the observations. T is not a sample period in the standard meaning.

The measured data at each time increment is $(x(t), x(t + T), \rho(t: t + T))$, where

$$\rho(t: t + T) = \int_{t}^{t+T} r(x(\tau), u(\tau)) \, d\tau, \qquad (11.6\text{-}19)$$

is the integral reinforcement measured on each time interval. This can be implemented by introducing an integrator $\dot{\rho} = r(x(t), u(t))$, as shown in Figure 11.6-1. That is, the integral reinforcement $\rho(t)$ is added as an extra continuous-time state. It functions as the memory or controller dynamics. The remainder of the controller is a sampled data controller.

Note that the control policy $\mu(x)$ is updated periodically after the critic weights have converged to the solution of (11.6-17) or (11.6-18). Therefore, the policy is piecewise constant in time. On the other hand, the control varies continuously with the state between each policy update. IRL for CT systems is, in fact, a hybrid CT/DT adaptive controller that converges to the optimal control solution in real time without knowing the drift dynamics $f(x)$. Due to the fact that the policy update (11.6-15) for CT systems does not involve the drift dynamics $f(x)$, no actor NN is needed in IRL. Only a critic NN is needed for VFA.

Online Solution of the Algebraic Riccati Equation without Full Plant Dynamics

It can be shown that the integral reinforcement form (11.6-11) is equivalent to the nonlinear Lyapunov (11.6-5) (Vrabie et al. 2009). Thus, the IRL controller solves the Lyapunov equation online without knowing the drift dynamics $f(x)$. Moreover, it converges to the optimal control so that it solves the HJB equation (11.6-7).

In the CT LQR case (11.6-3), (11.6-4) we have linear state feedback control policies $u = -Kx$. Then, equation (11.6-5) is

$$(A - BK)^{\mathsf{T}} P + P(A - BK) + Q + K^{\mathsf{T}} RK = 0, \qquad (11.6\text{-}20)$$

which is a Lyapunov equation. The HJB equation (11.6-7) becomes the CT ARE

$$A^{\mathsf{T}} P + PA + Q - PBR^{-1} B^{\mathsf{T}} P = 0. \qquad (11.6\text{-}21)$$

Thus, IRL solves both the Lyapunov equation and the ARE online in real time, using data measured along the system trajectories, without knowing the A matrix.

For the CT LQR, (11.6-13) is equivalent to a Lyapunov equation at each step, so that policy iteration is exactly the same as Kleinman's algorithm (Kleinman 1968) for solving the CT Riccati equation. This is a Newton method for finding the optimal value. CT value iteration, on the other hand, is a new algorithm that solves the CT ARE based on iterations on certain discrete-time Lyapunov equations that are equivalent to (11.6-16).

Example 11.6-1.   Continuous-time Optimal Adaptive Control Using IRL

This example shows the hybrid control nature of the IRL optimal adaptive controller. Consider the DC motor model

$$\dot{x} = Ax + Bu = \begin{bmatrix} -10 & 1 \\ -0.002 & -2 \end{bmatrix} x + \begin{bmatrix} 0 \\ 2 \end{bmatrix} u,$$

with cost weight matrices $Q = I$, $R = I$. The solution to the CT ARE is computed to be

$$P = \begin{bmatrix} 0.05 & 0.0039 \\ 0.0039 & 0.2085 \end{bmatrix}.$$

In this simulation we used the IRL-based CT VI algorithm. This algorithm does not require knowledge of the system A matrix. For the CT LQR, the value is quadratic in the states. Therefore, the basis functions for the critic NN are selected as the quadratic polynomial vector in the state components, $\phi(x) = \begin{bmatrix} x_1^2 & x_1 x_2 & x_2^2 \end{bmatrix}$. The IRL time interval was selected as $T = 0.04$ sec. To update the 3 critic weights $\bar{p}_{j+1} \equiv W_{j+1}$ (that is, the ARE solution elements) using (11.6-18), a batch LS solution can be obtained. Measurements of the data set $(x(t), x(t+T), \rho(t:t+T))$ are taken over 3 time intervals of $T = 0.04$ sec. Then, provided that there is enough excitation in the system, after each 0.12 sec enough data are collected from the system to solve for the value of the matrix P. Then a greedy policy update is performed using (11.6-15), that is, $u = -R^{-1} B^T P x \equiv -Kx$.

The states are shown in Figure 11.6-2, which shows the good regulation achieved. The control input and feedback gains are shown in Figure 11.6-3. Note that the control

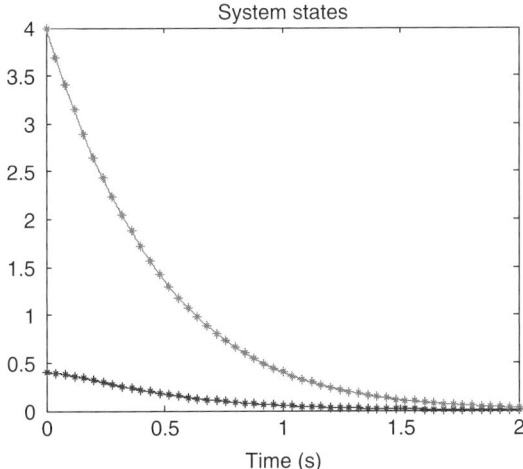

System states

FIGURE 11.6-2   System states during the first 2 sec, showing that the continuous time IRL adaptive controller regulates the states to zero without knowing the system A matrix.

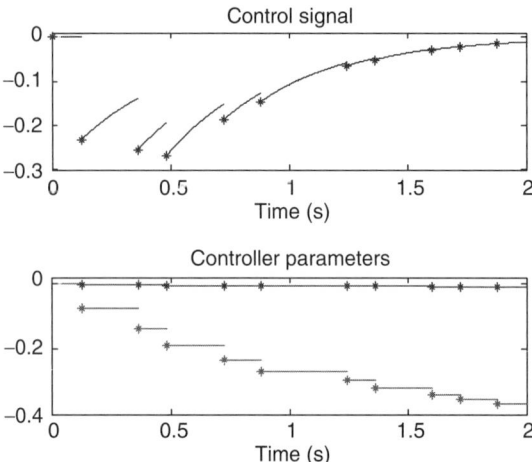

FIGURE 11.6-3   Control input and feedback gains, showing the hybrid nature of the IRL optimal adaptive controller. The controller gain parameters are discontinuous and piecewise constant, while the control signal itself is continuous between the gain parameter updates.

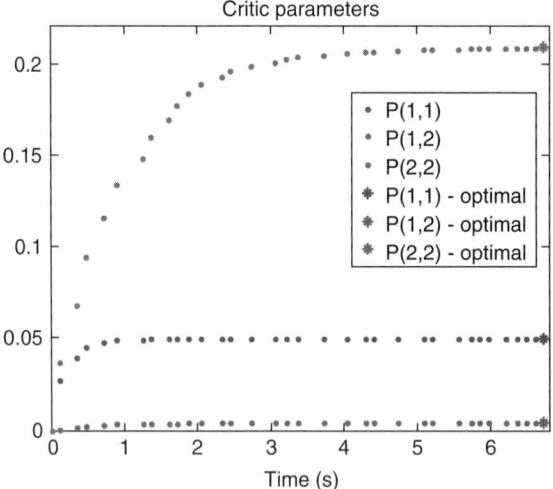

FIGURE 11.6-4 P matrix parameter estimates, showing that the IRL adaptive controller converges online to the optimal control Riccati equation solution without knowing the system A matrix.

gains are piecewise constant, while the control input is a continuous function of the state between policy updates. The critic NN parameter estimates are shown in Figure 11.6-4. They converge almost exactly to the entries in the Riccati solution matrix P. Thus, the ARE has been solved online without knowing the system A matrix.   ∎

Example 11.6-2.   Continuous-time Optimal Adaptive Control for Power System Using IRL

In this example we simulate the CT IRL optimal adaptive control based on VI for the electric power system in Example 11.5-1. The same system matrices and performance index were used. The solution to the CT ARE is computed to be

$$P_{\text{ARE}} = \begin{bmatrix} 0.4750 & 0.4766 & 0.0601 & 0.4751 \\ 0.4766 & 0.7831 & 0.1237 & 0.3829 \\ 0.0601 & 0.1237 & 0.0513 & 0.0298 \\ 0.4751 & 0.3829 & 0.0298 & 2.3370 \end{bmatrix}.$$

This is close to the DT ARE solution presented in Example 11.5-1, since the sample period used there is small.

The VI IRL algorithm was simulated, which does not require knowledge of the system A matrix. The IRL time interval was taken as $T = 0.1$ sec. (Note the IRL interval is not related at all to the sample period used to discretize the system in Example 11.5-1.) Fifteen data points $(x(t), x(t+T), \rho(t: t+T))$ were taken to compute each batch LS update for the critic parameters $\overline{p}_{j+1} \equiv W_{j+1}$ (e.g., the elements of the ARE solution P) using (11.6-18). Hence, the value estimate was updated every 1.5 sec. Then the policy was computed using (11.6-15), that is, $u = -R^{-1}B^{T}Px \equiv -Kx$.

The state trajectories are similar to those presented in Example 11.5-1. The critic parameter estimates for the P matrix entries are shown in Figure 11.6-5. They converge to the true solution to the CT ARE. Thus, the ARE has been solved online without knowing the system A matrix.

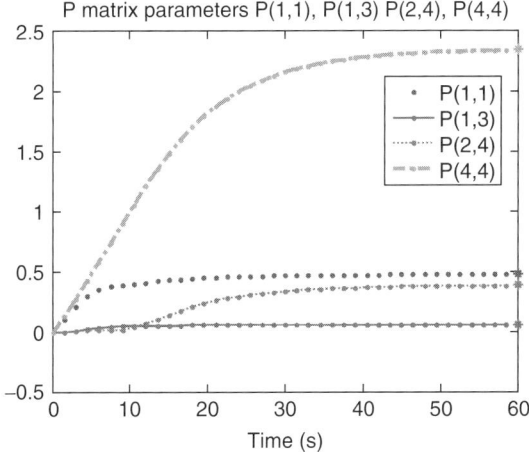

FIGURE 11.6-5 P matrix parameter estimates, showing that the IRL adaptive controller converges online to the optimal control Riccati equation solution without knowing the system A matrix.

Note that far less computation is needed using this IRL algorithm on the CT dynamics than was used in Example 11.5-1 for the DT optimal adaptive control algorithm. There, the critic parameter estimates were updated every 0.15 sec. Yet, the parameter estimates for the P matrix entries almost overlay each other. ∎

*Example 11.6-3.   Continuous-time IRL Optimal Adaptive Control
for Nonlinear System*

In this example we show that IRL can solve the HJB equation for nonlinear CT systems by using data measured along the trajectories in real time. This example was developed using the converse HJB approach (Nevistic and Primbs 1996), which allows construction of nonlinear systems starting from the known optimal cost function.

Consider the nonlinear system given by the equations

$$\begin{cases} \dot{x}_1 = -x_1 + x_2 + 2x_2^3 \\ \dot{x}_2 = f(x) + g(x)u \end{cases}, \qquad (11.6\text{-}22)$$

with $f(x) = -\frac{1}{2}(x_1 + x_2) + \frac{1}{2}x_2(1 + 2x_2^2)\sin^2(x_1)$, $g(x) = \sin(x_1)$. If we define $Q(x) = x_1^2 + x_2^2 + 2x_2^4$, $R = 1$, then the optimal cost function for this system is $V^*(x) = \frac{1}{2}x_1^2 + x_2^2 + x_2^4$ and the optimal controller is $u^*(x) = -\sin(x_1)(x_2 + 2x_2^3)$. It can be verified that for these choices the HJB equation (11.6-7) and the Bellman equation (11.6-5) are both satisfied.

The cost function was approximated by the smooth function $V_j(x(t)) = W_j^T \phi(x(t))$ with $L = 8$ neurons and $\phi(x) = \begin{bmatrix} x_1^2 & x_1 x_2 & x_2^2 & x_1^4 & x_1^3 x_2 & x_1^2 x_2^2 & x_1 x_2^3 & x_2^4 \end{bmatrix}^T$. The PI IRL algorithm (11.6-17), (11.6-15) was used. This does not require knowledge of the drift dynamics $f(x)$.

To ensure exploration so that the HJB solution is found over a suitable region, data were taken along five trajectories defined by five different initial conditions chosen randomly in the region $\Omega = \{-1 \le x_i \le 1; i = 1, 2\}$. The IRL time period was taken as $T = 0.1$ sec. At each iteration step we set up a batch least-squares problem to solve for the eight NN weights using 40 data points measured on each of the 5 trajectories in Ω. Each data point consists of $(x(t), x(t + T), \rho(t: t + T))$, with $\rho(t: t + T)$ the measured integral

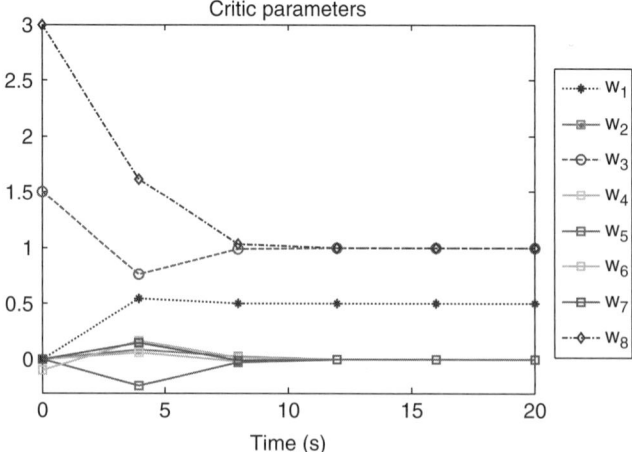

FIGURE 11.6-6 Convergence of critic NN parameters, showing that the IRL adaptive controller converges online to the optimal control Riccati equation solution without knowing the system A matrix.

reinforcement cost. In this way, at every 4 sec, the value was solved for and then a policy update was performed.

The result of applying the algorithm is presented in Figure 11.6-6, which shows that the parameters of the critic neural network converged to the coefficients of the optimal cost function $V^*(x) = \frac{1}{2}x_1^2 + x_2^2 + x_2^4$, that is, $W = \begin{bmatrix} 0.5 & 0 & 1 & 0 & 0 & 0 & 0 & 1 \end{bmatrix}^T$. We observe that after 3 iteration steps (that is, after 12 sec) the critic NN parameters have effectively converged. Then the controller is close to the optimal controller $u^*(x) = -\sin(x_1)(x_2 + 2x_2^3)$. The approximate solution to the HJB equation has been determined online and the optimal control has been found without knowing the system drift dynamics $f(x)$. Note that analytic solution of the HJB equation in this example would be intractable. ∎

11.7   SYNCHRONOUS OPTIMAL ADAPTIVE CONTROL FOR CONTINUOUS-TIME SYSTEMS

The integral reinforcement learning controller just given tunes the critic NN to determine the value while holding the control policy fixed, then a policy update is performed. Now we develop an adaptive controller that has two NN, one for value function approximation and one to approximate the control. We could call these the critic NN and actor NN. These two NN are tuned simultaneously, that is, synchronously in time (Vamvoudakis and Lewis 2010a and b). This is more nearly in line with accepted practice in adaptive control. Though this synchronous controller does require knowledge of the dynamics, it converges to the approximate local solutions to the HJB equation and the Bellman equation online, yet does not require explicitly solving either one. The HJB is generally impossible to solve for nonlinear systems.

Based on the CT Hamiltonian (11.6-6) and the stationarity condition $0 = \partial H(x, u, \nabla V^\mu)/\partial u$, we could write a PI algorithm for CT systems based on the policy evaluation step

$$0 = H(x, \mu_j(x), \nabla V_{j+1}) = r(x, \mu_j(x)) + (\nabla V_{j+1})^T(f(x) + g(x)\mu_j(x)),$$
$$V_{j+1}(0) = 0 \qquad (11.7\text{-}1)$$

and the policy improvement step

$$\mu_{j+1} = \arg\min_\mu H(x, \mu, \nabla V_{j+1}). \qquad (11.7\text{-}2)$$

Unfortunately, the full dynamics $f(x), g(x)$ are needed to implement this algorithm. Moreover, (11.7-1) is a nonlinear equation and cannot generally be solved.

However, this algorithm provides the structure needed to develop another adaptive control algorithm that can be implemented online using measured data along the trajectories and converges to the optimal control. Specifically, select a value function approximation (VFA), or critic NN, structure as

$$V(x) = W_1^T \phi(x) \qquad (11.7\text{-}3)$$

and a control action approximation structure or actor NN as

$$u(x) = -\tfrac{1}{2}R^{-1}g^{\mathrm{T}}(x)\nabla\phi^{\mathrm{T}}W_2, \tag{11.7-4}$$

which could be, for instance, two neural networks with unknown parameters (weights) W_1, W_2, and $\phi(x)$ the basis set (activation functions) of the first NN. The structure of the second action NN comes from (11.6-15). Then, it can be shown that tuning the NN weights as

$$\dot{W}_1 = -\alpha_1 \frac{\sigma}{(\sigma^{\mathrm{T}}\sigma + 1)^2}[\sigma^{\mathrm{T}}W_1 + Q(x) + u^{\mathrm{T}}Ru], \tag{11.7-5}$$

$$\dot{W}_2 = -\alpha_2\{(F_2 W_2 - F_1\bar{\sigma}^{\mathrm{T}}W_1) - \tfrac{1}{4}D(x)W_2 m^{\mathrm{T}}(x)W_1\}, \tag{11.7-6}$$

guarantees system stability as well as convergence to the optimal value and control (Vamvoudakis and Lewis, 2010a,b).

In these parameter estimation algorithms, $\alpha_1, \alpha_2, F_1, F_2$ are algorithm tuning parameters, $D(x) = \nabla\phi(x)g(x)R^{-1}g^{\mathrm{T}}(x)\nabla\phi^{\mathrm{T}}(x)$, $\sigma = \nabla\phi(f + gu)$, $\bar{\sigma} = \sigma/(\sigma^{\mathrm{T}}\sigma + 1)$, and $m(x) = \sigma/(\sigma^{\mathrm{T}}\sigma + 1)^2$. A PE condition on $\bar{\sigma}(t)$ is needed to get convergence to the optimal value.

This is an adaptive control algorithm that requires full knowledge of the system dynamics $f(x), g(x)$, yet converges to the optimal control solution. That is, it solves (locally approximately) the HJB equation, which is generally intractable for general nonlinear systems. In the CT LQR case, it solves the ARE using data measured along the trajectories (and knowledge of A,B). The importance of this algorithm is that it can approximately solve the HJB equation for nonlinear systems using data measured along the system trajectories in real time. The HJB is generally impossible to solve for nonlinear systems.

The VFA tuning algorithm for W_1 is based on gradient descent, while the control action tuning algorithm is a form of backpropagation (Werbos 1989), which is, however, also tuned by the VFA weights W_1. The similarity to the actor–critic RL structure in Figure 11.1-1 is clear. However, in contrast to IRL, this algorithm is a CT optimal adaptive controller with two parameter estimators tuned simultaneously, that is, synchronously and continuously in time.

Example 11.7-1.   Continuous-time Synchronous Optimal Adaptive Control

In this example we show that the synchronous optimal adaptive control algorithm can approximately solve the HJB equation for nonlinear CT systems by using data measured along the trajectories in real time. This example was developed using the method of Nevistic and Primbs (1996).

Consider the affine in the control input nonlinear system $\dot{x} = f(x) + g(x)u$, $x = [x_1 \ x_2]^{\mathrm{T}} \in R^2$, where

$$f(x) = \begin{bmatrix} -x_1 + x_2 \\ -x_1^3 - x_2 - \frac{x_1^2}{x_2} + 0.25x_2(\cos(2x_1 + x_1^3) + 2)^2 \end{bmatrix}$$

$$g(x) = \begin{bmatrix} 0 \\ \cos(2x_1 + x_1^3) + 2 \end{bmatrix}.$$

We select $Q = I$, $R = 1$. Then the optimal value function that solves the HJB equation is $V^*(x) = \frac{1}{4}x_1^4 + \frac{1}{2}x_2^2$ and the optimal control policy is $u^*(x) = -\frac{1}{2}(\cos(2x_1 + x_1^3) + 2)x_2$.

We select the critic NN vector activation function as $\phi(x) = \begin{bmatrix} x_1^2 & x_2^2 & x_1^4 & x_2^4 \end{bmatrix}$. The tuning algorithms (11.7-5), (11.7-6) were run for the critic NN and control actor NN, respectively, simultaneously in time. A probing noise was added to the control to guarantee persistence of excitation. This noise was decayed exponentially during the simulation. The evolution of the states is given in Figure 11.7-1. They are stable and approach zero as the probing noise decays to zero.

Figure 11.7-2 shows the critic parameters, denoted by $W_1 = [W_{c1} \ W_{c2} \ W_{c3} \ W_{c4}]^{\mathrm{T}}$ After 80 sec the critic NN parameters converged to $W_1(t_f) = [0.0033 \ 0.4967 \ 0.2405 \ 0.0153]^{\mathrm{T}}$, which is close to the true weights corresponding to the optimal value $V^*(x)$ that solves the HJB equation. The actor NN parameters converge to $W_2(t_f) = [0.0033 \ 0.4967 \ 0.2405 \ 0.0153]^{\mathrm{T}}$. Thus, the control policy converges to

$$\hat{u}_2(x) = -\frac{1}{2}\begin{bmatrix} 0 \\ \cos(2x_1 + x_1^3) + 2 \end{bmatrix}^{\mathrm{T}} \begin{bmatrix} 2x_1 & 0 & 4x_1^3 & 0 \\ 0 & 2x_2 & 0 & 4x_2^3 \end{bmatrix} \hat{W}_2(t_f).$$

This is the optimal control.

Figure 11.7-3 shows the 3-D plot of the difference between the approximated value function, by using the online synchronous adaptive algorithm, and the optimal value. The errors are small relative to the magnitude of the optimal value. Figure 11.7-4 shows the

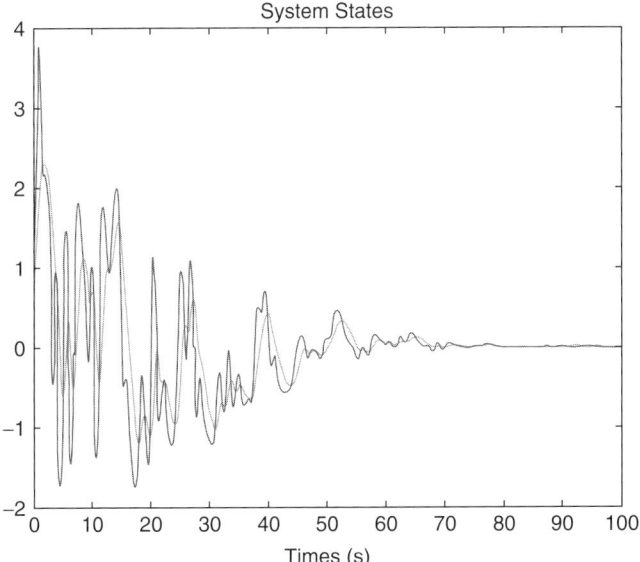

FIGURE 11.7-1 Evolution of the states, showing that the synchronous optimal adaptive controller ensures stability and regulates the states to zero.

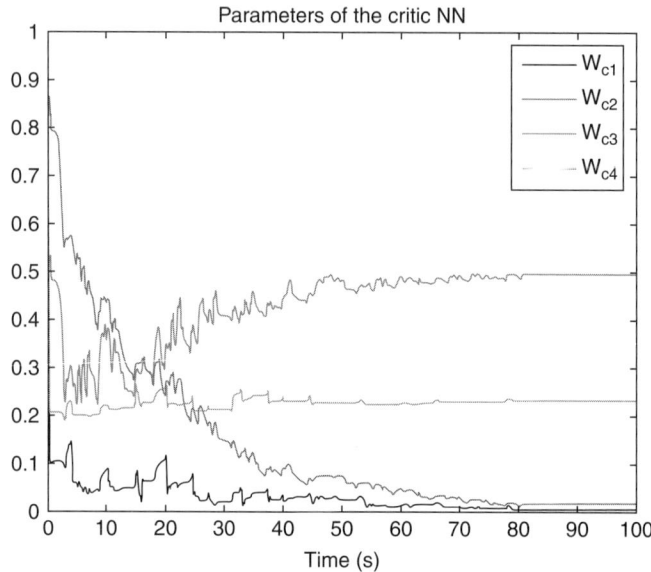

FIGURE 11.7-2   Convergence of critic NN parameters, showing that the optimal adaptive controller converges to the approximate solution of the nonlinear HJB equation.

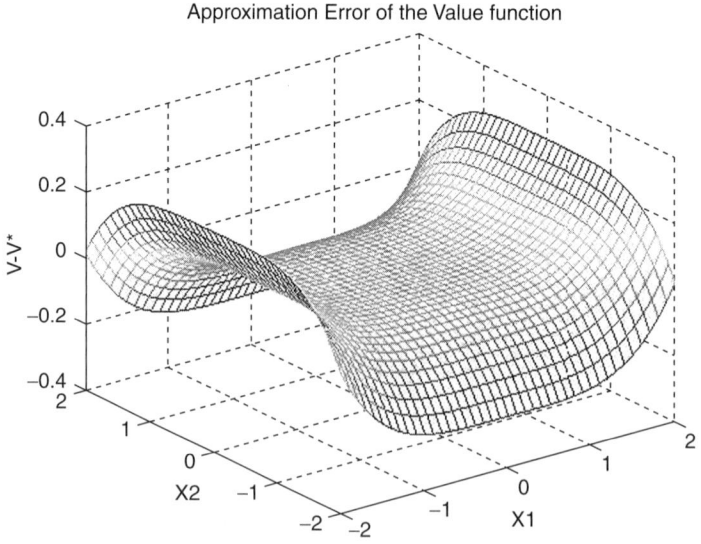

FIGURE 11.7-3   Error between optimal and approximated value function. This 3-D plot of the value function error shows that the synchronous optimal adaptive controller converges to a value function that is very close to the true solution of the HJB equation.

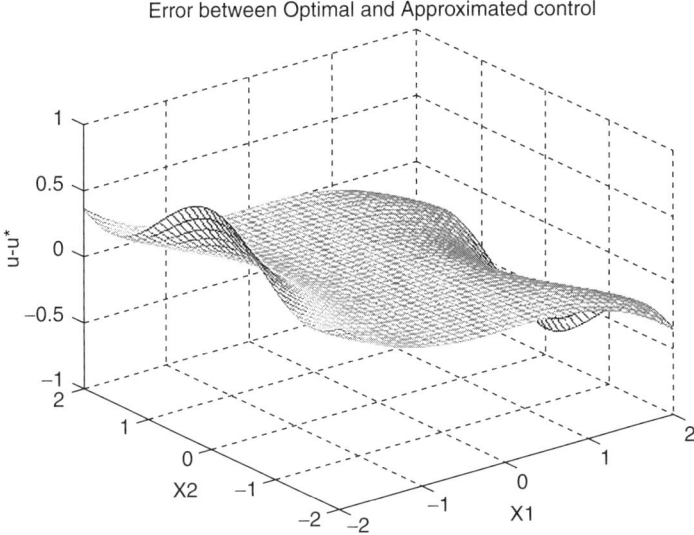

FIGURE 11.7-4   Error between optimal and approximated control input. This 3-D plot of the feedback control policy error shows that the synchronous optimal adaptive controller converges very close to the true optimal control policy.

3-D plot of the difference between the approximated feedback control policy found by using the online algorithm and the optimal control.

   This example demonstrates that the synchronous optimal adaptive controller is capable of approximately solving the HJB equation online by using data measured along the system trajectories. The HJB equation for this example is intractable to solve analytically.   ■

APPENDIX A

REVIEW OF MATRIX ALGEBRA

We present here a brief review of some concepts that are assumed as background for the text. Good references include Gantmacher (1977), Brogan (1974), and Strang (1980).

A.1 BASIC DEFINITIONS AND FACTS

The *determinant* of an $n \times n$ matrix is symbolized as $|A|$. If A and B are both square, then

$$|A| = |A^{\mathrm{T}}|, \tag{A.1-1}$$

$$|AB| = |A| \cdot |B|, \tag{A.1-2}$$

where the superscript T represents transpose. If $A \in C^{m \times n}$ and $B \in C^{n \times m}$ (where n can equal m), then

$$\mathrm{trace}(AB) = \mathrm{trace}(BA) \tag{A.1-3}$$

$$|I_m + AB| = |I_n + BA|. \tag{A.1-4}$$

(C represents the complex numbers.)
For any matrices A and B,

$$(AB)^{\mathrm{T}} = B^{\mathrm{T}} A^{\mathrm{T}} \tag{A.1-5}$$

and if A and B are nonsingular, then

$$(AB)^{-1} = B^{-1} A^{-1}. \tag{A.1-6}$$

The *Kronecker product* of two matrices $A = [a_{ij}] \in C^{m \times n}$ and $B = [b_{ij}] \in C^{p \times q}$ is

$$A \otimes B = [a_{ij} B] \in C^{mp \times nq}. \tag{A.1-7}$$

(It is sometimes defined as $A \otimes B = [A b_{ij}]$.) If $A = [a_1 a_2 \cdots a_n]$, where a_i are the columns of A, the *stacking operator* is defined by

$$s(A) = \begin{bmatrix} a_1 \\ a_2 \\ \vdots \\ a_n \end{bmatrix}. \tag{A.1-8}$$

It converts $A \in C^{m \times n}$ into a vector $s(A) \in C^{mn}$. An identity that is often useful is

$$s(ABD) = (D^{\mathrm{T}} \otimes A) s(B). \tag{A.1-9}$$

If $A \in C^{m \times m}$ and $B \in C^{p \times p}$, then

$$|A \otimes B| = |A|^p \cdot |B|^m. \tag{A.1-10}$$

See Brewer (1978) for other results.

If λ_i is an eigenvalue of A with eigenvector v_i, then $1/\lambda_i$ is an eigenvalue of A^{-1} with the same eigenvector, for

$$A v_i = \lambda_i v_i \tag{A.1-11}$$

implies that

$$\lambda_i^{-1} v_i = A^{-1} v_i. \tag{A.1-12}$$

If λ_i is an eigenvalue of A with eigenvector ω_i, and μ_j is an eigenvalue of B with eigenvector w_j, then $\lambda_i \mu_j$ is an eigenvalue of $A \otimes B$ with eigenvector $v_i \otimes w_j$ (Brewer 1978).

A.2 PARTITIONED MATRICES

If

$$D = \begin{bmatrix} A_{11} & 0 & 0 \\ 0 & A_{22} & 0 \\ 0 & 0 & A_{33} \end{bmatrix}, \tag{A.2-1}$$

where A_{ij} are matrices, then we write $D = \mathrm{diag}(A_{11}, A_{22}, A_{33})$ and say that D is *block diagonal*. If the A_{ii} are square, then

$$|D| = |A_{11}| \cdot |A_{22}| \cdot |A_{33}|, \tag{A.2-2}$$

and if $|D| \neq 0$, then

$$D^{-1} = \mathrm{diag}(A_{11}^{-1}, A_{22}^{-1}, A_{33}^{-1}). \tag{A.2-3}$$

If

$$D = \begin{bmatrix} A_{11} & A_{12} & A_{13} \\ 0 & A_{22} & A_{23} \\ 0 & 0 & A_{33} \end{bmatrix}, \tag{A.2-4}$$

where A_{ij} are matrices, then D is *upper block triangular* and (A.2-2) still holds. *Lower block triangular* matrices have the form of the transpose of (A.2-4).

If

$$A = \begin{bmatrix} A_{11} & A_{12} \\ A_{21} & A_{22} \end{bmatrix}, \tag{A.2-5}$$

we define the *Schur complement of A_{22}* as

$$D_{22} = A_{22} - A_{21} A_{11}^{-1} A_{12} \tag{A.2-6}$$

and the *Schur complement of A_{11}* as

$$D_{11} = A_{11} - A_{12} A_{22}^{-1} A_{21}. \tag{A.2-7}$$

The inverse of A can be written

$$A^{-1} = \begin{bmatrix} A_{11}^{-1} + A_{11}^{-1} A_{12} D_{22}^{-1} A_{21} A_{11}^{-1} & -A_{11}^{-1} A_{12} D_{22}^{-1} \\ -D_{22}^{-1} A_{21} A_{11}^{-1} & D_{22}^{-1} \end{bmatrix}, \tag{A.2-8}$$

$$A^{-1} = \begin{bmatrix} D_{11}^{-1} & -D_{11}^{-1} A_{12} A_{22}^{-1} \\ -A_{22}^{-1} A_{21} D_{11}^{-1} & A_{22}^{-1} + A_{22}^{-1} A_{21} D_{11}^{-1} A_{12} A_{22}^{-1} \end{bmatrix}, \tag{A.2-9}$$

or

$$A^{-1} = \begin{bmatrix} D_{11}^{-1} & -A_{11}^{-1} A_{12} D_{22}^{-1} \\ -A_{22}^{-1} A_{21} D_{11}^{-1} & D_{22}^{-1} \end{bmatrix}, \tag{A.2-10}$$

depending, of course, on whether $|A_{11}| \neq 0$, $|A_{22}| \neq 0$, or both. These can be verified by checking that $AA^{-1} = A^{-1}A = I$. By comparing these various forms, we obtain the well-known *matrix inversion lemma*

$$(A_{11}^{-1} + A_{12} A_{22} A_{21})^{-1} = A_{11} - A_{11} A_{12} (A_{21} A_{11} A_{12} + A_{22}^{-1})^{-1} A_{21} A_{11}. \tag{A.2-11}$$

The Schur complement arises naturally in the solution of linear simultaneous equations, for if

$$\begin{bmatrix} A_{11} & A_{12} \\ A_{21} & A_{22} \end{bmatrix} \begin{bmatrix} X \\ Y \end{bmatrix} = \begin{bmatrix} 0 \\ Z \end{bmatrix}, \tag{A.2-12}$$

then from the first equation

$$X = -A_{11}^{-1} A_{12} Y,$$

and using this in the second equation yields

$$(A_{22} - A_{21}A_{11}^{-1}A_{12})Y = Z. \tag{A.2-13}$$

If A is given by (A.2-5), then

$$|A| = |A_{11}| \cdot |A_{22} - A_{21}A_{11}^{-1}A_{12}| = |A_{22}| \cdot |A_{11} - A_{12}A_{22}^{-1}A_{21}|. \tag{A.2-14}$$

Therefore, the determinant of A is the product of the determinant of A_{11} (or A_{22}) and the determinant of the Schur complement of A_{22} (or A_{11}).

A.3   QUADRATIC FORMS AND DEFINITENESS

If $x \in R^n$ is a vector, then the square of the Euclidean norm is

$$\|x\|^2 = x^T x. \tag{A.3-1}$$

If S is any nonsingular transformation, the vector Sx has a norm squared of $(Sx)^T Sx = x^T S^T Sx$. Letting $P = S^T S$, we write

$$\|x\|_P^2 = x^T P x \tag{A.3-2}$$

as the norm squared of Sx. We call $\|x\|_P$ the *norm of* x with respect to P. We call

$$x^T Q x \tag{A.3-3}$$

a *quadratic form*. We shall assume Q is real.

Every real square matrix Q can be decomposed into a *symmetric part* Q_s (i.e., $Q_s^T = Q_s$) and an *antisymmetric part* Q_a (i.e., $Q_a^T = -Q_a$):

$$Q = Q_s + Q_a, \tag{A.3-4}$$

where

$$Q_s = (Q + Q^T)/2, \tag{A.3-5}$$

$$Q_a = (Q - Q^T)/2. \tag{A.3-6}$$

If the quadratic form $x^T A x$ has A antisymmetric, then it must be equal to zero since $x^T A x$ is a scalar, so that $x^T A x = (x^T A x)^T = x^T A^T x = -x^T A x$. For a general real square Q, then

$$x^T Q x = x^T (Q_s + Q_a)x = x^T Q_s x. \tag{A.3-7}$$

We can therefore assume without loss of generality that Q in (A.3-3) is symmetric. Let us do so.

We say Q is:

Positive definite $(Q > 0)$ if $x_T Q x > 0$ for all nonzero x.

Positive semi-definite ($Q \geq 0$) if $x^T Q x \geq 0$ for all nonzero x.

Negative semi-definite ($Q \leq 0$) if $x^T Q x \leq 0$ for all nonzero x.

Negative definite ($Q < 0$) if $x^T Q x < 0$ for all nonzero x.

Indefinite if $x^T Q x > 0$ for some x, $x^T Q x < 0$ for other x.

We can test for definiteness independently of the vectors x. If λ_i are the eigenvalues of Q, then

$$
\begin{aligned}
Q > 0 &\quad \text{if all} \quad \lambda_i > 0, \\
Q \geq 0 &\quad \text{if all} \quad \lambda_i \geq 0, \\
Q \leq 0 &\quad \text{if all} \quad \lambda_i \leq 0, \\
Q < 0 &\quad \text{if all} \quad \lambda_i < 0.
\end{aligned}
\tag{A.3-8}
$$

Another test is provided as follows. Let $Q = [q_{ij}] \in R^{n \times n}$. The *leading minors* or Q are

$$
m_1 = q_{11},
$$

$$
m_2 = \begin{vmatrix} q_{11} & q_{12} \\ q_{21} & q_{22} \end{vmatrix},
\tag{A.3-9}
$$

$$
m_3 = \begin{vmatrix} q_{11} & q_{12} & q_{13} \\ q_{21} & q_{22} & q_{23} \\ q_{31} & q_{32} & q_{33} \end{vmatrix}, \ldots,
$$

$$
m_n = |Q|.
$$

In terms of the minors, we have

$$
Q > 0 \quad \text{if } m_i > 0, \text{ all } i,
$$

$Q \geq 0$ if *all* principal minor not only leading minors) are nonnegative.

$$
Q \leq 0 \quad \text{if} - Q \geq 0,
\tag{A.3-10}
$$

$$
Q < 0 \quad \text{if } \begin{cases} m_i < 0, & \text{all odd } i \\ m_i > 0, & \text{all even } i \end{cases}
$$

Any positive semidefinite matrix Q can be factored into *square roots* either as

$$
Q = \sqrt{Q}\sqrt{Q}^T
\tag{A.3-11}
$$

or as

$$
Q = \sqrt{Q}^T \sqrt{Q}.
\tag{A.3-12}
$$

The ("left" and "right") square roots in (A.3-11) and (A.3-12) are not in general the same. Indeed, Q may have several roots since each of these factorizations is not even unique. If $Q > 0$, then all square roots are nonsingular.

If $P > 0$, then (A.3-2) is a norm. If $P \geq 0$, it is called a *seminorm* since $x^T P x$ may be zero even if x is not.

A.4   MATRIX CALCULUS

Let $x \in C^n = [x_1 \, x_2 \cdots x_n]^T$ be a vector, $s \in C$ be a scalar, and $f(x) \in C^m$ be an m-vector function of x. The differential in x is

$$dx = \begin{bmatrix} dx_1 \\ dx_2 \\ \vdots \\ dx_n \end{bmatrix}, \tag{A.4-1}$$

and the derivative of x with respect to s (which could be time) is

$$\frac{dx}{ds} = \begin{bmatrix} dx_1/ds \\ dx_2/ds \\ \vdots \\ dx_n/ds \end{bmatrix}. \tag{A.4-2}$$

If s is a function of x. Then the *gradient* of s with respect to x is the *column* vector

$$s_x \stackrel{\Delta}{=} \frac{\partial s}{\partial x} = \begin{bmatrix} \partial s/\partial x_1 \\ \partial s/\partial x_2 \\ \vdots \\ \partial s/\partial x_n \end{bmatrix}. \tag{A.4-3}$$

(The gradient is defined as a row vector in some references.) Then the total differential in s is

$$ds = \left(\frac{\partial s}{\partial x}\right)^T dx = \sum_{i=1}^{n} \frac{\partial s}{\partial x_i} dx_i. \tag{A.4-4}$$

If s is a function of two vectors x and y, then

$$ds = \left(\frac{\partial s}{\partial x}\right)^T dx + \left(\frac{\partial s}{\partial y}\right)^T dy. \tag{A.4-5}$$

The *Hessian* of s with respect to x is the second derivative

$$s_{xx} \stackrel{\Delta}{=} \frac{\partial^2 s}{\partial x^2} = \left[\frac{\partial^2 s}{\partial x_i \partial x_j}\right], \tag{A.4-6}$$

which is a symmetric $n \times n$ matrix. In terms of the gradient and the Hessian, the *Taylor series expansion* of $s(x)$ about x_0 is

$$s(x) = s(x_0) + \left(\frac{\partial s}{\partial x}\right)^T (x - x_0) + \frac{1}{2}(x - x_0)^T \frac{\partial^2 s}{\partial x^2}(x - x_0) + O(3), \tag{A.4-7}$$

where $O(3)$ represents terms of order 3, and s_x and s_{xx} are evaluated at x_0.

The *Jacobian* of f with respect to x is the $m \times n$ matrix

$$f_x \triangleq \frac{\partial f}{\partial x} = \left[\frac{\partial f}{\partial x_1} \frac{\partial f}{\partial x_2} \cdots \frac{\partial f}{\partial x_n} \right], \tag{A.4-8}$$

so that the total differential of f is

$$df = \frac{\partial f}{\partial x} dx = \sum_{i=1}^{n} \frac{\partial f}{\partial x_i} dx_i. \tag{A.4-9}$$

We shall use the shorthand notation

$$\frac{\partial f^T}{\partial x} \triangleq \left(\frac{\partial f}{\partial x} \right)^T \in C^{n \times m}. \tag{A.4-10}$$

If y is a vector and A, B, D, Q are matrices, all with dimensions so that the following expressions make sense, then we have the following results:

$$\frac{d}{dt}(A^{-1}) = -A^{-1}\dot{A}A^{-1}. \tag{A.4-11}$$

Some useful gradients are

$$\frac{\partial}{\partial x}(y^T x) = \frac{\partial}{\partial x}(x^T y) = y, \tag{A.4-12}$$

$$\frac{\partial}{\partial x}(y^T A x) = \frac{\partial}{\partial x}(x^T A^T y) = A^T y, \tag{A.4-13}$$

$$\frac{\partial}{\partial x}(y^T f(x)) = \frac{\partial}{\partial x}(f^T(x)y) = f_x^T y, \tag{A.4-14}$$

$$\frac{\partial}{\partial x}(x^T A x) = A x + A^T x, \tag{A.4-15}$$

and if Q is symmetric, then

$$\frac{\partial}{\partial x}(x^T Q x) = 2Qx, \tag{A.4-16}$$

$$\frac{\partial}{\partial x}(x - y)^T Q(x - y) = 2Q(x - y). \tag{A.4-17}$$

The chain rule for two vector functions becomes

$$\frac{\partial}{\partial x}(f^T y) = f_x^T y + y_x^T f. \tag{A.4-18}$$

Some useful Hessians are

$$\frac{\partial^2 x^T A x}{\partial x^2} = A + A^T, \tag{A.4-19}$$

and if Q is symmetric

$$\frac{\partial^2 x^\mathrm{T} Q x}{\partial x^2} = 2Q, \tag{A.4-20}$$

$$\frac{\partial^2}{\partial x^2}(x - y)^\mathrm{T} Q(x - y) = 2Q. \tag{A.4-21}$$

Some useful Jacobians are

$$\frac{\partial}{\partial x}(Ax) = A \tag{A.4-22}$$

(contrast this with (A.4-12)), and the chain rule

$$\frac{\partial}{\partial x}(sf) = \frac{\partial}{\partial x}(fs) = sf_x + fs_x^\mathrm{T} \tag{A.4-23}$$

(contrast this with (A.4-18)).

Some useful derivatives involving the trace and determinant are

$$\frac{\partial}{\partial A}\mathrm{trace}(A) = I, \tag{A.4-24}$$

$$\frac{\partial}{\partial A}\mathrm{trace}(BAD) = B^\mathrm{T} D^\mathrm{T}, \tag{A.4-25}$$

$$\frac{\partial}{\partial A}\mathrm{trace}(ABA^\mathrm{T}) = 2AB, \text{ if } B = B^\mathrm{T} \tag{A.4-26}$$

$$\frac{\partial}{\partial A}|BAD| = |BAD|A^{-\mathrm{T}}, \tag{A.4-27}$$

where $A^{-\mathrm{T}} \stackrel{\Delta}{=} (A^{-1})^\mathrm{T}$.

A.5   THE GENERALIZED EIGENVALUE PROBLEM

Consider the generalized eigenvalue problem

$$Gz = \mu Fz, \tag{A.5-1}$$

where

$$\det(\mu F - G) \equiv 0. \tag{A.5-2}$$

Then the finite generalized eigenvalues are the roots of $\det(\mu F - G)$. Let μ_i be the roots of $\det(\mu F - G)$ and define

$$\eta_i = \mathrm{dimker}(\mu_i F - G). \tag{A.5-3}$$

Then the rank 1 finite generalized eigenvectors are defined by

$$(\mu_i F - G)z_{ij}^1 = 0, \, j \in \hat{\eta}_i \tag{A.5-4}$$

(where $\hat{\eta}_i = \{1, 2, \ldots, \eta_i\}$) and the rank k finite eigenvectors for $k > 1$ and each i and j by

$$(\mu_i F - G)z_{ij}^{k+1} = -F z_{ij}^k, \quad k \geq 1. \tag{A.5-5}$$

If F is nonsingular, the above equation can be used to solve recursively for the z_{ij}^k beginning with the highest rank eigenvector in each chain. In that case this construction provides the eigenstructure of $F^{-1}G$. In the case where F in singular, the above equation cannot generally be used to recursively generate the z_{ij}^k. Furthermore, there exist eigenvalues at infinity and corresponding eigenvectors that can be constructed as follows. Define $\eta = \dim \ker(F)$. Then the rank 1 infinite eigenvectors are defined by

$$F z_{\infty j}^1 = 0, \quad j = \hat{\eta} \tag{A.5-6}$$

and the rank k infinite eigenvectors for $k > 1$ and each j by

$$F z_{\infty j}^{k+1} = G z_{\infty j}^k, \quad k \geq 1. \tag{A.5-7}$$

By arranging the eigenvectors as the columns of two nonsingular matrices according to

$$Z = [z_{ij}^k | z_{\infty j}^k], \ W = [F z_{ij}^k | G z_{\infty j}^k] \tag{A.5-8}$$

with i, j, k incrementing in odometer order, then

$$W^{-1}FV = \begin{bmatrix} I & 0 \\ 0 & N \end{bmatrix}, W^{-1}GV = \begin{bmatrix} M & 0 \\ 0 & I \end{bmatrix}, \tag{A.5-9}$$

where M is a Jordan form matrix containing the finite generalized eigenvalues of (G, F) and N is a nilpotent Jordan matrix representing the infinite generalized eigenvalues. The above canonical form is also known as the *Weierstrass form*.

REFERENCES

Abu-Khalaf, M., and F. L. Lewis, "Nearly optimal control laws for nonlinear systems with saturating actuators using a neural network HJB approach," *Automatica*, **41**, 779–791 (2005).

Abu-Khalaf, M., F. L. Lewis, and Jie Huang, "Policy iterations on the Hamilton-Jacobi-Isaacs equation for H_∞ state feedback control with input saturation," *IEEE Trans. Automatic Control*, **51** (12), 1989–1995 (2006).

Abu-Khalaf, M., J. Huang, and F. L. Lewis, *Nonlinear H2/H-Infinity Constrained Feedback Control: A Practical Design Approach Using Neural Networks*, Berlin: Springer-Verlag, 2006.

Abu-Khalaf, M., F. L. Lewis, and J. Huang, "Neurodynamic programming and zero-sum games for constrained control systems," *IEEE Trans. Neural Networks*, **19** (7), 1243–1252 (2008).

Al-Tamimi, A., F. L. Lewis, and M. Abu-Khalaf, "Discrete-time nonlinear HJB solution using approximate dynamic programming: convergence proof," IEEE *Trans. Systems, Man, Cybernetics, Part B*, **38** (4), 943–949 (2008).

Anderson, B. D. O., and Y. Liu, "Controller reduction: concepts and approaches," *IEEE Trans. Automatic Control*, AC-34, 802–812 (1989).

Anderson, B. D. O., and J. B. Moore, *Linear Optimal Control*, Englewood Cliffs, NJ: Prentice-Hall, 1971.

Armstrong, E. S., *ORACLS, A Design System for Linear Multivariable Control*, New York: Dekker, 1980.

Åström, K. J., and B. Wittenmark, *Computer Controlled Systems*, Englewood Cliffs, NJ: Prentice-Hall, 1984.

Athans, M., "A tutorial on the LQG/LTR method," *Proc. Am. Control Conf.*, 1289–1296 (1986).

Athans, M., and P. Falb, *Optimal Control*, New York: McGraw-Hill, 1966.

Athans, M., P. Kapsouris, E. Kappos, and H. A. Spang III, "Linear quadratic Gaussian with loop-transfer recovery methodology for the F-100 engine," *J. Guid.*, 9, 45–52 (1986).

Baird, L., "Reinforcement learning in continuous time: advantage updating," *Proc. International Conference on Neural Networks*, Orlando, FL, June 1994.

Balakrishnan, S. N., J. Ding, and F. L. Lewis, "Issues on stability of ADP feedback controllers for dynamical systems," *IEEE Trans. Systems, Man, Cybernetics, Part B*, **38** (4), 913–917 (2008).

Bardi, M., and I. Capuzzo-Dolcetta, *Optimal Control and Viscosity Solutions of Hamilton-Jacobi-Bellman Equations*, Boston: Birkhauser, 1997.

Bartels, R. H., and G. W. Stewart, "Solution of the matrix equation $AX + XB = C$," *Commun. ACM* 15 (6), 820–826 (1984).

Barto, A. G., R. S. Sutton, and C. Anderson. "Neuron-like adaptive elements that can solve difficult learning control problems," *IEEE Trans. Systems, Man Cybernetics*, **SMC-13**, 834–846 (1983).

Başar, T., and G. J. Olsder, *Dynamic Noncooperative Game Theory*, 2nd ed., Philadelphia, PA: SIAM, 1999.

Bell, R. F., E. W. Johnson, R. V. Whitaker, and R. V. Wilcox, "Head positioning in a large disk drive," *Hewlett Packard J.*, pp. 14–20, Jan. 1984.

Bellman, R. E., *Dynamic Programming*, Princeton, NJ: Princeton University Press, 1957.

Bellman, R. E., and S. E. Dreyfus, *Applied Dynamic Programming*, Princeton, NJ, Princeton University Press, 1962.

Bellman, R. E., and R. E. Kalaba, *Dynamic Programming and Modern Control Therapy*, Orlando, FL Academic Press, 1965.

Bertsekas, D. P., and J. N. Tsitsiklis, *Neuro-dynamic Programming*, Athena Scientific, Cambridge, MA, 1996.

Bierman, G. J., *Factorization Methods for Discrete Sequential Estimation*, Orlando FL: Academic Press, 1977.

Bittanti, S., A. J. Laub, and J. C. Willems, *The Riccati Equation*, New York: Springer-Verlag, 1991.

Blakelock, J. H., *Automatic Control of Aircraft and Missiles*, New York: Wiley, 1965.

Bradtke, S., B. Ydstie, and A. Barto, *Adaptive Linear Quadratic Control Using Policy Iteration*, report CMPSCI-94-49, University of Massachusetts, June 1994.

Brewer, J. W., "Kronecker products and matrix calculus in system theory," *IEEE Trans. Circuits Systems*, CAS-25 (9), 772–781 (1978).

Brogan, W. L., *Modern Control Theory*, New York: Quantum, 1974.

Broussard, J., and N. Halyo, "Active flutter control discrete optimal constrained dynamic compensators," *Proc. Am. Control Conf.*, 1026–1034 (1983).

Bryson, A. E., Jr., and Y. C. Ho, *Appl. Optimal Control*, New York: Hemisphere, 1975.

Businger, P., and G. H. Golub, "Linear least squares solution by householder transformations," *Numer. Math.*, 7, 269–276 (1965).

Busoniu, L., R. Babuska, B. De Schutter, and D. Ernst, *Reinforcement Learning and Dynamic Programming Using Function Approximators*, Boca Raton, FL: CRC, 2009.

Cao, X., *Stochastic Learning and Optimization*, Berlin: Springer-Verlag, 2007.

Casti, J., *Dynamical Systems and Their Applications: Linear Theory*, Orlando, FL: Academic Press, 1977.

Casti, J., "The linear quadratic control problem: some recent results and outstanding problems," *SIAM Rev.*, 22 (4), 459–485 (1980).

Chang, S. S. L., *Synthesis of Optimum Control Systems*, New York: McGraw-Hill, 1961.

Chen, B. M., Z. Lin, and Y. Shamash, *Linear Systems Theory: a Structural Decomposition Approach*, Boston: Birkhauser, 2004.

Clarke, D. W., and P. J. Gawthrop, "Self-tuning controller," *Proc. IEE*, 122 (9), 929–934 (1975).

Darwin, C., *On the Origin of Species by Means of Natural Selection*, London: J. Murray, 1859.

Davison, E. J., and I. J. Ferguson, "The design of controllers for the multivariable robust servomechanism problem using parameter optimization methods," *IEEE Trans. Automatic Control*, AC-26, 93–110 (1981).

Doya, K. "Reinforcement learning in continuous time and space," *Neural Computation*, vol. 12, pp. 219–245, MIT Press, 2000.

Doya, K., H. Kimura, and M. Kawato, "Neural mechanisms for learning and control," *IEEE Control Systems Magazine*, 42–54 (2001).

Doyle, J. C., "Guaranteed margins for LQG regulators," *IEEE Trans. Automatic Control*, AC-23, 756–757 (1978).

Doyle, J. C., and G. Stein, "Robustness with observers," *IEEE Trans. Automatic Control*, AC-24, 607–611 (1979).

Doyle, J. C., and G. Stein, "Multivariable feedback design: concepts for a classical/modern synthesis," *IEEE Trans. Automatic Control*, AC-26, 4–16 (1981).

Doyle, J. C., K. Glover, P. P. Khargonekar, and B. A. Francis, "State-space solutions to standard H_2 and H_∞ control problems," *IEEE Trans. Automatic Control*, AC-34, 831–847 (1989).

Dyer, P., and S. R. McReynolds, "Extension of square root filtering to include process noise," *J. Optimiz. Theory Applic.*, 3 (6), 444 (1969).

Elbert, T. F., *Estimation and Control of Systems*, New York: Van Nostrand Reinhold, 1984.

Francis, B. A., *A Course in H_∞ Control Theory*, Springer Verlag, Lecture notes in *Control and Info. Sci.*, 88, (1986).

Francis, B. A., and J. C. Doyle, "Linear control theory with an H_∞ optimality criterion," *SIAM J. Control Optim.*, 815–844 (1987).

Francis, B. A., J. W. Helton, and G. Zames, "H_∞-optimal feedback controllers for linear multivariable systems," *IEEE Trans. Automatic Control*, AC-29, 888–900 (1984).

Franklin, G. F., and J. D. Powell, *Digital Control of Dynamic Systems*, Reading, MA: Addison-Wesley, 1980.

Franklin, G. F., J. D. Powell, and A. Emami-Naeini, *Feedback Control of Dynamic Systems*, Reading, MA: Addison-Wesley, 1986.

Fulks, W., *Advanced Calculus*, New York: Wiley, 1967.

Gangsaas, D., K. R. Bruce, J. D. Blight, and U.-L. Ly, "Application of modern synthesis to aircraft control: three case studies," *IEEE Trans. Automatic Control*, AC-31, 995–1014 (1986).

Gantmacher, F. R., *The Theory of Matrices*, New York: Chelsea, 1977.

Gawthrop, P. J., "Some interpretations of the self-tuning controller," *Proc. IEEE Control Sci.*, 124 (10), 889–894 (1977).

Gelb, A., ed., *Applied Optimal Estimation*, Cambridge, MA: MIT Press, 1974.

Golub, G. H., S. Nash, and C. Van Loan, "A Hessenberg-Schur method for the matrix problem $AX + XB = C$," *IEEE Trans. Automatic Control*, AC-24, 909–913 (1979).

Green, M., and D. Limebeer, *Robust Control Theory*, Englewood Cliffs, NJ: Prentice-Hall, 1993.

Grimble, M. J., and M. A. Johnson, *Optimal Control and Stochastic Estimation: Theory and Applications*, vol. 1, New York: Wiley, 1988.

Hanselmann, T., L. Noakes, and A. Zaknich, "Continuous-time adaptive critics," *IEEE Trans. Neural Networks*, 18 (3), 631–647 (2007).

Harvey, C. A., and G. Stein, "Quadratic weights for asymptotic regulator properties," *IEEE Trans. Automatic Control*, AC-23, 378–387 (1978).

Hewer, G. A., "An iterative technique for the computation of steady state gains for the discrete optimal regulator," *IEEE Trans. Automatic Control*, 16 (4), 382–384 (1971).

IMSL, *Library Contents Document*, 8th ed., International Mathematical and Statistical Libraries, Inc., 7500 Bellaire Blvd., Houston, Texas, 77036, 1980.

Ioannou, P., and B. Fidan, *Adaptive Control Tutorial*, Philadelphia: SIAM Press, 2006.

Jadbabaie, A., J. Lin, and S. Morse, "Coordination of groups of mobile autonomous agents using nearest neighbor rules," *IEEE Trans. Automatic Control*, **48** (6), 988–1001 (2003).

Kailath, T., *Linear Systems*, Englewood Cliffs, NJ: Prentice-Hall, 1980.

Kalman, R. E., "New methods in Wiener filtering Theory," *Proceedings of the Symposium on Engineering Applications of Random Function Theory and Probability*, New York: Wiley, 1963.

Kalman, R. E., and R. S. Bucy, "New results in linear filtering and prediction theory," *Trans. ASME J. Basic Eng.*, 83, 95–108 (1961).

Kaminski, P. G., A. E. Bryson, and S. F. Schmidt, "Discrete square root filtering: a survey of current techniques," *IEEE Trans. Automatic Control*, AC-16 (6), 727–736 (1971).

Kimura, H., Y. Lu, and R. Kawatani, "On the structure of H_∞ control systems and related extensions," *IEEE Trans. Automatic Control*, AC-36, 653–667 (1991).

Kirk, D. E., *Optimal Control Theory*, Englewood Cliffs, NJ: Prentice-Hall, 1970.

Kleinman, D. L., "On an iterative technique for Riccati equation computations," *IEEE Trans. Automatic Control*, **AC-13** (1), 114–115. (1968).

Knobloch, H. W., A. Isidori, and D. Flokcerzi, *Topics in Control Theory*, Berlin: Springer-Verlag, 1993.

Koivo, H. N., "A multivariable self-tuning controller," *Automatica*, 16, 351–366 (1980).

Kreindler, E., and D. Rothschild, "Model-following in linear quadratic regulator," *AIAA J.*, 14 (7), 835–842 (1976).

Kučera, V., *Discrete Linear Control, The Polynomial Equation Approach*, New York: Wiley, 1979.

Kwakernaak, H., and R. Sivan, *Linear Optimal Control Systems*, New York: Wiley-Interscience, 1972.

Lancaster, P., and L. Rodman, *Algebraic Riccati Equations*, Oxford University Press, UK, 1995.

Laub, A. J., "A Shur Method for Solving Algebraic Riccati Equations," *IEEE Trans. Automatic Control*, AC-24, 913–921 (1979).

Laub, A. J., "Efficient Multivariable Frequency Response Computations," *IEEE Trans. Automatic Control*, AC-26, 407–408 (1981).

Letov, A. M., "Analytical Controller Design, I, II," *Autom. Remote Control*, 21, 303–306 (1960).

Levine, W. S., and M. Athans, "On the Determination of the Optimal Constant Output Feedback Gains for Linear Multivariable Systems," *IEEE Trans. Automatic Control*, AC-15, 44–48 (1970).

Lewis, F. L., *Optimal Estimation*, New York: Wiley, 1986.

Lewis, F. L., and D. Vrabie, "Reinforcement learning and adaptive dynamic programming for feedback control," *IEEE Circuits Systems Mag.*, 32–38 (2009).

Lewis, F. L., L. Xie, and D. Popa, *Optimal & Robust Estimation: With an Introduction to Stochastic Control Theory*, 2nd ed., Boca Raton, FL: CRC Press, 2007.

Lewis, F. L., G. Lendaris, and Derong Liu, "Special issue on approximate dynamic programming and reinforcement learning for feedback control," *IEEE Trans. Systems, Man Cybernetics, Part B*, 38 (4) (2008).

Li, Z. H., and M. Krstic, "Optimal design of adaptive tracking controllers for nonlinear systems," *Automatica*, **33** (8), 1459–1473 (1997).

Ljung, L., *System Identification*, Englewood Cliffs, NJ: Prentice-Hall, 1999.

Luenberger, D. G., *Optimization by Vector Space Methods*, New York: Wiley, 1969.

Luenberger, D. G., *Introduction to Dynamic Systems*, New York: Wiley, 1979.

MacFarlane, A. G. J., "Return difference and return-ratio matrices and their use in the analysis and design of multivariable feedback control systems," *Proc. IEE*, 117, 2037–2049 (1970).

MacFarlane A. G. J., and B. Kouvaritakis, "A Design Technique for Linear Multivariable Feedback Systems," *Int. J. Control*, 25, 837–874 (1977).

Marion, J. B., *Classical Dynamics of Particles and Systems*, Orlando, FL: Academic Press, 1965.

MATLAB, The MathWorks, Inc., Cochituate Place, 24 Prime Parkway, Natick, MA 01760, 1992.

McClamroch, N. H., *State Models of Dynamic Systems*, New York: Springer-Verlag, 1980.

McFarlane, D., and K. Glover, "A loop shaping design procedure using H_∞ synthesis," *IEEE Trans. Automatic Control* AC-37, 759–769 (1992).

McReynolds, S. R., Ph.D. thesis, Harvard University, Cambridge, MA, 1966.

Medanic, J., "Closed-loop Stackelberg strategies in linear quadratic problems," *IEEE Trans. Automatic Control*, AC-23, 632–637 (1978).

Mehta, P., and S. Meyn, "Q-learning and Pontryagin's minimum principle," *Proc. IEEE Conf. Decision and Control*, 3598–3605. (2009).

Mendel, J. M., and R. W. MacLaren, "Reinforcement learning control and pattern recognition systems," in *Adaptive, Learning, and Pattern Recognition Systems: Theory and Applications*, ed. Mendel, J. M., and K. S. Fu, pp. 287–318, New York: Academic Press, 1970.

Mil. Spec. 1797, *Flying Qualities of Piloted Vehicles*, 1987.

Moerder, D. D., and A. J. Calise, "Convergence of a numerical algorithm for calculating optimal output feedback gains," *IEEE Trans. Automatic Control*, AC-30, 900–903 (1985).

Moore, B. C., "Principal component analysis in linear systems: controllability, observability and model reduction," *IEEE Trans. Automatic Control*, AC-26, 17-32 (1982).

Moore, K. L., *Iterative Learning Control for Deterministic Systems*, London: Springer-Verlag, 1993.

Morari, M., and E. Zafiriou, *Robust Process Control*, Englewood, NJ: Prentice-Hall, 1989.

Morf, M., and T. Kailath, "Square root algorithms for least-squares estimation," *IEEE Trans. Automatic Control*, AC-20 (4), 487–497 (1975).

Murray, J., C. Cox, R. Saeks, and G. Lendaris, "Globally convergent approximate dynamic programming applied to an autolander," *Proc. Am. Control Conf.*, pp. 2901–2906, Arlington, VA, 2001.

Nelder, J. A., and R. Mead, "A simplex method for function minimization," *Comput. J.* 7, 308–313 (1964).

Nevistic, V., and J. Primbs, *Constrained Nonlinear Optimal Control: A Converse HJB Approach*, Technical Report 96-021, California Institute of Technology, 1996.

O'Brien, M. J., and J. B. Broussard, "Feedforward control to track the output of a forced model," *Proc. IEEE Conference on Decision and Control*, Dec. 1978.

Olfati-Saber, R., and R. M. Murray, "Consensus problems in networks of agents with switching topology and time-delays," *IEEE Trans. Automatic Control*, **49** (9), 1520–1533 (2004).

Papavassilopoulos, G. P., and J. B. Cruz, Jr., "On the existence of solutions to coupled matrix Riccati differential equations in linear quadratic Nash games," *IEEE Trans. Automatic Control*, AC-24, 127–129 (1979).

Papoulis, A., *Probability, Random Variables, and Stochastic Processes*, 2nd ed., New York: McGraw Hill, 1984.

Pappas, T., A. J. Laub, and N. R. Sandell, "On the numerical solution of the discrete-time algebraic Riccati equation," *IEEE Trans. Automatic Control*, AC-25, 631–641 (1980).

Pontryagin, L. S., V. G. Boltyanskii, R. V. Gamkrelidze, and E. F. Mishchenko, *The Mathematical Theory of Optimal Processes*, New York: Wiley-Interscience, 1962.

Postlethwaite, I., J. M. Edmunds, and A. G. J. MacFarlane, "Principal gains and principal phases in the analysis of linear multivariable systems," *IEEE Trans. Automatic Control*, AC-26, 32–46 (1981).

Powell, W. B., *Approximate Dynamic Programming*, Hoboken, NJ: Wiley, 2007.

Press, W. H., B. P. Flanerry, S. A. Teukolsky, and W. T. Vetterling, *Numerical Recipes: The Art of Scientific Computing*, New York: Cambridge University Press, 1986.

Rosenbrock, H. H., *Computer-aided Control System Design*, New York: Academic Press, 1974.

Safanov, M. G., and M. Athans, "Gain and phase margin for multiloop LQG regulators," *IEEE Trans. Automatic Control*, AC-22, 173–178 (1977).

Safanov, M. G., A. J. Laub, and G. L. Hartmann, "Feedback properties of multivariable systems: the role and use of the return difference matrix," *IEEE Trans. Automatic Control*, AC-26, 47–65 (1981).

Sandell, W. R., "Decomposition vs. decentralization in large-scale system theory," *Proc. Conf. Dec. Control*, 1043–1046 (1976).

Schmidt, S. F., "Estimation of state with acceptable accuracy constraints," TR 67-16, Analytical Mechanics Assoc., Palo Alto, California, 1967.

Schmidt, S. F., "Computational techniques in Kalman filtering," *Theory and Applications of Kalman Filtering*, Chap. 3, NATO Advisory Group for Aerospace Research and Development, AGARDograph 139, Feb. 1970.

Schultz, D. G., and J. L. Melsa, *State Functions and Linear Control Systems*, New York: McGraw-Hill, 1967.

Schultz, W,. "Neural coding of basic reward terms of animal learning theory, game theory, microeconomics and behavioral ecology," *Neurobiology*, **14**, 139–147 (2004).

Sewell, G., "IMSL software for differential equations in one space variable," IMSL Tech. Report Series, No. 8202, 1982. IMSL, Inc., 7500 Bellaire Blvd., Houston, TX 77036.

Shin, V., and C. Chen, "On the Weighting Factors of the Quadratic Criterion in Optimal Control," *Int. J. Control*, 19, 947–955 (1974).

Si, J., A. Barto, W. Powell, and D. Wunsch, *Handbook of Learning and Approximate Dynamic Programming*, IEEE Press, USA, 2004.

Söderström, T., "On some algorithms for design of optimal constrained regulators," *IEEE Trans. Automatic Control*, AC-23, 1100–1101 (1978).

Southworth, R. W., and S. L. Deleeuw, *Digital Computation and Numerical Methods*, New York: McGraw-Hill, 1965.

Stein, G., and M. Athans, "The LQR/LTR procedure for multivariable feedback control design," *IEEE Trans. Automatic Control*, AC-32, 105–114 (1987).

Stevens, B. L., and F. L. Lewis, *Aircraft Control and Simulation*, New York: Wiley, 1992.

Strang, G., *Linear Algebra and Its Applications*, 2nd ed. Orlando, FL: Academic Press, 1980.

Sutton, R. S., and A. G. Barto, *Reinforcement Learning—An Introduction*, Cambridge, MA: MIT Press, 1998.

Tsitsiklis, J., *Problems in Decentralized Decision Making and Computation*, Ph.D. dissertation, Dept. Elect. Eng. and Comput. Sci., Cambridge, MA: MIT, 1984.

Vamvoudakis, K. G., and F. L. Lewis, "Online actor-critic algorithm to solve the continuous-time infinite horizon optimal control problem," *Automatica*, **46** (5), 878–888 (2010a).

Vamvoudakis, K. G., and F. L. Lewis, "Online solution of nonlinear two-player zero-sum games using synchronous policy iteration", *Proc. IEEE Conf. Decision & Control*, 3040–3047 (2010b).

Vamvoudakis, K. G., and F. L. Lewis, "Multi-player non-zero sum games: online adaptive learning solution of coupled Hamilton-Jacobi equations," *Automatica*, **47** (8), 1556–1 2011.

van der Schaft, A. J., "L_2-Gain analysis of nonlinear systems and nonlinear state feedback H_∞ control," *IEEE Trans. Automatic. Control*, **37** (6), 770–784 (1992).

Vaughan, D. R., "A nonrecursive algebraic solution to the discrete Riccati equation," *IEEE Trans. Automatic Control*, AC-15, 597–599 (1970).

Verriest, E. I., and F. L. Lewis, "On the linear quadratic minimum-time problem," *IEEE Trans. Automatic Control*, AC-36, pp. 859–863, July 1991.

Vrabie, D., and F. L. Lewis, "Neural network approach to continuous-time direct adaptive optimal control for partially-unknown nonlinear systems," *Neural Networks*, **22** (3), 237–246 (2009).

Vrabie, D., and F. L. Lewis, "Adaptive dynamic programming algorithm for finding online the equilibrium solution of the two-player zero-sum differential game," *Proc. Int. Joint Conf. Neural Networks*, 1–8 (2010a).

Vrabie, D., and F. L. Lewis, "Integral reinforcement learning for online computation of feedback Nash strategies of nonzero-sum differential games," *Proc. IEEE Conf. Decision Control*, 3066–3071 (2010b).

Vrabie, D., O. Pastravanu, M. Abu-Khalaf, and F. L. Lewis, "Adaptive optimal control for continuous-time linear systems based on policy iteration," *Automatica*, **45**, 477–484 (2009).

Wang, F. Y., H. Zhang, D. Liu, "Adaptive dynamic programming: an introduction," *IEEE Computational Intelligence Magazine*, 39–47 (2009).

Watkins, C., *Learning from Delayed Rewards*, Ph.D. Thesis, Cambridge, UK: Cambridge University, 1989.

Watkins, C., and P. Dayan, "Q-learning," *Machine Learning*, 8, 279–292 (1992).

Werbos, P. J., "Neural networks for control and system identification," *Proc. IEEE Conf. Decision and Control*, Florida, 1989.

Werbos., P. J., "A menu of designs for reinforcement learning over time," *Neural Networks for Control*, pp. 67–95, ed. W. T. Miller, R. S. Sutton, and P. J. Werbos, Cambridge, MA: MIT Press, 1991.

Werbos, P. J., "Approximate dynamic programming for real-time control and neural modeling," *Handbook of Intelligent Control*, ed. D. A. White and D. A. Sofge, New York: Van Nostrand Reinhold, 1992.

Wheeler, R. M., and K. S. Narendra, "Decentralized learning in finite Markov chains," *IEEE Trans. Automatic Control*, 31 (6), 1986.

Wolovich, W. A., *Linear Multivariable Systems*, New York: Springer-Verlag, 1974.

Zames, G., "Feedback and optimal sensitivity: model reference transformations, multiplicative seminorms and approximate inverses," *IEEE Trans. Automatic Control*, 26, 301–320 (1981).

Zhang, H., J. Huang, and F. L. Lewis, "Algorithm and stability of ATC receding horizon control," *Proc. IEEE Symp. ADPRL*, pp. 28–35, Nashville, TN, Mar. 2009.

INDEX